Lecture Notes in Computer Sci

Commenced Publication in 1973
Founding and Former Series Editors:
Gerhard Goos, Juris Hartmanis, and Jan van Leeuwen

Editorial Board

David Hutchison
Lancaster University, UK

Takeo Kanade
Carnegie Mellon University, Pittsburgh, PA, USA

Josef Kittler
University of Surrey, Guildford, UK

Jon M. Kleinberg
Cornell University, Ithaca, NY, USA

Alfred Kobsa
University of California, Irvine, CA, USA

Friedemann Mattern
ETH Zurich, Switzerland

John C. Mitchell
Stanford University, CA, USA

Moni Naor
Weizmann Institute of Science, Rehovot, Israel

Oscar Nierstrasz
University of Bern, Switzerland

C. Pandu Rangan
Indian Institute of Technology, Madras, India

Bernhard Steffen
TU Dortmund University, Germany

Madhu Sudan
Microsoft Research, Cambridge, MA, USA

Demetri Terzopoulos
University of California, Los Angeles, CA, USA

Doug Tygar
University of California, Berkeley, CA, USA

Gerhard Weikum
Max Planck Institute for Informatics, Saarbruecken, Germany

For further volumes:
http://www.springer.com/series/7407

Giuseppe Nicosia · Panos Pardalos (Eds.)

Learning and Intelligent Optimization

7th International Conference, LION 7
Catania, Italy, January 7–11, 2013
Revised Selected Papers

 Springer

Editors

Giuseppe Nicosia
Department of Mathematics
and Computer Science
University of Catania
Catania
Italy

Panos Pardalos
Department of Industrial and Systems
Engineering
University of Florida
Gainesville, FL
USA

ISSN 0302-9743 ISSN 1611-3349 (electronic)
ISBN 978-3-642-44972-7 ISBN 978-3-642-44973-4 (eBook)
DOI 10.1007/978-3-642-44973-4
Springer Heidelberg New York Dordrecht London

Library of Congress Control Number: 2013955073

CR Subject Classification (1998): F.2, G.1.6, G.2, F.1, I.2, G.1

LNCS Sublibrary: SL1 – Theoretical Computer Science and General Issues

© Springer-Verlag Berlin Heidelberg 2013
This work is subject to copyright. All rights are reserved by the Publisher, whether the whole or part of the material is concerned, specifically the rights of translation, reprinting, reuse of illustrations, recitation, broadcasting, reproduction on microfilms or in any other physical way, and transmission or information storage and retrieval, electronic adaptation, computer software, or by similar or dissimilar methodology now known or hereafter developed. Exempted from this legal reservation are brief excerpts in connection with reviews or scholarly analysis or material supplied specifically for the purpose of being entered and executed on a computer system, for exclusive use by the purchaser of the work. Duplication of this publication or parts thereof is permitted only under the provisions of the Copyright Law of the Publisher's location, in its current version, and permission for use must always be obtained from Springer. Permissions for use may be obtained through RightsLink at the Copyright Clearance Center. Violations are liable to prosecution under the respective Copyright Law.
The use of general descriptive names, registered names, trademarks, service marks, etc. in this publication does not imply, even in the absence of a specific statement, that such names are exempt from the relevant protective laws and regulations and therefore free for general use.
While the advice and information in this book are believed to be true and accurate at the date of publication, neither the authors nor the editors nor the publisher can accept any legal responsibility for any errors or omissions that may be made. The publisher makes no warranty, express or implied, with respect to the material contained herein.

Printed on acid-free paper

Springer is part of Springer Science+Business Media (www.springer.com)

Preface

By bringing together scientists working in optimization and machine learning, LION aims to provide delegates with the opportunity to learn more about other research areas, where the algorithms, methods and theories on show are likely to be relevant to their own research.

Optimization and machine learning researchers are now forming their own community and identity. The International Conference on Learning and Optimization is proud to be the premiere conference in the area. As its organizers, we are honored to have such a variety of innovative and original scientific articles presented this year.

LION 2013 is the seventh international conference dedicated entirely to the field of optimization and machine learning. It was held in Catania - Italy, during January 7–11, 2013.

There were three plenary lectures:

Youssef Hamadi, Microsoft Research - UK
Mauricio G.C. Resende, AT&T Labs Research - USA
Qingfu Zhang, University of Essex - UK,

and four tutorial speakers:

Paola Festa, University of Napoli "Federico II" - Italy,
Mario Guarracino, CNR - Italy,
Boris Goldengorin, National Research University, Moscow, Russia,
Yaroslav D. Sergeyev, University of Calabria, Italy.

We had more submissions than ever this year, each manuscript was independently reviewed by at least three members of the Programme Committee in a blind review process. So, in these proceedings there are 49 research articles written by leading scientists in the field, from 47 different countries on 5 continents, describing an impressive array of ideas, technologies, algorithms, methods and applications in Optimization and Machine Learning.

We couldn't have organized this conference without these researchers, so we thank them all for coming. We also couldn't have organized LION without the excellent work of all of the Programme Committee members, the session chair Giovanni Stracquadanio, the publicity chair, and the chair of the local Organizing Committee, Patrizia Nardon.

We would like to express our appreciation to the keynote and tutorial speakers who accepted our invitation, and to all authors who submitted research papers to LION 2013.

January 2013

Giuseppe Nicosia
Panos Pardalos

Organization

LION 2013 Committees

Conference and Technical Program Committee Co-chairs
Giuseppe Nicosia   University of Catania, Italy
Panos Pardalos   University of Florida, USA

Special Session Chair

Giovanni Stracquadanio   Johns Hopkins University, USA

Local Organization

Patrizia Nardon   Reactive Search srl, Trento, Italy

Liaison with Springer

Thomas Stuetzle   University Libre de Bruxelles, Belgium

LION 2013 Website

Marco Dallariva   Reactive Search SrL

Publicity Chair

Mauro Brunato   Reactive Search srl, Catania, Italy

Technical Program Committee

Hernan Aguirre	Shinshu University, Japan
Ethem Alpaydin	Bogazici University, Turkey
Dirk Arnold	Dalhousie University, Canada
Luigi Barone	SolveIT Software, Adelaide, Australia
Julio Barrera	CINVESTAV-IPN, San Pedro Zacatenco, Mexico
Roberto Battiti	University of Trento, Italy
Mauro Birattari	University Libre de Bruxelles, Belgium
Christian Blum	Universitat Politecnica de Catalunya, Spain
Juergen Branke	University of Warwick, UK
Dimo Brockhoff	Inria, Lille, France
Mauro Brunato	University of Trento, Italy
Philippe Codognet	Université Pierre et Marie Curie, Paris 6, France
Carlos Coello Coello	CINVESTAV-IPN, San Pedro Zacatenco, Mexico
Pierre Collet	Université de Strasbourg, France
Carlos Cotta	Universidad de Malaga, Spain
Clarisse Dhaenens Flipo	Laboratoire LIFL/Inria Villeneuve d'Ascq, France
Luca Di Gaspero	University of Udine, Italy
Federico Divina	Pablo de Olavide University, Seville, Spain
Karl F. Doerner	University of Vienna, Austria
Marco Dorigo	Université Libre de Bruxelles, Belgium
Talbi El-Ghazali	Polytech Lille, France
Michael Emmerich	LIACS Leiden University, The Netherlands
Andries Engelbrecht	University of Pretoria, South Africa
Valerio Freschi	University of Urbino, Italy
Xavier Gandibleux	The University of Nantes, France
Pablo Garcia Sanchez	University of Granada, Spain
R. Ruiz Garcia	Universidad Politècnica de Valencia, Spain
Deon Garrett	Icelandic Institute for Intelligent Machines, Reykjavík, Iceland
Michel Gendreau	Ecole Polytechnique de Montreal, Canada
Tobias Glasmacher	Ruhr-University Bochum, Germany
Martin C. Golumbic	CRI Haifa, Israel
Salvatore Greco	University of Catania, Italy
Walter J. Gutjahr	University of Vienna, Austria
Youssef Hamadi	Microsoft Research, UK
Jin-Kao Hao	University of Angers, France
Simon Harding	University of Bristol, UK
Richard Hartl	University of Vienna, Austria
Geir Hasle	SINTEF Applied Mathematics, Norway
A. G. Hernandez-Diaz	Pablo de Olavide University, Spain
Francisco Herrera	University of Granada, Spain
Tomio Hirata	Nagoya University, Japan
Frank Hutter	University of British Columbia, Canada

Hisao Ishibuchi	Osaka Prefecture University, Japan
Yaochu Jin	University of Surrey, UK
Laetitia Jourdan	LIFL University of Lille 1, France
Narendra Jussien	Ecole des Mines de Nantes, France
Tanaka Kiyoshi	Shinshu University, Nagano, Japan
Zeynep Kiziltan	University of Bologna, Italy
Dario Landa-Silva	University of Nottingham, UK
A. J. Fernandez Leiva	Universidad de Malaga, Spain
Arnaud Liefooghe	Inria, Villeneuve d'Ascq, France
Manuel Lopez-Ibanez	Université Libre de Bruxelles, Belgium
Antonio Lopez-Jaimes	CINVESTAV-IPN, San Pedro Zacatenco, Mexico
Thibaut Lust	Université Catholique de Louvain, Belgium
Dario Maggiorini	University of Milan, Italy
Ogier Maitre	University of Strasbourg, France
Vittorio Maniezzo	University of Bologna, Italy
Francesco Masulli	University of Genoa, Italy
Basseur Matthieu	LERIA Angers, France
J. J. Merelo	Universidad de Granada, Spain
Bernd Meyer	Monash University, Australia
Zbigniew Michalewicz	University of Adelaide, Australia
Nenad Mladenovic	Brunel University, London, UK
M. A. Montes de Oca	IRIDIA, Université Libre de Bruxelles, Belgium
Antonio M. Mora Garcia	University of Granada, Spain
Amir Nakib	Université Paris Este Creteil, France
Giuseppe Nicosia	University of Catania, Italy
Gabriela Ochoa	University of Nottingham, UK
Yew-Soon Ong	Nanyang Technological University, Singapore
Djamila Ouelhadj	University of Portsmouth, UK
Patricia Paderewski	University of Granada, Spain
Natalia Padilla-Zea	LIVE - GEDES, University of Granada, Spain
Luis Paquete	CISUC, University of Coimbra, Spain
Panos M. Pardalos	University of Florida, USA
Andrew Parkes	University of Nottingham, UK
Marcello Pelill	University of Venice, Italy
Diego Perez	University of Essex, UK
Vincenzo Piuri	University of Milan, Italy
Silvia Poles	Enginsoft Srl, Trento, Italy
Mike Preuss	TU Dortmund, Germany
Gunther R. Raidl	Vienna University of Technology, Austria
Franz Rendl	University of Klagenfurt, Austria
Celso C. Ribeiro	Universidade Federal Fluminense, Brazil
Florian Richoux	University of Nantes, France
Laura Anna Ripamonti	University of Milan, Italy
Andrea Roli	Alma Mater Studiorum University of Bologna, Italy
E. Rodriguez-Tello	CINVESTAV-Tamaulipas, Mexico
Samuel Rota Bulò	Ca' Foscari University of Venice, Italy

Wheeler Ruml	University of New Hampshire, USA
Ilya Safro	Argonne National Laboratory, USA
Horst Samulowitz	National ICT Australia, Sydney, Australia
Hiroyuki Sato	The University of Electro-Communications, Tokyo, Japan
Frederic Saubion	University of Angers, France
Andrea Schaerf	University of Udine, Italy
Marc Schoenauer	Inria Saclay, France
Oliver Schettze	CINVESTAV-IPN, San Pedro Zacatenco, Mexico
Yaroslav D. Sergeyev	University of Calabria, Italy
Patrick Siarry	University Paris-Est Creteil, France
Ankur Sinha	Aalto University, Helsinki, Finland
Christine Solnon	University of Lyon, France
Theo Stewart	University of Cape Town, South Africa
Giovanni Stracquadanio	Johns Hopkins University, Baltimore, USA
Thomas Stutzle	Université Libre de Bruxelles, Belgium
Ke Tang	University of Science and Technology of China, China
Julian Togelius	IDSIA, Lugano, Switzerland
Shigeyoshi Tsutsui	Hannan University, Osaka, Japan
Pascal Van Hentenryck	Brown University, Providence, USA
Sebastien Verel	Inria, Calais Cedex, France
Stefan Voss	University of Hamburg, Germany
Markus Wagner	University of Adelaide, Australia
Toby Walsh	NICTA and UNSW, Australia
David L. Woodruff	University of California, Davis, USA
Petros Xanthopoulos	University of Central Florida, USA
Ning Xiong	Mälardalen University, Västerås, Sweden

Contents

Interleaving Innovization with Evolutionary Multi-Objective Optimization in Production System Simulation for Faster Convergence

Amos H.C. Ng[1](✉), Catarina Dudas[1], Henrik Boström[2],
and Kalyanmoy Deb[1,3]

[1] Virtual Systems Research Centre, University of Skövde, Skövde, Sweden
amos.ng@his.se
[2] Department of Computer and Systems Sciences, Stockholm University,
Stockholm, Sweden
[3] Department of Electrical and Computer Engineering, Michigan State University,
East Lansing, USA

Abstract. This paper introduces a novel methodology for the optimization, analysis and decision support in production systems engineering. The methodology is based on the *innovization* procedure, originally introduced to unveil new and innovative design principles in engineering design problems. The innovization procedure stretches beyond an optimization task and attempts to discover new design/operational rules/principles relating to decision variables and objectives, so that a deeper understanding of the underlying problem can be obtained. By integrating the concept of innovization with simulation and data mining techniques, a new set of powerful tools can be developed for general systems analysis. The uniqueness of the approach introduced in this paper lies in that decision rules extracted from the multi-objective optimization using data mining are used to modify the original optimization. Hence, faster convergence to the desired solution of the decision-maker can be achieved. In other words, faster convergence and deeper knowledge of the relationships between the key decision variables and objectives can be obtained by interleaving the multi-objective optimization and data mining process. In this paper, such an interleaved approach is illustrated through a set of experiments carried out on a simulation model developed for a real-world production system analysis problem.

Keywords: Innovization · Multi-objective optimization · Data mining · Production system simulation

1 Introduction

Optimization involves the process of finding one or more solutions which correspond to the minimization or maximization of one or more objectives. In a single optimization problem, a single optimal solution is sought to optimize a single objective function and in a multi-objective optimization (MOO) problem the optimization involves more than one objective function. In most MOO problems, especially those

G. Nicosia and P. Pardalos (Eds.): LION 7, LNCS 7997, pp. 1–18, 2013.
DOI: 10.1007/978-3-642-44973-4_1, © Springer-Verlag Berlin Heidelberg 2013

found in the real world, these objective functions are in conflict with each other. Thus, seeking one single best solution that optimizes all of them simultaneously is impossible, because improving one objective would deteriorate the others [1]. This scenario gives rise to a set of optimal compromised (trade-off) solutions, largely known as Pareto-optimal solutions. The so-called Pareto Front (PF) in the objective space consists of solutions in the Pareto-optimal solution set.

Despite the existence of multiple trade-off solutions, in most cases, only one of them will be chosen as the solution for implementation, for example, in a product or system design. Therefore, two equally important tasks are usually involved in solving an MOO problem: (1) searching for the PF solution set, so that the decision-maker can acquire an idea on the extent and nature of the trade-off among the objectives, and (2) choosing a particular preferred solution from the Pareto-optimal set. While the first task is computationally intensive, and can be fully automated using an MOO algorithm, the second task usually necessitates a manual decision-making process using the preference information of the decision-maker. It is interesting to note that an MOO problem can easily be converted into a single-objective optimization problem, by formulating a weighted-sum objective function which is composed of the multiple objectives, so that a single trade-off optimal solution can effectively be sought. However, the major drawback is that the trade-off solution obtained by using this procedure is very sensitive to the relative preference vector. Therefore, the choice of the preference weights and thus the obtained trade-off solution is highly subjective to the particular decision-maker. Firstly, without detailed knowledge about the product or system under study, selecting the appropriate preference vector can be a very difficult task. Secondly, converting an MOO problem into a simplistic single-objective problem puts decision-making ahead of knowing the best possible trade-offs. In other words, thanks to the generation of multiple trade-off solutions, an MOO procedure can contribute to support the decision-making, in comparison to a single-objective optimization procedure. On one hand, the decision-maker is provided with multiple "optimal" (or precisely near-optimal) alternatives for consideration before making the final choice. On the other hand, since these optimal solutions are "high-performing" with respect to at least one objective, conducting an analysis that answers "What makes these solutions optimal?" can provide the decision-maker with very important information, or knowledge, which cannot be obtained if only one single solution is sought in the optimization task. The idea of deciphering knowledge, or knowledge discovery, using the post-optimality analysis of Pareto-optimal solutions from an MOO, was first proposed by Deb and Srinivasan [2]. They coined the term *innovization* (innovation via optimization) to describe the task of discovering the salient common principles present in the Pareto-optimal solutions, in order to obtain deeper knowledge/insights regarding the behavior/nature of the problem. The innovization task employed in earlier publications involved the manual identification of the important relationships among decision variables and objectives that are common to the obtained trade-off solutions. Recent studies have shown that using data mining (DM) techniques to enable innovization procedures to be performed automatically [3, 4] can be promising for various engineering problems. In these innovization tasks, the efficient evolutionary multi-objective optimization (EMO) algorithm, NSGA-II [5], has been applied to generate the Pareto-optimal solutions.

Due to their population-based approach and wide-spread applicability, EMO algorithms are in general very suitable for the optimization task in an automated innovization procedure.

Research in combining MOO and DM has attracted increasing interest in the last decade. Obayashi and Sasaki [6] used Self-Organizing Maps (SOM) to cluster the design space, in order to gain more information about design trade-offs in the objective space. Jeong et al. [7] applied a combination of SOM and analysis of variance (ANOVA) in the design process for aerodynamic optimization problems. SOM was used to analyze the key design variables found in the ANOVA for further examination, in order to gain insight into how they influence the objective functions. Sugimura et al. [8] explored the use of decision trees and rough sets to analyze the optimal solutions, in order to extract the design rules for the blower efficiency and stability of the inflow for a diffuser. In [9], the dominant design features were extracted by decomposing the shape and flow data into a set of orthogonal base vectors describing the optimal design of an aerodynamic transonic airfoil. In addition to data-mining methods, data visualization techniques like 4D-plots, Parallel coordinates, and hyper-radial visualization have been used to analyze the Pareto-optimal solutions applied to a range of automotive engineering problems [10].

In summary, most of the above-mentioned related studies were focused on engineering or product design problems. As a matter of fact, by integrating the concept of innovization with simulation and DM techniques, the innovization procedure can be applied to systems design and analysis in general. In particular, by using MOO on discrete-event simulation models, the innovization task can be effectively employed for analysis and decision-making support in the system design/development of industrial-scale production or supply-chain systems. Such a so-called Simulation-based Innovization (SBI) procedure has been proposed in some of our previous work [11–13]. In contrast to other automated innovization procedures using data-mining techniques [3, 4], the uniqueness of our proposed SBI approach lies in:

- Using decision trees/rules as the induction techniques, instead of mathematical formulae, to capture the relationship between decision variables and objectives, is believed to make understanding the extracted knowledge easier for the decision-maker.
- The SBI approach focusing on combining the data set in DM with both optimal and non-optimal solutions, so that research on "What distinguishes the Pareto-optimal solutions and non-optimal solutions?" can be conducted. This is different from other existing innovization procedures, which only focus on unveiling the salient common principles of the Pareto-optimal solutions.

Related to the latter point, it is logical to assume that a non-optimal solution, which is closer to the PF, possesses the attributes that are closer to a Pareto-optimal solution than one which is far away from it. This is particularly apparent for an MOO problem with continuous decision variables and objective functions. Therefore, a distance-based approach that performs pre-processing on the data set generated from MOO has been proposed [14]. With such a distance-based pre-processing approach, the subsequent DM task becomes a regression problem in which the distances of the solutions in the data set to the PF are used as the dependent continuous variable. The overall aim of this procedure is therefore to decipher attributes or patterns in what distinguishes a solution with a short distance to the PF from a solution with a long distance to the PF, in order to portray the optimality of the PF solutions and acquire deeper knowledge of the designed system, before a final decision is made.

Very recently, the EMO literature has highlighted the importance of using a local search procedure along with an EMO procedure [15]. The article includes a proposal of a serial innovization and local search approach, in which the common principles present in EMO solutions are first deciphered and then used as heuristics in the local search operation. The basic idea of this approach is that the relationships between the decision variables and the objectives derived from an innovization task can be used as heuristics in the local search procedure, in order to obtain a faster convergence than a single application of EMO to the problem would achieve. We believe that this concept can be extended when we consider that a common decision-making scenario would require the decision-maker to go through the following process iteratively, before a decision could be made: (1) run MOO to gain an approximate idea of the extent of the PF; (2) select some specific region(s) of interest on the PF; (3) discover the attributes of the solutions in the selected region(s); (4) perform a local search procedure to further explore other possible solutions in the interested region(s), e.g., using the reference point-based approach [16]. In other words, an advanced SBI procedure should be able to support the decision-maker, so that optimization, knowledge discovery and decision-making tasks can be applied in an interleaving manner, before a final, well-informed/confident decision is made.

The aim of this paper is to introduce such an advanced SBI procedure in which MOO and DM are interleaved. The usefulness of such a procedure to solve real-world system optimization problems is illustrated through a case study of an industrial production system analysis problem. The paper continues as follows. Section 2 introduces the original SBI process and then the interleaving approach. Section 3 presents the industrial case study and results of the experiments using the proposed approach to converge to the different preferred regions chosen by the decision-maker. A presentation of the conclusions is found in Sect. 4.

2 Simulation-Based Innovization (SBI)

As argued in [14], a SBI process does not deviate much from a standardized process model of an ordinary, knowledge discovery in databases (KDD) process. The major difference, when comparing SBI with a KDD process, is that the data used for the pattern detection is not from a data source containing historical data, but from experimental data generated from simulation-based multi-objective optimization (SMO). Simulation is a common approach to solve industrial problems. The detail level of the simulation model can vary from a conceptual model to one with detailed operation logics, but stochastic processes are commonly used in almost all production simulation models to model variability, e.g., due to machine failures. It is such a challenge, imposed by stochastic simulation, commonly found in production system simulation that leads to the design of some distance-based DM approaches specifically for these types of innovization applications.

2.1 Distance-Based SBI

Similar to any innovization process, the core goal of the SBI process is to discover what properties of the decision variables are in common with the solutions in the non-dominated set. At the same time, by analyzing the design variables together with the objective function values, the distinguishing factors that separate good solutions from lesser ones can also be determined. There are two issues that complicate the latter analysis:

- Due to the stochastic behavior of the simulation model, the binary classification of a solution as either non-dominated or dominated is not entirely trustworthy.
- A solution closer to the PF is more likely to possess attribute values that resemble a non-dominated solution more than a solution which is far away from the PF would. This is particularly true for an MOO problem with continuous decision variables and objective functions.

To address these issues, a distance-based approach to perform pre-processing on the dataset generated from SMO, which differs significantly from other innovization approaches, has been devised. Instead of treating dominated and non-dominated solutions as belonging to two different classes, and hence finding distinguishing factors by building classification models, the problem is converted into a regression problem, in which the distance to the PF is used as the dependent continuous variable in the subsequent DM process. The task is therefore to find factors to distinguish solutions with short distances to the PF from solutions with long distances, as mentioned earlier. As an example, a 2-D interpolation curve for a two-objective MOO problem, in which the minimum Euclidian distances of the solutions to the interpolated curve of the PF are represented by the color scale, is illustrated in Fig. 1 (see [14] for more details of the calculations).

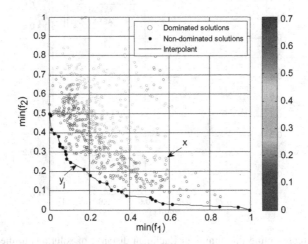

Fig. 1. Color representing the minimum Euclidian distance of solutions to the interpolation curve in a 2-objective MOO problem.

2.2   An ASF-Based Distance Calculation

While the above distance calculation is easy to apply and has been combined with K-means clustering so that the decision-maker can choose a preferred segment for the DM analysis, a better alternative for the decision-maker is using a reference point (RP) approach to explicitly specify the preferred region, which is perhaps the most important way to provide preference information in an interactive optimization procedure [16]. Suggested by Wierzbicki [17–19], the RP approach can be used to find a weakly Pareto-optimal solution closest to a supplied aspiration level RP, reflecting desirable values for the objective function, based on solving an achievement scalarizing problem. Given a RP z for an MOO problem of minimizing $(f_1(x), \ldots, f_M(x))$ with $x \in S$, the following single-objective optimization is solved:

$$\text{Minimize} \quad max_{i=1}^{M}[w_i(f_i(x) - z_i)]$$

where w_i is the ith component of a chosen weight vector used for scalarizing the objectives. If the decision-maker is interested in biasing some objectives more than others, a suitable weight vector can be used with the RP selected.

The use of RP to guide an EMO algorithm was proposed earlier in [20] and later in [21] and [22]. Here, we are interested in investigating the use of RP for the decision-maker to specify the preference for the distance-based SBI analysis. This concept is illustrated in Fig. 2. For a 2-objectives problem, a Pareto-optimal solution can be found by solving the above achievement scalarizing function (ASF), using the RP and weight vector supplied by the decision-maker. Therefore, the Euclidean distances for all solutions with respect to this so-called ASF solution can be calculated as shown by the color scale in Fig. 2.

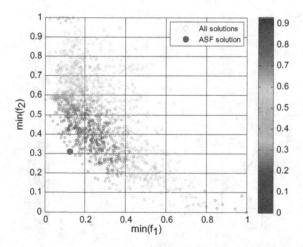

Fig. 2. Color representing the minimum Euclidian distance of solutions to the ASF point.

2.3 Interleaving MOO, Knowledge Discovery and Decision-Making

With the introduction of ASF-based distance calculation, a general framework of an interactive MOO, knowledge discovery and decision-making procedure is proposed as shown in Fig. 3. Many powerful algorithms can be used for the knowledge discovery to generate the decision rules. Among the various DM models, decision trees [23] are particularly appealing for the purpose of innovization, because they provide interpretable (non-opaque) predictive models. In the context of the distance-based approach, the extracted decision tree models can provide insights into how the decision variables should be configured in order to obtain a close-to-optimal and close-to-preferred solution based on the ASF solution. In other words, the purpose of using DM in this context is to find relationships of the decision variables to explain why certain solutions are closer than others to the ASF solution. It is also important to note that, unlike other innovization approaches in the literature, e.g. [3], by feeding the DM algorithms with all the solutions generated in an MOO, both Pareto and non-Pareto ones, the extracted models can also provide information about the values and relationships of the decision variables constituent in the poor solutions.

The main purpose of SBI is to find innovative principles and to present novel information to the user. It is hence of great interest to pay attention to the visualization step of the SBI process. This step can be divided into two parts: the first is to interpret the results of the decision tree analysis and the second is to present the discovered information to the user in a comprehensible way. In terms of extracting important rules from the decision tree analysis, each node in the decision tree corresponds to one

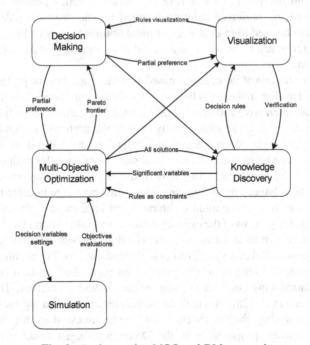

Fig. 3. An interactive MOO and DM approach.

rule, which is represented by two parts: an antecedent set of conditions and the consequent averaged regression value (r_v). The elements in the antecedent set of conditions consist of a design vector (x_1, ..., x_n) and its corresponding values (v_1, ..., v_n) which are linked by an operator ($op1$, ..., opn), as shown in the following form:

$$\text{Rule j:} \mathbf{IF}(x_1 \, \mathbf{op1} \, v_1) \mathbf{AND} \ldots \mathbf{AND}(x_i \, \mathbf{opn} \, v_i) \mathbf{THEN} \, r_v = d$$

For a rule to be "interesting", $r_v=d$ (d is the predicted Euclidian distance) must be sufficiently small. Therefore, all nodes are checked in order to determine all the rules with d below a certain threshold of interestingness. Such a threshold ensures that the high interestingness of the selected rules can be predetermined or determined at a later stage by the decision-maker in conjunction with the visualization of the extracted rules.

In general, since EMO algorithms do not use any mathematical optimality conditions in their operators, the solutions obtained after a finite number of computations are not guaranteed to be optimal, although an asymptotic convergence for EMOs with certain restrictions has been proven in the past [24]. To enhance the convergence properties of EMO algorithms, one common approach is to first apply an EMO and the solutions obtained are then modified one at a time by using a local search procedure. Although this hybrid method is commonly employed, the overall computational effort needed to execute the local search for each EMO solution can be burdensome. Therefore, an alternative strategy, which involves the hybrid use of MOO and DM, also schematically illustrated in Fig. 3, is proposed in this paper. First, an MOO is run to generate a data set of sufficient size for the innovization study using DM techniques. Since the derived rules can reveal salient properties present in the MOO solutions close to the preferred solution selected by the decision-maker, the rules obtained can then be used to re-define the original optimization MOO problem, so that a faster convergence can be achieved, compared to a single application of EMO to the original problem.

With the introduction of the distance-based approach based on the preference region selected by the decision-maker, it is believed that faster and "better" optimization can be achieved in an interactive and interleaved manner. An interactive manner means the decision-maker can select and subsequently change the preferred region by choosing different reference points. With interleaving, it implies that several iterations of the MOO-DM cycles can be repeated in order to obtain important rules with respect to the preference of the decision-maker. In addition, the efficiency of the optimization can simultaneously be enhanced by having the optimization converge faster to the preferred region. While there is a vast amount of literature on interactive multi-objective optimization (see e.g. [25]), most of the existing methods in Multi-criteria Decision Making (MCDM) literature aim to assist the user in selecting the best solution through some interaction processes. Related visualization techniques in [26, 27] are also targeted to help the decision-maker analyze the Pareto solutions, particularly in the objective space, but not the relationships between decision variables and objectives. The approach proposed by Greco et al. [28], in which decision rules are used to represent user preferences and also to describe the Pareto front, is very relevant to this current work. Similar to the concept proposed here, the Dominance-based Rough Set Approach

(DRSA) described in the article can also be used for the progressive exploration of the Pareto optimal set, which is interesting from the point of view of the decision-maker's preferences as well as using the extracted rules to refine the original constraints of the optimization problem. Nevertheless, there are some key contrasts between SBI and other approaches like DRSA in the MCDM literature. In practical problems like production systems engineering, decision-makers usually have a strong interest in the decision space. While they indicate their preference in the objective space, they would prefer to acquire information about the values of the decision variables that put the solutions closer to their targeted region in the objectives. The preference model using, e.g., the pairwise comparison of solutions is also less suitable for this type of problem with stochastic continuous objectives, because a large number of solutions can be close to the preferred solution. In order to illustrate the applicability of the proposed interleaved SBI approach for practical production system problems, we are more interested in the results when it is applied in studies of real-world production optimization problems, instead of some theoretical benchmarking functions.

3 Industrial Application Study

The application case study presented here was part of an industrial-based research project conducted at some automotive manufacturers in Sweden. As a matter of fact, many industrial, production systems engineering problems can be effectively formulated as some optimization problems, because their aim is to find the optimal setting, through re-allocations of workforce, work-in-process and/or buffers, to improve the performance of the system [29]. One objective of the project is to verify and evaluate the combined use of SMO and innovization, as a new and innovative manufacturing management toolset to support decision-making [30]. For the SMO and the subsequent SBI methodology to be useful in solving real-world manufacturing decision-making problems, the optimization of production systems via simulation models that take into account both productivity and environmental factors related to the decision-making process is essential. In other words, formulating the optimization objectives related to both productivity and environmental issues, such as energy efficiency, was the first step.

The production system considered in this industrial case study is a truck gears machining cell comprising seven machines connected by conveyors and industrial robots. The cell produces five types of truck gears with different cycle (processing) times on different machines and is usually staffed by 2 operators who perform manual tasks including tool changes, setups, and measurements. In the case study, some new processing plans involving gear machining and cutting-time changes had been carried out, which required the cell to be re-configured. The goal of the study was therefore to investigate how changes in capacities, conveyor (buffer) sizes, setup sequences, operator utilization, and planning schemes would affect the productivity and energy efficiency cost, in order to optimally re-configure the cell. Figure 4 schematically illustrates the product flow of the cell and its major components. There are five lathes (L1-5) and two parallel milling machining centers (M1-2). Work-pieces flow from the raw material inventory (RMI) to L1, L2 and L3 through some long conveyors (CB1-3) which are not only used for material handling, but also serve as in-process buffers for temporarily storing the work-pieces. The buffer sizes are therefore determined by the lengths of the conveyors. An identical type of conveyor buffer is located before L5 (CB5). Operators are located in two regions and they serve different machine groups. As later indicated, the number of workers in each region (W1 or W2) is also a decision variable in the MOO study.

A discrete-event simulation model of the production line was developed for the machining cell and used in the simulation-based optimization of this study. Readers are referred to [31] for the full details of the cell and the simulation model.

3.1 Optimization Objectives and Decision Variables

The key purpose of the simulation-based optimization is to seek the optimum parameter selection to optimize the following key performance indicators:

- Maximize the number of gears produced, i.e., maximize production of the cell.
- Maximize the machine and worker utilizations.
- Minimize the number of work-in-process (WIP).
- Minimize the energy consumption.

Since, in general, high utilization is only a secondary objective, it is usually correlated with higher production. Therefore, in the case study, only three objectives were considered in the SMO, namely, *Maximize (Production), Minimize (WIP), Minimize (Energy Usage)*. In the simulation model, the energy consumption (E) of a machine is calculated according to the following formula using the runtime and down-time proportion of that machine during the entire simulation period:

$$E = \left(P_{run} \times E_{run} + P_{down} \times E_{down} + (1 - P_{run} - P_{down})E_{standby}\right) \times SimTime$$

where

P_{run}	= Proportion of the time the machine is in working mode.
P_{down}	= Proportion of the time the machine is in down mode, including setup and tool changes.
E_{run}	= Energy consumption per unit time when the machine is in working mode.
E_{down}	= Energy consumption per unit time when the machine is in down mode.
$E_{standby}$	= Energy consumption per unit time when the machine is in standby mode.

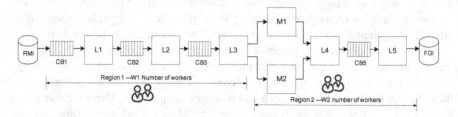

Fig. 4. Schematic illustration of the components and product flow of the machining cell.

SimTime   = Total simulation time.

A total of nine decision variables were considered in the SMO:

- Conveyor (buffer) lengths before L1, 2, 3 and 5, i.e., CB_1, CB_2, CB_3 and CB_5.
- Production scheme, P_S [1...5]; 5 different pre-defined production plans, batch sizes and product mix that the manufacturer wanted to select.
- Lathe setting (L_S); the company provided two sets of lathe settings with different cycle times and change-over intervals.
- Milling m/c setting, M_S.
- Number of workers in region 1 (W_1) and region 2 (W_2), as shown in Fig. 4.

3.2 Optimization Results and SBI Analysis

Five thousand simulation evaluations were run with NSGA-II [5] as the MOO algorithm. Figure 5 plots the optimization results (all 5000 evaluations) and highlights the non-dominated solutions.

In Fig. 5, it is very clearly indicated that there are several clusters in the objective space. By using an interactive user interface, the decision-maker can browse the color change in the 4D plot by adjusting each decision variable. The color change during the browsing allows the relationship between the selected decision variable and the objectives in the 3D plot to be visualized and interpreted, as shown in Fig. 6. Applying this procedure to the data set generated from the SMO has revealed that there are mainly two decision variables that determine which objective cluster a solution will fall into; high production can only be achieved when L_S is 1 and the P_s is the parameter which further divides the solutions into 8 clusters in the entire 3D objective space. The variables controlling the conveyor lengths, which indirectly control the

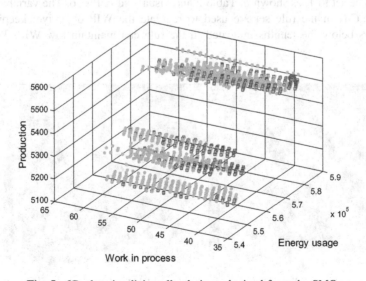

Fig. 5. 3D plot visualizing all solutions obtained from the SMO.

buffer sizes between workstations, contribute to controlling the final average WIP of the whole cell, which is logical and not being shown here.

Apart from the visualization of the solutions generated from the MOO, we used the ASF-based distance SBI data mining process to extract decision rules from the entire MOO data set. In this study, the decision-maker had chosen the reference point (RP) to be [WIP = 35, EnergyUsage = 550000 (kW), Production = 5700]. This is an ideal performance indicator, due to the relatively low WIP and low energy usage, but it can achieve the highest possible total production, which also reflects a typical decision strategy of production managers. In Fig. 5, this RP is already highlighted as the red dot. A set of weight vectors had been tried with such a RP, which projected to different non-dominated solutions that were presented to the decision-maker:

$$ASF[1, 1, 1] = [47, 553566.1, 5402]$$
$$ASF[1, 1, 10] = [46, 576910.1, 5587]$$
$$ASF[1, 1, 10] = [46, 552152, 5380]$$
$$ASF[1, 1, 10] = [36, 551462.1, 5342]$$
$$ASF[1, 10, 10] = [36, 576840.3, 5585]$$

The weight vector [10, 1, 10] was finally chosen because of the desirable high production and very low WIP attributes, even though energy usage would be sacrificed somewhat. The rule set in Table 1 was obtained by using the ASF point [36, 576840.3, 5585] to calculate the Euclidean distance for all the solutions in the solution set.

The rules generated have clearly verified the importance of the key decision variable, L_S, as the main splitting variable which divides the solution set into the two distinct regions, one with lower energy usage and less production and the other with higher energy usage but also higher production. In order to maximize the production, L_S has to be set to 1, as shown in Table 2 and visualized in Fig. 6. The variables CB_1, CB_2 and CB_5 in the rule set are used to regulate the WIP objective; keeping the conveyors below the lengths indicated in the rule can maintain low WIP. The rule

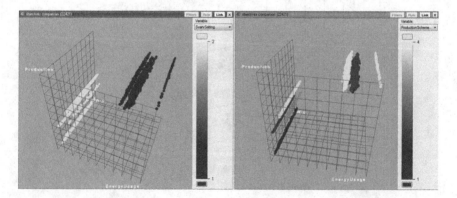

Fig. 6. 4D plot visualizing how key decision variables divide the solutions into clusters.

extracted may conclude that the other decision variables, i.e., CB_3, W_1 and W_2 have less influence, with respect to the ASF point chosen by the decision-maker.

The accuracy of the extracted rule can be partially verified by highlighting the solutions that have attributes represented by the rule "$L_S = 1 \vee CB_1 < 5.61 \vee CB_2 < 5.43 \vee P_S = 2 \vee CB_5 < 1.705$" in the objective space with all solutions, as shown in Fig. 7 (the ASF point indicated by the blue dot).

3.3   Local Search Using the Extracted Rule as Constraints

Another way to verify that the extracted rule set is reliable is by re-running the optimization using the selected rule as the constraints for decision variables. Table 3 shows the new upper boundaries for CB1, CB2 and CB5 when they are limited according to Rule no. 5 in Table 1, whereby L_S and P_S are restricted to 1 and 2 respectively.

By considering the visualization of the comparison between the original solution set (the grey dots) and the solution set of the second optimization run (the blue triangles) in Fig. 8, it is clear that the rule set which was used as the new constraints does capture the main attributes of the solutions closer to the reference point.

With an aim to verify that the ASF-based approach can be used to control the convergence of the optimization towards the preference region, DM was run with the ASF [1, 10, 1], which resulted in a PF solution with the lowest WIP and very low energy usage. In other words, the Production objective is sacrificed. DM results are

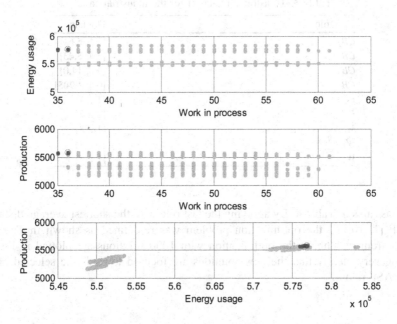

Fig. 7. 2D plots visualizing how close the solutions represented by the rule are to the ASF point for the problem with 3 objectives.

Table 1. The influence of the rules on the average distance.

Rule applied	No of solutions	Average distance
No rule, all solutions	5 000	0.815
$L_S = 1$	2 005	0.414
$L_S = 1 \vee CB_1 < 5.61$	743	0.204
$L_S = 1 \vee CB_1 < 5.61 \vee CB_2 < 5.43$	394	0.123
$L_S = 1 \vee CB_1 < 5.61 \vee CB_2 < 5.43 \vee P_S = 2$	222	0.068
$L_S = 1 \vee CB_1 < 5.61 \vee CB_2 < 5.43 \vee P_S = 2 \vee CB_5 < 1.705$	183	0.057

Table 2. Influence of the rule set on the objectives.

Rule applied	min / max WIP	min / max Production	min / max Energy
No rule, all solutions	35 / 61	5168 / 5587	548764 / 583340
$L_S = 1$	35 / 61	5500 / 5587	572190 / 583340
$L_S = 1 \vee CB_1 < 5.61$	35 / 51	5503 / 5585	572330 / 583220
$L_S = 1 \vee CB_1 < 5.61 \vee CB_2 < 5.43$	35 / 45	5519 / 5585	573020 / 583220
$L_S = 1 \vee CB_1 < 5.61 \vee CB_2 < 5.43 \vee P_S = 2$	35 / 41	5550 / 5585	575180 / 576840
$L_S = 1 \vee CB_1 < 5.61 \vee CB_2 < 5.43$ $\vee P_S = 2 \vee CB_5 < 1.705$	35 / 38	5550 / 5585	575180 / 576840

Table 3. Constrained data set for the industrial case.

Variable	Domain
CB_1	[5.1, **5.61**]
CB_2	[3.4, **5.43**]
CB_3	[3.0, 14.0]
CB_5	[0.5, **1.705**]
L_S	[**1** .. **1**]
P_S	[**2** .. **2**]
M_S	[1 .. 8]
W_1	[1 .. 3]
W_2	[1 .. 3]

shown as rules in Table 4. By applying the last rule with the shortest average distance to ASF [1, 10, 1], the optimization problem was re-defined as shown in Table 5. Results from re-running the optimization with 1000 solutions are plotted in Fig. 9, showing very clearly that the new solutions are focused towards the selected alternative ASF point.

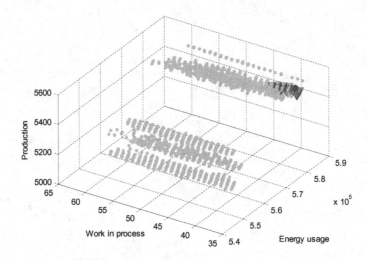

Fig. 8. Solutions from the second optimization run (blue triangles) with the refined constraints; grey solutions from previous optimization; red dot=[36, 576840.3, 5585].

Table 4. The influence of the rules on the average distance.

Rule applied	No of solutions	Average distance
No rule, all solutions	5 000	0.649
$L_S = 2$	2 995	0.441
$L_S = 2 \lor CB_1 < 5.33$	705	0.245
$L_S = 2 \lor CB_1 < 5.33 \lor CB_2 < 5.59$	397	0.195
$L_S = 2 \lor CB_1 < 5.33 \lor CB_2 < 5.59 \lor P_S = 2$	186	0.089

Table 5. Constrained data set for the industrial case.

Variable	Domain
CB_1	[5.1, **5.33**]
CB_2	[3.4, **5.59**]
CB_3	[3.0, 14.0]
CB_5	[0.5, 7]
L_S	[**2..2**]
P_S	[**2..2**]
M_S	[1..8]
W_1	[1..3]
W_2	[1..3]

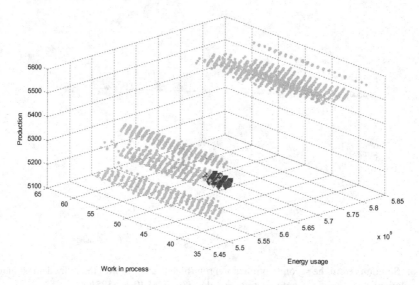

Fig. 9. Solutions from the optimization run (blue triangles) with the refined constraints based on new selected rule; grey solutions from first optimization; red dot=[36, 551462.1, 5342].

4 Conclusions

In this paper, we describe an extension of the SBI procedure to extract knowledge from SMO. The SBI process is based on the post-optimality analysis of Pareto-optimal solutions, to discover knowledge, in terms of rules/principles that relate key influencing decision variables and objectives. Recent work in using DM techniques to automate the post-optimality analysis of Pareto-optimal solutions has shown that some engineering design problems can be successfully handled. Not long ago, we proposed a distance-based data pre-processing approach specifically to generate high-quality rules from stochastic simulation models for real-world production systems. In the paper, this approach is further extended by the introduction of interleaving some MOO-DM cycles, in order to enhance the efficiency of the optimization, in terms of faster convergence to the preferred region in the objective space selected by the decision-maker. Such an enhanced MOO procedure has been demonstrated with a simulation model developed in an industrial production system optimization study. In our future work, we will apply the methodology to address complex production systems engineering problems with more decision variables, as well as continue some quantitative comparisons with other innovization approaches.

References

1. Deb, K.: Multi-Objective Optimization using Evolutionary Algorithms. Wiley, Chichester (2004)
2. Deb, K., Srinivasan, A., Innovization: innovating design principles through optimization. In: Proceedings of the 8th Annual Conference on Genetic and Evolutionary Computation, Seattle, USA, July 2006, pp. 162–1636 (2006)
3. Bandaru, S., Deb, K.: Automated discovery of vital knowledge from Pareto optimal solutions: first results from engineering design. In: IEEE Congress on Evolutionary Computation, CEC '10, pp. 1–8 (2010)
4. Bandaru, S., Deb, K.: Towards automating the discovery of certain innovative design principles through a clustering based optimization technique. Eng. Optim. **43**(9), 911–941 (2011)
5. Deb, K., Pratap, A., Agarwal, S., Meyarivan, T.: A fast and elitist multi-objective genetic algorithm: NSGA-II. IEEE Trans. Evol. Comput. **6**(2), 181–197 (2002)
6. Obayashi, S., Sasaki, D.: Visualization and data mining of Pareto solutions using self-organizing map. In: Fonseca, C.M., Fleming, P.J., Zitzler, E., Deb, K., Thiele, L. (eds.) EMO 2003. LNCS, vol. 2632, pp. 796–809. Springer, Heidelberg (2003)
7. Jeong, S., Chiba, K., Obayashi, S.: Data mining for aerodynamic design space. J. Aerosp. Comput. Inf. Commun. **2**(11), 452–469 (2005)
8. Sugimura, K., Obayashi, S., Jeong, S.: Multi-objective design exploration of a centrifugal impeller accompanied with a vaned diffuser. In: Proceedings of FEDSM2007, 5th Joint AME/JSME Fluids Engineering Conference, San Diego, USA, 30 July–2 August, pp. 939–946 (2007)
9. Oyama, A., Nonomura, T., Obayashi, S.: Data mining of Pareto optimal transonic airfoil shapes using proper orthogonal decomposition. In: Proceedings of 19th AIAA Computational Fluid Dynamics, San Antonio, USA, 22–25 June, pp. 1514–1523 (2009)
10. Liebscher, M., Witowski, K, Goel, T.: Decision making in multi-objective optimization for industrial applications – data mining and visualization of Pareto data. In: The 8th World Congress on Structural and Multidisciplinary Optimization, Lisbon, Portugal, 1–5 June (2009)
11. Ng, A.H.C, Deb, K., Dudas, C.: Simulation-based innovization for production systems improvement: an industrial case study. In: Proceedings of the International 3rd Swedish Production Symposium, Göteborg, Sweden, December 2009, pp. 278–286 (2009)
12. Dudas, C., Frantzén, M., Ng, A.H.C.: A synergy of multi-objective optimization and data mining for the analysis of a flexible flow shop. Robot. Comput. Integr. Manuf. **27**(4), 687–695 (2011)
13. Ng, A.H.C., Dudas, C., Nießen, J., Deb, K.: Simulation-based innovization using data mining for production systems analysis. In: Wang, L., Ng, A., Deb, K. (eds.) Evolutionary Multi-objective Optimization in Product Design and Manufacturing, pp. 401–430. Springer, London (2011)
14. Ng, A.H.C., Dudas, C., Pehrsson, L., Deb, K.: Knowledge discovery in production simulation by interleaving multi-objective optimization and data mining. In: Proceedings of the 5th Swedish Production Symposium (SPS'12), Linköping, Sweden, 6–8 November 2012, pp. 461–471 (2012)
15. Deb, K., Datta, R.: Hybrid evolutionary multiobjective optimization and analysis of machining operations. Eng. Optim. **44**(6), 685–706 (2011)

16. Branke, J.: Consideration of partial user preferences in evolutionary multiobjective optimization. In: Branke, J., Deb, K., Miettinen, K., Słowiński, R. (eds.) Multiobjective Optimization. LNCS, vol. 5252, pp. 157–178. Springer, Heidelberg (2008)
17. Wierzbicki, A.P.: The use of reference objectives in multiobjective optimization. In: Fandel, G., Gal, T. (eds.) Multiple Criteria Decision Making Theory and Applications, pp. 468–486. Springer, Berlin (1980)
18. Wierzbicki, A.P.: A mathematical basis for satisficing decision making. Math. Model. **3**, 391–405 (1982)
19. Wierzbicki, A.P.: On the completeness and constructiveness of parametric characterizations to vector optimization problems. OR Spektrum **8**, 73–87 (1986)
20. Fonseca, C.M., Fleming, P.J.: Genetic algorithms for multiobjective optimization: formulation, discussion, and generalization. In: International Conference on Genetic Algorithms, pp. 416–423 (1993)
21. Fonseca, C.M., Fleming, P.J.: Multiobjective optimization and multiple constraint handling with evolutionary algorithms - Part I: a unified formulation. IEEE Trans. Syst. Man Cybern. Part A **28**(1), 26–37 (1998)
22. Deb, K., Sundar, J., Reddy, U.B., Chaudhuri, S.: Reference point based multiobjective optimization using evolutionary algorithms. Int. J. Comput. Intell. Res. **2**(3), 273–286 (2006)
23. Breiman, L., Friedman, J.H., Olshen, R.A., Stone, C.J.: Classification and Regression Trees. Wadsworth, Belmont (1984)
24. Rudolph, G., Agapie, A.: Convergence properties of some multi-objective evolutionary algorithms. In: Proceedings of the 2000 congress on evolutionary computation (CEC2000), 16–19 July 2000, San Diego, CA, pp. 1010–1016. IEEE Press, Piscataway (2000)
25. Miettinen, K., Ruiz, F., Wierzbicki, A.P.: Introduction to multiobjective optimization: interactive approaches. In: Branke, J., Deb, K., Miettinen, K., Słowiński, R. (eds.) Multiobjective Optimization. LNCS, vol. 5252, pp. 27–57. Springer, Heidelberg (2008)
26. Korhonen, P., Wallenius, J.: Visualization in the multiple objective decision-making framework. In: Branke, J., Deb, K., Miettinen, K., Słowiński, R. (eds.) Multiobjective Optimization. LNCS, vol. 5252, pp. 195–212. Springer, Heidelberg (2008)
27. Lotov, A.V., Miettinen, K.: Visualizing the Pareto frontier. In: Branke, J., Deb, K., Miettinen, K., Słowiński, R. (eds.) Multiobjective Optimization. LNCS, vol. 5252, pp. 213–243. Springer, Heidelberg (2008)
28. Greco, S., Matarazzo, B., Słowiński, R.: Dominance-based rough set approach to interactive multiobjective optimization. In: Branke, J., Deb, K., Miettinen, K., Słowiński, R. (eds.) Multiobjective Optimization. LNCS, vol. 5252, pp. 121–155. Springer, Heidelberg (2008)
29. Li, S., Meerkov, S.M.: Production Systems Engineering. Springer, New York (2008)
30. Pehrsson, L., Ng, A.H.C., Bernedixen, J.: Multi-objective production system optimisation including investment and running costs. In: Wang, L., Ng, A., Deb, K. (eds.) Evolutionary Multi-objective Optimization in Product Design and Manufacturing, pp. 431–454. Springer, London (2011)
31. Hossain, M., Harari, N., Semere, D., Mårtensson, P., Ng, A.H.C., Andersson, M.: Integrated modeling and application of standardized data schema. In: Proceedings of the 5th Swedish Production Symposium (SPS'12), Linköping, Sweden, 6–8 November 2012

Intelligent Optimization for the Minimum Labelling Spanning Tree Problem

Sergio Consoli[1]([⊠]), José Andrés Moreno Pérez[2], and Nenad Mladenović[3]

[1] Joint Research Centre, European Commission, Via Enrico Fermi 2749,
21027 Ispra, VA, Italy
[2] DEIOC, IUDR, Facultad de Matemáticas, Universidad de La Laguna,
4a planta Astrofisico F. Sánchez s/n, 38271 Santa Cruz de Tenerife, Spain
[3] School of Information Systems, Computing and Mathematics,
Brunel University, Uxbridge, Middlesex UB8 3PH, UK
sergio.consoli@istc.cnr.it, jamoreno@ull.es,
nenad.mladenovic@brunel.ac.uk

Abstract. Given a connected, undirected graph whose edges are labelled (or coloured), the minimum labelling spanning tree (MLST) problem seeks a spanning tree whose edges have the smallest number of distinct labels (or colours). In recent work, the MLST problem has been shown to be NP-hard and some effective heuristics have been proposed and analysed. In this paper we present preliminary results of a currently ongoing project regarding the implementation of an intelligent optimization algorithm to solve the MLST problem. This algorithm is obtained by the basic Variable Neighbourhood Search heuristic with the integration of other complements from machine learning, statistics and experimental algorithmics, in order to produce high-quality performance and to completely automate the resulting optimization strategy.

Keywords: Combinatorial optimization · Graphs and networks · Minimum labelling spanning trees · Intelligent optimization · Hybrid local search

1 Preliminary Discussion

In a currently ongoing project, we investigate a new possibility for solving the *minimum labelling spanning tree* (MLST) by an intelligent optimization algorithm. The minimum labelling spanning tree problem is a challenging combinatorial problem [1]. Given an undirected graph with labelled (or coloured) edges as input, with each edge assigned with a single label, and a label assigned to one or more edges, the goal of the MLST problem is to find a spanning tree with the minimum number of labels (or colours).

The MLST problem can be formally formulated as a network or graph problem [2]. We are given a labelled connected undirected graph $G = (V, E, L)$, where V is the set of nodes, E is the set of edges, and L is the set of labels. The purpose is to find a spanning tree T of G such that $|L_T|$ is minimized, where L_T

G. Nicosia and P. Pardalos (Eds.): LION 7, LNCS 7997, pp. 19–23, 2013.
DOI: 10.1007/978-3-642-44973-4_2, © Springer-Verlag Berlin Heidelberg 2013

is the set of labels used in T. Although a solution to the MLST problem is a spanning tree, it is easier to work firstly in terms of feasible solutions. A feasible solution is defined as a set of labels $C \subseteq L$, such that all the edges with labels in C represent a connected subgraph of G and span all the nodes in G. If C is a feasible solution, then any spanning tree of C has at most $|C|$ labels. Moreover, if C is an optimal solution, then any spanning tree of C is a minimum labelling spanning tree. Thus, in order to solve the MLST problem we first seek a feasible solution with the least number of labels [3].

The MLST problem was first introduced in [1]. The authors also proved that it is an NP-hard problem and provided a polynomial time heuristic, the Maximum Vertex Covering Algorithm (MVCA), successively improved in [4]. Other heuristics for the MLST problem have been proposed in the literature [2, 3, 5–9].

The aim of this paper is to present preliminary results concerning the design of a novel heuristic solution approach for the MLST problem, with the goal of obtaining high-quality performance. The proposed optimization strategy is an intelligent hybrid metaheuristic, obtained by combining Variable Neighbourhood Search (VNS) [10] and Simulated Annealing (SA) [11], with the integration of other complements in order to improve the effectiveness and robustness of the optimization process, and to completely automate the resulting solution strategy.

2   Complementary Variable Neighbourhood Search

The first extension that we introduce for the MLST problem is a local search mechanism that is inserted at top of the Variable Neighbourhood Search metaheuristic [10]. The resulting local search method is referred to as *Complementary Variable Neighbourhood Search* (COMPL).

For our implementation, given a labelled graph $G = (V, E, L)$, with n vertices, m edges, ℓ labels, each solution is encoded as a binary string, i.e. $C = (c_1, c_2, \ldots, c_\ell)$ where $c_i = 1$ if label i is in solution C, $c_i = 0$ otherwise, $\forall i = 1, \ldots, \ell$.

Given a solution C, COMPL extracts a solution from the *complementary space* of C, and then replaces the current solution with the solution extracted. The complementary space of a solution C is defined as the set of all the labels that are not contained in C, that is $(L \Delta C)$. To yield the solution, COMPL applies a constructive heuristic, such as the MVCA [1, 4], to the subgraph of G with labels in the complementary space of the current solution. Note that COMPL stops if either a feasible solution is obtained (i.e. a single connected component is obtained), or the set of unused labels contained in the complementary space is empty, (i.e. $(L \Delta C) = \emptyset$), producing a final infeasible solution. Then, the basic VNS is applied in order to improve the resulting solution. At the starting point of VNS, it is required to define a suitable neighbourhood structure of size k_{max}. The simplest and most common choice is a structure in which the neighbourhoods have increasing cardinality: $|N_1(\cdot)| < |N_2(\cdot)| < \ldots < |N_{k_{max}}(\cdot)|$. In order to impose a neighbourhood structure on the solution space S, comprising all possible solutions, we define the distance between any two such solutions $C_1, C_2 \in S$, as the Hamming distance: $\rho(C_1, C_2) = |C_1 \Delta C_2| = \sum_{i=1}^{\ell} \lambda_i$, where $\lambda_i = 1$ if label i is included in one of the solutions but not in the other, and 0 otherwise, $\forall i = 1, ..., \ell$. VNS starts from an initial solution C with k increasing from 1 up to the maximum neighborhood size, k_{max}, during the progressive execution.

The basic idea of VNS to change the neighbourhood structure when the search is trapped at a local minimum, is implemented by the shaking phase. It consists of the random selection of another point in the neighbourhood $N_k(C)$ of the current solution C. Given C, we consider its kth neighbourhood, $N_k(C)$, as all the different sets having a Hamming distance from C equal to k labels, where $k \leftarrow 1, 2, \ldots, k_{max}$. In order to construct the neighbourhood of a solution C, the algorithm first proceeds with the deletion of labels from C. In other words, given a solution C, its kth neighbourhood, $N_k(C)$, consists of all the different sets obtained from C by removing k labels, where $k \leftarrow 1, 2, ..., k_{max}$. In a more formal way, given a solution C, its kth neighbourhood is defined as $N_k(C) = \{S \subset L : (|C \Delta S|) = k\}$, where $k \leftarrow 1, 2, ..., k_{max}$.

The iterative process of selection of a new incumbent solution from the complementary space of the current solution if no improvement has occurred, is aimed at increasing the diversification capability of the basic VNS for the MLST problem. When the local search is trapped at a local minimum, COMPL extracts a feasible complementary solution which lies in a very different zone of the search domain, and is set as new incumbent solution for the local search. This new starting point allows the algorithm to escape from the local minimum where it is trapped, producing an immediate peak of diversification.

3 The Intelligent Optimization Algorithm

In order to seek further improvements and to automate on-line the search process, Complementary Variable Neighbourhood Search has been modified by replacing the inner local search based on the deterministic MVCA heuristic with a *probability-based local search* inspired by a "Simulated Annealing cooling schedule" [11], with the view of achieving a proper balance between intensification and diversification capabilities. The strength of this probabilistic local search is tuned by an automated process which allows the intelligent strategy to adapt on-line to the problem instance explored and to react in response to the search algorithm's behavior [12]. The resulting metaheuristic represents the intelligent optimization algorithm that we propose for the MLST problem.

The probability-based local search is another version of the MVCA heuristic, but with a probabilistic choice of the next label to be added. It extends the basic greedy construction criterion of the MVCA by allowing moves to worse solutions. Starting from an initial solution, successively a candidate move is randomly selected; this move is accepted if it leads to a solution with a better objective function value than the current solution, otherwise the move is accepted with a probability that depends on the deterioration, Δ, of the objective function value.

Following the SA criterion, the acceptance probability is computed according to the Boltzmann function as $\exp(-\Delta/T)$, using the temperature (T) as control parameter. The value of T is initially high, which allows many worse moves to be accepted, and is gradually reduced following a specific geometric cooling schedule:

$$T_{k+1} = \alpha \cdot T_k \qquad \text{where} \begin{cases} T_0 = |Best_C|, \\ \alpha = 1/|Best_C| \in [0,1], \end{cases} \qquad (1)$$

with $Best_C$ being the current best solution, and $|Best_C|$ its number of labels. This cooling law is very fast for the MLST problem, yielding a good balance between intensification and diversification. Furthermore, thanks to its self-tuning parameters setting, which is guided automatically by the best solution $Best_C$ without requiring any user-intervention, the algorithm is allowed to adapt on-line to the problem instance explored and to react in response to the search algorithm's behavior [12].

The aim of the probabilistic local search is to allow, with a specified probability, worse components with a higher number of connected components to be added to incomplete solutions. Probability values assigned to each label are inversely proportional to the number of components they give. So the labels with a lower number of connected components will have a higher probability of being chosen. Conversely, labels with a higher number of connected components will have a lower probability of being chosen. Thus, the possibility of choosing less promising labels is allowed. Summarizing, at each step the probabilities of selecting labels giving a smaller number of components will be higher than the probabilities of selecting labels with a higher number of components. Moreover, these differences in probabilities increase step by step as a result of the reduction of the temperature for the adaptive cooling schedule. It means that the difference between the probabilities of two labels giving different numbers of components is higher as the algorithm proceeds. The probability of a label with a high number of components will decrease as the algorithm runs and will tend to zero. In this sense, the search becomes MVCA-like.

A simple VNS implementation which uses the probabilistic local search as constructive heuristic has been tested. However, the best results were obtained by combining Complementary Variable Neighbourhood Search with the probabilistic local search, resulting in the hybrid intelligent algorithm that we propose. Note that the probabilistic local search is applied both in COMPL, to obtain a solution from the complementary space of the current solution, and in the inner local search phase, to restore feasibility by adding labels to incomplete solutions.

4   Summary and Outlook

Concerning the achieved optimization strategy, the whole approach seems to be highly promising for the MLST problem. Ongoing investigation will consist in a statistical comparison of the resulting strategy against the best MLST algorithms in the literature, in order to quantify and qualify the improvements obtained by the proposed intelligent optimization algorithm.

References

1. Chang, R.S., Leu, S.J.: The minimum labelling spanning trees. Inf. Process. Lett. **63**(5), 277–282 (1997)
2. Consoli, S., Darby-Dowman, K., Mladenović, N., Moreno-Pérez, J.A.: Greedy randomized adaptive search and variable neighbourhood search for the minimum labelling spanning tree problem. Eur. J. Oper. Res. **196**(2), 440–449 (2009)
3. Xiong, Y., Golden, B., Wasil, E.: A one-parameter genetic algorithm for the minimum labelling spanning tree problem. IEEE Trans. Evol. Comput. **9**(1), 55–60 (2005)
4. Krumke, S.O., Wirth, H.C.: On the minimum label spanning tree problem. Inf. Process. Lett. **66**(2), 81–85 (1998)
5. Xiong, Y., Golden, B., Wasil, E.: Improved heuristics for the minimum labelling spanning tree problem. IEEE Trans. Evol. Comput. **10**(6), 700–703 (2006)
6. Cerulli, R., Fink, A., Gentili, M., Voß, S.: Metaheuristics comparison for the minimum labelling spanning tree problem. In: Golden, B.L., Raghavan, S., Wasil, E.A. (eds.) The Next Wave on Computing, Optimization, and Decision Technologies, pp. 93–106. Springer-Verlag, New York (2005)
7. Brüggemann, T., Monnot, J., Woeginger, G.J.: Local search for the minimum label spanning tree problem with bounded colour classes. Oper. Res. Lett. **31**, 195–201 (2003)
8. Chwatal, A.M., Raidl, G.R.: Solving the minimum label spanning tree problem by ant colony optimization. In: Proceedings of the 7th International Conference on Genetic and Evolutionary Methods (GEM 2010), Las Vegas, Nevada (2010)
9. Consoli, S., Moreno-Pérez, J.A.: Solving the minimum labelling spanning tree problem using hybrid local search. In: Proceedings of the mini EURO Conference XXVIII on Variable Neighbourhood Search (EUROmC-XXVIII-VNS), Electronic Notes in Discrete Mathematics, vol. 39, pp. 75–82 (2012)
10. Hansen, P., Mladenović, N.: Variable neighbourhood search: principles and applications. Eur. J. Oper. Res. **130**, 449–467 (2001)
11. Kirkpatrick, S., Gelatt, C.D., Vecchi, M.P.: Optimization by simulated annealing. Science **220**(4598), 671–680 (1983)
12. Osman, I.H.: Metastrategy simulated annealing and tabu search algorithms for the vehicle routing problem. Ann. Oper. Res. **41**, 421–451 (1993)

A Constraint Satisfaction Approach
to Tractable Theory Induction

John Ahlgren and Shiu Yin Yuen[✉]

City University of Hong Kong, Hong Kong, China
ahlgren@ee.cityu.edu.hk, kelvin.ee@cityu.edu.hk
http://www.cityu.edu.hk

Abstract. A novel framework for combining logical constraints with
theory induction in Inductive Logic Programming is presented. The constraints are solved using a boolean satisfiability solver (SAT solver) to
obtain a candidate solution. This speeds up induction by avoiding
generation of unnecessary candidates with respect to the constraints.
Moreover, using a complete SAT solver, search space exhaustion is
always detectable, leading to faster small clause/base case induction. We
run benchmarks using two constraints: input-output specification and
search space pruning. The benchmarks suggest our constraint satisfaction approach can speed up theory induction by four orders of magnitude
or more, making certain intractable problems tractable.

Keywords: Inductive Logic Programming · Theory induction · Constraint satisfaction

1 Introduction

Inductive Logic Programming (ILP) is a branch of machine learning that
represent knowledge using predicate logic. Examples are generalized into a theory
that cover all positive and no negative examples [1,2]. Thus, positive examples
provide observables, and negative examples provide instances of what should
never be observed. The induced theories are typically given in Prolog code (first
order Horn clauses), which has both declarative and procedural interpretations
[3,4]. Thus ILP is simultaneously capable of first order concept learning and
Turing complete program synthesis. For a summary of ILP as a research field,
and its applications, we refer the reader to [1].

State of the art ILP systems—such as Progol and Aleph—use the method
of Inverse Entailment to compute a bottom clause from a positive example,
background knowledge, and a non-mandatory set of mode declarations specifying
the input-output requirements of predicates [5].

The bottom clause is intended to be a most specific clause for the example
(relative to background knowledge and mode declarations) and is used to constrain the search space by providing a bottom element in the lattice generated
by the subsumption order [2,6,7].

G. Nicosia and P. Pardalos (Eds.): LION 7, LNCS 7997, pp. 24–29, 2013.
DOI: 10.1007/978-3-642-44973-4_3, © Springer-Verlag Berlin Heidelberg 2013

A search through the lattice is then performed to find a suitable candidate solution, and all examples made redundant by the new candidate are removed. The process of generating a bottom clause from an example and searching for a candidate continues until all positive examples have been covered [8].

As for the mode declarations, they specify the data type of each variable, and whether it is an input or output. For our purposes in this paper, we are only interested in the mode declaration's specification of which variables are inputs and outputs.

Example 1. If the bottom clause is $h(X, Y) \leftarrow b_1(X, Z), b_2(Z, Y)$, there are four candidate solutions, given by (ordered) subsets of the bottom clause's literals: $h(X, Y)$ (the top element), $h(X, Y) \leftarrow b_1(X, Z)$, $h(X, Y) \leftarrow b_2(X, Z)$, and the bottom clause itself. In general, there are 2^n candidates whenever the bottom clause has n body literals.

Searching for a viable candidate in the lattice bounded by a bottom clause amounts to selecting and evaluating candidates for positive and negative example coverage. If a candidate covers a negative example, it is said to be *inconsistent*, otherwise *consistent*. Intuitively, a good candidate should be consistent, cover as many positive examples as possible, and contain as few literals as possible.

Example 2. Continuing Example 1, assume we evaluate all four candidates, obtaining positive and negative example coverage as pairs: $(20, 5)$, $(17, 2)$, $(6, 0)$, and $(2, 0)$, respectively. As can be seen, $h(X, Y)$ and $h(X, Y) \leftarrow b_1(X, Z)$ are inconsistent (they cover at least one negative example), so we discard them. The candidate $h(X, Y) \leftarrow b_2(X, Z)$ and the bottom clause are consistent, but the former covers more positive examples (as well as fewer literals), so it is preferable. Adding the candidate, we remove the 6 positive examples it covers. There is at least 14 remaining examples (as can be seen by the top element's coverage), so the process of picking a new positive example, computing bottom clause, and searching for best candidate continues in an iterative manner.

There are other approaches to ILP using constraints. Constraint logic programming with ILP [9] converts negative examples into constraints in order to deal with numerical constraints. Theta-subsumption with constraint satisfaction [10] speeds up subsumption testing [2, 6, 7] using constraints. Our approach differs from these in that our constraints are propositional and explicitly used to construct candidate solutions by describing existential relations between the literals of bottom clauses.

2   The NrSample Framework

We present NrSample (non-redundant sampling), a novel constraint satisfaction framework where the search space is constrained not only by the bottom clause, but also by arbitrary propositional constraints [11]. The propositional variables refer to literals of the bottom clause: interpreting the variable as true means that the literal is to be included in a candidate, false means it is to be omitted. Thus, if the bottom clause has n body literals, there are n propositional variables, which we name $b_1, \ldots, b_n$. A boolean satisfiability solver (SAT solver) is then invoked to obtain a model for the constraints [11,12], which in turn corresponds to a constraint-satisfied candidate. For high performance, we have used the Chaff algorithm [13] for SAT solving.

Example 3. Using the bottom clause from Example 1, $n = 2$. The propositional clause $b_1 \vee b_2$ specifies that at least one of b_1 and b_2 must be included in the candidate. A model for this propositional clause is setting b_1 to false and b_2 to true, which corresponds to the candidate obtained by including the second, but not the first, literal from the bottom clause: $h(X, Z) \leftarrow b_2(Y, Z)$.

We call candidates that satisfy all constraints *valid*. Candidates that are not valid are said to be *invalid*.

We have tested our framework using two particular constraints: mode declarations (*mode constraints*) and search space pruning (*pruning constraints*), which we present next.

2.1   Mode Constraints

Mode constraints are constraints that require correct input-output chaining as a candidate's literals are computed from left to right: in order for a literal to appear, all its inputs must have been instantiated by previous outputs (or occur as inputs to the head, which means they are instantiated by the query itself). Moreover, the candidate's head outputs must be instantiated, either by appearing as outputs in body literals or as inputs to the head.

Example 4. Given bottom clause

$$h(A, B) \leftarrow b_1(A, C), b_2(A, D), b_3(C, B), b_4(C, D, B)$$

and mode declarations specifying that for each literal, the last argument is an output variable and all other arguments are input variables, we see that b_4 requires both b_1 and b_2 (to instantiate C and D), b_3 requires b_1 (b_3 also needs C), and instantiating h's output variable B means at least one of b_3 and b_4 must be present. As propositional clauses, this is $b_4 \rightarrow b_1$, $b_4 \rightarrow b_2$, $b_3 \rightarrow b_1$, and $b_3 \vee b_4$. One model is $\{b_1, b_3\}$, which specifies the candidates containing precisely the first and third body literals: $h(A, B) \leftarrow b_1(A, C), b_3(C, B)$. The constraints prevent the invalid candidate $h(A, B) \leftarrow b_1(A, C)$ from being generated: intuitively, because it does not instantiate B; logically, because the constraint $b_3 \vee b_4$ is not satisfied.

In general, for each input variable X of a bottom clause's body literal b_i (which does not also occur as input to the head), we find all preceding literals which have X as output, say $b_{X_1}, \ldots b_{X_k}$. Then the propositional clause $b_i \rightarrow b_{X_1} \vee \ldots \vee b_{X_k}$ is added as a constraint, stating that at least one of $b_{X_1}, \ldots b_{X_k}$

must occur for b_i to occur. Moreover, for each output Z of the head (and where it does not also occur as input), we find all body literals where Z occurs as output, $b_{Z_1}, \ldots b_{Z_k}$. Since at least one of these body literals must be present to instantiate Z, we add the clause $b_{Z_1} \vee \ldots \vee b_{Z_k}$.

2.2 Pruning Constraints

Pruning constraints arise due to the generality order of the search space (lattice). Note that a clause C containing a subset of a clause D's literals necessarily means that C logically implies D (C is a generalization of D). In particular, C will cover at least all the examples of D. Thus, if a candidate is consistent, a constraint preventing generation of its specializations is added, as these can never cover more positive examples. Conversely, if a candidate is inconsistent, so must its generalizations be, and a constraint preventing their generation is added.

Example 5. Continuing from the previous example, if candidate $h(A, B) \leftarrow b_1(A, C), b_3(C, B)$ is consistent, we prune all specializations, given by all candidates containing at least both the first and third literals. The pruning constraint is hence the clause $\neg(b_1 \wedge b_3) = \neg b_1 \vee \neg b_3$. Note that generation of the candidate itself, as well as the bottom clause, is prevented by this constraint as both b_1 and b_3 can no longer occur simultaneously. Conversely, if the candidate is inconsistent, we prune all generalizations, given by all candidates containing only a subset of $\{b_1, b_3\}$. The pruning constraint is now the (complement) clause $\neg(\neg b_2 \wedge \neg b_4) = b_2 \vee b_4$. In particular, this prevents generation of the candidate itself, as well as the top element.

In general, all specializations of a candidate C are given by all candidates containing a superset of C's body literals (and the same head). The propositional clause is $\neg(\bigwedge_{l \in C_-} l) = \bigvee_{l \in C_-} \neg l$, where C_- denotes the set of body literals in C. All generalizations are given by subsets of C (which means no other literals may be included). The propositional clause is $\neg(\bigwedge_{l \in C_-^c} \neg l) = \bigvee_{l \in C_-^c} l$, where C_-^c is the set of literals in the bottom clause that are not in C.

3 Experimental Results

We benchmark NrSample against a best-first and enumeration search on 9 problems. Best-first search is an implementation of Progol's well established A^* algorithm [5], whereas enumeration search is the default search method in the state-of-the-art Aleph ILP system.

First, we borrow some concept learning problems from the Progol distribution[1]: animal, train, grammar, member, and append. Second, we add our own problems: sorted, sublist, reverse, and add.[2] Moreover, append is benchmarked

---

[1] Available at http://www.doc.ic.ac.uk/~shm/Software/progol4.4/
[2] Available with our source code distribution upon request.

Table 1. Execution time in seconds.

Test	T_S	T_H	T_E	T_H/T_S	T_E/T_S	V_H	V_E
member	0.019	0.03	0.018	1.579	0.947	44.6	33.8
animal	0.026	0.041	0.025	1.577	0.962	100	42.2
train	0.079	0.09	0.019	1.139	0.241	29.1	7
grammar	0.022	0.088	0.032	4	1.455	0.7	0.3
sorted	0.008	0.028	0.008	3.5	1	19.7	19.3
sublist	0.13	0.846	0.154	6.507	1.185	3.3	2.8
append1	0.092	7.236	1.584	78.652	17.217	< 0.1	< 0.1
append2	0.11	> 600	45.351	> 2985	412.282	?	< 0.1
append3	0.238	> 600	> 600	> 1295	> 1295	?	?
reverse	0.009	0.195	0.036	21.667	4	0.4	0.3
add1	0.027	1.825	0.271	67.593	10.037	0.1	0.1
add2	0.04	> 600	> 600	> 16216	> 16216	< 0.1	< 0.1

at three difficulty levels, which we call append1, append2, and append3. Similarly, add is benchmarked as add1 and add2. The problems are made more difficult by altering default settings. By default, an upper limit of $i = 3$ iterations are used in bottom clause construction; at most $n = 200$ candidates are then explored. The maximum number of body literals in any body candidate is $c = 4$. For append1, we use $c = 3$; for append3 and add2, we use $c = \infty$.

Table 1 shows the average execution time over 30 trials for NrSample (T_S), best-first search (T_H), and enumeration (T_E). All benchmarks were performed on an Intel Dual Core i5 (2×2.30 GHz) with 4 GB RAM. NrSample outperformed best-first on all experiments. Compared to enumeration, NrSample performed approximately equal for the smaller problems (member, animal, grammar, sorted, sublist), but much better for the larger (reverse, add1, add2, all versions of append). The only exception is train, where enumeration outperformed NrSample. We put an execution time limit of 10 minute per run on all tests. This limit was reached in append2 for best-first. In append3 and add2, both best-first and enumeration timed out.

The growing performance difference between NrSample and its competing algorithms for larger problems—ranging from nothing to more than 1200 for append3 and 16000 for add2—can be explained as follows: A more complex problem is characterized by a longer bottom clause, which makes the search space grow exponentially larger (in the number of bottom clause literals, see Example 1). As the search space grows larger, there will be more ways of violating the input-output specification given by the mode declarations. The percentage of valid candidates during best-first and enumeration search is given by V_H and V_E in Table 1, respectively. As can be seen, the probability of blindly generating a valid candidate approaches zero for the larger tests, having less than 1 in 1000 valid candidates for append2, append3, and add2.[3] By contrast, with NrSample, 100 % of the candidates are valid.

---

[3] We could not measure the exact proportion for the tests that timed out, but it is estimated to be even less than its easier variants, thus always less than 0.1 %.

4 Conclusions

We have presented NrSample, a novel framework for theory induction using constraints. To the best of our knowledge, we are the first to use constraints to construct valid candidates rather than to eliminate already constructed candidates. Our benchmarks indicate that, as the problems become larger, it becomes increasingly beneficial to use mode and pruning constraints, with observed increases of four orders of magnitude.

Given that top-down and bottom-up algorithms also prune the search space, pruning constraints do not by themselves provide an advantage for NrSample. Its SAT solving overhead can only be compensated for by providing additional (non-pruning) constraints. We used mode constraints, but other alternatives may be to use domain specific knowledge. We plan to do more research into using other kinds of constraints within our framework.

Acknowledgments. The work described in this paper was supported by a grant from the Research Grants Council of the Hong Kong Special Administrative Region, China [Project No. CityU 124409].

References

1. Muggleton, S., Raedt, L.D., Poole, D., Bratko, I., Flach, P.A., Inoue, K., Srinivasan, A.: ILP turns 20 - biography and future challenges. Mach. Learn. **86**(1), 3–23 (2012)
2. Nienhuys-Cheng, S.H., de Wolf, R.: Foundations of Inductive Logic Programming. Springer-Verlag New York Inc., Secaucus (1997)
3. Blackburn, P., Bos, J., Striegnitz, K.: Learn Prolog Now!. College Publications, London (2006)
4. Sterling, L., Shapiro, E.: The art of Prolog: advanced programming techniques, 2nd edn. MIT Press, Cambridge (1994)
5. Muggleton, S.H.: Inverse entailment and progol. New Gener. Comput. **13**, 245–286 (1995)
6. Plotkin, G.D.: A note on inductive generalization. Mach. Intell. **5**, 153–163 (1970)
7. Plotkin, G.D.: A further note on inductive generalization. Mach. Intell. **6**, 101–124 (1971)
8. Lavrac, N., Dzeroski, S.: Inductive Logic Programming: Techniques and Applications. Ellis Horwood, New York (1994)
9. Sebag, M., Rouveirol, C.: Constraint inductive logic programming. In: De Raedt, L. (ed.) Advances in ILP. IOS Press, Amsterdam (1996)
10. Maloberti, J., Sebag, M.: Fast theta-subsumption with constraint satisfaction algorithms. Mach. Learn. **55**(2), 137–174 (2004)
11. Harrison, J.: Handbook of Practical Logic and Automated Reasoning. Cambridge University Press, Cambridge (2009)
12. Davis, M., Logemann, G., Loveland, D.: A machine program for theorem-proving. Commun. ACM **5**, 394–397 (1962)
13. Moskewicz, M.W., Madigan, C.F., Zhao, Y., Zhang, L., Malik, S.: Chaff: engineering an efficient SAT solver. In: Proceedings of the 38th Annual Design Automation Conference, DAC '01, pp. 530–535. ACM, New York (2001)

Features for Exploiting Black-Box Optimization Problem Structure

Tinus Abell[1], Yuri Malitsky[2]([⊠]), and Kevin Tierney[1]

[1] IT University of Copenhagen, Copenhagen, Denmark
{tmab,kevt}@itu.dk
[2] Cork Constraint Computation Centre, Cork, Ireland
y.malitsky@4c.ucc.ie

Abstract. Black-box optimization (BBO) problems arise in numerous scientific and engineering applications and are characterized by computationally intensive objective functions, which severely limit the number of evaluations that can be performed. We present a robust set of features that analyze the fitness landscape of BBO problems and show how an algorithm portfolio approach can exploit these general, problem independent, features and outperform the utilization of any single minimization search strategy. We test our methodology on data from the GECCO Workshop on BBO Benchmarking 2012, which contains 21 state-of-the-art solvers run on 24 well-established functions.

1 Introduction

This paper tackles the challenge of crafting a set of features that can capture the structure of black-box optimization (BBO) problem fitness landscapes for use in portfolio algorithms. BBO problems involve the minimization of an objective function $f(x_1, \ldots, x_n)$, subject to the constraints $l_i \leq x_i \leq u_i$, over the variables $x_i \in \mathbb{R}, \forall 1 \leq i \leq n$. These types of problems are found throughout the scientific and engineering fields, but are difficult to solve due to their oftentimes expensive objective functions. This complexity can arise when the objective involves difficult to compute expressions or that are too complicated to be defined by a simple mathematical expression. Even though BBO algorithms do not guarantee the discovery of the optimal solution, they are an effective tool for finding approximate solutions. However, different BBO algorithms vary greatly in performance across a set of problems. Thus, deciding which solver to apply to a particular problem is a difficult task.

Portfolio algorithms, such as Instance Specific Algorithm Configuration (ISAC), which uses a clustering approach to identify groups of similar instances, provide a way to automatically choose a solver for a particular BBO instance using offline learning. However, such methods require a set of features that consolidate

Yuri Malitsky is partially supported by the EU FET grant ICON (project 284715).
Kevin Tierney is supported by the Danish Council for Strategic Research as part of the ENERPLAN project.

G. Nicosia and P. Pardalos (Eds.): LION 7, LNCS 7997, pp. 30–36, 2013.
DOI: 10.1007/978-3-642-44973-4_4, © Springer-Verlag Berlin Heidelberg 2013

the relevant attributes of a BBO instance into a vector that can then be used for learning. The only way to generate these features for BBO problems is by evaluating expensive queries to the black box, which contrasts with most non-black-box problems, e.g. SAT or the set covering problem, where many features can be quickly inferred from the problem definition itself.

In this paper, we propose a novel set of features that are fast to compute and are descriptive enough of the instance structure to allow a portfolio algorithm like ISAC to accurately cluster and tune for the benchmark. These features are based on well-known fitness landscape measures and are learned through sampling the black box. They allow for the analysis and classification of BBO problems so that anybody can take advantage of the recent advances in the ISAC framework in order to more efficiently solve their BBO problems. This paper is a short version of [1].

Related Work. There has been extensive research studying the structure of BBO problems, and copious measures have been proposed for determining the hardness of local search problems by sampling their fitness landscape [9], such as the search space diameter, optimal solution density/distribution [6], fitness-distance correlation (FDC) [10], the correlation length [16,19], epistasis measures [14], information analysis [17], modality and neutrality measures [15], and fitness-distance analysis [13]. Difficulty measures for BBO problems in particular were studied by [8], who concluded that in the worst case building predictive difficulty measures for BBO problems is not possible to do in polynomial time.[1] Most recently, Watson introduced several cost models for combinatorial landscapes in order to try to understand why certain algorithms perform well on certain landscapes [18].

In [12], the authors identify six "low-level feature classes" to classify BBO problems into groups. In [4], algorithm selection for BBO problems is considered with a focus on minimizing the cost of incorrect algorithm selections, unlike our approach, which minimizes a score based on the penalized expected runtime. Our approach also differs from online methods [5] and reactive techniques [3] that attempt to guide algorithms based on information from previously explored states because ISAC performs all of its work offline.

2   BBO Dataset and Solver Portfolio

We evaluate the effectiveness and robustness of our features on a dataset from the GECCO 2012 Workshop on Black-Box Optimization Benchmarking (BBOB) [2]. The dataset contains the number of evaluations required to find a particular objective value within some precision on one of 24 continuous, noise-free, optimization functions from [7] in 6 different dimension settings for 27 solvers. The solvers are all run on the data 15 times, each time with a different target value set as the artificial global optimum. Note that the BBOB documentation refers

---

[1] Our results do not contradict this, as we are not predicting the hardness of instances.

to each of these target values as an "instance". To avoid confusion with the instances that ISAC uses to train and test on, we will only refer to BBOB targets. Removing 7 instances from the dataset for which no solver was able to find a solution, the dataset consists of 1289 instances.

We use the 21 solvers of the BBOB dataset with full solution data for all instances. This portfolio consists of a diverse set of continuous optimizers, including 10 covariance matrix adaptation (CMA) variants, 8 differential evolution (DE) variants, an ant colony optimization (ACO) algorithm, a genetic algorithm (GA), and a particle swarm optimization (PSO) algorithm.[2]

3   Features

Computing features for BBO problems is difficult because evaluating the objective function of a BBO problem is expensive, and there is scarce information about a problem instance in its definition, other than the number of dimensions and the desired solver accuracy. In the absence of any structure in the problem definition, we have to sample the fitness landscape. However, such sampling is expensive, and on our dataset performing more than 600 objective evaluations removes all benefits of using a portfolio approach. We therefore introduce a set of 10 features that are based on well-studied aspects of search landscapes in the literature [18]. Our features are drawn from three information sources: the problem definition, hill climbs, and random points. Table 1 summarizes our features.

The *problem definition features* contain the desired accuracy of the continuous variables (Feature 1), and the number of dimensions that the problem has (Feature 2), which, together, describe the size of the problem.

The *hill climbing features* are based off of a number of hill climbs that are initiated from random points and continued until a local optimum or a fixed number of evaluations is reached. We then calculate the average and standard deviation of the distance between optima (Features 3 and 4), which describes the density of optima in the landscape. Using the best optimum found, we then

Table 1. BBO problem features.

Problem definition features
1.   Solver accuracy
2.   Number of dimensions
Hill climbing features
3–4. Average distance between optima (average, std. dev.)
5–6. Distance between best optima and other optima (average, std. dev.)
7.   Percent of optima that are the best optimum
Random point features
8–9. Distance to local optimum (average, std. dev.)
10.   Fitness-distance correlation (FDC)

---

[2] Full details about the algorithms are available in [2].

compute the average and standard deviation of the distance between the optima and the best optimum (Features 5 and 6), using the nearest to each non-best optimum for these features if multiple optima qualify as the best. Feature 7 describes what percentage of the optima are equal to the best optimum, giving a picture of how spread out the optima are throughout the landscape.

The *random point features* 8 and 9 contain the average and standard deviation of the distance of each random point to the nearest optimum, which describes the distribution of local optima around the landscape. Feature 10 computes the fitness-distance correlation, a measure of how effectively the fitness value at a particular point can guide the search to a global optimum [10]. In feature 10, we compute an approximation to the FDC.

4   Numerical Results

In this section we describe the results of using our features, in full and in various combinations, to train a portfolio solver using the ISAC method on the BBOB 2012 dataset. We measure the performance of each solver using a penalized score that takes into account the relative performance of each solver on an instance. We do not directly use the expected running time (ERT) value because the amount of evaluations can vary greatly between instances, and too much focus would be placed on instances where a large number of evaluations is required. The penalized score of solver s on an instance i is given by:

$$score(s, i) = \frac{PERT(s, i) - best(i)}{worst(i) - best(i)}$$

where $PERT(s, i)$ is the penalized ERT defined by

$$PERT(s, i) = \begin{cases} ERT(s, i) & \text{if } ERT(s, i) < \infty \\ worst(i) \cdot 10 & \text{otherwise,} \end{cases}$$

$best(i)$ refers to the lowest ERT score on instance i, and $worst(i)$ refers to the highest non-infinity ERT score on the instance. The penalized ERT therefore returns ten times the worst ERT on an instance for solvers that were unable to find the global optimum. We are forced to use a penalized measure because if a solver cannot solve a particular instance, it becomes impossible to calculate its performance over the entire dataset.

4.1   ISAC Results

Table 2 shows the results of training and testing ISAC on the BBOB 2012 dataset. For each entry in the table, we run a 10-fold cross validation using features from each of the 15 BBOB target values. The scores of each of the cross-validation folds are accumulated for each instance, and the entries in the table are the average and standard deviation across all instances in the dataset.

Table 2. The average and standard deviation of the scores across all instances for various minimum cluster sizes, numbers of hill climbs and hill climb lengths for the best single solver and ISAC using various features.

κ		10/10				50/20				200/400			
		Test		Train		Test		Train		Test		Train	
		⊘	σ	⊘	σ	⊘	σ	⊘	σ	⊘	σ	⊘	σ
50	BSS	2.23	5.29	2.23	5.29	2.23	5.29	2.23	5.29	2.23	5.29	2.23	5.29
	F_1	2474.63	3×10^5	2.04	5.08	2474.66	3×10^5	2.04	5.08	2474.64	3×10^5	2.04	5.08
	F_2	1.24	4.02	1.24	4.02	1.24	4.02	1.24	4.02	1.24	4.02	1.24	4.02
	$F_{1,2}$	189.11	6743.81	1.27	4.07	189.10	6743.81	1.27	4.07	189.10	6743.81	1.26	4.07
	All	51.32	1801.27	1.21	3.96	96.15	3105.76	0.79	2.94	13.41	452.79	0.82	3.30
	All*	51.42	1801.33	1.32	4.05	97.15	3110.46	1.82	9.90	95.25	1161.92	83.12	760.37
	LSF	1.25	4.01	1.24	4.00	88.18	3137.52	0.82	3.03	**0.53**	2.73	0.55	2.75
	LSF*	1.35	4.09	1.34	4.08	89.18	3142.23	1.85	9.93	99.44	1323.68	82.86	760.40
100	BSS	2.23	5.29	2.23	5.29	2.23	5.29	2.23	5.29	2.23	5.29	2.23	5.29
	F_1	2474.63	3×10^5	2.04	5.08	2474.66	3×10^5	2.04	5.08	2474.64	3×10^5	2.04	5.08
	F_2	1.24	4.02	1.24	4.02	1.24	4.02	1.24	4.02	1.24	4.02	1.24	4.02
	$F_{1,2}$	189.11	6743.81	1.27	4.07	189.11	6743.81	1.27	4.07	189.10	6743.81	1.27	4.07
	All	1.25	4.02	1.24	4.00	1.25	4.03	1.23	4.00	1.16	3.86	1.12	3.80
	All*	1.35	4.10	1.34	4.08	2.28	10.21	2.26	10.20	83.46	760.60	83.43	760.34
	LSF	1.25	4.02	1.24	4.01	1.22	3.99	1.19	3.93	1.20	3.85	1.15	4.00
	LSF*	1.35	4.10	1.34	4.09	2.25	10.19	2.22	10.17	97.31	1223.98	83.45	760.34

We compare our results against the best single solver (BSS) on the dataset, which is the best performing solver across all instances, which is MVDE [11].

We train using several subsets of our features; only feature 1 (F_1), only feature 2 (F_2), and only features 1 and 2 ($F_{1,2}$). We then train using all features (All), and only landscape features (LSF), i.e., features 3 through 10. All* and LFS* include the evaluations necessary to compute the features, whereas all other entries do not include the feature computation in the results. We used several different settings of the number of hill climbs and maximum hill climb length based on our feature robustness experiments in [1]: 10 hill climbs of maximum length 10, 50 hill climbs of maximum length 20, and 200 hill climbs of maximum length 400. The closer a score is to 0 (the score of the virtual best solver) the better the performance of an approach.

Based on results for F_1, F_2 and $F_{1,2}$, the easy to compute BBO features alone are only able to give ISAC some information about the dataset, and that a landscape analysis is justified. On the other hand, F_2 outperforms BSS. In fact, F_2 performs equally well to All and LSF for cluster 100 with 10 hill climbs of length 10 and for 50 hill climbs of length 20. In addition, F_2 significantly outperforms All on cluster size 50, where it is clear that it overfits the training data. This is a clear indication that 10 hill climbs of length 10, or 50 hill climbs of length 20, do not provide enough information to train ISAC to be competitive with simply using the number of dimensions of a problem.

The fact that LSF* is able to match the performance of F_2 on 10 hill climbs of length 10 for both cluster size 50 and 100 an important accomplishment. With so little information learned about the landscape, the fact that ISAC can learn such an effective model indicates that our features are indeed effective.

Once we move up to 200 hill climbs of length 400, LSF significantly outperforms F_2, and even outperforms All, which suffers from overfitting. In fact, LSF is able to cut the total score to under a fourth of BSS's score, and to one half of F_2's score, indicating that the fitness landscape can indeed be used for a portfolio. In addition, LSF has a lower standard deviation than BSS. LSF's score on the training set of 0.53 and 0.55 on the test set are surprisingly close to the virtual best solver, which has a score of zero, indicating that ISAC is able to exploit the landscape features to nearly always choose the best or second best solver for each instance. On the downside, 200 hill climbs of length 400 requires too many evaluations to be used in a competitive portfolio, and All* needs 50 times the evaluations of BSS. However, the 200/400 features are still useful for classifying instances into groups and analyzing the landscape.

5   Conclusion and Future Work

We introduced a set of features based on accepted and well-studied properties and measures of fitness landscapes to categorize BBO problems for use in algorithm portfolios, like ISAC, that can greatly improve the ability of practitioners to solve BBO problems. We experimentally validated our features within the ISAC framework, showing that ISAC is able to exploit problem structure learned during feature computation to choose the fastest solver for an unseen instance. The success of the features we introduced clearly indicates that selecting algorithms from a portfolio based on the landscape structure is possible. For future work, features analyzing landscape structure could be incorporated into problems, providing an alternative view of problem structure.

References

1. Abell, T., Malitsky, Y., Tierney, K.: Fitness landscape based features for exploiting black-box optimization problem structure. Technical report TR-2012-163, IT University of Copenhagen (2012)
2. Auger, A., Hansen, N., Heidrich-Meisner, V., Mersmann, O., Posik, P., Preuss, M.: In: GECCO 2012 Workshop on Black-Box Optimization Benchmarking (BBOB). http://coco.gforge.inria.fr/doku.php?id=bbob-2012 (2012)
3. Battiti, R., Brunato, M.: Reactive search optimization: learning while optimizing. In: Gendreau, M., Potvin, J.-Y. (eds.) Handbook of Metaheuristics, vol. 146, pp. 543–571. Springer, New York (2010)
4. Bischl, B., Mersmann, O., Trautmann, H., Preuß, M.: Algorithm selection based on exploratory landscape analysis and cost-sensitive learning. In: GECCO'12, pp. 313–320. ACM, New York (2012)
5. Boyan, J., Moore, A.W.: Learning evaluation functions to improve optimization by local search. J. Mach. Learn. Res. 1, 77–112 (2001)

6. Brooks, C., Durfee, E.: Using landscape theory to measure learning difficulty for adaptive agents. In: Alonso, E., Kudenko, D., Kazakov, D. (eds.) AAMAS 2000 and AAMAS 2002. LNCS (LNAI), vol. 2636, pp. 291–305. Springer, Heidelberg (2003)

7. Finck, S., Hansen, N., Ros, R., Auger, A.: Real-parameter black-box optimization benchmarking 2010: presentation of the noisy functions. Technical report 2009/21, Research Center PPE (2010)

8. He, J., Reeves, C., Witt, C., Yao, X.: A note on problem difficulty measures in black-box optimization: classification, realizations and predictability. Evol. Comput. **15**, 435–443 (2007)

9. Hoos, H.H., Stützle, T.: Stochastic Local Search: Foundations & Applications. Morgan Kaufmann Publishers Inc., San Francisco (2004)

10. Jones, T., Forrest, S.: Fitness distance correlation as a measure of problem difficulty for genetic algorithms. In: ICGA-95, pp. 184–192 (1995)

11. Melo, V.V.: Benchmarking the multi-view differential evolution on the noiseless bbob-2012 function testbed. In: GECCO'12, pp. 183–188. ACM (2012)

12. Mersmann, O., Bischl, B., Trautmann, H., Preuß, M., Weihs, C., Rudolph, G.: Exploratory landscape analysis. In: GECCO'11, pp. 829–836. ACM (2011)

13. Merz, P., Freisleben, B.: Fitness landscapes, memetic algorithms, and greedy operators for graph bipartitioning. Evol. Comput. **8**, 61–91 (2000)

14. Naudts, B., Kallel, L.: A comparison of predictive measures of problem difficulty in evolutionary algorithms. IEEE Trans. Evol. Comp. **4**(1), 1–15 (2000)

15. Smith, T., Husbands, P., Layzell, P., O'Shea, M.: Fitness landscapes and evolvability. Evol. Comput. **10**(1), 1–34 (2002)

16. Stadler, P.F., Schnabl, W.: The landscape of the traveling salesman problem. Phys. Lett. A **161**(4), 337–344 (1992)

17. Vassilev, V.K., Fogarty, T.C., Miller, J.F.: Information characteristics and the structure of landscapes. Evol. Comput. **8**, 31–60 (2000)

18. Watson, J.: An introduction to fitness landscape analysis and cost models for local search. In: Gendreau, M., Potvin, J. (eds.) Handbook of Metaheuristics, vol. 146, pp. 599–623. Springer, New York (2010)

19. Weinberger, E.: Correlated and uncorrelated fitness landscapes and how to tell the difference. Biol. Cybern. **63**, 325–336 (1990)

MOCA-I: Discovering Rules and Guiding Decision Maker in the Context of Partial Classification in Large and Imbalanced Datasets

Julie Jacques[1,2,3], Julien Taillard[1], David Delerue[1],
Laetitia Jourdan[2,3(✉)], and Clarisse Dhaenens[2,3]

[1] Société ALICANTE, 50 Rue Philippe de Girard, 59113 Seclin, France
[2] INRIA Lille Nord Europe, 40 Av. Halley, 59650 Villeneuve d'Ascq, France
[3] LIFL, Université Lille 1, Bât. M3, 59655 Villeneuve d'Ascq cedex, France
{Laetitia.jourdan,clarisse.dhaenens}@lifl.fr
{julie.jacques,julien.taillard,david.delerue}@alicante.fr

Abstract. This paper focuses on the modeling and the implementation as a multi-objective optimization problem of a Pittsburgh classification rule mining algorithm adapted to large and imbalanced datasets, as encountered in hospital data. We associate to this algorithm an original post-processing method based on ROC curve to help the decision maker to choose the most interesting rules. After an introduction to problems brought by hospital data such as class imbalance, volumetry or inconsistency, we present MOCA-I - a Pittsburgh modelization adapted to this kind of problems. We propose its implementation as a dominance-based local search in opposition to existing multi-objective approaches based on genetic algorithms. Then we introduce the post-processing method to sort and filter the obtained classifiers. Our approach is compared to state-of-the-art classification rule mining algorithms, giving as good or better results, using less parameters. Then it is compared to C4.5 and C4.5-CS on hospital data with a larger set of attributes, giving the best results.

1 Introduction

Data mining on real datasets can lead to handling imbalanced data. It occurs when many attributes are available for each observation, but only a few are actually entered. This is especially the case with medical data: ICD-10[1] – a medical coding system – allows encoding up to 14,199 diseases and symptoms. However in hospital data, for each patient, only a very small subset of these codes will be used: up to 100 symptoms and diseases. This implies that most frequent symptoms, like *high blood pressure*, are found on at best 10 % of the patients. For less common diseases, like *transient ischemic stroke* it can be lower to less than 0.5 % of patients. This can also happen with market-basket data: many different

---

[1] International classification of diseases; http://www.who.int/classifications/icd/en/

G. Nicosia and P. Pardalos (Eds.): LION 7, LNCS 7997, pp. 37–51, 2013.
DOI: 10.1007/978-3-642-44973-4_5, © Springer-Verlag Berlin Heidelberg 2013

items are available in the store but only a few are actually bought by a single customer. Additionally, more and more information is available and collected nowadays: algorithms must be able to deal with larger datasets. According to Fernández et al. dealing with large datasets is still a challenge that needs to be addressed [1]. This work is a part of the OPCYCLIN project – an industrial project involving Alicante company, hospitals and academics as partners – that aims at providing a tool to optimize screening of patients for clinical trials.

Different tasks are available in datamining, this paper focuses on the *classification* task, useful to predict or explain a given *class* (e.g.: *cardiovascular risk*) on unseen observations. Classification will use known data, composed of a set of N known observations $i_1, i_2, ..., i_N$ to build a model. Each observation can be described by M attributes $a_1, a_2, ..., a_M$ and a class c. Therefore each observation i is associated with a set of values $v_{i1}, v_{i2}, ..., v_{iM}$ where $v_{ij} \in V_j\{val_1, val_2, ... val_p\}$; V_j being the set of possible values for attribute a_j. In the same manner, each observation i is associated to a class value $c_i \in C$, C being the set of all possible values for the class. A classification algorithm will be able to generate a model that describes how to determine c_v on an unseen observation v, using its values $v_{v1}, v_{v2}, ..., v_{vM}$. This paper focuses on models able to give a good interpretability: they allow medical experts to give a feed back about them. Decision trees and classification rules give easy-to-interpret models, by generating trees or rules — like "$a_j = val_j$ and $a_g = val_g \Rightarrow class$", where $val_j \in V_j$, $val_g \in V_g$ and $class \in C$; using combinations of attributes $a_1, a_2, ..., a_M$ and one of their possible values $val_1, val_2, ..., val_M$ to lead to the decision.

Decision trees – like C4.5 [2] or CART (classification and regression trees) – are popular and efficient solutions for knowledge extraction. However the tree representation is composed of conjunctions, not allowing expressing classes explained by different contexts (e.g.: presence of *overweight* or *high blood pressure* implies an increased *cardiovascular risk*, having both increases the risk more). Separate and conquer strategy, frequently implemented in tree algorithms often contribute to miss rules issued from different contexts: each sub-tree is constructed using a sub-part of data (observations not corresponding to the top of the tree are removed from learning). To avoid this problem we will focus on classification rule mining approaches.

The majority of state-of-the-art classification algorithms will have trouble to deal with imbalanced data because they use *Accuracy* to build their predictive model [1]. *Accuracy* focuses on counting good classifications obtained by a given algorithm: *true positives* and *true negatives*. However, when predicting a class available on only 1 % of the observations, an algorithm can get a very good classification *Accuracy* – 99 % – while predicting each observation as negative (99 % of observations) and missing each positive observation. Some resampling methods exist to pre-process the data and convert it into balanced data, an overview can be found in [3]. Jo and Japkowicz showed that combining data resampling and an algorithm able to deal with class imbalance is more effective than using resampling alone [4]. Moreover, in addition to class imbalance and a huge amount of data, hospital data is subject to uncertainty. When data is missing on one patient, two cases can happen: the patient does not have the disease or the patient has the disease but was not diagnosed yet (or the diagnosis was not entered in the system). This is difficult to predict the consequences of resampling on such data.

The remaining of this paper is organised as follows: Sect. 2 will introduce some common rule interestingness measures, and will show how the classification rule mining problem can be seen as a multi-objective problem. Section 3 will propose the modeling as a multi-objective local search optimization problem. Then, the *Dominance-based local search* algorithm will be presented, as well as the associated implementation details such as neighborhood. This section will conclude by the description of an original post-processing method to select rules based on ROC curve. In Sect. 4, we will assess the performance of our approach. At first we will compare our results to those gathered by Fernandez et al. with 22 state-of-the-art classifiers in the context of imbalanced data [1], showing our approach can be applied on more general datasets. Secondly, we will compare our approach to C4.5 – a state-of-the-art decision tree algorithm – and C4.5-CS – an adaptation of the C4.5 algorithm to imbalanced data – on real hospital data. Finally, Sect. 5 gives conclusions and perspectives for future works.

2   A Multi-Objective Model to Discover Partial Classification Rules in Imbalanced Data

This section will present some rule interestingness measures and their meaning. Then it presents the 3 objectives that will be used to find rules.

2.1   Rule Interestingness Measures

When mining rules, an important question will raise: how can we assess that a rule is better than another? Over 38 common rule interestingness measures are referenced by Geng and Hamilton in their review [5], while Ohsaki et al. studied measures used in medical domain [6] and Greco et al. studied Bayesian confirmation measures [7].

Table 1. Confusion matrix

	P	$\overline{P}$
C	TP	FP
$\overline{C}$	FN	TN
		N

The majority of rule interestingness measures are based on a confusion matrix, like the one provided in Table 1. For a given rule $C \rightarrow P$, TP (*true positives*) will represent count of observations having both C and P; TN (*true negatives*) count of observations not having C and not having P. FN (*false negatives*) and FP (*false positives*) count observations on which C and P do not match. When dealing with imbalanced data

$$\overline{P} = FP + TN >> P = TP + FN, \tag{1}$$

therefore problems may rise with some measures like previously seen with the *Accuracy*.

To ease the conception of a rule mining algorithm we must focus on a subset of these measures. Indeed, handling too many measures will add complexity and will increase computational time. An analysis of these measures showed that *Confidence* and *Sensitivity*

$$Confidence = \frac{TP}{TP + FP}, Sensitivity = \frac{TP}{TP + FN}, \tag{2}$$

are two interesting complementary measures. Increasing *Confidence* decreases the number of *false positives* while increasing *Sensitivity* decreases the number of *false negatives*. However, increasing *Confidence* often decreases *Sensitivity* while increasing *Sensitivity* decreases *Confidence*. To the medical domain point of view, only rules having both good *Confidence* and *Sensitivity* are interesting. Moreover, Bayardo and Agrawal showed that mining rules optimizing both *Confidence* and *Support* leads to obtain rules optimizing several other measures including *Gain, Chi-squared value, Gini, Entropy gain, Laplace, Lift*, and *Conviction* [8]. Since in classification, *Sensitivity* and *Support* measures are proportional, optimizing *Confidence* and *Support* will bring the same rules than optimizing *Confidence* and *Sensitivity*.

When mining variable-length rules, *bloat* can happen: rules endlessly grow with no predictive enhancement. Because of *bloat*, a rule $R_1 : C \Rightarrow P$ can turn into $R_2 : C$ OR $C \Rightarrow P$, then $R_3 : C$ OR C OR $C \Rightarrow P$, increasing computational time and preventing the algorithm to stop. Most of all, R_3 is needlessly complex and harder to interpret than R_1. Rissanen introduced the *Minimum Description Length (MDL) principle* that can be used to overcome this problem [9]. Given two equivalent rules, the simplest must be preferred. The addition of one objective promoting simpler rules is a common solution, successfully applied in Reynolds and Iglesia work [10]. In addition to this, Barcadit et al. used rule deletion operators [11]. In application of this principle, we introduce a third objective to promote simpler rulesets: minimizing the count of terms of each solution. Finally, we choose to find rules optimizing the 3 following objectives:

– maximize *Confidence*
– maximize *Sensitivity*
– minimize number of terms

3   A Multi-Objective Model to Discover Partial Classification Rules in Imbalanced Data

In addition to class imbalance, our hospital data raises another problem: more than 10,000 attributes are available for each patient. This implies a huge number of possible rules to explore. As a rule may be seen as a combination of pairs <attribute, value>, the rule mining problem is a combinatorial one. Moreover, regarding the large number of attributes, it requires methods dedicated to deal with very large search spaces such as combinatorial optimization methods and metaheuristics. Moreover, some datamining tasks contain NP-hard problems; in their review, Corne et al. explain how operations research and metaheuristics can help solving these problems that may be seen as combinatorial optimization problems [12].

The three objectives identified in the previous section highlight the need of methods able to deal with several objectives. Multi-objective optimization can handle such problems; Srinivasan and Ramkrishnan made a review of rule mining approaches using multi-objective optimization [13].

As explained later, in our work we will adopt a Dominance-based approach: each objective will be treated separately. Metaheuristics working on a population of solutions are particularly well suited for this type of problems [14] and we will adopt one of them to our classification rules problem. In the following, the solution modeling and the algorithm proposed are detailed.

3.1   Solution Modeling

Solution Representation. Two main solution representations exist for rule mining in metaheuristics: *Michigan* and *Pittsburgh*. Michigan is the widely used one, where each solution represents a single rule. However algorithms using this representation can miss some rules: an interesting rule will be removed if a slightly better rule is found, even if that rule targets different observations from the dataset.

This problem does not appear in Pittsburgh representation where each solution is a set of rules. This more complex representation will impact the size of the search space – now larger than the one of Michigan – and introduce new problems such as rule redundancy or conflicts between rules: which prediction must be chosen when different rules in the same ruleset have conflicting predictions? This is only an overview of problems that may happen. Casillas et al. identified more possible inconsistencies risen by Pittsburgh modeling [15].

We propose to use a Pittsburgh representation where each solution is a ruleset. Each ruleset is composed only of rules predicting the positive class, called partial rules. A rule is a conjunction of terms; a term is the expression of a test made on an attribute, for a given observation. Attributes can be binary (e.g: *hasHighBloodPressure?*), or associated to an operator $(=, <, >)$ and a value, where the value is taken from a list of values, ordered or not. Any observation that triggers at least one rule from the ruleset will be labeled as positive class. Our representation is designed to handle binary classes, thus observations not triggering any rules are labeled as negative class. Since a ruleset groups together only partial classification rules predicting the positive class, there is no need to store the right part of each rule. Moreover, there are no inconsistencies since rules in a same ruleset cannot predict different classes. Rule redundancy is avoided by minimizing the size of each ruleset. Thus, a rule will be added only if it improves *Sensitivity* (*true positives rate*) or *Confidence* of the ruleset.

Evaluation Function. Previously we saw three objectives can be used to find rulesets: maximizing *Confidence*, maximizing *Sensitivity* and minimizing the number of terms. The third objective corrects one drawback of Pittsburgh representation that brings *bloat*. We use a dominance-based approach to handle the different objectives, in opposition to some classification rule mining algorithms like GAssist using scalar approaches [11]. Dominance-based approaches use a dominance relation to compare solutions over several objectives, avoiding searching the good adjustment of weights to combine the different objectives, needed in scalar approaches. Moreover, with a weighted fitness function two solutions having different objective values can have the same fitness score. Our method is based on Pareto Dominance. This will generate a population of rulesets, in our case rulesets with very high *Confidence* but relatively low *Sensitivity*, rulesets with middle *Confidence* and *Sensitivity*, rulesets with high *Sensitivity* and relatively low *Confidence*, etc. These rule sets are stored in an *archive*, thus we need an algorithm able to handle populations.

3.2   DMLS Algorithm

Dominance-based multi-objective local search (DMLS) is a local search algorithm, based on a dominance relation [16]. It needs the definition of a neighborhood function that associates to each solution a set of solutions – called neighbors – by applying a small modification on it. A neighborhood of a rule can be, for example, all rules having one more or one less term.

Most of multi-objective rule mining contributions presented in the review of Srinivasan and Ramkrishnan are based on the metaheuristic NSGA-II (genetic algorithm dedicated to multi-objective problems) [13]. DMLS has previously proven to give at least as good results as NSGA-II on several problems [16]. Moreover, DMLS is easier to parameter than a genetic algorithm as we only have to define a neighborhood operator. Therefore we used DMLS implemented by Liefooghe et al. [16] in ParadisEO framework [17], with an unbounded archive, using the natural stopping criterion. DMLS algorithm is detailed in Algorithm 1. It evolves a population of non dominated rulesets. At first, all rulesets are marked as unvisited. While unvisited rulesets exist, DMLS will chose randomly one of them from the archive, visit its neighborhood and add all the non-dominated neighbors to the archive for future visits.

Initialization. DMLS is initialized with a population of 100 rulesets. Each is made of two rules, whose attributes are randomly picked in a same observation. This ensures the chosen attributes appears together at least on one observation. Then, one or two random attributes are replaced to add some diversity.

Algorithm 1. Dominance-based multi-objective local search

---

generates 100 rulesets RS_a, composed of 2 initial rules
$RS_a.setVisited(false)$
$archive.add(RS_a)$
while $RS_{current} \leftarrow archive.selectRandomUnvisitedSol()$ **do**
  $RS_n \leftarrow generateNeighbors(RS_{current})$
  for $RS_{neighbor} \in RS_n$ **do**
    /** add non dominated neighbor to archive, for future visits **/
    if $!RS_{current} \succ RS_{neighbor}$ **then**
      $RS_{neighbor}.setVisited(false)$
      $archive.add(RS_{neighbor})$
    end if
    /** stops when a dominating neighbor is found **/
    if $RS_{neighbor} \succ RS_{current}$ **then**
      break
    end if
  end for
  if all neighbors in RS_n were visited **then**
    $RS_{current}.setVisited(true)$
  end if
end while

Neighborhood. The neighborhood function is defined as a generator of all rulesets having a one-term difference: one term removed, one term added or one term modified. They are randomly visited. The neighborhood size is important, since a term addition can happen on each rule of the ruleset, for each available attribute. To reduce the neighborhood size on attributes taking values in ordered lists, we designed a simplified term neighborhood where only boundaries (adjacent values) are visited. Table 2 indicates for each operator $(=, <, >)$ the list of possible neighbors, assuming values are ordered $(v_{i-1} < v_i < v_{i+1})$. Ømeans remove the term.

Table 2. Neighborhood of list-valued terms

$a = v_i$	$a < v_i$	$a > v_i$
$\emptyset$	$\emptyset$	$\emptyset$
$a > v_{i-1}$	$a = v_{i-1}$	$a = v_{i+1}$
$a < v_{i+1}$	$a < v_{i-1}$	$a > v_{i+1}$
$a = v_{i-1}$	$a < v_{i+1}$	$a < v_{i+2}$
$a = v_{i+1}$	$a > v_{i-2}$	$a > v_{i-1}$

On a dataset with mixed attributes (ordered and not) (heart dataset, introduced in results section), this simplified neighborhood decreases computational time (in average by 14 %), while not degrading too much the classification performance (less than 1 %). Another optimization on the neighborhood exploration is done on rules with $Confidence = 1$: adding one term can only result in decreasing $Sensitivity$ because the obtained rule will be more specific and will concern less observations. In this case, we restrict neighbors to modification or removing of one random term.

3.3   Post-processing Using ROC Curve

Multi-objective optimization finds a population of rulesets, corresponding to compromise solutions between the different objectives. Some datasets may obtain up to 400 rulesets after one single run. As it is hard to choose one ruleset, we propose a post-processing method based on ROC curve which combines all obtained rules to take advantage of the diversity issued from the multi-objective algorithm, and a tool to help choosing rules. Then we present how the result can be used by the decision maker to choose the final ruleset. In addition, we propose a method to determine automatically the final ruleset.

ROC Curve is often used in data mining to assess the performance of classification algorithms, especially ranking algorithms. It is plotted using *true positive rate* (*TPR*) (known as *Sensitivity*) and *false positives rate* (*FPR*) (also called *1 - Specificity*) as axes and allows comparing algorithms. Fawcett presented different ROC curve usages [18]. In our case, we use ROC curve to select which rules to keep. Since the objective is to use the developed method in a medical context, it can also be used to help our medical users to calibrate classifier or choosing rules using a tool they are familiar with. Algorithm 2 describes how ROC curve can be generated for a given ruleset. Rules are first ordered from the highest *Confidence* score to the lower; rules having the same *Confidence* are ordered by descending order according to *Sensitivity*. Then, *TPR* and *FPR* are computed and drawn for each subruleset $\{R_1\}, \{R_1, R_2\}, \ldots, \{R_1, R_2 \ldots R_i\}$.

Algorithm 2. Draw ROC curve of a ruleset RS

order rules of RS by *Confidence* DESC, *Sensitivity* DESC
create an empty ruleset RS_{roc}
for rule $R_i \in RS$ $\{R_1, R_2, \ldots, R_n\}$ **do**
  /* get TPR and FPR for sub-rules-set $R_1, R_2, \ldots, R_i$ */
  $RS_{roc}.add(R_i)$
  tpr $\leftarrow RS_{roc}.computeTruePositiveRate()$
  fpr $\leftarrow RS_{roc}.computeFalsePositiveRate()$
  plot(fpr,tpr)
end for

Rule Selection Using ROC Curve. One drawback of dominance-based methods lies in obtaining a set of compromise solutions, that are difficult to handle by the decision maker. The following post-processing method is proposed to cope with this problem. Figure 1 shows on the right, one sample ruleset containing rules $R_1 \ldots R_{10}$, ordered from the highest *Confidence* to the lower and in descending order of *Sensitivity* for rules having the same *Confidence* score. On the left the matching ROC curve is drawn. Each point on this curve depicts the performance of a subruleset, e.g.: R_1, R_2, R_3. The higher is the point, the more observations of positive class it detects. Additionally, the more a point is on the right, the more false positive it brings. Consequently, point $(0, 1)$ is the ideal point where all positive observations are found, without bringing any false positive. This figure shows the performance of the ruleset when cut at different places, allowing to choose the subruleset giving the most interesting performance according to decision maker's needs. On this curve we can see that between point a and point b, and after point c there is only a small improvement of *True positives rate*, but it brings much more *false positives*. Performance is more interesting before point a, matching ruleset R_1, R_2, R_3. Subruleset $R_1 \ldots R_4$ (cut b) does not seem to be a good choice because it brings much more false positives than positives cases. Point c brings more true positives cases, giving ruleset $R_1, R_2, ..., R_7$. Cutting at point d finds a ruleset able to detect all positive observations: $R_1, R_2, ..., R_9$, keeping rule R_{10} is useless and will only increase false positives. Depending on how many false positive are tolerable, the cut point can be changed. In medical context, cut points bringing less false positives will be preferred (like cut a). To the contrary, an advertising campaign will accept more false positives, to deal with a larger audience. In fraud detection, the cut point can be moved until a given number of positive observations are found.

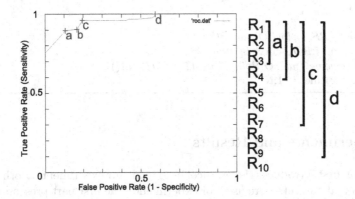

Fig. 1. Example of ROC curve obtained from one ruleset. $R_1 \ldots R_{10}$ is one ruleset obtained after the post-processing. a, b, c and d represent different cuts and their associated position on the roc curve.

Ruleset Post-processing. Regarding the OPCYCLIN project, the decision maker will have to deal with up to 30 different predictions for each clinical trial, leading to 30 rulesets and 30 ROC curves. It makes the manual rule selection harder. In addition to the previous rule selection method, we propose a solution to determine automatically the best cut point. Thus, we can obtain a classifier with good performance without the intervention of the decision maker. A final ruleset classifier is generated from all obtained rulesets coming from archive, as described in Algorithm 3. After merging all rules into one ruleset, the ROC curve is drawn. The subruleset giving the point closest to the ideal point (0,1) according to the Euclidean distance is chosen. All rules after this point are removed (or disabled if we want to allow the decision maker to change the *Sensitivity* accordingly to his needs). Once this ruleset is obtained, common data mining measures can be computed on the entire ruleset: *Confidence*, *Support*, etc. Diverse cut conditions have been tested but only the above presented one gives classifiers with interesting performance. An improvement of this condition could consist in weighting *true positives rate* and *false positives rate* according to decision maker's needs, since *false positives* can be more or less important than *true positives*, depending on the context.

Algorithm 3. Obtain a ruleset RS_{all} from a set of rulesets RS_i

---

create an empty ruleset RS_{all}
/* merge all obtained rules into RS_{all} */
for obtained ruleset RS_i **do**
   for each rule $R_j \in RS_i \ \{R_1, R_2, \ldots, R_n\}$ **do**
      /* avoid duplicates */
      if $R_j \notin RS_{all}$ **then**
         RS_{all}.add(R_j)
      end if
   end for
end for
$rocCurve \leftarrow RS_{all}.plotROCcurve()$
/* best point of ROC curve is (0,1) */
$i \leftarrow rocCurve.getIndexOfPointClosestToBestPoint()$
RS_{all}.removeRules(i+1,N)

---

4   Experiments and Results

This section first introduces the protocol used in all our experiments. Both benchmarks and real datasets were used for experiments; the first part presents results obtained on benchmarks and compares them to the ones obtained by algorithms of literature. In the last part we compare *C4.5* and *C4.5-CS* decision tree algorithms and our approach on a real dataset having an important imbalance and a large number of attributes.

4.1   Protocol

According to the protocol proposed by Fernandez et al., our algorithm was run 25 times for each dataset. We use 5-fold cross-validation: datasets are split into 5 parts, each containing 20 % of observations. Then 4 parts are used for training, 1 for evaluation. For each available partition, as the algorithm contains some stochastic components, it was run 5 times. So we obtain for stochastic algorithms 25 Pareto fronts for each dataset. For each partition, solutions are evaluated on both training and test partitions. In our case, the objective is to maximize the results on test data, because it shows the ability of the algorithm to handle unseen data. A discretization of data was applied with Weka when necessary (*weka.filters.unsupervised.attribute.Discretize; bins=10, findNumBins=true*) to allow our algorithm to handle datasets containing continuous attributes.

Generally *accuracy* measure is used to assess the performance of classification. Previously we saw that *accuracy* is not effective to handle class imbalance. Therefore Fernandez et al. proposed to use *Geometric mean of the true rates* (GM):

$$GM = \sqrt{\frac{TP}{TP+FN} \times \frac{TN}{FP+TN}}. \tag{3}$$

In order to have a good score, a classifier has now to classify correctly both classes: positive and negative. GM has one drawback though: when a classifier is not able to predict one class, score is worth 0. Here, when two classifiers failed to predict the negative class, there is no difference between the classifier able to find 50 % of positive observations and an other classifier predicting 70 % of positive observations: both have a score of 0.

For each dataset we computed the average of GM values obtained in each 25 runs. In order to get a single GM value from the rulesets proposed by our algorithm, we generated a ruleset, its ROC curve and cut it automatically as shown previously. Then we computed GM on the resulting ruleset: if an observation matches a rule from the ruleset it is considered as positive class. Observations not matching any rule are considered as negative class.

Tests were carried out on a computer with a Xeon 3500 quad core and 8 GB of memory, under Ubuntu 12. We used Weka software version 3.6 for discretization of datasets and for running C4.5 tests. Our approach is implemented in C++, using metaheuristics from ParadisEO framework [17]. In our experimentations we set *MOCA-I* max ruleset size = 5, max rule size = 9 for each dataset.

4.2 Experiments on Imbalanced Benchmarks Datasets

Fernandez et al. performed a comparison of 22 classification rule mining algorithms on imbalanced datasets and provided material to compare to their results [1]. Since our algorithm is designed to handle discrete attributes, datasets with less continuous attributes were preferred. We selected 6 imbalanced datasets in those proposed by Fernandez et al. Their details are available in Table 3. The degree of class imbalance varies from 0.77 % to 27.42 %: in the *abalone19* dataset the positive class happens on only 0.77 % of observations. In addition to these datasets, *tia* dataset - a real dataset - will be used in the next experiments. The *tia* dataset comes from hospital data. It is composed of 10,000 patients taking values in 10,250 available attributes: medical procedures and diagnoses. The

Table 3. Datasets main attributes

Name	#ind.	#feat.	% repar.
haberman	306	3	27.42
ecoli1	336	7	22.92
ecoli2	336	7	15.48
yeast3	1484	8	10.38
yeast2vs8	482	8	4.15
abalone19	4174	8	0.77
tia	10,000	699	0.74

#ind.: count of observations; #feat.: count of attributes; % repar.: Percentage of observations having the class to be predicted

objective is to predict the presence of the diagnosis *Transient cerebral ischemic attack*, available on 0.74 % of the observations. In order to allow some state-of-the-art algorithms processing this dataset, the number of attributes is reduced. Only attributes available on at least one observation having the class are kept, leading to 699 attributes.

Results are available in Table 4. Our approach is denoted *MOCA-I* (Multi-Objective Classifier Algorithm for Imbalanced data). We compared only to algorithms giving the best results regarding the average of GM over the 25 runs. In addition, we will also compared to *C4.5-CS* – a cost-sensitive version of *C4.5* available in KEEL Framework [19]. For each dataset the best average of GM is recorded. Then we computed relative error to the obtained best: a score of 0 indicates the algorithm got the best result on the dataset. As an indication, the last line indicates the average of the relative errors, over the 6 datasets.

We can observe that the majority of algorithms had some difficulties to handle *abalone19* dataset, which has a high imbalance. Our model outperforms other algorithms on 3 datasets. On the 3 remaining datasets, *C4.5-CS* outperforms all algorithms. However, when outperformed the model still gives interesting results.

4.3   Experiments on a Real Dataset

In addition to literature datasets, we tested the scalability of our method on one real large dataset: the previously presented *tia* dataset. We compared to results obtained by J48 – the *C4.5* algorithm implementation of Weka; well-known by some medical users. Since *C4.5* may encounters trouble to deal with imbalanced data, we compared to results obtained by *C4.5-CS* algorithm that obtained good results on the benchmark datasets. Each algorithm was run 5 times using 5-fold cross-validation to obtain 25 Pareto fronts for *MOCA-I*. *C4.5* and *C4.5-CS* uses default parameters provided by Weka and KEEL.

As observed in Table 5, reporting average and standard deviation of GM, on training and test data, our approach obtains a better GM score than *C4.5* and *C4.5-CS*, on test datasets. *C4.5-CS* is more effective than *C4.5* and *MOCA-I* on training data but is subject to over-fitting: it encounters problems when dealing with unknown observations, like on the test data. In addition to its

Table 4. Relative error to the best average of GM: algorithms with 0 obtained the best average of GM

	MOCA-I	XCS	O-DT	SIA	CORE	GAssist	OCEC	DT-GA	HIDER	C4.5	C4.5-CS
haberman	**0.00**	0.41	0.04	0.16	0.40	0.27	0.27	0.42	0.48	0.42	0.19
ecoli1	0.05	0.04	0.10	0.19	0.03	0.05	0.36	0.07	0.16	0.06	**0.00**
ecoli2	**0.00**	0.66	0.06	0.05	0.15	0.04	0.42	0.15	0.35	0.07	0.03
yeast3	0.01	0.81	0.10	0.10	0.23	0.06	0.01	0.10	0.43	0.07	**0.00**
yeast2vs8	0.11	0.18	0.18	0.88	0.14	0.26	0.16	0.88	0.18	0.88	**0.00**
abalone19	**0.00**	1.00	0.88	1.00	1.00	1.00	0.03	1.00	1.00	1.00	0.52
err. average	0.03	0.52	0.40	0.23	0.32	0.28	0.21	0.44	0.43	0.42	0.12

MOCA-I: Multi-Objective Classifier Algorithm for Imbalanced data; SIA: Supervised Inductive Algorithm, O-DT: Oblique Decision Tree, CORE: CO-Evolutionary Rule Extractor, GAssist: Genetic Algorithms based claSSIfier sySTem, OCEC: Organizational Co-Evolutionary algorithm for Classification, DT-GA: Hybrid Decision Tree - Genetic Algorithm, HIDER: HIerarchical DEcision Rules

Table 5. Comparison to *C4.5* and *C4.5-CS* on the real dataset *tia*

	GM on Training	GM on Test
MOCA-I	0.92 ±0.02	**0.74 ±0.07**
C4.5	0.58 ±0.03	0.52 ±0.11
C4.5-CS	**0.99 ±0.0009**	0.47 ±0.11

Table 6. Impact of rule selection

	Cf tra	Cf tst	Se tra	Se tst	GM tra	GM tst
MOCA-I cut 1	1	1	0.36	0.14	0.60	0.38
MOCA-I cut 2	1	0.75	0.53	0.21	0.72	0.46
MOCA-I cut 3	0.10	0.06	0.86	0.71	0.90	0.81
C4.5	1	0.75	0.39	0.21	0.62	0.46

Confidence (Cf), Sensitivity (Se) and Geometric mean of the true rates (GM) on one fold, with different cuts after ROC post-processing. Tra=training data, Tst=test data

best performance on test data, *MOCA-I* uses the previously presented postprocessing method to output a ruleset with different cut possibilities. In Table 5 the ruleset is cut to improve GM; Table 6 shows results obtained with different cut points over the ROC curve. *Cut 1* is a cut where there is no false positive. *Cut 3* is the cut presented previously in our post-processing method. *Cut 2* is between these two cuts on the ROC curve. The cut point can be adapted depending on the cost of a *false positive*. When no error is tolerable, cut points bringing less false positives will be preferred (like *Cut 1* or *Cut 2* in Table 6).

5 Conclusion and Further Research

We proposed and implemented a Pittsburgh classification rule mining system using partial classification rules, adapted to imbalanced data. Our method based on a multi-objective local search using *Confidence*, *Sensitivity* and rule length was shown to be effective in this context, compared to state-of-the-art rule mining classification algorithms. Moreover, it was proven to be more effective than *C4.5* and *C4.5-CS* on real hospital data to predict unknown observations. The use of partial classification rules avoids some common inconsistencies brought by Pittsburgh modeling, simplifying the conception of neighborhood operators. Thanks to DMLS algorithm, parameters are easier to configure than in other approaches based on genetic algorithm, while giving best results. To overcome one of the drawback of dominance-based algorithms, obtaining an archive of 400 and more compromise solutions, we proposed two methods based on ROC curve. The first helps the final user to choose the rules to keep, while the second automatically generates one single solution without the intervention of the user. Further research may include the development of new neighborhood operators, like a covering operator that introduces rules concerning uncovered individuals. Operators dealing with attribute granularity can be interesting, like the one presented in Plantevit et al. work [20]. They will allow generalizing or specializing rules, defining new rule neighbors:

- *R1* : *juvenile diabetes → increased risk of stroke*
- *R1′* : *diabetes → increased risk of stroke*

With enhanced computational power another interesting approach would be to use the *area under the ROC curve* (AUC) as an optimization criterion, instead of *Sensitivity* and *Confidence*. Or the left-most portion of the area under the curve (LAUC) as defined by Zhang et al. [21], if we want to allow a fixed number of false positives.

References

1. Fernández, A., Garciá, S., Luengo, J., Bernadó-Mansilla, E., Herrera, F.: Genetics-based machine learning for rule induction: state of the art, taxonomy, and comparative study. IEEE Trans. Evol. Comput. **14**(6), 913–941 (2010)
2. Quinlan, J.R.: C4.5: Programs for Machine Learning. Morgan Kaufmann Publishers Inc., San Francisco (1993)
3. Chawla, N.V.: Data mining for imbalanced datasets: an overview. In: Maimon, O., Rokach, L. (eds.) Data Mining and Knowledge Discovery Handbook, 2nd edn, pp. 875–886. Springer, New York (2010)
4. Jo, T., Japkowicz, N.: Class imbalances versus small disjuncts. ACM SIGKDD Explor. Newsl. **6**(1), 40–49 (2004)
5. Geng, L., Hamilton, H.J.: Interestingness measures for data mining: a survey. ACM Comput. Surv. (CSUR) **38**(3), 1–32 (2006)
6. Ohsaki, M., Abe, H., Tsumoto, S., Yokoi, H., Yamaguchi, T.: Evaluation of rule interestingness measures in medical knowledge discovery in databases. Artif. Intell. Med. **41**, 177–196 (2007)
7. Greco, S., Pawlak, Z., Slowiński, R.: Can bayesian confirmation measures be useful for rough set decision rules? Eng. Appl. Artif. Intell. **17**(4), 345–361 (2004)
8. Bayardo, J., Agrawal, R.: Mining the most interesting rules. In: Proceedings of the Fifth ACM SIGKDD, ser. KDD '99, pp. 145–154 (1999)

9. Rissanen, J.: Modeling by shortest data description. Automatica **14**(5), 465–471 (1978)

10. Reynolds, A., de la Iglesia, B.: Rule induction for classification using multi-objective genetic programming. In: Obayashi, S., Deb, K., Poloni, C., Hiroyasu, T., Murata, T. (eds.) EMO 2007. LNCS, vol. 4403, pp. 516–530. Springer, Heidelberg (2007)

11. Bacardit, J., Stout, M., Hirst, J.D., Sastry, K., Llorà, X., Krasnogor, N.: Automated alphabet reduction method with evolutionary algorithms for protein structure prediction. In: GECCO, pp. 346–353 (2007)

12. Corne, D., Dhaenens, C., Jourdan, L.: Synergies between operations research and data mining: the emerging use of multi-objective approaches. Eur. J. Oper. Res. **221**(3), 469–479 (2012)

13. Srinivasan, S., Ramakrishnan, S.: Evolutionary multi objective optimization for rule mining: a review. Artif. Intell. Rev. **36**(3), 205–248 (2011)

14. Coello Coello, C.A, Dhaenens, C., Jourdan, L. (eds.): Advances in Multi-Objective Nature Inspired Computing. SCI, vol. 272. Springer, Heidelberg (2010)

15. Casillas, J., Martínez, P., Benítez, A.: Learning consistent, complete and compact sets of fuzzy rules in conjunctive normal form for regression problems. Soft Comput. (A Fusion of Foundations, Methodologies and Applications) **13**, 451–465 (2009)

16. Liefooghe, A., Humeau, J., Mesmoudi, S., Jourdan, L., Talbi, E.-G.: On dominance-based multiobjective local search: design, implementation and experimental analysis on scheduling and traveling salesman problems. J. Heuristics **18**, 317–352 (2012)

17. Liefooghe, A., Jourdan, L., Talbi, E.-G.: A software framework based on a conceptual unified model for evolutionary multiobjective optimization: paradiseo-moeo. Eur. J. Oper. Res. **209**(2), 104–112 (2011)

18. Fawcett, T.: An introduction to roc analysis. Pattern Recogn. Lett. **27**(8), 861–874 (2006)

19. Alcalá-Fdez, J., et al.: Keel: a software tool to assess evolutionary algorithms for data mining problems. Soft Comput. (A Fusion of Foundations, Methodologies and Applications) **13**, 307–318 (2009)

20. Plantevit, M., Laurent, A., Laurent, D., Teisseire, M., Choong, Y.W.: Mining multidimensional and multilevel sequential patterns. ACM TKDD **4**(1), 1–37 (2010)

21. Zhang, J., Bala, J.W., Hadjarian, A., Han, B.: Learning to rank cases with classification rules. In: Fürnkranz, J., Hüllermeier, E. (eds.) Preference Learning, pp. 155–177. Springer, Heidelberg (2011)

Sharing Information in Parallel Search with Search Space Partitioning

Davide Lanti and Norbert Manthey[✉]

Knowledge Representation and Reasoning Group,
Technische Universität Dresden, 01062 Dresden, Germany
norbert@janeway.inf.tu-dresden.de

Abstract. In this paper we propose a new approach to share information among the computation units of an iterative search partitioning parallel SAT solver by approximating validity. Experimental results show the streh of the approach, against both existing sharing techniques and absence of sharing. With the improved clause sharing, out of 600 instances we could solve 13 more than previous sharing techniques.

1 Introduction

Search problems arise from various domains, ranging from small logic puzzles over scheduling problems like railway scheduling [1], to large job shop scheduling problems [2]. As long as the answers need not to be optimal, these problems can be translated into a constraint satisfaction problem [3], or into satisfiability testing (SAT) [4]. SAT approach is often successful, e.g. scheduling railway trains has been improved by a speedup up to 10000 compared to the state-of-the-art domain specific solver [1]. With the advent of parallel architectures, the interest moved towards parallel SAT solvers [5–7]. Most relevant parallel SAT solvers can be divided in two families: *portfolio* solvers, where several sequential solvers compete each other over the same formula, and *iterative partitioning*, where each solver is assigned a partition of the original problem and partitions are created iteratively. Portfolio solvers received much attention from the community, leading to enhancements by means of sharing according to some filter heuristics [5], or by controlling the diversification and intensification among the solvers [8]. The same cannot be said for iterative partitioning: for a grid implementation of the parallel solver, only a study on how to divide the search space [9] and on limited sharing has been done [10]. As for portfolio solvers [5,11], Hyvärinen et.al report that in average even this limited sharing results in a speedup. In this paper we present an improved clause sharing mechanism for the iterative partitioning approach. Our evaluation reveals interesting insights: first, sharing clauses introduces almost no overhead in computation. Furthermore, the performance of the overall search is increased. One of the reasons for this improved behavior is that the number

---

Davide Lanti was supported by the European Master's Program in Computational Logic (EMCL).

G. Nicosia and P. Pardalos (Eds.): LION 7, LNCS 7997, pp. 52–58, 2013.
DOI: 10.1007/978-3-642-44973-4_6, © Springer-Verlag Berlin Heidelberg 2013

of shared clauses increases, strengthening the cooperation among the parallel running solvers. Finally, the approach scales with more cores, that is crucial as increasingly parallel architectures become available.

After giving some preliminaries on SAT solving in Sect. 2, we present our new clause sharing approach in Sect. 3. An empirical evaluation is performed in Sect. 4, and finally we draw conclusions in Sect. 5.

2   Preliminaries

Let V be a finite set of Boolean variables. The set of *literals* $V \cup \{\overline{x} \mid x \in V\}$ consists of positive and negative Boolean variables. A *clause* is a finite disjunction of literals and a *formula* (in *conjunctive normal form* (CNF)) is a finite conjunction of clauses. We sometimes consider clauses and formulae as sets of literals and sets of clauses, respectively, because duplicates can be removed safely. We denote clauses with square brackets and formulae with angle brackets, so that $((a \vee b) \wedge (\overline{a} \vee c \vee \overline{d}))$ is written as $\langle [a, b], [\overline{a}, c, \overline{d}] \rangle$. An *interpretation* J is a (partial or total) mapping from the set of variables to the set $\{\top, \bot\}$ of truth values; the interpretation is represented by a set of literals, also denoted by J, with the understanding that a variable x is mapped to $\top$ if $x \in J$ and is mapped to $\bot$ if $\overline{x} \in J$. One should observe that $\{x, \overline{x}\} \not\subseteq J$ for any x and J.

A clause C is *satisfied* by an interpretation J if $l \in J$ for some literal $l \in C$. An interpretation *satisfies* a formula F, if it satisfies every clause in F. If there exists an interpretation that satisfies F, then F is *satisfiable*, otherwise F is *unsatisfiable*. An interpretation J that satisfies a formula F is called *model* of F ($J \models F$). Given two formulae F and G, we say that F models G ($F \models G$) if and only if every model of F is also a model of G. Two formulae F and G are *equivalent* ($F \equiv G$), if they have the same set of models. Observe that if $F \models G$, then $F \cup G \equiv F$. Let $C = [x, c_1, \ldots, c_m]$ and $D = [\overline{x}, d_1, \ldots, d_n]$ be two clauses. We call the clause $E = [c_1, \ldots, c_m, d_1, \ldots, d_n]$ the *resolvent* of C and D, which has been produced by *resolution* on variable x. We write $E = C \otimes D$. Note, that $\langle C, D \rangle \models \langle E \rangle$, and therefore $\langle C, D \rangle \equiv \langle C, D, E \rangle$.

2.1   Parallel SAT Solving

Satisfiability testing answers the question whether a propositional formula is satisfiable. Structured sequential SAT solvers create a partial interpretation based on the *Davis-Putnam-Loveland-Logemann* (*DPLL*) algorithm [12]. *Conflict Driven Clause Learning* (CDCL) sequential SAT solvers are an enhancement of DPLL where backtracking is performed according to *conflict analysis* [13]. Conflict analysis produces new *learnt* clauses based on resolution. Katebi et.al. show in [14] that among all the major improvements to sequential SAT solvers clause learning is the most beneficial technique.

Studies on parallel SAT solvers started in 1994 [15]. An overview of these studies since that time is given in [16,17]. Parallelizing the search process inside the DPLL algorithm has been done in [18], however this approach does not scale

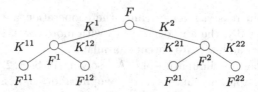

Fig. 1. The tree shows how a formula can be partitioned iteratively by using a partitioning function that creates two child formulae.

beyond two cores. Since modern hardware provides many more cores, we focus on techniques that are more promising, namely:

- **parallel portfolio search** [5], where different sequential solvers solve the same input formula in parallel.
- **iterative partitioning** [9], where a formula is partitioned iteratively into a tree of subproblems and each subproblem is solved in parallel.

Portfolio parallelization is the most common approach and many parallel SAT solvers rely on this technique, e.g. [5,19]. Iterative partitioning is a partitioning scheme that does not suffer of the theoretical slow down common to other partitioning approaches [20]. Since [20] reports that iterative partitioning is the most scalable algorithm, we focus on improving it further by allowing more communication.

3   Sharing Information in Parallel Search

The partitioning of the search space of a formula F is illustrated by the *partition tree* in Fig. 1. A *partition function* splits a formula F into n sub-problems $F_1, \ldots, F_n$ meeting the following constraints: $F \equiv F_1 \vee \cdots \vee F_n$ and $F_i \wedge F_j \equiv \bot$, for each $1 \leq i < j \leq n$. W.l.o.g one can assume that every partition F_i is of the form $F \wedge K_i$, for some *partitioning constraint* K_i in CNF. A partition tree for a formula F w.r.t. a partition function ϕ is a tree $\mathcal{T}$ rooted in F such that, for every node F' in $\mathcal{T}$ the set of its direct successors is $\phi(F')$. A more convenient notation for nodes in a tree is given by marking them with their *positions*: the root node has the empty position ϵ, whereas the node at position pi is the *i-th* successor of the node at position p. The set of all positions in $\mathcal{T}$ is $pos(\mathcal{T})$. With F^p we denote the node at position p of a tree rooted in F. Observe that for every position $p \in pos(\mathcal{T})$, it holds $F^p = F \cup K^{i_1} \cup K^{i_1 i_2} \cup \ldots \cup K^{i_1 \cdots i_n}$, if $p := i_1 \ldots i_n$ and each $i_j \in \{1, \ldots, |\phi(F^{i_1 \cdots i_{j-1}})|\}$. Since a partition tree is created upon a partition function, clearly $F^p \equiv \bigvee_i F^{pi}$ and $\forall_{i \neq j} F^{pi} \wedge F^{pj} \equiv \bot$, for every $p \in pos(\mathcal{T})$, $i, j \in \{1, \ldots, |\phi(F_p)|\}$. Sharing learnt clauses among solvers that solve child formulae has been considered in [10]. There, Hyvärinen et.al. introduce an expensive mechanism called *assumption-based (learnt) clause tagging* and a fast approximation method *flag-based (learnt) clause tagging*. In the following, we present an extension of the latter that allows more information to be shared without introducing significant overhead to the solver.

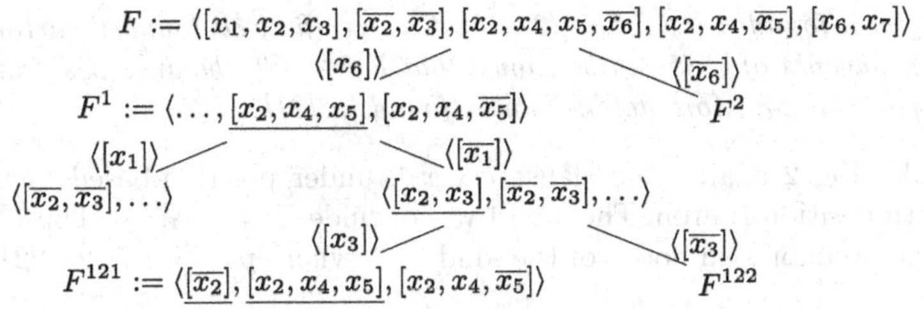

$$F := \langle [x_1, x_2, x_3], [\overline{x_2}, \overline{x_3}], [x_2, x_4, x_5, \overline{x_6}], [x_2, x_4, \overline{x_5}], [x_6, x_7] \rangle$$

$\langle [x_6] \rangle$ $\langle [\overline{x_6}] \rangle$

$$F^1 := \langle \ldots, [x_2, x_4, x_5], [x_2, x_4, \overline{x_5}] \rangle \qquad\qquad F^2$$

$\langle [x_1] \rangle$ $\langle [\overline{x_1}] \rangle$

$\langle [\overline{x_2}, \overline{x_3}], \ldots \rangle$ $\langle [x_2, x_3], [\overline{x_2}, \overline{x_3}], \ldots \rangle$

$\langle [x_3] \rangle$ $\langle [\overline{x_3}] \rangle$

$$F^{121} := \langle [\overline{x_2}], [x_2, x_4, x_5], [x_2, x_4, \overline{x_5}] \rangle \qquad\qquad F^{122}$$

Fig. 2. Partition tree for F, where unsafe clauses are underlined. Each node has been applied resolution w.r.t. its incoming constraints. Clause $[x_4, x_2]$, that could be learnt by the solver working on F^{121}, is unsafe because it depends on constraint $\langle [x_6] \rangle$. However, it could safely be shared among the children of F^1.

3.1 Flag-Based Clause Tagging

The idea of *flag-based* clause tagging is to share only those clauses that do not depend on a partitioning constraint. Each clause is assigned a Boolean flag that represents being *safe* or *unsafe*. Clauses are tagged *unsafe* if they either belong to a partitioning constraint or have been obtained in a resolution derivation involving some unsafe clause. All other clauses are tagged as *safe*. Thus, every safe clause either belongs to the original problem, or it can be obtained by applying resolution over clauses of the original problem. Therefore, a safe clause is a semantic consequence of every node in a partition tree.

Due to limited space, we provide examples in [21] where flag-based sharing effectively permits to speed-up the local computation on nodes in the tree. However, this approach is only an approximation. Indeed, if a clause is a semantic consequence only of a strict subset of the nodes in a partition tree, then this clause cannot be shared at all. However, it would be safe to share it at least among the nodes belonging to that subset. Figure 2 depicts such a situation: clause $[x_4, x_2]$ could be shared among all the children of F^{123}, but this is not allowed by the flag-based approach. We now propose an improvement to clause sharing that addresses this problem without introducing overhead on the parallel solver.

3.2 Position-Based Clause Tagging

Instead of flags, we tag each clause with a position in the partition tree. A clause C tagged with a position p is indicated as C^p. Given a partition tree T for a formula F, clauses belonging to F are tagged with the empty position ϵ. Clauses in a partitioning constraint K^p are tagged with p. A clause obtained from a resolution derivation $(R_1^{p_1}, \ldots, R_n^{p_n})$ is tagged with $arg\,max_{p_i} |p_i|$, where $1 \leq i \leq n$. The following theorem clarifies the main contribution of this work:

Theorem 1. *Let T be a partition tree for a formula F, and consider two nodes F^{pq} and F^r in T. Let C^p be a clause learnt at some point in the computation over the node F^r. Then $F^{pq} \models C^p$.*

Proof. By well-founded induction [22] w.r.t. a specific well-founded partial order over the resolvents of C it can be shown that $F^{pq} \models C^p$, because position p is a prefix of position pq. More details can be found in [21].

Consider Fig. 2 again. The clause $[x_4, x_2]$, under position-based tagging, is tagged with position 1. From Theorem 1 we conclude $F^1 \models [x_4, x_2]$. Thus, $[x_4, x_2]$ can be shared among all nodes of the kind F^{1p}, where $p \in \{\epsilon, 1, 2, 21, 22\}$.

4   Empirical Evaluation

The experiments have been run on AMD Opteron 6274 CPUs with 2.2 GHz and 16 cores, so that 16 local solvers run in parallel. Each instance is assigned a time-out of 1 h (wall clock) and a total of 16 GB main memory. Each sharing approach has been tested over 600 instances of the instances of SAT challenge 2012. Our iterative-partitioning solver is based on Minisat [7], following the ideas of [20]. As in [9] the resources of the local solvers are restricted: a branch is created after 8096 conflicts, and a local solver is allowed to search until 512000 conflicts are reached. The partitioning function uses VSIDS scattering [9]. We tested our solver with 4 different configurations: "POS" and "FLAG" use position-based and flag-based clause sharing, respectively. Here, a clause can be shared only if its size is less than or equal to 2. "RAND" uses position-based sharing where each learnt clause can be shared with a probability of 5 %. The last configuration, "NONE", does not share any clauses.

Table 1. Number of solved instances

Approach	Solved	SAT	UNSAT	AVG	CPU ratio	Score	Shared
POS	**430**	239	**191**	377.397	11.5	**78**	17202
RAND	380	232	148	**374.445**	11.5	−50	**209199**
FLAG	417	234	183	378.969	11.4	30	6557
NONE	418	**244**	174	383.785	**12.1**	−58	0

The results are depicted in Table 1. For satisfiable instances, sharing no clauses seems to be the best option, allowing the parallel solvers to diverse. On the other hand, for unsatisfiable instances the level based sharing gives the best results. The *CPU ratio* shows how many cores have been used in average to solve all the instances. Observe that sharing does not introduce significant overhead w.r.t. a setting where solvers do not communicate. As expected, position-based tagging permits to share (thrice) more clauses than flag-based. We expect even better results by using more sophisticated filters than fixed clause size.

5   Conclusion

We presented a new position-based clause sharing technique that allows to share clauses for subsets of a parallel iterative-partitioning SAT solver. Position-based clause sharing improves the intensification of parallel searching SAT solvers by identifying the search space in which a shared clause is valid so that the total number of shared clauses can be increased compared to previous work [10].

Future work could improve shared clauses further. By rejecting resolution steps, the sharing position of learnt clauses can be improved. Moreover, a filter on the receiving solver should be considered as well. Also, it is not trivial to decide what shared clauses are important and if these should actively drive the search. Additionally, parallel resources should be exploited further, for example by using different partitioning strategies or by replacing the local sequential solver by another parallel SAT solver. Finally, improvements to the local solver, as for example restarts and advanced search direction techniques, could also be incorporated into the search space partitioning.

References

1. Großmann, P., Hölldobler, S., Manthey, N., Nachtigall, K., Opitz, J., Steinke, P.: Solving periodic event scheduling problems with SAT. In: Jiang, H., Ding, W., Ali, M., Wu, X. (eds.) IEA/AIE 2012. LNCS, vol. 7345, pp. 166–175. Springer, Heidelberg (2012)
2. Carlier, J., Pinson, E.: An algorithm for solving the job-shop problem. Manage. Sci. 35(2), 164–176 (1989)
3. Rossi, F., Beek, P.V., Walsh, T.: Handbook of Constraint Programming (Foundations of Artificial Intelligence). Elsevier Science Inc, New York (2006)
4. Biere, A., Heule, M., van Maaren, H., Walsh, T. (eds.): Handbook of Satisfiability. Frontiers in Artificial Intelligence and Applications, vol. 185. IOS Press, Amsterdam (2009)
5. Hamadi, Y., Jabbour, S., Sais, L.: Manysat: a parallel sat solver. JSAT 6(4), 245–262 (2009)
6. Biere, A.: Lingeling, Plingeling, PicoSAT and PrecoSAT at SAT Race 2010. FMV Report Series Technical Report 10/1. Johannes Kepler University, Linz, Austria (2010)
7. Eén, N., Sörensson, N.: An extensible SAT-solver. In: Giunchiglia, E., Tacchella, A. (eds.) SAT 2003. LNCS, vol. 2919, pp. 502–518. Springer, Heidelberg (2004)
8. Guo, L., Hamadi, Y., Jabbour, S., Sais, L.: Diversification and intensification in parallel SAT solving. In: Cohen, D. (ed.) CP 2010. LNCS, vol. 6308, pp. 252–265. Springer, Heidelberg (2010)
9. Hyvärinen, A.E.J., Junttila, T., Niemelä, I.: Partitioning SAT instances for distributed solving. In: Fermüller, C.G., Voronkov, A. (eds.) LPAR-17. LNCS, vol. 6397, pp. 372–386. Springer, Heidelberg (2010)
10. Hyvärinen, A.E.J., Junttila, T., Niemelä, I.: Grid-based SAT solving with iterative partitioning and clause learning. In: Lee, J. (ed.) CP 2011. LNCS, vol. 6876, pp. 385–399. Springer, Heidelberg (2011)
11. Arbelaez, A., Hamadi, Y.: Improving parallel local search for SAT. In: Coello, C.A.C. (ed.) LION 2011. LNCS, vol. 6683, pp. 46–60. Springer, Heidelberg (2011)
12. Davis, M., Logemann, G., Loveland, D.: A machine program for theorem-proving. Commun. ACM 5, 394–397 (1962)
13. Marques Silva, J.P., Sakallah, K.A.: Grasp: a search algorithm for propositional satisfiability. IEEE Trans. Comput. 48(5), 506–521 (1999)

14. Katebi, H., Sakallah, K.A., Marques-Silva, J.: Empirical study of the anatomy of modern sat solvers. In: Sakallah, K.A., Simon, L. (eds.) SAT 2011. LNCS, vol. 6695, pp. 343–356. Springer, Heidelberg (2011)
15. Böhm, M., Speckenmeyer, E.: A fast parallel sat-solver - efficient workload balancing, (1994)
16. Martins, R., Manquinho, V., Lynce, I.: An overview of parallel sat solving. Constraints 17(3), 304–347 (2012)
17. Hölldobler, S., Manthey, N., Nguyen, V., Stecklina, J., Steinke, P.: A short overview on modern parallel SAT-solvers. In: Wasito, I., et al. (ed.) ICACSIS, pp. 201–206 (2011)
18. Manthey, N.: Parallel SAT solving - using more cores. In: Pragmatics of SAT(POS'11) (2011)
19. Audemard, G., Hoessen, B., Jabbour, S., Lagniez, J.-M., Piette, C.: Revisiting clause exchange in parallel SAT solving. In: Cimatti, A., Sebastiani, R. (eds.) SAT 2012. LNCS, vol. 7317, pp. 200–213. Springer, Heidelberg (2012)
20. Hyvärinen, A.E.J., Manthey, N.: Designing scalable parallel SAT solvers. In: Cimatti, A., Sebastiani, R. (eds.) SAT 2012. LNCS, vol. 7317, pp. 214–227. Springer, Heidelberg (2012)
21. Lanti, D., Manthey, N.: Sharing information in parallel search with search space partitioning. Technical Report 1, Knowledge Representation and Reasoning Group, Technische Universität Dresden, 01062 Dresden, Germany (2013)
22. Baader, F., Nipkow, T.: Term rewriting and all that. Cambridge University Press, New York (1998)

Fast Computation of the Multi-Points Expected Improvement with Applications in Batch Selection

Clément Chevalier[1,2] and David Ginsbourger[2(✉)]

[1] Institut de Radioprotection et de Sûreté Nucléaire (IRSN),
31, avenue de la Division Leclerc, 92260 Fontenay-aux-Roses, France
[2] IMSV, University of Bern, Alpeneggstrasse 22, 3012 Bern, Switzerland
{clement.chevalier,ginsbourger}@stat.unibe.ch

Abstract. The Multi-points Expected Improvement criterion (or q-EI) has recently been studied in batch-sequential Bayesian Optimization. This paper deals with a new way of computing q-EI, without using Monte-Carlo simulations, through a closed-form formula. The latter allows a very fast computation of q-EI for reasonably low values of q (typically, less than 10). New parallel kriging-based optimization strategies, tested on different toy examples, show promising results.

Keywords: Computer experiments · Kriging · Parallel optimization · Expected improvement

1 Introduction

In the last decades, *metamodeling* (or *surrogate modeling*) has been increasingly used for problems involving costly computer codes (or "black-box simulators"). Practitioners typically dispose of a very limited evaluation budget and aim at selecting evaluation points cautiously when attempting to solve a given problem.

In global optimization, the focus is usually put on a real-valued function f with d-dimensional source space. In this settings, Jones et al. [1] proposed the now famous *Efficient Global Optimization* (EGO) algorithm, relying on a kriging metamodel [2] and on the Expected Improvement (EI) criterion [3]. In EGO, the optimization is done by sequentially evaluating f at points maximizing EI. A crucial advantage of this criterion is its fast computation (besides, the analytical gradient of EI is implemented in [4]), so that the hard optimization problem is replaced by series of much simpler ones.

Coming back to the decision-theoretic roots of EI [5], a Multi-points Expected Improvement (also called "q-EI") criterion for batch-sequential optimization was defined in [6] and further developed in [7,8]. Maximizing this criterion enables choosing batches of $q > 1$ points at which to evaluate f in parallel, and is of particular interest in the frequent case where several CPUs are simultaneously available. Even though an analytical formula was derived for the 2-EI in [7], the

G. Nicosia and P. Pardalos (Eds.): LION 7, LNCS 7997, pp. 59–69, 2013.
DOI: 10.1007/978-3-642-44973-4_7, © Springer-Verlag Berlin Heidelberg 2013

Monte Carlo (MC) approach of [8] for computing q-EI when $q \geq 3$ makes the criterion itself expensive-to-evaluate, and particularly hard to optimize.

A lot of effort has recently been paid to address this problem. The pragmatic approach proposed by Ginsbourger and Le Riche [8] consists in circumventing a direct q-EI maximization, and replacing it by simpler strategies where batches are obtained using an offline q-points EGO. In such strategies, the model updates are done using dummy response values such as the kriging mean prediction (Kriging Believer) or a constant (Constant Liar), and the covariance parameters are re-estimated only when real data is assimilated. In [9] and [10], q-EI optimization strategies were proposed relying on the MC approach, where the number of MC samples is tuned online to discriminate between candidate designs. Finally, Frazier [11] proposed a q-EI optimization strategy involving stochastic gradient, with the crucial advantage of *not* requiring to evaluate q-EI itself.

In this article we derive a formula allowing a fast and accurate approximate evaluation of q-EI. This formula may contribute to significantly speed up strategies relying on q-EI. The main result, relying on Tallis' formula, is given in Sect. 2. The usability of the proposed formula is then illustrated in Sect. 3 through benchmark experiments, where a brute force maximization of q-EI is compared to three variants of the Constant Liar strategy. In particular, a new variant (CL-mix) is introduced, and is shown to offer very good performances at a competitive computational cost. For self-containedness, a slightly revisited proof of Tallis' formula is given in appendix.

2   Multi-Points Expected Improvement Explicit Formulas

In this section we give an explicit formula allowing a fast and accurate deterministic approximation of q-EI. Let us first give a few precisions on the mathematical settings. Along the paper, f is assumed to be one realisation of a Gaussian Process (GP) with known covariance kernel and mean known up to some linear trend coefficients, so that the conditional distribution of a vector of values of the GP conditional on past observations is still Gaussian (an improper uniform prior is put on the trend coefficients when applicable). This being said, most forthcoming derivations boil down to calculations on Gaussian vectors. Let $\mathbf{Y} := (Y_1, \ldots, Y_q)$ be a Gaussian Vector with mean $\mathbf{m} \in \mathbb{R}^q$ and covariance matrix Σ. Our aim in this paper is to explicitly calculate expressions of the following kind:

$$\mathbb{E}\left[\left(\max_{i \in \{1, \ldots, q\}} Y_i - T \right)_+ \right] \tag{1}$$

where $(.)_+ := \max(.,0)$. In Bayesian optimization (say maximization), expectations and probabilities are taken conditional on response values at a given set of n points $(\mathbf{x}_1, \ldots, \mathbf{x}_n) \in \mathbb{X}^n$ where $\mathbb{X}$ is the input set of f (often, a compact subset of $\mathbb{R}^d$, $d \geq 1$), the threshold $T \in \mathbb{R}$ is usually the maximum of those n available response values, and $\mathbf{Y}$ is the vector of unknown responses at a given batch of q points, $\mathbf{X}^q := (\mathbf{x}_{n+1}, \ldots, \mathbf{x}_{n+q}) \in \mathbb{X}^q$. In such framework, the vector $\mathbf{m}$ and the matrix Σ are the so-called "Kriging mean" and "Kriging covariance" at $\mathbf{X}^q$ and can be calculated relying on classical Kriging equations (see, e.g., [12]).

In order to obtain a tractable analytical formula for Expression (1), not requiring any Monte-Carlo simulation, let us first give a useful formula obtained by [13], and recently used in [14] for GP modeling with inequality constraints:

Proposition 1 (Tallis' formulas). *Let* $\mathbf{Z} := (Z_1, \ldots, Z_q)$ *be a Gaussian Vector with mean* $\mathbf{m} \in \mathbb{R}^q$ *and covariance matrix* $\Sigma \in \mathbb{R}^{q \times q}$. *Let* $\mathbf{b} = (b_1, \ldots, b_q) \in \mathbb{R}^q$. *The expectation of any coordinate* Z_k *under the linear constraint* $(\forall j \in \{1, \ldots, q\},\ Z_j \leq b_j)$ *denoted by* $\mathbf{Z} \leq \mathbf{b}$ *can be expanded as follows:*

$$\mathbb{E}(Z_k | \mathbf{Z} \leq \mathbf{b}) = m_k - \frac{1}{p} \sum_{i=1}^{q} \Sigma_{ik}\, \varphi_{m_i, \Sigma_{ii}}(b_i)\, \Phi_{q-1}(\mathbf{c}_{\cdot i}, \Sigma_{\cdot i}) \tag{2}$$

where:

- $p := \mathbb{P}(\mathbf{Z} \leq \mathbf{b}) = \Phi_q(\mathbf{b} - \mathbf{m}, \Sigma)$
- $\Phi_q(\mathbf{u}, \Sigma)$ $(\mathbf{u} \in \mathbb{R}^q, \Sigma \in \mathbb{R}^{q \times q}, q \geq 1)$ *is the c.d.f. of the centered multivariate Gaussian distribution with covariance matrix* Σ.
- $\varphi_{m,\sigma^2}(.)$ *is the p.d.f. of the univariate Gaussian distribution with mean* m *and variance* σ^2
- $\mathbf{c}_{\cdot i}$ *is the vector of* $\mathbb{R}^{q-1}$ *with general term* $(b_j - m_j) - (b_i - m_i)\frac{\Sigma_{ij}}{\Sigma_{ii}}$, $j \neq i$
- $\Sigma_{\cdot i}$ *is a* $(q-1) \times (q-1)$ *matrix obtained by computing* $\Sigma_{uv} - \frac{\Sigma_{iu}\Sigma_{iv}}{\Sigma_{ii}}$ *for* $u \neq i$ *and* $v \neq i$. *This matrix corresponds to the conditional covariance matrix of the random vector* $\mathbf{Z}_{-i} := (Z_1, \ldots, Z_{i-1}, Z_{i+1}, \ldots, Z_q)$ *knowing* Z_i.

For the sake of brevity, the proof of this Proposition is sent in the Appendix. A crucial point for the practical use of this result is that there exist very fast procedures to compute the c.d.f. of the multivariate Gaussian distribution. For example, the work of Genz [15,16] have been used in many R packages (see, e.g., [17,18]). The Formula (2) above is an important tool to efficiently compute Expression (1) as shown with the following Property:

Proposition 2. *Let* $\mathbf{Y} := (Y_1, \ldots, Y_q)$ *be a Gaussian Vector with mean* $\mathbf{m} \in \mathbb{R}^q$ *and covariance matrix* Σ. *For* $k \in \{1, \ldots, q\}$ *consider the Gaussian vectors* $\mathbf{Z}^{(k)} := (Z_1^{(k)}, \ldots, Z_q^{(k)})$ *defined as follows:*

$$Z_j^{(k)} := Y_j - Y_k,\ j \neq k$$

$$Z_k^{(k)} := -Y_k$$

Denoting by $\mathbf{m}^{(k)}$ *and* $\Sigma^{(k)}$ *the mean and covariance matrix of* $\mathbf{Z}^{(k)}$, *and defining the vector* $\mathbf{b}^{(k)} \in \mathbb{R}^q$ *by* $b_k^{(k)} = -T$ *and* $b_j^{(k)} = 0$ *if* $j \neq k$, *the EI of* $\mathbf{X}^q$ *writes:*

$$EI(\mathbf{X}^q) = \sum_{k=1}^{q} \left((m_k - T)p_k + \sum_{i=1}^{q} \Sigma_{ik}^{(k)} \varphi_{m_i^{(k)}, \Sigma_{ii}^{(k)}}(b_i^{(k)}) \Phi_{q-1}\left(\mathbf{c}_{\cdot i}^{(k)}, \Sigma_{\cdot i}^{(k)}\right) \right) \tag{3}$$

where:

- $p_k := \mathbb{P}(\mathbf{Z}^{(k)} \le \mathbf{b}^{(k)}) = \Phi_q(\mathbf{b}^{(k)} - \mathbf{m}^{(k)}, \Sigma^{(k)})$.
  p_k *is actually the probability that* Y_k *exceeds* T *and* $Y_k = \max_{j=1,\dots,q} Y_j$.
- $\Phi_q(.,\Sigma)$ *and* $\varphi_{m,\sigma^2}(.)$ *are defined in Proposition 1*
- $\mathbf{c}_{.i}^{(k)}$ *is the vector of* $\mathbb{R}^{q-1}$ *constructed like in Proposition 1, by computing*
  $(b_j^{(k)} - m_j^{(k)}) - (b_i^{(k)} - m_i^{(k)}) \frac{\Sigma_{ij}^{(k)}}{\Sigma_{ii}^{(k)}}$, *with* $j \ne i$
- $\Sigma_{.i}^{(k)}$ *is a* $(q-1) \times (q-1)$ *matrix constructed from* $\Sigma^{(k)}$ *like in Proposition 1. It corresponds to the conditional covariance matrix of the random vector* $\mathbf{Z}_{-i}^{(k)} := (Z_1^{(k)}, \dots, Z_{i-1}^{(k)}, Z_{i+1}^{(k)}, \dots, Z_q^{(k)})$ *knowing* $Z_i^{(k)}$.

Proof 1. Using that $\mathbb{1}_{\{\max_{i \in \{1,\dots,q\}} Y_i \ge T\}} = \sum_{k=1}^q \mathbb{1}_{\{Y_k \ge T,\, Y_j \le Y_k\ \forall j \ne k\}}$, we get

$$EI(\mathbf{X}^q) = \mathbb{E}\left[\left(\max_{i \in \{1,\dots,q\}} Y_i - T \right) \sum_{k=1}^q \mathbb{1}_{\{Y_k \ge T,\, Y_j \le Y_k\ \forall j \ne k\}} \right]$$

$$= \sum_{k=1}^q \mathbb{E}\left((Y_k - T) \mathbb{1}_{\{Y_k \ge T,\, Y_j \le Y_k\ \forall j \ne k\}} \right)$$

$$= \sum_{k=1}^q \mathbb{E}\left(Y_k - T \,\Big|\, Y_k \ge T,\, Y_j \le Y_k\ \forall j \ne k \right) \mathbb{P}\left(Y_k \ge T, Y_j \le Y_k\ \forall j \ne k \right)$$

$$= \sum_{k=1}^q \left(-T - \mathbb{E}\left(Z_k^{(k)} \Big| \mathbf{Z}^{(k)} \le \mathbf{b}^{(k)} \right) \right) \mathbb{P}\left(\mathbf{Z}^{(k)} \le \mathbf{b}^{(k)} \right)$$

Now the computation of $p_k := \mathbb{P}\left(\mathbf{Z}^{(k)} \le \mathbf{b}^{(k)} \right)$ simply requires one call to the Φ_q function and the proof can be completed by applying Tallis' formula (2) to the random vectors $\mathbf{Z}^{(k)}$ ($1 \le k \le q$).

Remark 1. From Properties (1) and (2), it appears that computing q-EI requires a total of q calls to Φ_q and q^2 calls to Φ_{q-1}. The proposed approach performs thus well when q is moderate (typically lower than 10). For higher values of q, estimating q-EI by Monte-Carlo might remain competitive. Note that, when q is larger (say, $q = 50$) and when q CPUs are available, one can always distribute the calculations of the q^2 calls to Φ_{q-1} over these q CPUs.

Remark 2. In the particular case $q = 1$ and with the convention $\Phi_0(.,\Sigma) = 1$, Eq. (3) corresponds to the classical EI formula proven in [1,5].

Remark 3. The Multi-points EI can be used in a batch-sequential strategy to optimize a given expensive-to-evaluate function f, as detailed in the next Section. Moreover, a similar criterion can also be used to perform optimization based on a Kriging model *with linear constraints*, such as the one developed by Da Veiga and Marrel [14]. For example expressions like: $\mathbb{E}\left[\left(\max_{i \in \{1,\dots,q\}} Y_i - T \right)_+ | \mathbf{Y} \le \mathbf{a} \right], \mathbf{a} \in \mathbb{R}^q$, can be computed using Tallis' formula and the same proof.

3   Batch Sequential Optimization Using Multi-Points EI

Let us first illustrate Proposition 2 and show that the proposed q-EI calculation based on Tallis' formula is actually consistent with a Monte Carlo estimation. From a kriging model based on 12 observations of the Branin-Hoo function [1], we generated a 4-point batch (Fig. 1, left plot) and calculated its q-EI value (middle plot, dotted line). The MC estimates converge to a value close to the latter, and the relative error after $5 * 10^9$ runs is less than 10^{-5}. 4-point batches generated from the three strategies detailed below are drawn on the right plot.

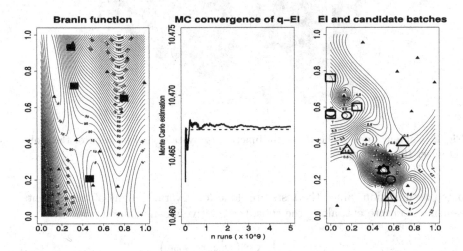

Fig. 1. Convergence (middle) of MC estimates to the q-EI value calculated with Proposition 2 in the case of a batch of four points (shown on the left plot). Right: candidate batches obtained by q-EI stepwise maximisation (squares), and the CL-min (circles) and CL-max (triangles) strategies.

We now compare a few kriging-based batch-sequential optimization methods on two different functions: the function $x \mapsto -\log(-\mathrm{Hartman6}(x))$ (see, e.g., [1]), defined on $[0,1]^6$ and the Rastrigin function [19,20] in dimension two restricted to the domain $[0, 2.5]^2$. The first function in dimension 6 is unimodal, while the second one has a lot of local optima (see: Fig. 2). The Rastrigin function is one of the 24 noiseless test function of the Black-Box Optimization Benchmark (BBOB) [19].

For each runs, we start with a random initial Latin hypercube design (LHS) of $n_0 = 10$ (Rastrigin) or 50 (Hartman6) points and estimate the covariance parameters by Maximum Likelihood (here a Matérn kernel with $\nu = 3/2$ is chosen). For both functions and all strategies, batches of $q = 6$ points are added at each iteration, and the covariance parameters are re-estimated after each

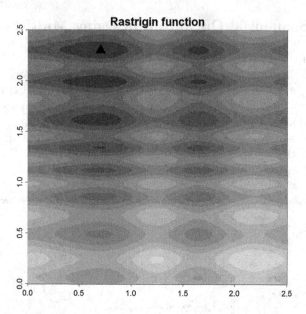

Fig. 2. Contour lines of the Rastrigin function (grayscale) and location of the global optimizer (black triangle)

batch assimilation. Since the tests are done for several designs of experiments, we chose to represent, along the runs, the relative mean squared error:

$$\text{rMSE} = \frac{1}{M} \sum_{i=1}^{M} \left(\frac{y_{\min}^{(i)} - y_{\text{opt}}}{y_{\text{opt}}} \right)^2 \tag{4}$$

where $y_{\min}^{(i)}$ in the current observed minimum in run number i and y_{opt} is the real unknown optimum. The total number M of different initial designs of experiments is fixed to 50. The tested strategies are:

– (1) q-EI stepwise maximization: q sequential d-dimensional optimizations are performed. We start with the maximization of the 1-point EI and add this point to the new batch. We then maximize the 2-point EI (keeping the first point obtained as first argument), add the maximizer to the batch, and iterate until q points are selected.
– (2) Constant Liar min (CL-min): We start with the maximization of the 1-point EI and add this point to the new batch. We then assume a dummy response (a "lie") at this point, and update the Kriging metamodel with this point and the lie. We then maximize the 1-point EI obtained with the updated kriging metamodel, get a second point, and iterate the same process until a batch of q points is selected. The dummy response has the same value over the $q - 1$ lies, and is here fixed to the minimum of the current observations.
– (3) Constant Liar max (CL-max): The lie in this Constant Liar strategy is fixed to the maximum of the current observations.

- (4) Constant Liar mix (CL-mix): At each iteration, two batches are generated with the CL-min and CL-max strategies. From these two "candidate" batches, we choose the batch with the best actual q-EI value, calculated based on Proposition 2.
- (5) Random sampling.

Note that CL-min tends to explore the function near the current minimizer (as the lie is a low value and we are minimizing f) while CL-max is more exploratory. Thus, CL-min is expected to perform well on unimodal functions. On the contrary, CL-max may perform better on multimodal functions. For all the tests we use the DiceKriging and DiceOptim packages [4]. The optimizations of the different criteria rely on a genetic algorithm using derivatives, available in the rgenoud package [21]. Figure 3 represents the compared performances of these strategies.

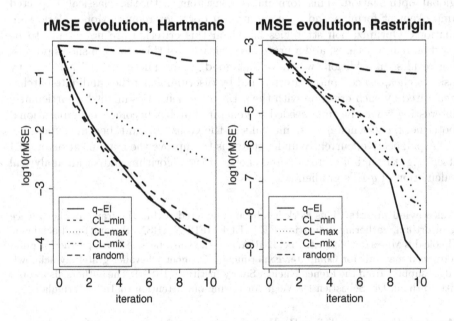

Fig. 3. Compared performances of the five considered batch-sequential optimization strategies, on two test functions.

From these plots we draw two main conclusions. From these plots we draw the following conclusions: first, the q-EI stepwise maximization strategy outperforms the strategies based on constant lies, CL-min and CL-max. However, the left graph of Fig. 3 points out that the CL-min strategy seems particularly well-adapted to the Hartman6 function. Since running a CL is computationally much cheaper than a brute fore optimization of q-EI, it is tempting to recommend the CL-min strategy for Hartman6. However, it is not straightforward to know in advance which of CL-min or CL-max will perform better on a given test case. Indeed, for example, CL-max outperforms CL-min on the Rastrigin function.

Now, we observe that using q-EI in the CL-mix heuristic enables very good performances in both cases without having to select one of the two lie values in advance. For the Hartman6 function, CL-mix even outperforms both CL-min and CL-max and has roughly the same performance as a brute force q-EI maximization. This suggests that a good heuristic might be to generate, at each iteration, candidate batches obtained with different strategies (e.g. CL with different lies) and to discriminate those batches using q-EI.

4   Conclusion

In this article we give a closed-form expression enabling a fast computation of the Multi-points Expected Improvement criterion for batch sequential Bayesian global optimization. This formula is consistent with the classical Expected Improvement formula and its computation does not require Monte Carlo simulations. Optimization strategies based on this criterion are now ready to be used on real test cases, and a brute maximization of this criterion shows promising results. In addition, we show that good performances can be achieved by using a cheap-to-compute criterion and by discriminating the candidate batches generated by such criterion with the q-EI. Such heuristics might be particularly interesting when the time needed to generate batches becomes a computational bottleneck, e.g. when $q \geq 10$ and calls to the Gaussian c.d.f. become expensive.

A perspective, currently under study, is to improve the maximization of q-EI itself, e.g. through a more adapted choice of the algorithm and/or an analytical calculation of q-EI's gradient.

Acknowledgments. This work has been conducted within the frame of the ReDice Consortium, gathering industrial (CEA, EDF, IFPEN, IRSN, Renault) and academic (Ecole des Mines de Saint-Etienne, INRIA, and the University of Bern) partners around advanced methods for Computer Experiments. Clément Chevalier gratefully acknowledges support from the French Nuclear Safety Institute (IRSN). The authors also would like to thank Dr. Sébastien Da Veiga for raising our attention to Tallis' formula.

Appendix: Proof for Tallis' Formula (2)

The proof proposed here follows exactly the method given in [13] in the particular case of a centered Gaussian Vector with normalized covariance matrix (i.e. a covariance matrix equal to the *correlation* matrix). Here, the proof is slightly more detailed and applies in a more general case.

Let $\mathbf{Z} := (Z_1, \ldots, Z_q) \sim \mathcal{N}(\mathbf{m}, \Sigma)$ with $\mathbf{m} \in \mathbb{R}^q$ and $\Sigma \in \mathbb{R}^{q \times q}$. Let $\mathbf{b} = (b_1, \ldots, b_q) \in \mathbb{R}^q$. Our goal is to calculate: $\mathbb{E}(Z_k | \mathbf{Z} \leq \mathbf{b})$. The method proposed by Tallis consists in calculating the conditional joint moment generating function (MGF) of $\mathbf{Z}$ defined as follows:

$$M_{\mathbf{Z}}(\mathbf{t}) := \mathbb{E}(\exp(\mathbf{t}^\top \mathbf{Z}) | \mathbf{Z} \leq \mathbf{b}) \tag{5}$$

It is known (see, e.g., [22]) that the conditional expectation of Z_k can be obtained by deriving such MGF with respect to t_k, in $\mathbf{t} = \mathbf{0}$. Mathematically this writes:

$$\mathbb{E}(Z_k|\mathbf{Z} \leq \mathbf{b}) = \left.\frac{\partial M_{\mathbf{Z}}(\mathbf{t})}{\partial t_k}\right|_{\mathbf{t}=0} \tag{6}$$

The main steps of this proof are then to calculate such MGF and its derivative with respect to any coordinate t_k.

Let us consider the **centered** random variable $\mathbf{Z}^c := \mathbf{Z} - \mathbf{m}$. Denoting $\mathbf{h} = \mathbf{b} - \mathbf{m}$, conditioning on $\mathbf{Z} \leq \mathbf{b}$ or on $\mathbf{Z}^c \leq \mathbf{h}$ are equivalent. The MGF of $\mathbf{Z}^c$ can be calculated as follows:

$$M_{\mathbf{Z}^c}(\mathbf{t}) := \mathbb{E}(\exp(\mathbf{t}^\top \mathbf{Z}^c)|\mathbf{Z}^c \leq \mathbf{h})$$

$$= \frac{1}{p}\int_{-\infty}^{h_1}\ldots\int_{-\infty}^{h_q} \exp(\mathbf{t}^\top \mathbf{u})\varphi_{0,\Sigma}(\mathbf{u})d\mathbf{u}$$

$$= \frac{1}{p}(2\pi)^{-\frac{q}{2}}|\Sigma|^{-\frac{1}{2}}\int_{-\infty}^{h_1}\ldots\int_{-\infty}^{h_q} \exp\left(-\frac{1}{2}\left(\mathbf{u}^\top\Sigma^{-1}\mathbf{u} - 2\mathbf{t}^\top\mathbf{u}\right)\right)d\mathbf{u}$$

where $p := \mathbb{P}(\mathbf{Z} \leq \mathbf{b})$ and $\varphi_{\mathbf{v},\Sigma}(.)$ denotes the p.d.f. of the multivariate normal distribution with mean $\mathbf{v}$ and covariance matrix Σ. The calculation can be continued by noting that:

$$M_{\mathbf{Z}^c}(\mathbf{t}) = \frac{1}{p}(2\pi)^{-\frac{q}{2}}|\Sigma|^{-\frac{1}{2}}\exp\left(\frac{1}{2}\mathbf{t}^\top\Sigma\mathbf{t}\right)\int_{-\infty}^{h_1}\ldots\int_{-\infty}^{h_q}\exp\left(-\frac{1}{2}(\mathbf{u}-\Sigma\mathbf{t})^\top\Sigma^{-1}(\mathbf{u}-\Sigma\mathbf{t})\right)d\mathbf{u}$$

$$= \frac{1}{p}\exp\left(\frac{1}{2}\mathbf{t}^\top\Sigma\mathbf{t}\right)\Phi_q(\mathbf{h}-\Sigma\mathbf{t},\Sigma)$$

where $\Phi_q(.,\Sigma)$ is the c.d.f. of the centered multivariate normal distribution with covariance matrix Σ.

Now, let us calculate for some $k \in \{1,\ldots,q\}$ the partial derivative $\frac{\partial M_{\mathbf{Z}^c}(\mathbf{t})}{\partial t_k}$ in $\mathbf{t} = \mathbf{0}$, which is equal by definition to $\mathbb{E}(Z_k^c|\mathbf{Z}^c \leq \mathbf{h})$.

$$p\,\mathbb{E}(Z_k^c|\mathbf{Z}^c \leq \mathbf{h}) = p\left.\frac{\partial M_{\mathbf{Z}^c}(\mathbf{t})}{\partial t_k}\right|_{\mathbf{t}=0}$$

$$= 0 + 1.\left.\frac{\partial}{\partial t_k}\left(\Phi_q\left(\mathbf{h}-t_k\begin{pmatrix}\Sigma_{1k}\\ \vdots \\ \Sigma_{qk}\end{pmatrix},\Sigma\right)\right)\right|_{t_k=0}$$

$$= -\sum_{i=1}^{q}\Sigma_{ik}\int_{-\infty}^{h_1}\ldots\int_{-\infty}^{h_{i-1}}\int_{-\infty}^{h_{i+1}}\ldots\int_{-\infty}^{h_q}\varphi_{0,\Sigma}(\mathbf{u}_{-i},u_i=h_i)d\mathbf{u}_{-i}$$

The last step is obtained applying the chain rule to $\mathbf{x} \mapsto \Phi_q(\mathbf{x},\Sigma)$ at the point $\mathbf{x} = \mathbf{h}$. Here, $\varphi_{0,\Sigma}(\mathbf{u}_{-i},u_i = h_i)$ denotes the c.d.f. of the centered multivariate normal distribution at given points $(\mathbf{u}_{-i},u_i = h_i) := (u_1,\ldots,u_{i-1}, h_i,u_{i+1},\ldots,u_q)$. Note that the integrals in the latter Expression are in dimension $q-1$ and not q. In the ith term of the sum above, we integrate with respect to all the q components except the component i. To continue the calculation we can use the identity:

$$\forall \mathbf{u} \in \mathbb{R}^q, \varphi_{0,\Sigma}(\mathbf{u}) = \varphi_{0,\Sigma_{ii}}(u_i)\varphi_{\Sigma_{ii}^{-1}\Sigma_i u_i, \Sigma_{-i,-i} - \Sigma_i \Sigma_{ii}^{-1} \Sigma_i^\top}(\mathbf{u}_{-i}) \qquad (7)$$

where $\Sigma_i = (\Sigma_{1i}, \dots, \Sigma_{i-1i}, \Sigma_{i+1i}, \dots, \Sigma_{qi})^\top$ $(\Sigma_i \in \mathbb{R}^{q-1})$ and $\Sigma_{-i,-i}$ is the $(q-1) \times (q-1)$ matrix obtained by removing the line and column i from Σ. This identity can be proven using Bayes formula and Gaussian vectors conditioning formulas. Its use gives:

$$p\, \mathbb{E}(Z_k^c | \mathbf{Z}^c \le \mathbf{h}) = -\sum_{i=1}^{q} \Sigma_{ik}\varphi_{0,\Sigma_{ii}}(h_i)\Phi_{q-1}(\mathbf{h}_{-i} - \Sigma_{ii}^{-1}\Sigma_i h_i, \Sigma_{-i,-i} - \Sigma_i \Sigma_{ii}^{-1}\Sigma_i^\top)$$

$$= -\sum_{i=1}^{q} \Sigma_{ik}\varphi_{m_i,\Sigma_{ii}}(b_i)\Phi_{q-1}(\mathbf{h}_{-i} - \Sigma_{ii}^{-1}\Sigma_i h_i, \Sigma_{-i,-i} - \Sigma_i \Sigma_{ii}^{-1}\Sigma_i^\top)$$

which finally delivers Tallis' formula, see Eq. (2).

References

1. Jones, D.R., Schonlau, M., William, J.: Efficient global optimization of expensive black-box functions. J. Glob. Optim. **13**(4), 455–492 (1998)
2. Santner, T.J., Williams, B.J.: The Design and Analysis of Computer Experiments. Springer, New York (2003)
3. Mockus, J.: Bayesian Approach to Global Optimization. Theory and Applications. Kluwer Academic Publisher, Dordrecht (1989)
4. Roustant, O., Ginsbourger, D., Deville, Y.: DiceKriging, DiceOptim: Two R packages for the analysis of computer experiments by kriging-based metamodelling and optimization. J. Stat. Softw. **51**(1), 1–55 (2012)
5. Mockus, J., Tiesis, V., Zilinskas, A.: The application of Bayesian methods for seeking the extremum. In: Dixon, L., Szego, E.G. (eds.) Towards Global Optimization, pp. 117–129. Elsevier, Amsterdam (1978)
6. Schonlau, M.: Computer experiments and global optimization. PhD thesis, University of Waterloo (1997)
7. Ginsbourger, D.: Métamodèles multiples pour l'approximation et l'optimisation de fonctions numériques multivariables. PhD thesis, Ecole nationale supérieure des Mines de Saint-Etienne (2009)
8. Ginsbourger, D., Le Riche, R., Carraro, L.: Kriging is well-suited to parallelize optimization. Computational Intelligence in Expensive Optimization Problems. Adaptation Learning and Optimization, vol. 2, pp. 131–162. Springer, Heidelberg (2010)
9. Janusevskis, J., Le Riche, R., Ginsbourger, D.: Parallel expected improvements for global optimization: summary, bounds and speed-up (August 2011)
10. Janusevskis, J., Le Riche, R., Ginsbourger, D., Girdziusas, R.: Expected improvements for the asynchronous parallel global optimization of expensive functions: potentials and challenges. In: Hamadi, Y., Schoenauer, M. (eds.) LION 6. LNCS, vol. 7219, pp. 413–418. Springer, Heidelberg (2012)
11. Frazier, P.I.: Parallel global optimization using an improved multi-points expected improvement criterion. In: INFORMS Optimization Society Conference, Miami FL (2012)

12. Chilès, J.P., Delfiner, P.: Geostatistics: Modeling Spatial Uncertainty. Wiley, New York (1999)
13. Tallis, G.: The moment generating function of the truncated multi-normal distribution. J. Roy. Statist. Soc. Ser. B **23**(1), 223–229 (1961)
14. Da Veiga, S., Marrel, A.: Gaussian process modeling with inequality constraints. Annales de la Faculté des Sciences de Toulouse **21**(3), 529–555 (2012)
15. Genz, A.: Numerical computation of multivariate normal probabilities. J. Comput. Graph. Stat. **1**, 141–149 (1992)
16. Genz, A., Bretz, F.: Computation of Multivariate Normal and t Probabilities. Springer, Heidelberg (2009)
17. Genz, A., Bretz, F., Miwa, T., Mi, X., Leisch, F., Scheipl, F., Bornkamp, B., Hothorn, T.: Mvtnorm: Multivariate Normal and t Distributions. R package version 0.9-9992 (2012)
18. Azzalini, A.: mnormt: The multivariate normal and t distributions. R package version 1.4-5 (2012)
19. Finck, S., Hansen, N., Ros, R., Auger, A.: Real-parameter black-box optimization bencharking 2009: Presentation of the noiseless functions. Technical report, Research Center PPE, 2009 (2010)
20. Hansen, N., Finck, S., Ros, R., Auger, A.: Real-parameter black-box optimization benchmarking 2009: Noiseless functions definitions. Technical report, INRIA 2009 (2010)
21. Mebane, W., Sekhon, J.: Genetic optimization using derivatives: The rgenoud package for R. J. Stat. Softw. **42**(11), 1–26 (2011)
22. Cressie, N., Davis, A., Leroy Folks, J.: The moment-generating function and negative integer moments. Am. Stat. **35**(3), 148–150 (1981)

R2-EMOA: Focused Multiobjective Search Using R2-Indicator-Based Selection

Heike Trautmann[1]($\boxtimes$), Tobias Wagner[2], and Dimo Brockhoff[3]

[1] Department of Information Systems, University of Münster, Münster, Germany
trautmann@uni-muenster.de
[2] Institute of Machining Technology, TU Dortmund University, Dortmund, Germany
wagner@isf.de
[3] INRIA Lille Nord-Europe, Dolphin Team, Villeneuve d'Ascq, France
dimo.brockhoff@inria.fr

Abstract. An indicator-based evolutionary multiobjective optimization algorithm (EMOA) is introduced which incorporates the contribution to the unary R2-indicator as the secondary selection criterion. First experiments indicate that the R2-EMOA accurately approximates the Pareto front of the considered continuous multiobjective optimization problems. Furthermore, decision makers' preferences can be included by adjusting the weight vector distributions of the indicator which results in a focused search behavior.

Keywords: Multiobjective optimization · Performance assessment · EMOA · R2-indicator · Indicator-based selection · Preferences

1 Introduction

Throughout this paper, we consider multiobjective optimization problems consisting of d objectives Y_j and objective functions $f_j : \mathbb{R}^n \to \mathbb{R}$ with $1 \leq j \leq d$. In the context of performance assessment of multiobjective optimizers, the (binary) R-indicator family was introduced by Hansen and Jaszkiewicz [5]. It is based on a set of utility functions. In total, three different variants were proposed which differ in the way the utilities are evaluated and combined – the ratio of one set being better than the other (R1), the mean difference in utilities (R2), or the mean relative difference in utilities (R3). In particular, the second variant R2 is one of the most recommended performance indicators [8] together with the hypervolume (HV, [9]) which directly measures the dominated objective hypervolume bounded by a reference point dominated by all solutions. Recently, we defined an equivalent unary version of this R2 indicator [3]. In case the standard weighted Tchebycheff utility function with ideal point $\mathbf{i}$ is used, it is defined as

$$R2(A, \Lambda, \mathbf{i}) = \frac{1}{|\Lambda|} \sum_{\lambda \in \Lambda} \min_{a \in A} \left\{ \max_{j \in \{1,...,d\}} \{\lambda_j | \mathbf{i}_j - a_j | \} \right\}$$

for a solution set A and a given set of weight vectors $\lambda = (\lambda_1, \ldots, \lambda_d) \in \Lambda$.

G. Nicosia and P. Pardalos (Eds.): LION 7, LNCS 7997, pp. 70–74, 2013.
DOI: 10.1007/978-3-642-44973-4_8, © Springer-Verlag Berlin Heidelberg 2013

Theoretical and experimental comparisons to the HV for $d = 2$ revealed that, contrarily to common assumptions, the R2 indicator even has a stronger bias towards the center of the Pareto front than the HV [3]. Furthermore, it could be proven that for $d = 2$ the optimal placement of a point w.r.t. the R2-indicator solely depends on its two nearest neighbors and a subset of Λ. In [6], the influence of the R2-indicator parametrization on the optimal distribution of μ points on the true Pareto front (PF) regarding R2 was investigated. It was shown that this distribution heavily depends on the position of the ideal point, as well as on the domain and distribution of the weight vectors. Thus, preferences of the decision maker can be reflected by a specifically parametrized R2-indicator. In [1] a similar approach relying on linear utility functions was used to identify knees of Pareto fronts. In this paper, we will investigate whether the approximated optimal distributions of μ points regarding R2 based on different preference articulations [6] can be accurately reproduced by a greedy R2-EMOA.

2   R2-EMOA

The proposed R2-EMOA implements a steady state strategy based on the contribution to the unary R2-indicator (see Algorithm 1).

1: draw multiset P with μ elements $\in \mathbb{R}^n$ at random
2: **repeat**
3:    generate offspring $z \in \mathbb{R}^n$ from P by variation
4:    $P = P \cup \{z\}$
5:    non-dominated sorting:
      build ranking $R_1, \ldots, R_h$ from P
6:    $\forall x \in R_h : r(x) = R2(P \setminus \{x\}; \Lambda; \mathbf{i})$
7:    $x^* = \operatorname{argmin}\{r(x) : x \in R_h\}$
8:    $P = P \setminus \{x^*\}$
9: **until** stopping criterion fulfilled

Algorithm 1: Pseudo code of the R2-EMOA.

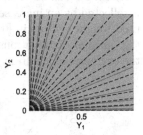

Fig. 1. $|\Lambda| = 19$ for $\gamma = 1$ (black dashed line) and $\gamma = 4$ (red solid line).

3   Experiments

Experiments were conducted to empirically show that the evolutionary procedure (selection pressure, variation) of the R2-EMOA is adequate to accurately approximate the R2-optimal distributions. This cannot be directly assumed, as the greedy strategy of the EMOA which only changes single solutions could be stuck in local optima of the objective functions or in suboptimal distributions.[1]

For the experiments, three bi-objective test functions with different problem characteristics were selected: ZDT1 (convex PF, $n = 30$) [7], DTLZ1 (linear PF,

---

[1] For the HV indicator, it has been, for example, theoretically proven that such a greedy strategy cannot always find a solution set with optimal HV value [2,10].

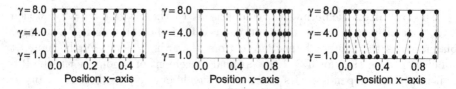

Fig. 2. Results of best R2-EMOA runs for increasing γ on DTLZ1 (left), DTLZ2 (middle) and ZDT1 (right). The movement of the x-axis positions for $\gamma \in \{1, 4, 8\}$ is shown. The optimal distributions regarding HV are reflected by dashed vertical lines.

$n = 6$), and DTLZ2 (concave PF, $n = 11$) [4]. On each function, ten independent runs were conducted using simulated binary crossover (SBX) and polynomial mutation ($p_c = 0.9$, $p_m = 1/n$, $\eta_c = 15$, $\eta_m = 20$), 150.000 function evaluations (FE), ideal point $\mathbf{i} = (0, 0)'$, and 501 weight vectors. A population size of $\mu = 10$ was chosen in order to allow a clear visualization of the results and the comparison to the reference distributions of [6].

The influence of restricted weight vector domains and altered weight vector distributions on the outcome of the R2-EMOA results is considered. Therefore, Algorithm 1 of [6] was used to generate weight vector distributions with increasing focus on the extremes of the weight vector domain (see Fig. 1). This is reflected by an increased value of γ while $\gamma = 1$ corresponds to equally distributed weight vectors in $[0, 1]^2$. The R2-EMOA is able to accurately approximate the optimal distributions. With increasing γ, the points tend to drift towards the extremes of the front (Fig. 2) which is perfectly in line with the results of [6].

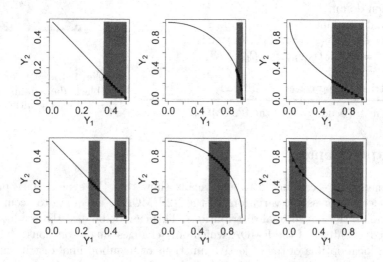

Fig. 3. Results of the best R2-EMOA runs (black dots) with restricted weight vector domains for DTLZ1 (left), DTLZ2 (middle) and ZDT1 (right). The areas within the intersections with the true PF (solid line) are highlighted.

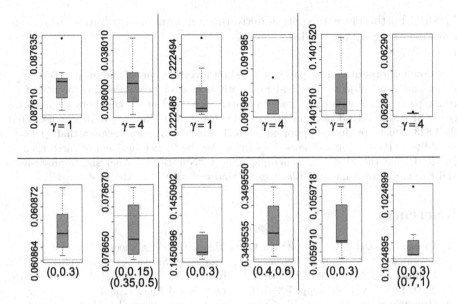

Fig. 4. Boxplots of R2 values at final R2-EMOA generation for DTLZ1 (left), DTLZ2 (middle) and ZDT1 (right) for altered weight distributions with parameter γ (top) or restricted weight space (bottom) corresponding to Fig. 3. The R2 value of the approximated optimal 10-distribution of R2 in [6] is visualized by a red horizontal line.

Individually for each problem, distributions close to the optimal ones regarding HV can be obtained for a specific choice of γ.

Moreover, the first component of the weight vector domain was restricted to one or two intervals within $[0, 1]$. From [6] it is known that in this setting the optimal solutions regarding R2 lie within the target cone defined by the two outmost weight vectors of the interval(s). This is reflected by the respective R2-EMOA results (Fig. 3).

Figure 4 relates the final R2 values of all experiments to the approximated optimal 10-distributions regarding R2 [6]. It can be observed that the variance of the R2-EMOA results is small. Sometimes even slightly better approximations of the optimal distributions are obtained than in [6]. This is rather surprising as these reference solutions were determined based on a global optimization on the front. The evolutionary mechanism and the greedy selection seem to provide efficient heuristics for the considered class of problems.

4   Conclusions and Outlook

First experiments show very promising results of the R2-EMOA regarding solution quality and the possibility of incorporating preferences of the decision maker. In future studies, the R2-EMOA will be theoretically and empirically compared to other EMOA optimizing the R2-indicator, such as MOEA/D and

MSOPS. Furthermore, theoretical derivations of optimal distributions of μ solutions regarding R2 are needed.

Acknowledgements. This paper is based on investigations of the project D5 "Synthesis and multi-objective model-based optimization of process chains for manufacturing parts with functionally graded properties" as part of the collaborative research center SFB/TR TRR 30 and the project B4 = C4 of the Collaborative Research Center SFB 823, which are kindly supported by the Deutsche Forschungsgemeinschaft (DFG). In addition, the authors acknowledge support by the French national research agency (ANR) within the Modèles Numérique project "NumBBO - Analysis, Improvement and Evaluation of Numerical Blackbox Optimizers" (ANR-12-MONU-0009-03).

References

1. Branke, J., Deb, K., Dierolf, H., Osswald, M.: Finding knees in multi-objective optimization. In: Yao, X., Burke, E.K., Lozano, J., Smith, J., Merelo-Guervós, J., Bullinaria, J.A., Rowe, J.E., Tino, P., Kabán, A., Schwefel, H.-P. (eds.) PPSN 2004. LNCS, vol. 3242, pp. 722–731. Springer, Heidelberg (2004)
2. Bringmann, K., Friedrich, T.: Convergence of hypervolume-based archiving algorithms I: effectiveness. In: Genetic and Evolutionary Computation Conference (GECCO 2011), pp. 745–752. ACM, New York (2011)
3. Brockhoff, D., Trautmann, H., Wagner, T.: On the properties of the $R2$ indicator. In: Genetic and Evolutionary Computation Conference (GECCO 2012), pp. 465–472. ACM, New York (2012)
4. Deb, K., Thiele, L., Laumanns, M., Zitzler, E.: Scalable multi-objective optimization test problems. In: Congress on Evolutionary Computation (CEC 2002), pp. 825–830. IEEE Press, New Jersey (2002)
5. Hansen, M.P., Jaszkiewicz, A.: Evaluating the quality of approximations of the nondominated set. Technical report, Institute of Mathematical Modeling, Technical University of Denmark (1998), IMM Technical Report IMM-REP-1998-7
6. Wagner, T., Trautmann, H., Brockhoff, D.: Preference articulation by means of the $R2$ indicator. In: Purshouse, R.C., Fleming, P.J., Fonseca, C.M., Greco, S., Shaw, J. (eds.) EMO 2013. LNCS, vol. 7811, pp. 81–95. Springer, Heidelberg (2013)
7. Zitzler, E., Deb, K., Thiele, L.: Comparison of multiobjective evolutionary algorithms: empirical results. Evol. Comput. **8**(2), 173–195 (2000)
8. Zitzler, E., Knowles, J.D., Thiele, L.: Quality assessment of Pareto set approximations. In: Branke, J., Deb, K., Miettinen, K., Słowiński, R. (eds.) Multiobjective Optimization. LNCS, vol. 5252, pp. 373–404. Springer, Heidelberg (2008)
9. Zitzler, E., Thiele, L.: Multiobjective optimization using evolutionary algorithms - a comparative case study. In: Eiben, A.E., Bäck, T., Schoenauer, M., Schwefel, H.-P. (eds.) PPSN 1998. LNCS, vol. 1498, pp. 292–301. Springer, Heidelberg (1998)
10. Zitzler, E., Thiele, L., Bader, J.: On set-based multiobjective optimization. IEEE Trans. Evol. Comput. **14**(1), 58–79 (2010)

A Heuristic Algorithm for the Set Multicover Problem with Generalized Upper Bound Constraints

Shunji Umetani[1]([⊠]), Masanao Arakawa[2], and Mutsunori Yagiura[3]

[1] Osaka University, Suita 565-0871, Japan
umetani@ist.osaka-u.ac.jp
[2] Fujitsu Limited, Kawasaki 211-8588, Japan
arakawa.masanao@jp.fujitsu.com
[3] Nagoya University, Nagoya 464-8601, Japan
yagiura@nagoya-u.jp

Abstract. We consider an extension of the set covering problem (SCP) introducing (i) multicover and (ii) generalized upper bound (GUB) constraints that arise in many real applications of SCP. For this problem, we develop a 2-flip neighborhood local search algorithm with a heuristic size reduction algorithm, in which a new evaluation scheme of variables is introduced taking account of GUB constraints. According to computational comparison with the latest version of a mixed integer programming solver, our algorithm performs quite effectively for various types of instances, especially for very large-scale instances.

1 Introduction

The set covering problem (SCP) is one of representative combinatorial optimization problems. We are given a ground set of m elements $i \in M = \{1, \ldots, m\}$, n subsets $S_j \subseteq M$ ($|S_j| \geq 1$) and costs $c_j(> 0)$ for $j \in N = \{1, \ldots, n\}$. We say that $X \subseteq N$ is a cover of M if $\bigcup_{j \in X} S_j = M$ holds. The goal of SCP is to find a minimum cost cover X of M. The SCP is formulated as a 0–1 integer programming (IP) problem as follows:

$$\min. \quad \sum_{j \in N} c_j x_j$$
$$\text{s.t.} \quad \sum_{j \in N} a_{ij} x_j \geq 1, \ i \in M, \tag{1}$$
$$x_j \in \{0, 1\}, \quad j \in N,$$

where $a_{ij} = 1$ if $i \in S_j$ holds and $a_{ij} = 0$ otherwise, and $x_j = 1$ if $j \in X$ holds and $x_j = 0$ otherwise, respectively.

The SCP is often referred in the literature that it has many important applications [2], e.g., crew scheduling, vehicle routing, facility location, and logical analysis of data. However, it is often difficult to formulate problems in real

G. Nicosia and P. Pardalos (Eds.): LION 7, LNCS 7997, pp. 75–80, 2013.
DOI: 10.1007/978-3-642-44973-4_9, © Springer-Verlag Berlin Heidelberg 2013

applications into the SCP, because they often have additional side constraints in practice. Most practitioners accordingly formulate them into general mixed integer programming (MIP) problem and apply general purpose solvers, which are usually less efficient compared to solvers specially tailored to SCP.

In this paper, we consider an extension of SCP introducing (i) multicover and (ii) generalized upper bound (GUB) constraints, which arise in many real applications of SCP. The multicover constraint is a generalization of covering constraint, in which each element $i \in M$ must be covered at least $b_i \in \mathbb{Z}_+$ ($\mathbb{Z}_+$ is the set of non-negative integers) times. GUB constraint is defined as follows. We are given a partition $\{G_1, \ldots, G_k\}$ of N ($\forall h \neq h', G_h \cap G_{h'} = \emptyset, \bigcup_{h=1}^k G_h = N$). For each block $G_h \subseteq N$ ($h \in K = \{1, \ldots, k\}$), the number of selected subsets S_j ($j \in G_h$) is constrained to be at most $d_h (\leq |G_h|)$. We call this problem the set multicover problem with GUB constraints (SMCP-GUB).

The SMCP-GUB is NP-hard, and the (supposedly) simpler problem of judging the existence of a feasible solution is NP-complete. We accordingly consider the following formulation of SMCP-GUB that allows violations of the multicover constraints and introduces a penalty function with a penalty weight vector $\boldsymbol{w} = (w_1, \ldots, w_m) \in \mathbb{R}_+^m$:

$$
\begin{aligned}
\text{min.} \quad & z(\boldsymbol{x}) = \sum_{j \in N} c_j x_j + \sum_{i \in M} w_i y_i \\
\text{s.t.} \quad & \sum_{j \in N} a_{ij} x_j + y_i \geq b_i, & i \in M, \\
& \sum_{j \in G_h} x_j \leq d_h, & h \in K, \\
& x_j \in \{0, 1\}, & j \in N, \\
& y_i \in \{0, \ldots, b_i\}, & i \in M.
\end{aligned}
\tag{2}
$$

For a given $\boldsymbol{x} \in \{0,1\}^n$, we can easily compute an optimal $\boldsymbol{y}$ by $y_i = \max\{b_i - \sum_{j \in N} a_{ij} x_j, 0\}$. We note that when $\boldsymbol{y}^* = \boldsymbol{0}$ holds for an optimal solution $(\boldsymbol{x}^*, \boldsymbol{y}^*)$ of SMCP-GUB under the soft multicover constraints, $\boldsymbol{x}^*$ is also optimal under the original (hard) multicover constraints. Moreover, for an optimal solution $\boldsymbol{x}^*$ under hard multicover constraints, $(\boldsymbol{x}^*, \boldsymbol{0})$ is also optimal with respect to soft multicover constraints if the values of w_i are sufficiently large, e.g., if $w_i > \sum_{j \in N} c_j$ holds for all $i \in M$. We accordingly set $w_i = \sum_{j \in N} c_j + 1$ for all $i \in M$.

In this paper, we proposes a 2-flip neighborhood local search algorithm with an efficient mechanism to find improved solutions. The above generalization of SCP substantially extends the variety of its applications. However, GUB constraints often make the pricing method less effective (which is known to be very effective for large-scale instances of SCP), because GUB constraints prevent solutions from containing highly evaluated variables together. To overcome this, we develop a heuristic size reduction algorithm, in which a new evaluation scheme of variables is introduced taking account of GUB constraints.

2 Lagrangian Relaxation and Subgradient Method

For a given vector $\boldsymbol{u} = (u_1, \ldots, u_m) \in \mathbb{R}_+^m$, called the Lagrangian multiplier vector, the Lagrangian relaxation of SMCP-GUB is defined as follows:

$$\min. \ z_{\mathrm{LR}}(\boldsymbol{u}) = \sum_{j \in N} c_j x_j + \sum_{i \in M} w_i y_i + \sum_{i \in M} u_i \left(b_i - \sum_{j \in N} a_{ij} x_j - y_i \right)$$

$$= \sum_{j \in N} \left(c_j - \sum_{i \in M} a_{ij} u_i \right) x_j + \sum_{i \in M} y_i (w_i - u_i) + \sum_{i \in M} b_i u_i \quad (3)$$

$$\text{s.t.} \quad \sum_{j \in G_h} x_j \le d_h, \quad h \in K,$$

$$x_j \in \{0, 1\}, \quad j \in N,$$

$$y_i \in \{0, \ldots, b_i\}, \quad i \in M,$$

where we call $\tilde{c}_j(\boldsymbol{u}) = c_j - \sum_{i \in M} a_{ij} u_i$ the Lagrangian cost associated with column $j \in N$. For any $\boldsymbol{u} \in \mathbb{R}_+^m$, $z_{\mathrm{LR}}(\boldsymbol{u})$ gives a lower bound on the optimal value of SMCP-GUB $z(\boldsymbol{x}^*)$. The problem of finding a Lagrangian multiplier vector $\boldsymbol{u}$ that maximizes $z_{\mathrm{LR}}(\boldsymbol{u})$ is called the Lagrangian dual problem.

A common approach to compute a near optimal Lagrangian multiplier vector $\boldsymbol{u}$ is the subgradient method. When huge instances of SCP are solved, the computing time spent on the subgradient method becomes very large if a naive implementation is used. Caprara et al. [1] developed a variant of pricing method on the subgradient method. They define a dual core problem consisting of a small subset of columns $C_d \subset N$ ($|C_d| \ll |N|$), chosen among those having the lowest Lagrangian costs $\tilde{c}_j(\boldsymbol{u})$ ($j \in C_d$), and iteratively update the dual core problem in a similar fashion to that used for solving large scale LP problems. In order to solve huge instances of SMCP-GUB, we also introduce their pricing method into the basic subgradient method (BSM) described in [3].

3 The 2-flip Neighborhood Local Search Algorithm

The local search (LS) starts from an initial solution $\boldsymbol{x}$ and repeats replacing $\boldsymbol{x}$ with a better solution $\boldsymbol{x}'$ in its neighborhood $\mathrm{NB}(\boldsymbol{x})$ until no better solution is found in $\mathrm{NB}(\boldsymbol{x})$. For a positive integer r, the r-flip neighborhood $\mathrm{NB}_r(\boldsymbol{x})$ is defined by $\mathrm{NB}_r(\boldsymbol{x}) = \{\boldsymbol{x}' \in \{0, 1\}^n \mid d(\boldsymbol{x}, \boldsymbol{x}') \le r\}$, where $d(\boldsymbol{x}, \boldsymbol{x}') = |\{j \in N \mid x_j \ne x_j'\}|$ is the Hamming distance between $\boldsymbol{x}$ and $\boldsymbol{x}'$. In other words, $\mathrm{NB}_r(\boldsymbol{x})$ is the set of solutions obtained from $\boldsymbol{x}$ by flipping at most r variables. In our LS, the r is set to 2. In order to improve efficiency, our LS searches $\mathrm{NB}_1(\boldsymbol{x})$ first, and $\mathrm{NB}_2(\boldsymbol{x}) \setminus \mathrm{NB}_1(\boldsymbol{x})$ only if $\boldsymbol{x}$ is locally optimal with respect to $\mathrm{NB}_1(\boldsymbol{x})$.

Yagiura et al. [4] developed an LS with the 3-flip neighborhood for SCP. They derived conditions that reduce the number of candidates in $\mathrm{NB}_2(\boldsymbol{x}) \setminus \mathrm{NB}_1(\boldsymbol{x})$ and $\mathrm{NB}_3(\boldsymbol{x}) \setminus \mathrm{NB}_2(\boldsymbol{x})$ without sacrificing the solution quality. However, those conditions are not applicable to the 2-flip neighborhood for SMCP-GUB because of GUB constraints. We therefore propose new conditions that reduce the number of candidates in $\mathrm{NB}_2(\boldsymbol{x}) \backslash \mathrm{NB}_1(\boldsymbol{x})$ taking account of GUB constraints. As a result, the number of solutions searched by our algorithm becomes $O(n + k\nu + n'\tau)$ while the size of NB_2 is $O(n^2)$, where $\nu = \max_{j \in N} |S_j|$, $n' = \sum_{j \in N} x_j$ and $\tau = \max_{j \in N} \sum_{i \in S_j} |N_i|$ for $|N_i| = \{j \in N | i \in S_j\}$.

Since the region searched in a single application of LS is limited, LS is usually applied many times. When a locally optimal solution is obtained, a standard strategy of our algorithm is to update penalty weights and to resume LS from the obtained locally optimal solution. We accordingly evaluate solutions with an alternative evaluation function $\hat{z}(\boldsymbol{x})$, where the original penalty weight vector $\boldsymbol{w}$ is replaced with $\hat{\boldsymbol{w}} = (\hat{w}_1, \ldots, \hat{w}_m) \in \mathbb{R}_+^m$. Our algorithm iteratively applies LS, updating the penalty weight vector $\hat{\boldsymbol{w}}$ after each call to LS.

Starting from the original penalty weight vector $\hat{\boldsymbol{w}} \leftarrow \boldsymbol{w}$, the penalty weight vector $\hat{\boldsymbol{w}}$ is updated as follows. Let $\boldsymbol{x}^{\mathrm{best}}$ denote the best feasible solution with respect to the original objective function $z(\boldsymbol{x})$. If the previous locally optimal solution $\boldsymbol{x}$ satisfies $\hat{z}(\boldsymbol{x}) \geq z(\boldsymbol{x}^{\mathrm{best}})$, our algorithm uniformly decreases the penalty weights $\hat{w}_i$ $(i \in M)$. Otherwise, our algorithm increases the penalty weights $\hat{w}_i$ $(i \in M)$ in proportion to the amount of violation of the ith multi-cover constraint.

4   Heuristic Reduction of Problem Sizes

For a near optimal Lagrangian multiplier vector $\boldsymbol{u}$, the Lagrangian costs $\tilde{c}_j(\boldsymbol{u})$ give reliable information on the overall utility of selecting columns $j \in N$ for SCP. Based on this property, the Lagrangian costs $\tilde{c}_j(\boldsymbol{u})$ are often utilized to solve huge instances of SCP, e.g., several heuristic algorithms successively solve a number of subproblems, called primal core problems, consisting of a small subset of columns $C_p \subset N$ ($|C_p| \ll |N|$), which are chosen among those having low Lagrangian costs $\tilde{c}_j(\boldsymbol{u})$ [1,2,4].

The Lagrangian costs $\tilde{c}_j(\boldsymbol{u})$ are unfortunately unreliable for selecting columns $j \in N$ for SMCP-GUB, because GUB constraints often prevent solutions from containing more than d_h variables x_j with the lowest Lagrangian costs $\tilde{c}_j(\boldsymbol{u})$. To overcome this, we develop an evaluation scheme of columns $j \in N$ for SMCP-GUB taking account of GUB constraints. The main idea of our algorithm is that we modify the Lagrangian costs $\tilde{c}_j(\boldsymbol{u})$ to reduce the number of redundant columns $j \in C_p$ resulting from GUB constraints.

For each block G_h $(h \in K)$, let γ_h be the value of the $(d_h + 1)$st lowest Lagrangian cost $\tilde{c}_j(\boldsymbol{u})$ among those for columns in G_h, where we set $\gamma_h \leftarrow 0$ if $d_h = |G_h|$ holds. We then define a score $\hat{c}_j(\boldsymbol{u})$ for a column $j \in G_h$ by $\hat{c}_j(\boldsymbol{u}) = \tilde{c}_j(\boldsymbol{u}) - \gamma_h$ if $\gamma_h < 0$ holds, and $\hat{c}_j(\boldsymbol{u}) = \tilde{c}_j(\boldsymbol{u})$ otherwise. That is, we normalize the Lagrangian costs $\tilde{c}_j(\boldsymbol{u})$ so that at most d_h columns have negative scores $\hat{c}_j(\boldsymbol{u}) < 0$ for each block G_h $(h \in K)$. Let $n' = \sum_{j \in N} x_j$ be the number of selected subsets for a solution $\boldsymbol{x}$. Given a solution $\boldsymbol{x}$ and a Lagrangian multiplier vector $\boldsymbol{u}$, a primal core problem is defined by a subset $C_p \subset N$ consisting of (i) columns $j \in N_i$ with the b_i lowest scores $\hat{c}_j(\boldsymbol{u})$ for each $i \in M$, and (ii) columns $j \in N$ with the $10n'$ lowest scores $\hat{c}_j(\boldsymbol{u})$.

Table 1. The benchmark instances for SMCP-GUB and time limits for our algorithm LS-SR and the MIP solver CPLEX (in seconds)

Instance	Rows	Columns	Density (%)	Instance types $(d_h/|G_h|)$				Time limit	
				Type1	Type2	Type3	Type4	LS-SR	CPLEX
G.1–G.5	1000	10,000	2.0	1/10	10/100	5/10	50/100	600	3600
H.1–H.5	1000	10,000	5.0	1/10	10/100	5/10	50/100	600	3600
I.1–I.5	1000	50,000	1.0	1/50	10/500	5/50	50/500	600	3600
J.1–J.5	1000	100,000	1.0	1/50	10/500	5/50	50/500	600	3600
K.1–K.5	2000	100,000	0.5	1/50	10/500	5/50	50/500	1200	7200
L.1–L.5	2000	200,000	0.5	1/50	10/500	5/50	50/500	1200	7200
M.1–M.5	5000	500,000	0.25	1/50	10/500	5/50	50/500	3000	18,000
N.1–N.5	5000	1,000,000	0.25	1/100	10/1000	5/100	50/1000	3000	18,000

5   Computational Results

We first prepared eight classes of random instances for SCP, where each class has five instances. We denote instances in class G as G.1, ..., G.5, and other instances in classes H–N similarly. The summary of these instances are given in Table 1, where the density is defined by $\sum_{i\in M}\sum_{j\in N} a_{ij}/mn$ and the costs c_j are random integers taken from interval $[1, 100]$. For each SCP instance, we generate four types of SMCP-GUB instances with different values of parameters d_h and $|G_h|$ as shown in Table 1, where all blocks G_h ($h \in K$) have the same size $|G_h|$ and upper bound d_h for each instance. Here, the right-hand sides of multicover constraints b_i are random integers taken from interval $[1, 5]$.

We compared our algorithm, called the local search algorithm with the heuristic size reduction (LS-SR), with one of the latest mixed integer program (MIP) solver called CPLEX12.3, where they were tested on an IBM-compatible personal computer (Intel Xeon E5420 2.5 GHz, 4 GB memory) and were run on a single thread. Table 1 also shows the time limits in seconds for LS-SR and CPLEX12.3, respectively. We tested two variants of LS-SR: LS-SR1 evaluates variables x_j with the proposed score $\hat{c}_j(\boldsymbol{x})$, and LS-SR2 uses the Lagrangian cost $\tilde{c}_j(\boldsymbol{x})$ in the heuristic reduction of problem sizes. We illustrate in Fig. 1 their comparison for each type of SMCP-GUB instances with respect to the relative gap $\frac{z(\boldsymbol{x})-z_{\mathrm{LP}}}{z_{\mathrm{LP}}} \times 100$, where z_{LP} is the optimal value of LP relaxation for SMCP-GUB. The horizontal axis shows the classes of instances G–N, and the vertical axis shows the average relative gap for five instances of each class.

We first observe that LS-SR1 and LS-SR2 achieve better upper bounds than CPLEX12.3 for types 3 and 4 instances, especially large instances with 10,000 variables or more. One of the main reasons for this is that the proposed algorithms evaluate a series of candidate solutions efficiently while CPLEX12.3 consumes much computing time for solving LP relaxation problems. We also observe that LS-SR1 achieves much better upper bounds than those of LS-SR2 and CPLEX12.3 for types 1 and 2 instances.

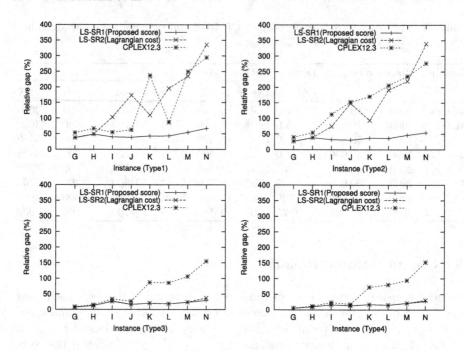

Fig. 1. Comparison of LS-SR and CPLEX12.3 on each instance type

6 Conclusion

In this paper, we considered an extension of SCP called the set multicover problem with the generalized upper bound constraints (SMCP-GUB). For this problem, we develop a 2-flip neighborhood local search algorithm with a heuristic size reduction algorithm, in which a new evaluation scheme of variables is introduced taking account of GUB constraints. According to computational comparison on benchmark instances with the latest version of a MIP solver called CPLEX12.3, our algorithm performs quite effectively for various types of instances, especially for very large-scale instances.

References

1. Caprara, A., Fischetti, M., Toth, P.: A heuristic method for the set covering problem. Oper. Res. **47**, 730–743 (1999)
2. Caprara, A., Toth, P., Fischetti, M.: Algorithms for the set covering problem. Ann. Oper. Res. **98**, 353–371 (2000)
3. Umetani, S., Yagiura, M.: Relaxation heuristics for the set covering problem. J. Oper. Res. Soc. Jpn. **50**, 350–375 (2007)
4. Yagiura, M., Kishida, M., Ibaraki, T.: A 3-flip neighborhood local search for the set covering problem. Eur. J. Oper. Res. **172**, 472–499 (2006)

A Genetic Algorithm Approach
for the Multidimensional Two-Way
Number Partitioning Problem

P.C. Pop[1](✉) and O. Matei[2]

[1] Department of Mathematics and Informatics, Technical University of Cluj-Napoca,
Cluj-Napoca, Romania
petrica.pop@ubm.ro
[2] Department of Electrical Engineering, Technical University of Cluj-Napoca,
Cluj-Napoca, Romania
oliviu.matei@holisun.com

Abstract. This paper addresses the problem of partitioning a set of
vectors into two subsets such that the sums per every coordinate should
be exactly or approximately equal. This problem, introduced by Kojic
[8], is called the multidimensional two-way number partitioning problem
(MDTWNPP) and generalizes the classical two-way number partitioning problem. We propose an efficient genetic algorithm based heuristic
for solving the multidimensional two-way number partitioning problem.
The performances of our genetic algorithm have been compared with
the existing numerical results obtained by CPLEX based on an integer
linear programming formulation of the problem. The obtained preliminary results, in the case of medium and large instances, reveal that our
proposed methodology performs very well in terms of both quality of
the solutions and the computational times compared with the previous
method of solving the MDTWNPP.

Keywords: Number partitioning problem · Genetic algorithms · Combinatorial optimization

1 Introduction

Number partitioning problem is a classical, challenging and surprisingly difficult
problem in combinatorial optimization. Given a set S of n integers, the two-way
number partitioning problem (TWNPP) asks for a division of S into two subsets
such that the sums of numbers in each subset should be equal or are close to be
equal.

Though the number partitioning problem is NP-complete (see [4]), there have
been proposed heuristic algorithms that solve the problem in many instances
either optimally or approximately: the set differencing heuristic introduced by
Karmarkar and Karp [7], a Simulated Annealing algorithm by Johnson et al. [6],

G. Nicosia and P. Pardalos (Eds.): LION 7, LNCS 7997, pp. 81–86, 2013.
DOI: 10.1007/978-3-642-44973-4_10, © Springer-Verlag Berlin Heidelberg 2013

genetic algorithm by Ruml et al. [9], GRASP by Arguello et al. [1], Tabu Search by Glover and Laguna [5], memetic algorithm by Berretta et al. [2], etc.

The problem has captioned a lot of attention due to its theoretical aspects and important real-world applications. For a more detailed description of the applications we refer to [3].

The multidimensional two-way number partitioning problem (MDTWNPP) was introduced by Kojic [8] and is a generalization of the TWNPP where instead of numbers we have a set of vectors we are looking for a partition of the vectors into two subsets such that the sums per every coordinate should be as close as possible.

The MDTWNPP is NP-hard, as it reduces when the vectors have dimension one to the TWNPP which is known to be an NP-hard problem. There is little research being done in mathematical modeling and solution methods for this problem. Kojic [8] described an integer programming formulation and tested the model on randomly generated sets using CPLEX, which as far as we know is the only approach to solve the problem. The obtained experimental results show that the MDTWNPP is very hard to solve even in the case of medium instances.

The aim of this paper is to describe a novel use of genetic algorithms with the goal of solving the multidimensional two-way number partitioning problem. The results of preliminary computational experiments are presented, analyzed and compared with the previous method introduced by Kojic [8]. The results reveal that our proposed methodology, in the case of medium and large instances, performs very well in terms of both quality of the solutions obtained and the computational times.

2   Definition of the Problem

Given a set of n vectors of dimension m

$$S = \{v_i \mid v_i = (v_{i1}, v_{i2}, ..., v_{im}), \ i \in \{1, ..., n\}, \ m \in \mathbb{N}\}$$

then according to Kojic [8] the multidimensional two-way number partitioning problem consists in splitting the elements of S into two sets, S_1 and S_2 such that

1. $S_1 \cup S_2 = S$ and $S_1 \cap S_2 = \emptyset$;
2. the sums of elements in the subsets S_1 and S_2 are equal or almost equal for all the coordinates.

If we introduce the variable t that denotes the greatest difference in sums per coordinate, i.e.

$$t = \max \left\{ \mid \sum_{i \in S_1} v_{ij} - \sum_{i \in S_2} v_{ij} \mid \ j \in \{1, ..., m\} \right\}$$

then the objective function of the MDTWNPP is to minimize t. If $\min t = 0$ then the partition will be called *perfect partition* for obvious reasons.

The MDTWNPP can be generalized easily to the case where a set of vectors is partitioned into a given number of subsets rather than into two subsets.

3   The Genetic Algorithm for Solving the MDMWNPP

In this section, we give the description of our genetic algorithm for solving the multidimensional two-way number partitioning problem.

We used a binary representation (encoding), where every chromosome is a fixed size (n-dimensional vector) ordered string of bits 0 or 1, identifying the set partition as assigned to the vectors. This representation ensures that the set of vectors belonging to the set S is partitioned into two subsets S_1 and S_2.

Concerning the initial population, experiments have been carried out with two different ways of generating the initial population: random generation and partially randomly and partially based on the problem structure. In the latter case, we picked randomly a number $q \in \{2, ..., n\}$ and then for the vectors belonging to $\{2, ..., q\}$ the genes are generated randomly and the other vectors are partitioned iteratively such that by adding each vector we reduce the greatest difference in sums per coordinate. Generating the population using as well the information about the problem structure permitted us to reduce the global fitness of the initial population with about 50 % in comparison to the randomly generation of the initial population.

The fitness value of the MDTWNPP, for a given partition of the vectors into two subsets is given by the greatest difference in sums per each coordinate and the aim of the problem is to find the partition that minimize this value.

Genetic operators are used in genetic algorithms to combine existing solutions into others (crossover-like operators) and to generate diversity (mutation-like operators). In our case we selected the two parents using the binary tournament method, where two solutions, called parents, are picked from the population, their fitness is compared and the better solution is chosen for a reproductive trial. In order to produce a child, two binary tournaments are held, each of which produces one parent. We have experimented both single and double point crossover. Since there was not a big difference in the results we got from both methods, we decided to use single point crossover. The crossover point is determined randomly by generating a random number between 1 and $n - 1$. We decided upon crossover rate of 85 % by testing the program with different values.

Mutation is a genetic operator that alters one or more gene values in a chromosome from its initial state. We consider a mutation operator that changes the new offspring by flipping bits from 1 to 0 or from 0 to 1. Mutation can occur at each bit position in the string with 10 % probability.

Computational experiments showed that our proposed GA involving just the crossover and the mutation operators is effective in producing high quality solutions in the case of medium and large size instances. However, we improved the GA algorithm by adding a problem specific heuristic operator involving the following local improvement step:

- let t be the greatest difference in sums per coordinate being achieved on coordinate j, $j \in \{1, ..., m\}$, then within the subset with higher sum we analyze the corresponding elements belonging to the coordinate j, choose the one closest to the value $\dfrac{t}{2}$ and finally reassigned it to the other subset.

Example. Considering the set of vectors: $S = \{(1,3),(5,5),(3,-2),(-3,12)\}$ and the partition:

$$S_1 = \{(5,5)\} \text{ and } S_2 = \{(1,3),(3,-2),(-3,12)\}$$

then the sums per coordinates are $(5,5)$ and $(1,13)$ and the difference is $(4,8)$ and therefore $t = 8$. The second component of the vector $(1,3)$ is the closest to $\frac{t}{2} = 4$ and we reassigned this vector to the subset S_1 getting the following partition:

$$S_1 = \{(1,3),(5,5)\} \text{ and } S_2 = \{(3,-2),(-3,12)\}$$

with the sums $(6,8)$, $(0,10)$ and the difference is $(6,2)$, $t = 6$.

In our algorithm we investigated and used the properties of (μ, λ) selection, where μ parent produce λ ($\lambda \gg \mu$) and only the offspring undergo selection. In other words, the lifetime of every individual is limited to only one generation. This may lead to short periods of recession, but it avoids long stagnation phases due to unadapted strategy parameters.

The genetic parameters are very important for the success of a GA. Based on preliminary experiments, we have chosen the following parameters: the population size μ has been set to 10 times the number of the vectors, the intermediate population size λ was chosen ten times the size of the population: $\lambda = 10 \cdot \mu$, mutation probability was set at 10 % and the maximum number of generations (epochs) in our algorithm was set to 10000.

In our algorithm the termination strategy is based on a maximum number of generations to be run if there is no improvement in the objective function for a sequence of 15 consecutive generations.

4   Preliminary Computational Results

In this section we present some computational results in order to asses the effectiveness of our proposed genetic algorithm for solving the multidimensional two-way number partitioning problem.

We conducted our computational experiments for solving the MDTWNPP on a set of instances generated randomly and following the general format $n-m$. We consider for each $n-m$ five instances denoted by a, b, c, d and e. These instances were used by Kojic [8] in her computational experiments. In our experiments we performed 10 independent runs for each instance. The testing machine was an Intel Dual-Core 1,6 GHz and 1 GB RAM with Windows XP Professional as operating system. The algorithm was developed in Java, JDK 1.6.

In Table 1 we present the computational results obtained with our GA: best solution and average solutions in comparison with those obtained by Kojic [8] using CPLEX for a set of medium instances containing 400 respectively 500 vectors with dimension between 10 and 20. Because CPLEX did not finish its work Kojic provided the best solutions achieved for a maximum of 30 min run.

The first column in the table gives the name of the instance, the second and third columns provide the results obtained by Kojic [8] using CPLEX: the best

Table 1. Computational results

Problem instance	Results of CPLEX		Results of GA		
	Best solution	Time (Sec.)	Best solution	Average solution	Time (Sec.)
400_10a	14836.579	1622.34	**7728.546**	12627.840	356.27
400_10b	17918.141	1215.03	**10918.141**	11829.286	376.89
400_10c	21213.818	1703.88	**9208.251**	14682.930	324.50
400_10d	15212.906	1283.81	**8212.906**	11025.633	352.28
400_10e	16369.531	1530.48	**13332.372**	15820.827	342.56
400_15a	37574.022	926.03	**29529.332**	31637.388	321.35
400_15b	34390.093	62.52	**21390.093**	28839.275	326.28
400_15c	37161.817	1463.43	**28621.829**	34527.229	378.29
400_15d	30019.198	1203.22	**16223.857**	24589.012	381.10
400_15e	32561.093	26.19	**30261.649**	33427.922	378.29
400_20a	41974.284	767.30	**28363.836**	36728.003	390.28
400_20b	46751.348	354.05	**38275.503**	43906.422	368.45
400_20c	47259.514	313.95	**32748.920**	45636.829	372.32
400_20d	51544.421	31.38	**36728.927**	45366.764	310.38
400_20e	48792.272	251.36	**43788.540**	50023.568	302.65
500_10a	19 183.301	1718.29	**12938.304**	16227.263	736.74
500_10b	12 161.350	128.48	**10393.382**	13427.589	701.29
500_10c	16 594.760	368.13	**10283.385**	12533.378	692.03
500_10d	20 284.381	1699.01	**16378.394**	18226.185	678.90
500_10e	15 548.670	1680.47	**14950.760**	16272.317	732.39
500_15a	30 316.775	1055.81	**20394.564**	26373.372	720.31
500_15b	31 878.383	1591.08	**28348.563**	33723.653	743.86
500_15c	32 792.472	803.77	**25484.567**	33526.279	783.09
500_15d	35 555.260	881.27	**27394.640**	34291.266	810.28
500_15e	30 806.719	455.06	**21849.570**	27382.387	807.62
500_20a	48 281.977	1000.12	**32934.495**	42638.251	843.30
500_20b	54 921.900	237.63	**38494.084**	48373.736	873.39
500_20c	41 578.884	1382.98	**39495.452**	42930.553	812.73
500_20d	54 293.200	1728.58	**43840.674**	48342.734	843.30
500_20e	41 092.622	1713.03	**40352.904**	42839.224	921.83

solution and the necessary computational time in order to get it and the last three columns provide the results obtained by our novel genetic algorithm: the best solutions, the average solutions and the required time to get these solutions. Because CPLEX did not finish its work in any considered instance in the table are provided the best solutions obtained.

Analyzing the computational results, we observe that our genetic algorithm based heuristic provides better solutions than the approach considered by Kojic [8] using CPLEX.

Regarding the computational times, it is difficult to make a fair comparison between algorithms, because they have been evaluated on different computers and they have different stopping criteria. The running time of our GA is proportional with the number of generations. From the computational experiments, it results that 10000 generations are enough to explore the solution space of the MDTWNPP. But generally speaking, it should be noted that the average CPU time of our GA heuristic is comparable with the average CPU time provided by CPLEX.

The proposed algorithm integrates a number of original features: we proposed a novel method of generating of the initial population and we considered a powerful local improvement step. The considered genetic operators and the local improvement step provide our algorithm with a good tradeoff between intensification and diversification.

In the future we plan to asses the generality and scalability of the GA by testing it on a larger number of instances, to improve it by considering as well other local search procedures and to explore the possibility of building a parallel implementation of the algorithm.

Acknowledgments. This work was supported by a grant of the Romanian National Authority for Scientific Research, CNCS - UEFISCDI, project number PN-II-RU-TE-2011-3-0113.

References

1. Arguello, M.F., Feo, T.A., Goldschmidt, O.: Randomized methods for the number partitioning problem. Comput. Oper. Res. **23**(2), 103–111 (1996)
2. Berretta, R.E., Moscato, P., Cotta, C.: Enhancing a memetic algorithms' performance using a matching-based recombination algorithm: results on the number partitioning problem. In: Resende, M.G.C., Souza, J. (eds.) Metaheuristics: Computer Decision-Making, pp. 65–90. Kluwer, Boston (2004)
3. Coffman, E., Lueker, G.S.: Probabilistic Analysis of Packing and Partitioning Algorithms. Wiley, New York (1991)
4. Garey, M.R., Johnson, D.S.: Computers and Intractability. A Guide to the Theory of NP-Completeness. W.H. Freeman, New York (1997)
5. Glover, F., Laguna, M.: Tabu Search. Kluwer Academic, Norwell (1997)
6. Johnson, D.S., Aragon, C.R., McGeoch, L.A., Schevon, C.: Optimization by simulated annealing: an experimental evaluation. Part II: Graph coloring and number partitioning. Oper. Res. **39**(3), 378–406 (1991)
7. Karmarkar, N., Karp, R.M.: The differencing method of set partitioning, Technical report UCB/CSD 82/113, University of California, Berkeley, Computer Science Division (1982)
8. Kojic, J.: Integer linear programming model for multidimensional two-way number partitioning problem. Comput. Math. Appl. **60**, 2302–2308 (2010)
9. Ruml, W., Ngo, J.T., Marks, J., Shieber, S.M.: Easily searched encodings for number partitioning. J. Optim. Theor. Appl. **89**(2), 251–291 (1996)

Adaptive Dynamic Load Balancing in Heterogeneous Multiple GPUs-CPUs Distributed Setting: Case Study of B&B Tree Search

Trong-Tuan Vu[✉], Bilel Derbel, and Nouredine Melab

DOLPHIN, INRIA Lille - Nord Europe, University Lille 1, Lille, France
{Trong-Tuan.Vu,Bilel.Derbel,Nouredine.Melab}@inria.fr

Abstract. The emergence of new hybrid and heterogenous multi-GPUs multi-CPUs large scale platforms offers new opportunities and poses new challenges when solving difficult optimization problems. This paper targets irregular tree search algorithms in which workload is unpredictable. We propose an adaptive distributed approach allowing to distribute the load dynamically at runtime while taking into account the computing abilities of either GPUs or CPUs. Using Branch-and-Bound and Flow-Shop as a case study, we deployed our approach using up to 20 GPUs and 128 CPUs. Through extensive experiments in different system configurations, we report near optimal speedups, thus providing new insights into how to take full advantage of both GPUs and CPUs power in modern computing platforms.

1 Introduction

Context and Motivation. The current trend in high performance computing is converging towards the development of new software tools which can be efficiently deployed over large scale hybrid platforms, interconnecting several hundreds to thousands of *heterogeneous* processing units (PUs) ranging from multiple distributed CPUs, multiple shared-memory cores, to multiple GPUs. Although the aggregation of those resources can in theory offer an impressive computing power, achieving high performance and scalability is still bound to the expertise of programmers in developing new parallel techniques and paradigms operating both at the algorithmic and at the system levels. The heterogeneity and incompatibility of resources in terms of computing power and programming models, make it difficult to parallelize a given application without significantly drifting away from the optimal and theoretically attainable performance. In particular, when parallelizing highly irregular applications producing unpredictable workload *at runtime*, mapping dynamically generated tasks into the hardware so that workload is distributed evenly is a challenging issue. In this context, adjusting the workload distributively is mandatory to maximize resource utilization and to optimize work balance over massively parallel and large scale distributed PUs.

G. Nicosia and P. Pardalos (Eds.): LION 7, LNCS 7997, pp. 87–103, 2013.
DOI: 10.1007/978-3-642-44973-4_11, © Springer-Verlag Berlin Heidelberg 2013

In the optimization field, irregular applications producing dynamic work-load do not stand for an exception. Many search algorithms operating in some decision space are essentially dynamic and irregular, meaning that neither the search trajectory nor the amount of work can be predicted in advance. For instance, while some search regions may require much computational efforts to be processed, some others may require only a few. This is typically the case of several tree search algorithms coming from discrete and combinatorial optimizations, artificial intelligence, expert systems, etc. Generally speaking, this paper is targeting tree search-like algorithms endowed with some splitting/selection, pruning/elimination and evaluation/bounding strategies to decide on what to explore/search next. Despite the possibly sophisticated and efficient strategies one can design, these kinds of algorithms still undergo a huge amount of process-ing time when tackling large scale and/or difficult problems. More importantly, the knowledge acquired during the search changes dynamically the shape of the tree. Hence, it ends up with an unpredictable search process producing a highly variable amount of work. From parallel computing and high performance per-spectives, these algorithms can be viewed as 'skillful' adversaries which are very difficult to counteract efficiently.

The goal of this paper is to push forward the design of parallel and distrib-uted optimization algorithms requiring dynamic load balancing, in order to run them efficiently on heterogenous systems consisting of multiple CPUs coupled with multiple GPUs. More precisely, we consider the case study of the Branch-and-Bound (B&B), viewed as a generic algorithm searching in a dynamic tree representing a set of candidate solutions built dynamically at runtime. Given that several distributed CPUs and GPUs coming from possibly different clusters connected through a network can be used to parallelize the B&B tree search, three major issues are addressed:

Q1. Can we benefit from the different degrees of parallelism available in the tree search procedure and map them efficiently into the different PUs?

Q2. Given no knowledge about the amount of work the search would pro-duce, can we distributively coordinate PUs so that parallelism dynamically unfolds, while communication cost and idle time of PUs are kept minimal?

Q3. Having PUs with different computing abilities, can we distribute the load evenly in order to attain optimal speedup while scaling the network?

Contribution Overview. In this paper, we answer the three previous ques-tions in the positive while giving new insights into how to fully benefit from het-erogenous computing systems and solve difficult optimization problems. More precisely, we describe a two-level and fully distributed parallel approach taking into account PU characteristics. Our approach incorporates an adaptive dynamic load balancing scheme based on distributed work stealing, in order to flow work-loads efficiently from overloaded PUs to idle ones at runtime. Furthermore, it does not require any parameter tuning or specific optimization operations so that it is adaptive to heterogeneous computing systems. We implemented and deployed our approach over a distributed system of up to 20 GPUs and 128 CPUs coming from three clusters. Different scales and configurations of PUs were experimented with the B&B algorithm and the well-known FlowShop combinatorial optimization problem [14] as a case study. Firstly, on one single GPU, we improve on the running time of the previous B&B GPUs implementations [4,11] by at least a factor of two on the considered instances (the speedup with respect to one CPU is around ×70). More importantly, independently of the scale and power of CPUs or GPUs, our approach provides a substantial speed-up which is *nearly optimal* compared to the ideal performance one could expect in theory. It is worth to notice that although our experimentations are conducted for the specific FlowShop problem, it is generic in the sense that it undergoes no specific optimization with respect to neither B&B nor FlowShop. Therefore, it can be appropriate to solve other optimization problems, *as far as a GPU parallel evaluation (bounding) of search nodes (viewed as a blackbox)* is available.

From the optimization perspective, relatively few investigations are known on heterogenous parallel tree search algorithms. Specific to B&B, some very recent GPU parallelizations are known for some specific problems [2–4,10,11]. The focus there is mainly on the parallelization of the bounding step which is known to very time-consuming. The only study we found on aggregating the power of multiple GPUs presents a Master/Slave-like model and an experimental scale of 2 GPUs [4]. The authors there stressed more on the parallel design issues and not on scalability nor performance optimality. They reported a good but sub-optimal speed-up when using 2 GPUs, which witness the difficulty of the problem. To the best of our knowledge, the new parallel approach presented in this paper is not only the first to scale near linearly up to 20 GPUs but also the first to address the joint use of multiple distributed CPUs in the system.

From the parallel perspective, very few works exist on the parallelization of highly irregular applications in heterogenous platforms. In particular, we found no in-depth and systematic studies of application speed-up at different CPU-GPU scales. Knowing that the adaptive workload distribution strategy adopted in this paper is generic and not specific to tree search or B&B, our study provides new insights into the scalability of distributed protocols harnessing *both* multiple GPUs and CPUs which have a substantial gap in their respective computing power.

Outline. In Sect. 2, we draw the main components underlying our distributed approach while motivating their design architecture. A more detailed and technical description then follows in Sect. 3. In Sect. 4, we report and discuss our experimental results. In Sect. 5, we conclude the paper and raise some open issues.

2   A Comprehensive Overview of Our Approach

In this section, we give the general design principles guiding our approach. The goal is to introduce different components of our approach in a comprehensive manner without going into system technicalities or implementation details.

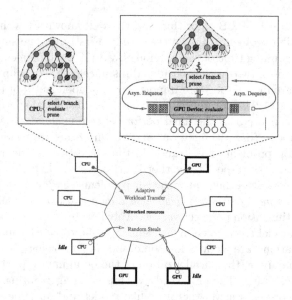

Fig. 1. Overview of our parallel approach

2.1   Application Model and Preliminaries

To simplify the presentation and clarify our contribution, let us model the B&B algorithm, as a tree search algorithm that starts from a root node representing an optimization problem. During the search, a parent node generates new child nodes (e.g., representing partial/complete candidate solutions) at runtime. The quality of these nodes is evaluated (bounding) using a given (heuristic) procedure. Then, according to the search state, some nodes are discarded (pruning) whether some others can be selected and the tree is expanded (branching) to push the search forward and so on. Having this in mind, the general architecture of our approach for distributing search computations is depicted in Fig. 1 and discussed in the following subsections. Each subsection will give an answer to one of the three questions addressed in the introduction.

2.2   A Two-Level Parallelism (Q1)

As shown in Fig. 1, our approach is based on two levels of parallelism mapping the search into possibly multiple CPUs and multiple GPUs. In *Level 1*, different CPUs or GPUs can explore different subtrees in parallel, i.e., select, branch, evaluate, prune, etc. As it will be discussed later, enabling the distribution of subtrees over PUs dynamically at runtime is at the heart of our approach. In *Level 2*, the evaluation of tree nodes (bounding for B&B) is done *inside* every GPU device, while the other search operations are performed *in parallel* by the GPU host, i.e., CPU. In fact, due to the irregularity and unpredictable shape of the tree, it is well understood that implementing the whole search operations inside GPU, could suffer from the thread divergence induced by the SIMD programming model of GPUs. The evaluation step, on the other side, can highly benefit from the parallelism offered by the many GPU cores. These design/model choices are essentially motivated by the fact that the evaluation phase of many combinatorial optimization problems is very time-consuming, e.g bounding for B&B, so that it dominates the other operations.

Although the GPU device can handle the evaluation of many tree nodes in parallel [4,10], the CPU host still has to prepare a data containing these nodes, copy them into GPU memory and copy back the result. This implies that while computations are carried out on the GPU device, the host is idle and vice-versa. In our Level 2 parallelism, the host and the device are managed to run computations *in parallel*, i.e., while the device is evaluating tree nodes, the host is preparing new data for the next evaluation in the device. Notice that the evaluation step of many tree nodes inside the GPU is of course implying another type of parallelism which we will not address in this paper, since our focus is on scalability and work distribution on multiple PUs.

2.3   Dynamic Work Stealing (Q2)

It is essential to fully explore the computing resources provided by a single CPU-GPU. However, it is more challenging to fully utilize the networked resources available in the distributed system. In fact, the irregularity generated at run-time can eventually lead to very poor performances because most computing nodes are underloaded and few others are highly overloaded, or because of the cost of synchronizing PUs and transferring work is so high. In this paper, we propose a distributed work stealing [7] based mechanism to tackle this issue. If a PU runs out of work, it acts like a thief and tries to steal work (i.e., subtree nodes) from another PU, called victim, *chosen uniformly at random*. This simple decentralized work stealing approach is motivated by two facts. Firstly, idle PUs acquire work cooperatively in parallel, thus eliminating the time required to synchronize, and to transfer/distribute data among them. In particular, no PU can constitute a communication bottleneck, so that the protocol would not suffer from scalability issues. Secondly, random work stealing (RWS) in shared memory is theoretically shown to give good performance under some application circumstances [1]. However, it has not been studied so far in a heterogenous networked setting involving the cooperation of both CPUs and GPUs at large scales in order to solve hard optimization problems.

2.4   Adaptive Work Balancing (Q3)

One crucial issue in RWS for efficient dynamic load balancing is the amount of work, denoted f, to be transferred between thieves and victims. Generally, the thief attempts to balance the load evenly between itself and the victim. In fact, when this amount of work is very small, the large overhead is observed since many load balancing operations are performed. At the opposite, when it is very large, too few load balancing operations will occur, thereby resulting in large idle times despite the fact that surplus work could be available. In classical RWS approaches, this is a hand-tuned parameter which depends on the distributed system and the application context [12]. In a theoretical study [1], the stability and optimality of RWS can be analytically guaranteed for $f \leq 1/2$. In practice, the so called steal-half strategy ($f = 1/2$) is often shown to perform efficiently using homogenous computing units. In a heterogenous and large scale scenario, this parameter is even more sensitive because of the wide variety of computing capabilities of different PUs. In this context, the community lacks relatively much knowledge to understand how to attain good performance for RWS based protocols.

To understand the issues we are facing when distributing tree search works over multiple CPUs and GPUs, one has to keep in mind that (i) a GPU is substantially faster in evaluating tree nodes than a CPU, (ii) nothing can be assumed about the amount of tree nodes initially. Hence, if GPUs run out of work and stay idle searching for work, the performance of the system can drop dramatically. If only few CPUs are available in the system, work stealing operations from CPUs to GPUs can cause a severe penalty to performance. This is because the few CPUs can only contribute very little to the overall performance but their stealing operations to GPUs can disturb the GPU computations and prevent them from reaching their maximal speed. In contrast, if work is scheduled more on GPUs, then a significant loss in performance can occur when a relatively large number of CPUs are available. To tackle these issues, we propose to configure RWS so that when performing a steal operation, the value of f is computed at runtime based on the normalized power of the thief and the victim, where the computing power of every PU is estimated continuously at runtime with respect to the application being tackled.

3   Parallel and Distributed Protocol Details

3.1   Concurrent Computations for Single CPU-GPU (*Level 2* Parallelism)

Generally speaking, an application is composed of multiple tasks and each task could be executed on a GPU or CPU depending on its characteristics. For each task running on a GPU, input data is transferred to GPU memory, a kernel is executed on the input and the outputs are copied back to the host for being processed. In other words, standard CPU/host GPU/device executions are synchronized sequentially. While the host is working, i.e., to prepare/process input/output data, the device sits idle, and vice versa. This can significantly slow down computations especially when the host and the device can perform concurrent operations *in parallel*.

With the rapid evolving of GPU devices, it is now possible to address the above issue by carefully exploiting the new available hardware and software technologies. For instance, NVIDIA GPUs with compute capability ≥ 1.1 are associated with a *compute* engine and a *copy* engine (DMA engine). NVIDIA's Fermi GPUs have up to 2 copy engines, one for uploading from host to device and one for downloading from device to host. Each engine is equipped with a queue to store pending data and kernels that will be processed by the engine shortly.

The *Level 2* host-device parallelism discussed in our approach can be enabled using CUDA primitives as sketched in Algorithm 1. Each ENQUEUE procedure dispatches CUDA operations into the GPU device *asynchronously*, i.e., pushes/retrieves data and launches the kernel. This is possible by wrapping those operations into a CUDA stream. All operations inside the same CUDA stream get automatically synchronized and executed sequentially, but the CUDA operations of different streams could overlap one with the other, e.g., execute the kernel of stream 1 and retrieve data from stream 2 concurrently in parallel. In our implementation, we use a maximum number of streams, i.e., variable r_{max}, which is the maximum number of elements (*data, kernel*) in the queue of GPU Copy engine and Compute engine. The maximum number of streams that a GPU can handle depends in general on GPU global memory characteristics. For B&B search, data is a pool of tree nodes and kernel is the bounding function. Asynchronously in parallel to the ENQUEUE procedure, the DEQUEUE procedure in Algorithm 1 waits for data copied back from the device on a given CUDA stream, and processes the output data. In our B&B implementation, this corresponds to the pruning operation. Notice also that Algorithm 1 is independent of the specific data or kernels being used, so that it can be customized with respect to the search operations or optimization problems at hand. In particular, any existing kernel implementing parallel tree node evaluation is applicable.

3.2 Distributed Work Stealing for Multiple CPUs/GPUs (*Level 1* Parallelism)

In this section, we describe how our *Level 1* parallelism is implemented, i.e., how tree nodes are distributed over PUs. As discussed previously, this is based on an adaptive work stealing paradigm and sketched in Algorithm 2 below — Notice that Algorithm 2 is to be executed *distributively by each* PU, i.e., v variable.

Algorithm 1: GPU *Level 2* parallelism — Concurrent host-device template

---

Data: q_host, q_device: queue of *task* in host and GPU; q_host_size: current size of q_host (0 initially); $stream[r_{max}]$: CUDA Stream of r_{max} elements; w_index, r_index: next index to write (resp. read) to (resp. from) the queues (0 initially).

1 **while** *tree nodes are available* **do in parallel:**
   // Push tree nodes for evaluation inside GPU
2   **Execute Procedure** ENQUEUE;
      // Retreive and process evaluated nodes from the GPU
3   **Execute Procedure** DEQUEUE;

Procedure Enqueue

---

1 **while** $q\_host\_size < r_{\max}$ **do**
2 $\quad q\_host[w\_index].task \leftarrow$ prepare a pool of tree nodes;
$\quad$ // Asynchronous Operations on stream[w_index]
3 $\quad$ cudaMemcpyAsyn($q\_device[w\_index], q\_host[w\_index]$,
$\quad$ sizeof($q\_host[w\_index].task$),
4 $\qquad\qquad\qquad$ ***cudaMemcpyHostToDevice, stream[w_index]***);
$\quad$ // Launch parallel evaluation (bounding) on device
5 $\quad$ KERNEL$<<<$ ***stream[w_index]*** $>>>$ ($q\_device[w\_index]$) ;
6 $\quad$ cudaMemcpyAsyn($q\_host[w\_index].bound, q\_device[w\_index].bound$,
7 $\qquad\qquad\qquad$ sizeof($q\_device[w\_index].bound$),
8 $\qquad\qquad\qquad$ ***cudaMemcpyDeviceToHost, stream[w_index]***) ;
9 $\quad w\_index \leftarrow (w\_index + 1) \pmod{r_{\max}}; \quad q\_host\_size \leftarrow q\_host\_size + 1$;

Procedure Dequeue

---

1 **if** $q\_host\_size > 0$ **then**
$\quad$ // Wait for results from device on stream[r_index]
2 $\quad$ cudaStreamSynchronize(***stream[r_index]***);
3 $\quad$ Process output data from $q\_host[r\_index]$, i.e., prune nodes ;
4 $\quad r\_index \leftarrow (r\_index + 1) \pmod{r_{\max}}; \quad q\_host\_size \leftarrow q\_host\_size - 1;$

Algorithm 2: *Level 1* Parallelism — Distributed Adaptive Work Stealing

---

1 **while** *Termination not detected* **do in parallel:**
2 $\quad$ **Execute Procedure** THIEF;
3 $\quad$ **Execute Procedure** VICTIM;

Procedure Thief

---

1 $x \leftarrow$ **runtime normalized** computing power;
2 **repeat**
3 $\quad u \leftarrow$ pick one PU victim **uniformly at random**;
$\quad$ // v denotes the actual thief PU executing the procedure
4 $\quad$ **Send** a steal request message (v, x) **to** u;
5 $\quad$ **Receive** u's response (reject or work) message;
6 **until** *some tree nodes are successfully transferred from victim u*;

Stealing Granularity. To efficiently balance the work load, stealing granularity, that is the amount of work to be transferred from victims to thieves, plays a crucial role. Depending on the hardware platform and the input application, there may exist a value of work granularity giving the best performance. For instance, for the Unbalanced Tree Search benchmark [6], which is often considered as an adversary application to load balancing [9,13], it was shown that steal

Procedure Victim

1	**if** *a steal request is pending* **then**
2	$\quad$ $y \leftarrow$ **runtime normalized** computing power ;
3	$\quad$ **if** *tree nodes are available* **then**
4	$\quad\quad$ $(v, x) \leftarrow$ pull the next pending thief request;
5	$\quad\quad$ $work \leftarrow$ share tree nodes in the proportion of $\dfrac{x}{x+y}$;
6	$\quad\quad$ **Send** back shared $work$ **to** v ;
7	$\quad$ **else**
8	$\quad\quad$ **Send** back a reject message **to** v ;

half works best for binomial trees. Instead, stealing a fixed amount of work items (i.e., 7 items) is shown to work well for geometric trees. Besides, in a heterogeneous and hybrid computing system, the hardware characteristics of PUs, e.g., clock speed, Cache, RAM, etc, can be highly needed to balance the work load evenly depending on the characteristics of every available PU. Because high variations in computing power among PUs can lead to high imbalance and idle times, one has also to manage this issue carefully when distributing work. One possible solution to the above issues could be to profile the system components/PUs and tune work granularity offline before application execution in order to get the best performance. It should be clear that such an approach is not reasonable nor feasible, for instance when the system may undergo a huge number of many different types of PUs, or when having many different applications at hand.

In our stealing approach, we make every PU maintain at runtime a measure reflecting its computing power, i.e., variable x in Algorithm 2. As the computations are running on, every PU adjusts its measure continuously with respect to the work processed in the previous iterations. In our approach, we simply use the average time needed for processing one tree node. More precisely, each PU sets its computing power to be $x = N/T$, where T is the (normalized) time elapsed since the PU has started the computation and N is the number of tree nodes explored locally by that PU. Notice that time T includes, in addition to tree node evaluation (i.e., B&B lower bounding), the time needed for other search operations (i.e., select, branch and prune) but *not* the time when a PU stays idle. When running out of work, a PU v then attempts to steal work by sending a request message to one other PU u chosen at random, while wrapping the value of x in the request. If a victim has some work to serve, then the amount of work (i.e., number of tree nodes) to be transferred is in the proportion of $x/(x+y)$, where y is the computing power maintained locally by the victim. Otherwise, a reject message is sent back to notify the thief and a new stealing round is performed. Initially, the value of x is normalized so that all PUs have the same computing ability. In other words, the system starts stealing half and then the stealing granularity is refined for each pairwise PU. Intuitively, each PU acts as a black-hole, so that the higher computing power of PUs is, the more available work are flowed to the black-hole. Furthermore, no knowledge about

PUs is needed so that any performance variation at system/application level would also be detected at runtime.

Termination Detection. One issue in the template of Algorithm 2 is how to decide on termination distributively (Line 1). For B&B, this occurs when all tree nodes are explored (explicitly or implicitly, i.e., pruned). However, since stealing is performed locally by idle PUs, the work remaining in the system is not maintained anywhere. This is a well understood issue for which an abundant literature can be found [5].

We use a fully distributed scheme, in which PUs are mapped into a tree overlay and the termination is detected in an 'Up-Down' distributed fashion. In the up phase, if a PU becomes idle and has not served any stealing request, it will then integrate a positive termination signal to its children signals. If a PU turns to idle and has served at least one stealing request, it will then integrate a negative termination signal to its children signals. Then the termination signal is forwarded to the parent and eventually to the root. In the down phase, if the root receives at least one negative termination signal from its children, it broadcasts a signal to restart a new round of termination detection. Otherwise, if only positive termination signals are received, the root broadcasts a message to announce global termination. The tree overlay used in our implementation is a binary one so that PU degrees and the overlay diameter are kept low. This allows us to scale out PUs while avoiding communication bottlenecks and performance degradation once a termination phase is performed.

Knowledge Exchange. An important ingredient missing to complete our approach, is the mechanism allowing PUs to exchange knowledge during the search. In B&B for instance, one important issue is to share the best upper bound found by any PU in order to avoid exploring unnecessary branches. We use the same tree overlay topology used in the above scheme for termination detection, to propagate search knowledge (new upper bounds) among PUs. Since the overlay diameter is logarithmic, propagating knowledge among PUs has a relatively limited communication cost.

4   Experimental Results

4.1   Experimental Setting

We consider the permutational FlowShop problem with the objective of minimizing the makespan (C_{max}) of scheduling n jobs over m machines as a case study in our experiments. The well-known Taillard' instances [14] of the family of 20 jobs and 20 machines are considered. To give an idea of the their difficulties, the time required for solving these instances on a standard modern CPU, starting from scratch (that is without any initial solution), can be several dozens of hours.

Our approach needs three major components to be experimented: (i) the distributed load balancing protocols (*Level 1* parallelism), (ii) the concurrent host-device computations (*Level 2* parallelism) and (iii) the GPU kernel for bounding w.r.t FlowShop. The GPU kernel was taken to be the one of [4,11] and used as a blackbox. *Level 1* (resp. *2*) was implemented using low level c++ libraries (resp. c++ concurrent threads and CUDA primitives). Three clusters C_1, C_2 and C_3 of the Grid'5000 French national platform [8] were involved in our experiments. Cluster C_1 contains 10 nodes, each equipped with 2 CPUs of 2.26 Ghz Intel Xeon processors with 4 cores per CPU. Besides, each node is coupled with two Tesla T10 GPUs. Each GPU contains 240 CUDA cores, a 4 GB global memory, a 16.38 KB shared memory and a warp size of 32 threads. Cluster C_2 (resp. C_3) were equipped with 72 nodes (resp. 34 nodes), each one equipped with 2 CPUs of 2.27 Ghz Intel Xeon processor with 4 cores per CPU (resp. 2 CPUs of 2.5 Ghz Intel Xeon processor having 4 cores) and a network card Infiniband-40G.

Let us point out that the GPU *kernel* implementation of [4,11] has a parameter s referring to the maximum number of B&B tree nodes that are pushed into GPU memory for parallel evaluation. It is shown in [11] that the parameter s has to be fixed to a value $s^\star$ so that the device memory is optimized and the performance is the best on a single GPU. In [4], it is shown how to tune the value of s online so that it converges to $s^\star$. Since we assume that the GPU kernel is provided as a black-box, and unless stated explicitly, the value of s is simply fixed to $s^\star$ in our experiments. In our experimental study, we are also interested in analyzing how our approach would perform when having GPU kernels allowing for different speed-ups in the evaluation phase. This can be typically the case for other type of problems, different hardware configurations, etc. Being able to understand whether our load balancing mechanism is efficient in such a heterogenous setting, independently of the considered scale or speedup gap between available CPUs and GPUs, is of great importance. In this paper, we additionally view the parameter s as allowing us to empirically reduce the intrinsic speed of a single GPU, and thus to experiment our approach while using different GPU and CPU configurations. In the remainder, we shall consider the following scenarios:

1. Enabling our *Level 2* parallelism within a single CPU-GPU.
2. Running our approach (*Level 1* + *Level 2*) at different scales with multiple GPUs.
3. Running our approach with a fixed number of GPUs, while scaling the CPUs.
4. Running our approach with a fixed number of CPUs, while scaling the GPUs.
5. Running our approach with CPUs and GPUs having different computational powers.

In all scenarios, a GPU device is launched with 1 CPU core taken from C_1. For the first four scenarios, CPUs are taken from cluster C_2. As for the fifth scenario, we mix CPUs of different hardware clock speeds, taken from C_2 and C_3, and GPUs launching kernels configured with different values of s. The previous scenarios aim at providing insights on how the system performs independently of the scale and/or the power of CPUs and GPUs. For all experiments, we measure

T and N, respectively the time needed to complete the B&B tree search and the number of B&B tree nodes that were effectively explored. All reported speedups are relative to the number of B&B tree nodes explored by time units, that is N/T.

4.2   Impact of Asynchronous Data Transfer on a Single GPU

We start our analysis by evaluating the impact of *Level 2* host device concurrent operations. For the ten instances in Taillard' family 20*20, we report in Fig. 2 execution time and speedup w.r.t. the baseline sequential host-device execution [11], for different number of concurrent CUDA streams (variable r_{max} in Algorithm 1) and different GPU kernel parameters s. One can clearly see that substantial improvements are obtained, i.e., our approach is at least 2 times faster. It also appears that the maximum number of concurrent CUDA streams r_{max}, which is the only parameter used in our approach, has only a marginal impact on performance. Figure 2 Right shows that the speed-up, w.r.t the sequential host-device execution, is substantial ($> 30\%$) but depends on kernel parameter s. This is because for lower values of s, the host spends more time pushing small amount of data, while the device is less efficient. In other words, *Level 2* parallelism performs better when the amount of data and computations on device is higher.

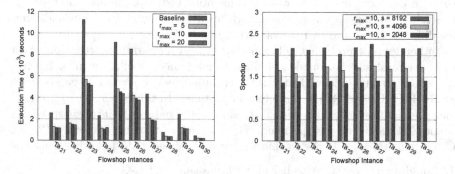

Fig. 2. *Level 2* parallelism *vs.* baseline sequential host-device execution [4]. **Left**: Execution time with different number r_{max} of CUDA streams and $s = s^\star$ (Lower is better). **Right**: Speedup w.r.t baseline for different values of s and $r_{max} = 10$ (Higher is better).

4.3   Scalability and Stealing Granularity for Multiple GPUs

In this section, we study the scalability of our approach when only multiple GPUs are available in the system. For this set of experiments we choose the first instance Ta_{21} to be our case study. In Fig. 3 Left, we report the speedup of our approach w.r.t one single GPU, and also the speedup obtained when using a static stealing granularity (with of course *Level 2* parallelism enabled). By static stealing, we mean that we initially fix the proportion of tree nodes to be stolen as a parameter $f \in \{1/2, 1/4, 1/8\}$. Two observations can be made. Firstly, our adaptive approach performs similar to the best static stealing, which is for $f = 1/2$ from our experiments. Other values of f in static stealing are in fact worse especially in high scales. Secondly, we are able to scale linearly with the number of GPUs. At scale 16, one can notice a slight decrease in speedup. We attribute this to two factors: (i) the communication cost of distributing work strategy to be not negligible in large scales, and (ii) sharable work becomes very fine grain so that it limits the maximal performance of GPUs. Actually, the results of Fig. 3 Left are obtained with parameter s being set to $s^\star$, i.e., the maximal (and best) amount of tree nodes such that a single GPU can handle. In Fig. 3 Right, we push our experiments further by taking other values for parameter s. We can clearly see that the speed-up (w.r.t. one single GPU running a kernel with the same value of s) is not impacted. The scalability is even slightly better when the kernels are less efficient. This can be interpreted as the scalability of our approach being not sensitive to other system/application settings with GPUs having possibly different processing powers.

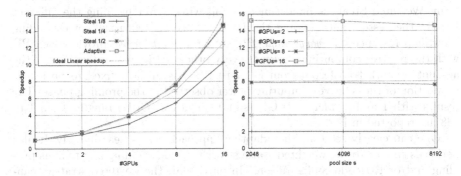

Fig. 3. Left: Scalability of our adaptive approach *vs.* static stealing $(s = s^\star)$. X-axis refers to the number of GPUs in log scale. Y-axis refers to the speed-up with respect to one GPU. **Right**: Speedups of our approach as a function of s. $r_{\max} = 10$.

4.4   Adaptive Stealing for Multiple GPUs Multiple CPUs

In this section, we study the properties of our approach when mixing both CPUs and GPUs. For that purpose, we proceed as following. Let α_i^j be the speedup obtained by a *single* PU j with respect to PU i. We naturally define the linear (ideal) normalized speedup *with respect to PU i*, to be $\sum_j \alpha_i^j$. For instance, having p *identical* GPUs and q *identical* CPUs, each GPU being β times faster than each CPU, our definition gives a linear speedup with respect to *one GPU* (resp. *one CPU*) of $p + q/\beta$ (resp. $q + \beta \cdot p$). The following sets of experiments shall allow us to appreciate the performance of our approach when varying substantially the ratio between the number of GPUs and CPUs.

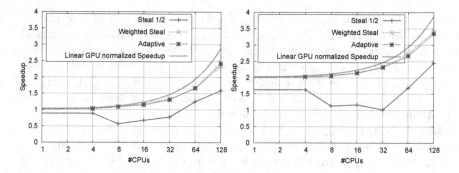

Fig. 4. Speedup of our approach *vs.* static steal when scaling CPUs and using 1 GPU (**Left**) and 2 GPUs (**Right**) X-axis is in the log scale. Speed-up are w.r.t. one GPU. $r_{max} = 10$.

CPU Scaling. In this set of experiments, we fix the number of GPUs and scale the number of CPUs. Besides, we experiment two other static baseline strategies. The first one is the standard steal half strategy. The second one, we term 'Weighted Steal', is hand tuned as following. After profiling the different PUs in the system and running the B&B tree search with the corresponding FlowShop instance on *every single PU until termination*, we provide each PU with the relative computing power of every other PU in the system. Then, the amount of work transferred from PU i to PU j is initially fixed to be in the proportion of the relative computing power observed in the profiling phase. The results with 1 and 2 (identical) GPUs and (identical) CPUs ranging from 1 to 128 are reported in Fig. 4.

One can clearly see that the adaptive approach scales near linearly. It also performs similar to the weighted static strategy while avoiding any tedious profiling and/or PU code configurations. In particular, the weighted strategy cannot be reasonable in production systems with different PU configurations since it requires much time to tune the systems. Turning to the steal half static strategy, it appears to perform substantially worse. When having relatively few CPUs, the performance of steal half is even worse than in a scenario where only GPUs are available (see Fig. 3). It is also getting worse as we push additional few CPUs in the system. Improvements over 1 or 2 GPUs are only observed when the number of CPUs is relatively high (w.r.t GPU power).

GPU Scaling. We now fix the number of CPUs and study how the behavior of the system when scaling the number of GPUs. Results with 128 (identical) CPUs and (identical) GPUs ranging from 1 to 16 are reported in Fig. 5. We can similarly see that our adaptive approach is still scaling in a linear manner while being near optimal. It is also substantially outperforming the static steal half strategy.

Mixed Scaling. Our last set of experiments is more complex since we manage to mix multiple GPUs with empirically different powers and multiple CPUs with different clock speeds. This scenario is in fact intended to reproduce a heterogenous setting where, even PUs in the same family do not have the same computing abilities. In this kind of scenario, where in addition the power of PUs can evolve, e.g., due to system maintenance constraints or hardware renewals/updates, even a weighted hand tuned steal strategy is not plausible nor applicable. In the results of Fig. 6, we fix the number of CPUs to be 128 with half of them taken from cluster C_2 and the other half from cluster C_3 (C_2 and C_3 have different CPU clock speeds as specified previously). For GPUs, we proceed as following. We use a variable number of GPUs in the range $p \in \{1, 4, 8, 12, 16, 20\}$. For $p > 1$, we configure the system so that $1/2$ of GPUs run a kernel with pool size $s^\star$, $1/4$ of them with pool size $s^\star/2$ and the last $1/4$ of them with pool size $s^\star/4$. Once again our approach is able to adapt the load for this complex heterogenous scenario and to obtain a nearly optimal speedup while outperforming the standard steal half strategy. From the previous set of experiments we can thus

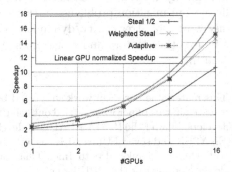

Fig. 5. Speedup (w.r.t. one GPU) when scaling GPUs and using 128 CPUs. $r_{\max} = 10$.

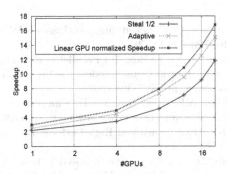

Fig. 6. Speedup when scaling heterogenous GPUs ($1/2$ with $s^\star$, $1/4$ with $s^\star/2$, $1/4$ with $s^\star/4$), and 128 heterogenous CPUs ($1/2$ from cluster C_2, $1/2$ from cluster C_3). Speedup is w.r.t. one GPU configured with $s^\star$. $r_{\max} = 10$.

conclude that our approach allows us to take full advantage of both GPU and CPU power independently of considered scales, or any hand tuned parameter.

5   Conclusion

In this paper, we proposed and experimented an adaptive load balancing distributed scheme for parallelizing computing intensive B&B-like tree search algorithms in heterogenous systems, where multiple CPUs and GPUs with possibly different properties are used. Our approach is based on a two-level parallelism allowing for (i) distributed subtree exploration among PUs and (ii) concurrent operations between every single GPU host and device. Through extensive experiments involving different PU configurations, we showed that the scalability of our approach is near optimal, which leaves very little space for further improvements. Besides being able to experiment our approach with other problem-specific GPU kernels, one interesting and challenging research direction would be to extend our approach in a *dynamic* distributed environment where: (i) processing units can join or leave the system, and (ii) different end-users can concurrently request the system for solving different optimization problems. In this setting, the load has to be balanced not only w.r.t. the irregularity/dynamicity of one single application, but also w.r.t many other factors and constraints that may affect the computing system at runtime.

Acknowledgments. This material is based on work supported by INRIA HEMERA project. Experiments presented in this paper were carried out using the Grid5000 experimental testbed, being developed under the INRIA ALADDIN development action with support from CNRS, RENATER and several Universities as well as other funding bodies (see https://www.grid5000.fr). Thanks also to Imen Chakroun for her precious contributions to the code development of the GPU kernel.

References

1. Blumofe, R.D., Leiserson, C.E.: Scheduling multithreaded computations by work stealing. J. ACM **46**, 720–748 (1999)
2. Boukedjar, A., Lalami, M.E., El-Baz, D.: Parallel branch and bound on a CPU-GPU system. In: 20th International Conference on Parallel, Distributed and Network-Based Processing, pp. 392–398 (2012)
3. Carneiro, T., Muritiba, A.E., Negreiros, M., De Campos, L., Augusto, G.: A new parallel schema for branch-and-bound algorithms using GPGPU. In: 23rd Symposium on Computer Architecture and High Performance Computing, pp. 41–47 (2011)
4. Chakroun, I., Melab, M.: An adaptative multi-GPU based branch-and-bound. a case study: the flow-shop scheduling problem. In: 14th IEEE Interernational Conference on High Performance Computing and Communications (2012)
5. Dijkstra, E.W.: Derivation of a termination detection algorithm for distributed computations. In: Broy, M. (ed.) Control Flow and Data Flow: Concepts of Distributed Programming, pp. 507–512. Springer, Berlin (1987)

6. Dinan, J., Olivier, S., Sabin, G., Prins, J., Sadayappan, P., Tseng, C.-W.: A message passing benchmark for unbalanced applications. Simul. Model. Pract. Theor. **16**(9), 1177–1189 (2008)
7. Matteo, F., Charles, E.L., Keith, H.R.: The implementation of the cilk-5 multithreaded language. SIGPLAN Not. **33**, 212–223 (1998)
8. Grid500 French national gird. https://www.grid5000.fr/
9. James, D., Brian, L.D., Sadayappan, P., Krishnamoorthy, S., Jarek, N.: Scalable work stealing. In: Proceedings of ACM Conference on High Performance Computing Networking, Storage and Analysis, pp. 53:1–53:11 (2009)
10. Lalami, M.E., El-Baz, D.: GPU implementation of the branch and bound method for knapsack problems. In: IPDPS Workshops, pp. 1769–1777 (2012)
11. Melab, N., Chakroun, I., Mezmaz, M., Tuyttens, D.: A GPU-accelerated b&b algorithm for the flow-shop scheduling problem. In: 14th IEEE Conference on Cluster Computing (2012)
12. Min, S.-J., Iancu, C., Yelick, K.: Hierarchical work stealing on manycore clusters. In: Proceedings of 5th Conference on Partitioned Global Address Space Programming Models (2011)
13. Saraswat, V.A., Kambadur, P., Kodali, S., Grove, D., Krishnamoorthy, S.: Lifeline-based global load balancing. In: 16th ACM Symposium on Principles and Practice of Parallel Programming (PPoPP '11), pp. 201–212 (2011)
14. Taillard, E.: Benchmarks for basic scheduling problems. Eur. J. Oper. Res. **64**(2), 278–285 (1993)

Multi-Objective Optimization for Relevant Sub-graph Extraction

Mohamed Elati, Cuong To, and Rémy Nicolle[✉]

iSSB CNRS UPS3509, University of Evry-Val-dEssonne EA4527,
Genopole Campus 1, 5 Rue Henri Desbruères, 91030 Evry CEDEX, France
{mohamed.elati,cuong.to,remy.nicolle}@issb.genopole.fr

Abstract. In recent years, graph clustering methods have rapidly emerged to mine latent knowledge and functions in networks. Most sub-graphs extracting methods that have been introduced fall into graph clustering. In this paper, a novel trend of relevant sub-graphs extraction problem was considered as multi-objective optimization. Genetic Algorithms (GAs) and Simulated Annealing (SA) were then used to solve the problem applied to biological networks. Comparisons between GAs, SA and Markov Cluster Algorithm (MCL) were carried out and the results showed that the proposed approach is superior.

Keywords: Sub-graph extraction · Genetic algorithms · Simulated annealing · Multi-objective optimization

1 Introduction

Nowadays, there are many types of data that can be represented as graphs (networks) such as web graphs, social networks, biological networks, communication networks, road networks, etc. Mining hidden knowledge in networks is a non-trivial task because networks are usually big (containing thousands of nodes and edges) and discrete (mathematics models are difficult to apply). A graph is a structure that consists of vertices (nodes) and edges. Vertices can be anything such as pages in web graphs, members in social networks, proteins or genes in biological networks, etc. Edges, which are links between two nodes, represent relationship between two nodes. There are many problems involving graphs such as the shortest paths, graph coloring, route, etc. In this work, we focus on graph clustering. Graph clustering is a problem where vertices are to be grouped into clusters satisfying some pre-defined criteria. Graph clustering methods have been applied in many fields, especially, in biological networks [1–3,10]. Most of introduced sub-graph extraction methods are based on graph clustering such as the Markov Cluster Algorithm (MCL) [5], a scalable clustering algorithm for graphs. The basic idea of MCL, is Markov Chains so that when starting at a node, and then randomly traveling to a connected node; you are more likely to stay within a cluster than travel between clusters. Spirin et al. [13] used three

G. Nicosia and P. Pardalos (Eds.): LION 7, LNCS 7997, pp. 104–109, 2013.
DOI: 10.1007/978-3-642-44973-4_12, © Springer-Verlag Berlin Heidelberg 2013

different approaches to identify connected sub-graphs. The first approach is complete enumeration. The second one is super-paramagnetic clustering. In the third
approach, the authors formulated the problem of finding highly connected sets
of nodes as an optimization problem, and then used Monte Carlo to solve it.
Although graph-clustering methods have been applied to many fields, they still
have some drawbacks. (1) The density (number of edges normalized by the number of nodes) of sub-graphs given by these methods are rather low. Yet, nodes of
a sub-graph (cluster) are desired to share similar traits which can be represented
by the den-sity in some circumstances. For instance, protein complexes in biological networks are described by their density [3], a cluster of social networks
is intuitively a collection of individuals with dense friendship patterns internally
and sparse friend-ships externally and in web graphs, clusters represent groups
of highly similar Web sites (similarity may be based on the degree of relationship
or association that exists between entities). Therefore, clusters with low density
can reflect groups of nodes with low similarity. (2) In large networks (containing
thousands of nodes and edges), graph clustering processes are resource-intensive
tasks and all identified sub-graphs are not all interesting for a given problem.
In other words, experts only want to extract dense sub-graphs connecting particular nodes (relevant subgraphs for short). However, in our knowledge there
have been very few studies directly addressing the problem of relevant subgraph
extraction; i.e, To find subgraph that can well capture the relationship among
k given nodes of interest in a large graph. The first work is done by Faloutsos
et al. [7] and then followed by that of [6]. Faloutsos et al. [7] present an electricity
analogue based method to find connection subgraph between two given nodes.
When there are more then two nodes of interest, the electricity analogues does
not work.

Based on the above drawbacks and inquiries, we propose an algorithm that
considers extracting reliable sub-graphs from networks as multi-objective optimization which is solved using Genetic algorithms or Simulated. The rest of this
work is organized as follow. Section 2 describes our algorithm including problem
statement, a brief introduction of Genetic algorithms and Simulated annealing,
and how to use them to solve a multi-objective optimization problem. Section 3
presents a case study of the application of our algorithm as a strategy to expand
the knowledge of protein complex by clustering the protein interaction network
in a semi-supervised fashion using affinity-purification/mass-spectrometry (AP-
MS) data.

2   Relevant Sub-graph Extraction Method

2.1   Problem Statement

Given an unweighted and undirected graph, $G(V, E)$ where V is a set of vertices
(nodes), $E \subseteq V \times V$ is a set of edges (links) and a list of interesting vertices called
ranked list, L. The goal of our algorithm is extracting sub-graphs (clusters),
which are dense and contain as many vertices of the ranked list as possible. The

density of a sub-graph is a fraction between number of edges and all possible edges of sub-graph given by (3).

The Multi-Objective Optimization Concept. The above problem has two objectives which need to be optimized (1) density of sub-graphs and (2) number of interesting nodes in sub-graphs. It considers the problem as a multi-objective optimization problem, which can be described as Finding sub-graphs, $G(V, E)$ which maximize

$$F(V', E') = (f_1(V', E'), f_2(V', E')) \tag{1}$$

Subject to

$$f_2(V', E') > 0 \tag{2}$$

where

- $f_1(V, E)$ represents density of sub-graph, given by

$$f_1(V', E') = \frac{2 \mid E' \mid}{\mid V' \mid (\mid V' \mid -1)} \tag{3}$$

with $\mid E' \mid$ and $\mid V' \mid$ are cardinality (number of items in a set) of E' and V', respectively.
- $f_2(V', E')$ counts number of vertices in the ranked list, L appearing in sub-graph, in form

$$f_2(V', E') = \frac{\sum\limits_{P \in V'} R(P)}{\sum\limits_{P \in L} R(P)} \tag{4}$$

with P is a node and $R(P)$ is the ranked value (a positive number or zero, if node is not in the ranked list) of node, P.
- The criterion (2) means that the result sub-graphs have to contain at least one node that belongs to the ranked list, L.

The ranges of f_1 and f_2 are within $[0, 1]$. Normally, the second objective gives high value if the sub-graph is big (containing a lot of vertices in the ranked list, L); but the density value of large sub-graph is usually low. Vice versa, if the sub-graph is small, the density value can become high then the number of vertices in the ranked list appearing in the sub-graph can be low. The two objectives of the of F (1) conflict with each other. So the algorithm has to find the sub-graphs which gives high values in both objectives.

Closed-form solutions of multi-objective optimization problems are difficult to obtain. There have been no closed-form solutions of global optimality for general multi-objective optimization problems [9]. While Genetic algorithms [15] can be considered as a "globalization technique" because they can handle a population of candidate solutions, Simulated annealing is a probabilistic method [11]. Another advantage of genetic algorithms and simulated annealing is that they do not use gradients or Hessians which may not exist or can be difficult to obtain. So we decided to use Genetic algorithms and Simulated annealing to solve the above multi-objective optimization.

2.2   Solving the Multi-Objective Optimization

In order to solve a multi-objective optimization problem, there are many types of methods such as: scalar approach, Pareto-based approach, normal boundary intersection, successive Pareto optimization, etc. [4,12]. We decided to use scalar approach [9], which gives one solution for each implementation, in the following formula:

$$F(V', E') = w_1 \times f_1(V', E') + w_2 \times f_2(V', E') \tag{5}$$

where w_i is a non-negative weight.

2.3   Representation of Solutions

Solutions of this multi-objective optimization are of course sub-graphs, $G(V, E)$. In other words, solutions are sets of vertices. Unfortunately, we do not know the number of proteins in each sub-graph in advance and these numbers are different, in principle. The number of vertices in sub-graph is intrinsically dimension of the search spaces. Consequently, the higher dimension the search space is, the more difficult the problem is.

In order to lessen the dimension, we represented a solution as a circle that is deter-mined by center and radius (see Fig. 1). Consequently, the dimension of search space is fixed of two, namely center (vertex) and radius (positive integers). Finally, each individual of GAs (SAs state) consists of center and radius; GAs and SA used (7) as the fitness function.

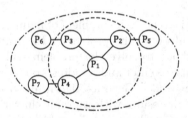

Fig. 1. Circle representation of solution. The dash circle has center, P1 and radius of one. The dash-dot circle has center, P1 and radius of two.

3   Results

Protein complex play an important role in cellular organization and function. The accumulation of large-scale protein interaction data on multiple organisms requires novel computational techniques to analyse biological data through these large-scale networks. Biological information can often be represented in the form of a ranked list.

The Biological General Repository for Interaction Datasets (BioGRID) [14] contains genetic and protein interaction data from model organisms and humans.

All the protein interactions corresponding to Saccharomyces cerevisiae (69,800 unique interactions) were downloaded and used as the network G to compute f_1. We used 20 ranked lists from the Affinity purification followed by mass spectrometry analysis (AP-MS) experiments from Gavin et al. [8]. These experiments produced lists of proteins (nodes) that potentially form a complex but very often capture non-specific proteins (false positive, FP) and miss core proteins (false negative, FN). Therefore we applied our algorithm to each of these lists to find the connected sub-graphs around these purified lists of proteins in order to eliminate FP and add highly connected proteins potentially missed by the experimental process.

We used Simulated annealing (with an initial temperature set to 10000) and Genetic algorithms to solve the multi-objective optimization problem. Based on the fitness function defined in Eq. 5, the Genetic Algorithm globally showed better results (mean 1.28, standard deviation 0.19) than simulated annealing (mean 1.22, standard deviation 0.20).

According to Brohee et al. [3], networks with large number of vertices are not well identified by the MCODE algorithm [1]. Moreover, one of the most successful clustering procedures that has been used so far in deriving clusters from PPI networks seems to be the MCL [5]. Those are the reasons that we selected MCL to do the comparisons. The parameters of MCL such as power and inflation are set to two.

The results obtained by the Genetic Algorithm had always a superior density f_1 (mean 0.57) than the results given by MCL (mean 0.1).

4   Conclusion

The objectif of this work is to pose the problem of finding a subgraph connecting n nodes in a graph as a multi-objective problem in the context of systems biology. We used stochastic methods to solve the problem and used a graph specific representation of solutions, a center and a radius, to lessen the search space. We used Simulated annealing and a Genetic Algorithm to solve the optimization problem and compared our results with a graph clustering method, the MCL algorithm. MCL does not actually solve this problem, it does not maximizes f_2, but graph clustering was the most logical choice for a comparison.

MCL produced very big sub-graphs (2000–6000 nodes) making the analysis of the results extremely difficult and adding many False Positive whereas the results produced by the Genetic Algorithm had 29–190 nodes which is much more realistic. The size of MCL clusters explains why the density is so low (average of 0.1) and why the number of nodes from L is so high therefore motivating the fact that the problem must be seen as a multi-objective problem. Moreover, the execution time of MCL on such large networks is on the order of hours or days (here 2 and a half hour on four cores) when solving the problem using a Genetic Algorithm took an average of 2 min.

Acknowledgment. We thank F. Radvanyi for fruitful discussions and the anonymous referees for their pertinent suggestions.This work is supported by the INCa (French

National Institute of Cancer) through the INCa project PL-2010-196. R. Nicolle is supported by a fellowship from the French Ministry of Higher Education and Research.

References

1. Bader, G., Hogue, C.: An automated method for finding molecular complexes in large protein interaction networks. BMC Bioinformatics **4**, 2 (2003)
2. Birmele, E., Elati, M., Rouveirol, C., Ambroise, C.: Identification of functional modules based on transcriptional regulation structure. BMC Proc. **2**(Suppl 4), S4 (2008)
3. Brohee, S., van Helden, J.: Evaluation of clustering algorithms for protein-protein interaction networks. BMC Bioinformatics **7**, 488 (2006)
4. Coello, C.A.C., Dhaenens, C., Jourdan, L. (eds.): Advances in Multi-Objective Nature Inspired Computing. Studies in Computational Intelligence, vol. 272. Springer, Heidelberg (2010)
5. van Dongen, S.: Graph clustering by flow simulation. Ph.D. thesis, University of Utrecht (May 2000)
6. Dupont, P., Callut, J., Dooms, G., Monette, J.N., Deville, Y.: Relevant subgraph extraction from random walks in a graph (2006)
7. Faloutsos, C., McCurley, K.S., Tomkins, A.: Fast discovery of connection subgraphs. In: Proceedings of the Tenth ACM SIGKDD International Conference on Knowledge Discovery and Data Mining, KDD '04,pp. 118–127 (2004)
8. Gavin, A., Aloy, P., Grandi, P., Krause, R., Boesche, M., Marzioch, M., Rau, C., Jensen, L., Bastuck, S., Dumpelfeld, B.: Proteome survey reveals modularity of the yeast cell machinery. Nature **440**, 631–636 (2006)
9. Ishibuchi, H., Yoshida, T., Murata, T.: Balance between genetic search and local search in hybrid evolutionary multi-criterion optimization algorithms. IEEE Trans. Evol. Comput. **7**, 204–223 (2002)
10. Jiang, P., Singh, M.: Spici: a fast clustering algorithm for large biological networks. Bioinformatics **26**(8), 1105–1111 (2010)
11. Kirkpatrick, S., Gelatt, C.D., Vecchi, M.P.: Optimization by simulated annealing. Science **220**, 671–680 (1983)
12. Mueller-Gritschneder, D., Graeb, H., Schlichtmann, U.: A successive approach to compute the bounded pareto front of practical multiobjective optimization problems. SIAM J. Optim. **20**, 915–934 (2009)
13. Spirin, V., Mirny, L.A.: Protein complexes and functional modules in molecular networks. Proc. Natl Acad. Sci. **100**, 12123–12128 (2003)
14. Stark, C., Breitkreutz, B.J., Reguly, T., Boucher, L., Breitkreutz, A., Tyers, M.: Biogrid: a general repository for interaction datasets. Nucleic Acids Res. **34**, D535–D539 (2006)
15. Whitley, D.: A genetic algorithm tutorial. Stat. Comput. **4**, 65–85 (1994)

PROGRESS: Progressive Reinforcement-Learning-Based Surrogate Selection

Stefan Hess[1]([✉]), Tobias Wagner[1], and Bernd Bischl[2]

[1] Institute of Machining Technology (ISF), TU Dortmund University,
Dortmund, Germany
{hess,wagner}@isf.de
[2] Faculty of Statistics, TU Dortmund University, Dortmund, Germany
bischl@statistik.tu-dortmund.de

Abstract. In most engineering problems, experiments for evaluating the performance of different setups are time consuming, expensive, or even both. Therefore, sequential experimental designs have become an indispensable technique for optimizing the objective functions of these problems. In this context, most of the problems can be considered as a black-box. Specifically, no function properties are known a priori to select the best suited surrogate model class. Therefore, we propose a new ensemble-based approach, which is capable of identifying the best surrogate model during the optimization process by using reinforcement learning techniques. The procedure is general and can be applied to arbitrary ensembles of surrogate models. Results are provided on 24 well-known black-box functions to show that the progressive procedure is capable of selecting suitable models from the ensemble and that it can compete with state-of-the-art methods for sequential optimization.

Keywords: Model-based optimization · Sequential designs · Black-box optimization · Surrogate models · Kriging · Efficient global optimization · Reinforcement learning

1 Introduction

The optimization of real-world systems based on expensive experiments or time-consuming simulations poses an important research area. Against the background of increasing flexibility and complexity of modern product portfolios, such kinds of problems have to be constantly solved. The use of surrogate (meta)-models $\hat{f}$ for approximating the expensive or time-consuming objective function $f : x \rightarrow y$ represents an established approach to this task. After determining the values of f for the points x of an initial design of experiments, the surrogate model $\hat{f}$ is computed and then used for the further analysis and optimization. Here, we consider deterministic, i.e., noise-free minimization problems. In such a scenario, the former approach has a conceptual drawback. The location of

G. Nicosia and P. Pardalos (Eds.): LION 7, LNCS 7997, pp. 110–124, 2013.
DOI: 10.1007/978-3-642-44973-4_13, © Springer-Verlag Berlin Heidelberg 2013

the optimum can only roughly be determined based on the initial design. A high accuracy of the optimization on the model does not necessarily provide improved quality with respect to the original objective function. As a consequence, the resources expended for the usually uniform coverage of the experimental region for the approximation of the global response surface may be spent more efficiently in order to increase the accuracy of the surrogate in the regions of the actual optimum.

A solution to this problem is provided by sequential techniques, called efficient global optimization (EGO) [16], sequential parameter optimization [1] and sequential designs [24] within the different disciplines. Sequential techniques do not focus on an approximation of a global response surface, but on an efficient way to obtain the global minimum of the objective function f. After evaluating a sparse initial design in the parameter space, much smaller than the actual experimental budget, the surrogate model is fitted and proposes a new point which is then evaluated on the original function f. The point is added to the design and the procedure is repeated until the desired objective value has been obtained or the experimental budget has been depleted.

For the design of a sequential technique, the choice of the surrogate model is a crucial decision. Whereas resampling techniques [3] can be used to estimate the global prediction quality in the classical approach, the optimization capabilities of a model have to be assessed in the sequential approach. Therefore, this capability is not necessarily static. Some models may be suited to efficiently identify the most promising basin in the beginning, whereas others are good for refining the approximation in the final stage of the optimization.

In this paper, we tackle the model selection problem for sequential optimization techniques. Hence, the proposed optimization algorithm utilizes a heterogeneous ensemble of surrogate models. An approach to solve the progressive model selection problem is proposed as a central scientific contribution. It is designed to identify models that are most promising at a certain stage of the optimization. Preference values are used to stochastically select a surrogate model, which in turn proposes a new design point in each iteration of the algorithm. Based on the quality of this design point, the rating of the model is adjusted by means of reinforcement learning techniques. The procedure is general and can be performed with arbitrary ensembles of surrogate models.

In the following, an overview of related research is provided by means of a brief review of the literature in Sect. 2. Details of the applied methods are presented in Sect. 3 and the actual PROGRESS algorithm is described in Sect. 4 based on these foundations. In Sect. 6, its results on the benchmark specified in Sect. 5 are presented and discussed. The main conclusions are summarized in Sect. 7 and an outlook for further research in this area is introduced.

2   Review

In the following review of the literature we mainly restrict the focus to sequential optimization techniques using ensembles of surrogate models in which the selection or combination of the models for the internal optimization is dynamically adjusted. A more general survey of sequential optimization algorithms has been recently provided by Shan et al. [31]. For the special class of kriging-based optimization algorithms exploiting the uncertainty estimation for a trade-off between local and global search, we refer to the taxonomy of Jones [15].

Ensemble-based sequential optimization procedures can be classified in three basic concepts:

1. All surrogate models are individually optimized and (subsets of) the design points are evaluated on the original objective f.
2. The predictions of all surrogate models are aggregated and their combined value is used for selecting the next design point.
3. A single surrogate model is used in each iteration. The selection of the model is based on dynamically adjusted scores or probabilities.

The first concept is particularly designed for applications in which speed-ups from parallelization can be expected. For instance, Viana [35] applied four different surrogate models in parallel to optimize engineering design problems. It was shown that the resulting procedure is often superior to a sequential optimization only relying on kriging. An application to surrogate-assisted evolutionary algorithms was reported by Lim et al. [20]. In their approach, the best solutions of each surrogate model are evaluated on the original function.

The second concept represents a general procedure combining predictions using an ensemble of surrogate models. Here, individual predictions are often aggregated via a weighted average. One major distinction of the techniques in this concept class is whether the technique can only be applied to surrogate models of the same basic type or whether a completely heterogeneous ensemble set is possible. The latter case is obviously preferable because of its increased generality. An early example for the former approach was presented by Hansen et al. [12] and relied upon combinations of different types of neural networks. In a similar manner, Ginsbourger et al. [9] proposed ensembles of kriging models based on different correlation functions.

An early approach for calculating weights for aggregating individual model predictions from a heterogeneous set was based on so-called Bayes factors [17], which from a Bayesian viewpoint denote the conditional probability of a surrogate being the true model given the current training data. More heuristic approaches for weight calculation based on cross-validated model accuracy were proposed and analyzed by Goel et al. [10]. The same means were also applied to select a subset of the ensemble within a tournament selection [37]. For each fold of a cross-validation, a representative surrogate was determined. The mean over the predictions of the selected models was used to guide the sequential optimization. An approach focusing on the combination and tuning of different surrogates was presented by Gorissen et al. [11]. They aggregated the active models of the ensemble by using the mean over the predictions. Since the key aspect of their work was the design of an evolutionary approach for tuning and selecting the models, the evaluation mainly focuses on the prediction quality of the resulting approach after the tuning. A comparison with other ensemble-based sequential techniques was not performed.

Algorithm 1: Model-based optimization

---

1 Let f be the black-box function that should be optimized;
2 Generate initial design set $\{x_1, \ldots, x_n\}$;
3 Evaluate f on design points: $y_i = f(x_i)$;
4 Let D=$\{(x_1, y_1), \ldots, (x_n, y_n)\}$;
5 **while** *stopping criterion not met* **do**
6 | Build surrogate model $\hat{f}$ based on D;
7 | Get new design point x^* by optimizing the infill criterion on $\hat{f}$;
8 | Evaluate new point $y^* = f(x^*)$;
9 | Extend design $D \leftarrow D \cup \{(x^*, y^*)\}$;

Currently, only a few related approaches exist for the third concept of ensemble techniques. Its main advantage is that it constitutes a very general approach which also allows many heterogeneous models to be integrated into the ensemble since only one model has to be fitted and optimized per iteration. Thus, the selected model can be subject to a more time-consuming tuning to specifically adapt it to the objective function. Friese et al. [8] applied and compared different strategies to assess their suitability for sequential parameter optimization, among them also ensemble-based methods using reinforcement learning. However, these methods were used in a rather out-of-the-box manner, without specifically adapting the generic reinforcement learning techniques to the problem at hand to exploit their full potential. Some of the potential problems, as well as enhancements to overcome them, will be discussed in this paper. Another variant of the approach was applied in the context of operator selection in evolutionary algorithms by Da Costa et al. [5].

3   Methods

In this section, the methodological foundations of our algorithm are introduced. First, the general concept of a model-based optimization (MBO) algorithm is described. Then the multi-armed bandit problem from reinforcement learning, which is later transferred to the progressive model selection problem in MBO, is presented.

3.1   Model-Based Optimization

Response surface models are a common approach in cases where the budget of evaluations available for the original function is severely limited. As this surface can be explored with a much higher amount of evaluations, the optimum of the so-called infill criterion can be accurately approximated using standard optimization methods. This generic MBO approach is summarized in Algorithm 1. The stages of proposing new points and updating the model are alternated in a sequential fashion.

In the following, the steps of the generic MBO algorithm are discussed and some details are provided.

1. For the initial design, many experimental design types are possible, but for nonlinear regression models usually space-filling designs like Latin hypercube sampling (LHS) are used, see [4] for an overview. Another important choice is the size of the initial design. Rules of thumb are usually somewhere between $4d$ and $10d$, the latter being recommended by [16].

2. As surrogate model, kriging was proposed in the seminal EGO paper [16] because it is especially suited for nonlinear, multimodal functions and allows local refinements to be performed, but basically any surrogate model is possible. As presented in Sect. 2, also more sophisticated approaches using ensemble methods have been applied within MBO algorithms.

3. The infill criterion is optimized in order to find the next design point for evaluation. It measures how promising the evaluation of a point x is according to the surrogate model. One obvious choice is the direct use of $\hat{f}(x)$. For kriging models, the expected improvement (EI) [23] is commonly used. It factors in the local model uncertainty in order to guarantee a reasonable trade-off between the exploration of the decision space and the exploitation of the already obtained information. These and other infill criteria have been proposed and assessed in several studies [15, 25, 30, 36].

4. As a stopping criterion, a fixed budget for function evaluations, the attainment of a specified y-level, or a combination of both is often used in practice.

3.2   Reinforcement Learning

The model selection problem in MBO can be considered as a "multi-armed-bandit" reinforcement learning problem. Here, in each iteration t an action a_t has to be chosen from a given set of finite choices $\mathcal{A} = \{v_1 \ldots v_m\}$ and we could envision those choices to be arms of a casino slot machine. In MBO, this choice corresponds to selecting a regression model which in turn is used to propose the next design point. Depending on the action a_t, we will receive a reward r_t according to an unknown probability distribution, and our aim is to maximize summed rewards over time.

After we have obtained some information regarding the pay-offs from the different slot machines, we face the fundamental exploration-exploitation dilemma in reinforcement learning: Should we try to gather more information regarding the expected pay-off of the actions or should we greedily select the action which currently seems most promising? The problem becomes even more difficult if we assume nonstationary rewards, i.e., reward distributions that change over time. In this scenario, we always have to allocate a significant proportion of selections for exploring the current situation.

Sutton and Barto [32] suggest several ways for balancing exploration and exploitation. One is a probability matching strategy called reinforcement comparison, where the actions are selected stochastically according to a vector $q_t \in \mathbb{R}^m$ of preference values. The main idea is that a high reward should strongly increase the preference value/selection probability of an action, whereas a low reward should decrease it.

Let us assume we are in iteration t and already have a preference vector $\boldsymbol{q}_t$. These preferences are then transformed to selection probabilities via a softmax function

$$\pi_{t,j} = \exp(q_{t,j}) \left(\sum_{k=1}^{m} \exp(q_{t,k}) \right)^{-1} , j \in \{1, \ldots, m\} . \qquad (1)$$

Based on these probabilities, we select action a_t and receive its stochastic reward r_t. Assuming we are in a general nonstationary scenario, we now have to decide whether r_t is favourable or not. For this, we compare it with a reference reward $\bar{r}_t$, which encodes the expected, average pay-off across all actions at iteration t. Assuming we already have such an $\bar{r}_t$, the element of the preference vector $\boldsymbol{q}_t$ for the chosen action $a_t = v_j \in \mathcal{A}$ is now updated, while the preferences for all other actions stay the same:

$$q_{t+1,k} = \begin{cases} q_{t,k} + \beta[r_t - \bar{r}_t], & \text{if } k = j \\ q_{t,k} & \text{else.} \end{cases} \qquad (2)$$

Here, the strategy parameter β encodes the strength of the adjustment, i.e., the desired trade-off between exploitation (high β) and exploration (low β).

Finally, we update the reference reward $\bar{r}_t$ via the following exponential smoothing formula

$$\bar{r}_{t+1} = \bar{r}_t + \alpha(r_t - \bar{r}_t) . \qquad (3)$$

The strategy parameter $\alpha \in (0, 1]$ determines how much influence the current reward has on the reference reward, i.e, how much we shift the reference reward towards r_t. It thus reflects the assumed degree of nonstationarity in the reward distributions.

4   Algorithm

In this section, we address how the action selection by means of the reinforcement comparison can be exploited for model selection in MBO. Regarding models as selectable actions seems straightforward, but apart from that many technical details of the basic method in Sect. 3.2 have to be clarified or adapted. The reward will be based on the improvement in objective value obtained by the proposed x^*. The sum of rewards over time then measures the progress made during optimization. The main idea is that models which generated larger improvements in the past should be preferred in the future.

Instead of using an expected improvement criterion, we directly optimize the response surface of the selected model, i.e., no local uncertainty estimation is used. Although this carries the risk of getting stuck in a local optimum for one model, it offers two important advantages: (a) It is possible to optimize this criterion for arbitrary regression models, and (b) by using a heterogeneous

---

Algorithm 2: PROGRESS

---

1 Let f be the black-box function that should be optimized;
2 Let $E = \{h_1, \ldots, h_m\}$ be the regression models in the ensemble;
3 Generate initial design $\{x_1, \ldots, x_n\}$;
4 Evaluate f on design $\forall i \in \{1, \ldots, n\} : y_i = f(x_i)$;
5 Let D=$\{(x_1, y_1), \ldots, (x_n, y_n)\}$;
6 **for** $j \in \{1, \ldots, m\}$ **do**
7   Build surrogate model $\hat{h}_j$ based on D;
8   Select next promising point $x^*(\hat{h}_j)$ by optimizing $\hat{h}_j$;
9   Evaluate new point $y^*(\hat{h}_j) = f(x^*(\hat{h}_j))$;
10   Extend design set $D \leftarrow D \cup \{(x^*(\hat{h}_j), y^*(\hat{h}_j))\}$;
11 Calculate vector of initial rewards $r_0 \in \mathbb{R}^m$ using equation (5);
12 Initialize preference vector $q_1 = r_0$ and reference reward $\bar{r}_1 = median(r_0)$;
13 Let $t = 1$;
14 **while** *stopping rule not met* **do**
15   Calculate model selection probabilities π_t from q_t using equation (1);
16   Sample model h_j according to π_t;
17   Build surrogate model $\hat{h}_j$ based on D;
18   Select next promising point $x^*(\hat{h}_j)$ by optimizing $\hat{h}_j$;
19   Evaluate new point $y^*(\hat{h}_j) = f(x^*(\hat{h}_j))$;
20   Extend design set $D \leftarrow D \cup \{(x^*(\hat{h}_j), y^*(\hat{h}_j))\}$;
21   Calculate reward r_t using equation (4);
22   Update preferences using equation (2) to obtain q_{t+1} ;
23   Update reference reward using equation (3) to obtain $\bar{r}_{t+1}$;
24   $t \leftarrow t+1$

---

ensemble the models are likely to focus on different basins. The exploration thus emerges from a successful adaptation of the model selection.

The complete procedure of the PROGressive REinforcement-learning-based Surrogate Selection (PROGRESS) is shown in Algorithm 2. It is an enhanced instance of the generic Algorithm 1, whereby the methodological contributions are: the initialization phase (lines 6 to 13), the stochastic model selection (lines 16 to 17) and the preference vector and reference reward updates (lines 22 to 24). Details are provided in the following.

4.1   Rewards

Let y_{min} be the minimum response value of the design before the integration of $x^*(\hat{h}_j)$. The reward of a chosen surrogate model h_j is then given by

$$r_t(\hat{h}_j) = \phi(y_{min} - f(x^*(\hat{h}_j))) , \tag{4}$$

where $x^*(\hat{h}_j)$ is the next point proposed by model $\hat{h}_j$ and ϕ is a simple linear rescaling function which will be detailed in the next section. Thereby, it is possible that a model produces negative rewards. We intentionally use all available information, also of iterations in which no improvements in the optimization of f are achieved, as this happened a considerable amount of times in our experiments. Surrogate models that poorly approximate interesting regions of the objective function are directly downgraded. Clipping rewards at zero would discard this information.

4.2 Initialization Phase and Reward Rescaling

Until now we have not defined how preference values and the reference reward are initialized before the first iteration $t = 1$. When using the generic reinforcement learning approach for MBO, we have to keep two peculiarities in mind: First, there is a high potential for large rewards in the first sequential step, as no optimization of the objective has been performed so far – this fact could cause a strong overrating of the first selected model. Second, it is hard to decide on a reasonable initial value for the reference reward – in particular, if independence with respect to the scaling of the objective function is desired.

In the initialization phase all models of the ensemble $E = \{h_1, \ldots, h_m\}$ are fitted to the initial design and we obtain all $\hat{h}_j$. Each one proposes a point $x^*(\hat{h}_j)$, which in turn is evaluated. We now define the vector initial rewards ρ by calculating the initial improvements of all models w.r.t. y_{min}, the best objective value we have observed on the initial design, and scale them to $[0, 1]$ by applying the linear transformation ϕ to obtain our initial rewards r_0:

$$\rho = (y_{min} - f(x^*(\hat{h}_1)), \ldots, y_{min} - f(x^*(\hat{h}_m)))^T$$
$$r_0 = (\phi(\rho_1), \ldots, \phi(\rho_m))^T, \quad \phi(x) = \frac{x - \min(\rho)}{\max(\rho) - \min(\rho)}. \tag{5}$$

This transformation ϕ (always with the initial values of ρ) is also applied to all upcoming rewards. The initial reference reward $\bar{r}_1$ is defined to be the median of the transformed rewards as a robust representative. The initial vector of probabilities π_1 is now simply the softmax transformation of q_1.

4.3 Sequential Update

In the sequential part of PROGRESS, a surrogate model is stochastically chosen according to the current probability vector π_t, fitted to the current design and proposes a new point $x^*(\hat{h}_j)$ by optimization of its response surface. After its reward has been determined based on Eq. 4, the original formulas of Sect. 3 can be used for updating the preferences and the reference reward. The algorithm usually stops when a given budget of function evaluations is exhausted.

5   Experimental Setup

PROGRESS and all experiments in this article have been implemented in the statistical programming language R [26]. We analyzed the performance of our

Table 1. Overview of the considered test functions and the problem features covered.

Separable	Low or moderate cond.	High cond. and unimodal	Adequate glob. struct.	Weak glob. struct.
1 Sphere	6 Attractive sector	10 Ellipsoidal function	15 Rastrigin	20 Schwefel
2 Ellipsoidal	7 Step ellipsoidal	11 Discus function	16 Weierstrass	21 Gallagher's Gaussian
3 Rastrigin	8 Rosenbrock	12 Bent cigar	17 Schaffers F7	(101-me Peaks)
4 Bueche-	(original)	13 Sharp ridge	18 Schaffers F7 (ill)	22 Gallagher's Gaussian
Rastrigin	9 Rosenbrock	14 Different powers	19 Composite Griewank	(21-hi Peaks)
5 Linear slope		(rotated)	Rosenbrock	23 Katsuura
				24 Lunacek bi-Rastrigin

algorithm on the 24 test functions of the BBOB noise-free test suite [13], which is a common benchmarking set for black-box optimization. It covers a variety of functions that differ w.r.t. problem features like separability, multi-modality, ill-conditioning and existence of global structure. A summary of these functions and their respective properties is provided in Table 1. Their function definitions, box constraints and global minima were taken from the soobench R package [21]. The dimension of the scalable test functions was set to $d = 5$.

The regression models used in our PROGRESS ensemble and their respective R packages are listed in Table 2: A second order (polynomial) response surface model (RSM) [14], a kriging model with power exponential covariance kernel [29], multivariate adaptive regression splines (MARS) [6], a feedforward neural network [14] with one hidden layer, a random forest [14], a gradient boosting machine (GBM) [7] and a regression tree (CART) [14]. Table 2 also lists (constant) parameter settings of the regression models which deviate from default values and box constraints for parameters which were tuned prior to model fitting in every iteration of PROGRESS based on all currently observed design points (lines 7 and 17 in Algorithm 2). To accomplish this, we used a 10-fold repeated cross-validation (5 repetitions) to measure the median absolute prediction error and minimized this criterion in hyperparameter space by CMAES with a low number of iterations. Integer parameters were determined by a rounding strategy and CMAES was always started at the point in hyperparameter space which was discovered as optimal during the previous tuning run of the same model (or a random point if no such point is available).

The response surfaces of the regression models were also optimized by CMAES with 100 iterations and $\lambda = 10d$ offspring. This setting deviates from the literature recommendation for technical reasons as more parallel model evaluations (predictions) reduce the computational overhead and lead to a better global search quality with the same number of iterations. We performed 10 random CMAES restarts and one additional restart at the currently best point of the observed design points for further exploration of the search space and to reduce the risk of getting stuck in a local optimum.

The learning parameters α and β of PROGRESS were manually tuned prior to the experiments. In order to not bias the results by overtuning on the problems, these parameters were fixed to $\alpha = 0.1$ and $\beta = 0.25$ for all of the test instances. Thereby, our aim was to find a global parametrization leading to robust results and a successful adaptation of the selection probabilities.

Table 2. Surrogate models, R packages, parameters and tuning ranges.

Model	R package	Parameter	Value/Range	Model	R package	Parameter	Value/Range
RSM	rsm [18]	-	-	RF	randomForest [19]	ntree	$\{50, \ldots, 500\}$
NNET	nnet [34]	size	5			mtry	$\{1, \ldots, 5\}$
		decay	$[0, 0.5]$	MARS	earth [22]	degree	$\{1, 2, 3\}$
GBM	gbm [27]	interaction.depth	$\{4, \ldots, 8\}$			penalty	$\{2, 3, 4\}$
		shrinkage	$[0.025, 0.05]$			nprune	$\{1, \ldots, 10\}$
		bag.fraction	$[0.5, 0.8]$	CART	rpart [33]	-	-
		n.minobsinnode	1	KM	DiceKriging [28]	nugget.estim	TRUE

We compared PROGRESS with an established global optimizer for expensive black-box functions, namely EGO based on a kriging with a power exponential kernel (implementation taken from the R package DiceOptim [28]; default parameters except for kernel). We also ran a random LHS to provide a baseline result. For PROGRESS we considered two variants, one with the above mentioned hyperparameter tuning and one without, where default settings for all regression models were set. All algorithms were allowed a budget of 200 function evaluations, from which 50 were allocated to an initial maximin LHS design. This is in accordance with the "$10 \cdot d$ rule" proposed in [16]. This initial design is shared by all MBO algorithms in our experiments within one replication to reduce variance in comparisons. All algorithm runs were replicated 20 times. To parallelize our experiments the BatchExperiments R package [2] was used.

6   Results

In the following results, we report the distance of the best visited design point to the global optimum in objective space as measure of algorithm performance. Because the optimization of the black-box function at hand is our major interest, we do not report global prediction quality indicators of the internal models. Neither do we calculate expected run times to obtain a certain target level, as we assume an a priori fixed budget of function evaluations. The performance distributions of the 20 runs of each algorithm on the 24 test functions of the benchmark are shown in Fig. 1 on a logarithmic scale.

The most obvious result is the superiority of the sequential designs over the static random LHS. In all cases, except for functions 12 and 23, at least one of the three sequential algorithms outperforms the static approach. Test functions 12 and 23 can apparently not be approximated well enough by any of the considered regression models to provide helpful guidance for the optimization process. This might be due to the high condition number (function 12) or the absence of global structure and an extreme number ($> 10^5$) of local optima (function 23). The comparison of the two versions of PROGRESS with and without hyperparameter tuning shows that both variants obtain similar results in most instances. There are cases (e.g., function 15, 17, 19), however, where hyperparameter tuning leads to a significantly better outcome.

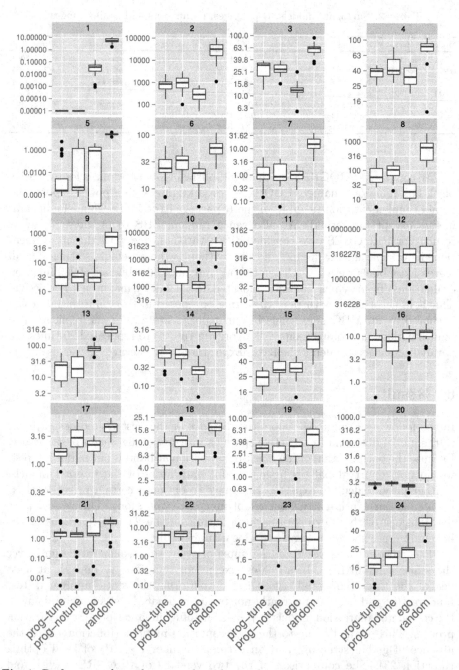

Fig. 1. Performance box plots per algorithm and test function. Displayed is the difference in objective space between best design point during optimization and global minimum on a log10 scale. PROGRESS is run in two variants, one with hyperparameter tuning of the selected surrogate model in each iteration and one without.

The EGO algorithm and the use of kriging models represent the state-of-the-art in sequential designs. If PROGRESS can in general compete with EGO, it can be considered as successful, as it manages to detect suitable surrogate models within the strictly limited budget. If we assign equal importance to all test functions in the benchmark, none of the two algorithms clearly dominates the other. Whereas PROGRESS outperforms EGO on the functions 1, 5, 13, 15, 16, 17, 21, and 24, the opposite holds on the functions 2, 3, 6, 8, 10, 14 and 20. On the remaining test cases, both algorithms show no significant difference in performance. Hence, PROGRESS is competitive with EGO and is the preferable choice on about one third of the benchmark set. For this reason, the proposed procedure can be considered as successful.

To obtain a more detailed understanding of the principles behind the surrogate selection, we analyzed the progression of the selection probabilities in three exemplary runs. These are shown in Fig. 2. We also display the total number of selections per model during the whole run.

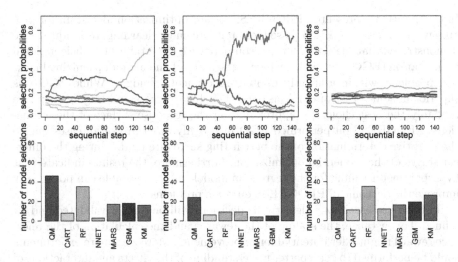

Fig. 2. Progression of selection probabilities and number of selections in PROGRESS (with tuning). From left to right: Lunacek bi-Rastrigin (24), Gallagher's Gaussian 21-hi Peaks (22) and Rastrigin (3).

In the first case shown in the left plot, a dynamic shift between a model capturing global trends (QM) and a more detailed model for local optimization (RF) is accomplished during the optimization. While the former efficiently identifies the basins of the Lunacek bi-Rastrigin Function,[1] the latter succeeds in guiding the search through the rugged area around the optimum. This synergy allows PROGRESS to significantly outperform EGO on this function. The center plot

---

[1] The weak global structure of this function is rooted in the existence of two basins of almost the same size.

shows an experiment on Gallagher's Gaussian 21-hi Peaks function. As "kriging" is merely a different term for "Gaussian process", it is unsurprising that this model is the best choice to approximate the 21 local optima of this test function. PROGRESS adapts its preference values after only a few iterations and learns to select the kriging model with high probability and therefore scores comparable to EGO. In the remaining plot on the right-hand side, the selection probabilities are shown for a test case where PROGRESS showed inferior results compared to EGO. Here, PROGRESS is not able to learn a superior model, but can only downgrade the apparently inappropriate models NNET and CART. A possible explanation for this problem might be the lack of balance between improvements and deteriorations on this function. For functions on which PROGRESS is inferior to EGO, such a problem can often be observed. We therefore consider this as one of the main starting points for future improvements.

7 Conclusions and Outlook

In this paper, we presented PROGRESS, a new optimization algorithm for progressive surrogate selection based on reinforcement learning techniques. We demonstrated that the algorithm can compete with the established efficient global optimization (EGO) algorithm, which is the state-of-the-art for optimizing blackbox problems within a strictly limited experimental budget. While EGO was superior in some cases of our considered benchmark cases and kriging therefore probably the best choice for approximating the response surface of these functions, PROGRESS outperformed EGO in roughly one third of all cases due to the adaptive determination of the best fitting surrogate model during the different stages of the sequential optimization. Furthermore, the results indicate that hyperparameter tuning for the regression models in the ensemble can potentially improve the outcome of PROGRESS on some problems.

In future research, our algorithmic decisions will be further validated and/or refined. For instance, the scaling of the rewards and the calculation of the reference reward might offer potential for improvements. Moreover, more experiments could be performed to get a better understanding of the effects and interactions of the algorithm parameters. Alternatively, other reinforcement learning techniques can be implemented and benchmarked with the reinforcement comparison with respect to their suitability for an adaptive model selection. Finally, in order to enhance the time-efficiency of PROGRESS, the necessity of a hyperparameter tuning in every sequential loop will be further examined.

Acknowledgements. This paper is based on investigations of the project D5 of the Collaborative Research Center SFB/TR TRR 30 and of the project C2 of the Collaborative Research Center SFB 823, which are kindly supported by the Deutsche Forschungsgemeinschaft (DFG).

References

1. Bartz-Beielstein, T., Lasarczyk, C.G., Preuss, M.: Sequential parameter optimization. In: McKay, B., et al. (eds.) Proceedings of the 2005 Congress on Evolutionary Computation (CEC'05), Edinburgh, Scotland, pp. 773–780. IEEE Press, Los Alamitos (2005)
2. Bischl, B., Lang, M., Mersmann, O., Rahnenfuehrer, J., Weihs, C.: BatchJobs and BatchExperiments: abstraction mechanisms for using R in batch environments. Submitted to Journal of Statistical Software (2012a)
3. Bischl, B., Mersmann, O., Trautmann, H., Weihs, C.: Resampling methods for meta-model validation with recommendations for evolutionary computation. Evol. Comput. **20**(2), 249–275 (2012b)
4. Bursztyn, D., Steinberg, D.M.: Comparison of designs for computer experiments. J. Stat. Planning Infer. **136**(3), 1103–1119 (2006)
5. DaCosta, L., Fialho, A., Schoenauer, M., Sebag, M.: Adaptive operator selection with dynamic multi-armed bandits. In: Proceedings of the 10th Conference Genetic and Evolutionary Computation (GECCO '08), pp. 913–920. ACM, New York (2008)
6. Friedman, J.: Multivariate adaptive regression splines. Ann. Stat. **19**(1), 1–67 (1991)
7. Friedman, J.: Greedy function approximation: a gradient boosting machine. Ann. Stat. **29**(5), 1189–1232 (2001)
8. Friese, M., Zaefferer, M., Bartz-Beielstein, T., Flasch, O., Koch, P., Konen, W., Naujoks, B.: Ensemble based optimization and tuning algorithms. In: Hoffmann, F., Hüllermeier, E. (eds.) Proceedings of the 21. Workshop Computational Intelligence, pp. 119–134 (2011)
9. Ginsbourger, D., Helbert, C., Carraro, L.: Discrete mixtures of kernels for kriging-based optimization. Qual. Reliab. Eng. Int. **24**(6), 681–691 (2008)
10. Goel, T., Haftka, R.T., Shyy, W., Queipo, N.V.: Ensemble of surrogates. Struct. Multidisc. Optim. **33**(3), 199–216 (2007)
11. Gorissen, D., Dhaene, T., Turck, F.: Evolutionary model type selection for global surrogate modeling. J. Mach. Learn. Res. **10**, 2039–2078 (2009)
12. Hansen, L., Salamon, P.: Neural network ensembles. IEEE Trans. Pattern Anal. Mach. Intell. **12**(10), 993–1001 (1990)
13. Hansen, N., Finck, S., Ros, R., Auger, A.: Real-Parameter Black-Box Optimization Benchmarking 2009: Noiseless Functions Definitions. Tech. Rep. RR-6829, INRIA (2009). http://hal.inria.fr/inria-00362633/en/
14. Hastie, T., Tibshirani, R., Friedman, J.: The Elements of Statistical Learning: Data Mining, Inference, and Prediction. Springer Series in Statistics. Springer, New York (2009)
15. Jones, D.R.: A taxonomy of global optimization methods based on response surfaces. J. Global Optim. **21**(4), 345–383 (2001)
16. Jones, D., Schonlau, M., Welch, W.: Efficient global optimization of expensive black-box functions. J. Global Optim. **13**(4), 455–492 (1998)
17. Kass, R.E., Raftery, A.E.: Bayes factors. J. Am. Stat. Assoc. **90**(430), 773–795 (1995)
18. Lenth, R.V.: Response-surface methods in R, using rsm. J. Stat. Softw. **32**(7), 1–17 (2009)
19. Liaw, A., Wiener, M.: Classification and regression by randomForest. R News **2**(3), 18–22 (2002)

20. Lim, D., Ong, Y.S., Jin, Y., Sendhoff, B.: A study on metamodeling techniques, ensembles, and multi-surrogates in evolutionary computation. In: Thierens, D., et al. (eds.) Proceedings of the 9th Annual Genetic and Evolutionary Computation Conference (GECCO 2007), pp. 1288–1295. ACM, New York (2007)
21. Mersmann, O., Bischl, B.: soobench: Single Objective Optimization Benchmark Functions (2012). http://CRAN.R-project.org/package=soobench, R package version 1.0-73
22. Milborrow, S.: earth: Multivariate Adaptive Regression Spline Models (2012). http://CRAN.R-project.org/package=earth, R package version 3.2-3
23. Mockus, J.B., Tiesis, V., Zilinskas, A.: The application of bayesian methods for seeking the extremum. In: Dixon, L.C.W., Szegö, G.P. (eds.) Towards Global Optimization 2, pp. 117–129. Elsevier North-Holland, New York (1978)
24. Myers, R.H., Montgomery, D.C., Anderson-Cook, C.M.: Response Surface Methodology, 3rd edn. Wiley, Hoboken (2009)
25. Picheny, V., Wagner, T., Ginsbourger, D.: A benchmark of kriging-based infill criteria for noisy optimization. Struct. Multidisc. Optim. 48(3), 607–626 (2013)
26. R Core Team: R: A Language and Environment for Statistical Computing. R Foundation for Statistical Computing, Vienna (2012). http://www.R-project.org/ ISBN 3-900051-07-0
27. Ridgeway, G.: gbm: Generalized Boosted Regression Models (2012). http://CRAN. R-project.org/package=gbm, R package version 1.6-3.2
28. Roustant, O., Ginsbourger, D., Deville, Y.: DiceKriging, DiceOptim: two R packages for the analysis of computer experiments by kriging-based metamodeling and optimization. J. Stat. Softw. 51(1), 1–55 (2012). http://www.jstatsoft.org/v51/i01/
29. Santner, T., Williams, B., Notz, W.: The Sesign and Analysis of Computer Experiments. Springer, New York (2003)
30. Sasena, M.J., Papalambros, P., Goovaerts, P.: Exploration of metamodeling sampling criteria for constrained global optimization. Eng. Optim. 34(3), 263–278 (2002)
31. Shan, S., Wang, G.G.: Survey of modeling and optimization strategies to solve high-dimensional design problems with computationally-expensive black-box functions. Struct. Multi. Optim. 41(2), 219–241 (2010)
32. Sutton, R., Barto, A.: Reinforcement Learning: An Introduction. Cambridge University Press, Cambridge (1998)
33. Therneau, T.M., port by Brian Ripley, B.A.R.: rpart: Recursive Partitioning (2012). http://CRAN.R-project.org/package=rpart, R package version 3.1-54
34. Venables, W.N., Ripley, B.D.: Modern Applied Statistics with S, 4th edn. Springer, New York (2002)
35. Viana, F.A.C.: Multiple Surrogates for Prediction and Optimization. Ph.D. thesis, University of Florida (2011)
36. Wagner, T., Emmerich, M., Deutz, A., Ponweiser, W.: On expected-improvement criteria for model-based multi-objective optimization. In: Schaefer, R., Cotta, C., Kołodziej, J., Rudolph, G. (eds.) PPSN XI. LNCS, vol. 6238, pp. 718–727. Springer, Heidelberg (2010)
37. Wichard, J.D.: Model selection in an ensemble framework. In: International Joint Conference on Neural Networks, pp. 2187–2192 (2006)

Neutrality in the Graph Coloring Problem

Marie-Eléonore Marmion[1,2(✉)], Aymeric Blot[2,3], Laetitia Jourdan[2],
and Clarisse Dhaenens[2]

[1] Université Libre de Bruxelles, Brussels, Belgium
`mmarmion@ulb.ac.be`
[2] Université Lille 1 - INRIA Lille Nord-Europe, Lille, France
`blot.aymeric@gmail.com`, {`laetitia.jourdan,clarisse.dhaenens`}`@lifl.fr`
[3] ENS Cachan/Bretagne, Université Rennes 1, Bruz, France

Abstract. In this paper, the neutrality of some hard instances of the graph coloring problem (GCP) is quantified. This neutrality property has to be detected as it impacts the search process. Indeed, local optima may belong to plateaus that represent a barrier for local search methods. Then, we also aim at pointing out the interest of exploiting neutrality during the search. Therefore, a generic local search dedicated to neutral problems, NILS, is performed on several hard instances.

Keywords: Graph coloring problem · Fitness landscape · Neutrality

1 Motivation

The graph coloring problem (GCP) consists in finding the minimal number of colors χ, called the chromatic number, that leads to a legal coloring of a graph. This is a $\mathcal{NP}$−hard problem [3] widely studied in the literature. Then, for large instances, approximate algorithms are used as local search methods [2] or evolutionary strategies [8]. The most efficient metaheuristic schemes include specific encodings and mechanisms for GCP. It requires a very good knowledge of the problem and a long time of experimental analysis to tune the best parameters.

Another way to design efficient algorithms is to analyze the problem structure. For example, the landscape analysis aims at understanding better the characteristics of the problems in order to design efficient algorithms [11]. Neutrality appears when neighboring solutions have the same fitness value. Thus, neutrality is a characteristic of the landscape [10].

Many insights about the neutrality of the GCP are raised when considering the number of edges with the same color at both ends. But, as far as we know, no deep analysis has ever been conducted in the literature. In this paper, we are interested in the χ-GCP. This problem aims at looking for a legal coloring with χ colors, while minimizing the number of conflicts. This paper analyses if the χ-GCP may be considered as a neutral problem and if the neutrality may be exploited to solve it. Therefore, Sect. 2 gives the results on the neutrality of hard GCP instances. Then, in Sect. 3, the benefit of exploiting the neutrality when

G. Nicosia and P. Pardalos (Eds.): LION 7, LNCS 7997, pp. 125–130, 2013.
DOI: 10.1007/978-3-642-44973-4_14, © Springer-Verlag Berlin Heidelberg 2013

solving the GCP is studied. Section 4 discusses the presented work and future research interests.

2   Characterizing the Neutrality

2.1   Definitions

A *neutral neighbor* of a solution is a neighboring solution having the same fitness value. The *neutral degree* of a given solution is the number of neutral solutions in its neighborhood. A fitness landscape is said to be *neutral* if there are "many" solutions with a high neutral degree. A *neutral fitness landscape* can be pictured by a landscape with many plateaus. A *portal* is a solution in a plateau of a local optimum, having at least one neighbor with a better fitness value.

2.2   Approach

The average or the distribution of neutral degrees over the landscape may be used to qualify the level of neutrality of a problem instance. This measure plays an important role in the dynamics of local search algorithms [12,13]. In the case a problem gets the neutrality property, Marmion et al. [6] suggested to characterize the plateaus found from the local optima. Then, they sampled plateaus using neutral walks from each local optimum found. The plateaus are classified in a three-class topology: (T1) the local optimum is the single solution of the plateau, (T2) no neighbor with a better fitness value was met for any solutions of the plateau encountered along the neutral walk and; (T3) a portal has been identified on the plateau. This latest type is the most interesting since it reveals some solutions that help to escape from the plateau by improving. Indeed, the number of solutions visited before finding a portal during a neutral random walk is a good indicator of the probability to find an improving solution. Then, Marmion et al. proposes to compute the distribution of the number of solutions on the plateau that are needed to visit before finding a portal. Also, to study the cost/quality trade-off, they compare this distribution with the one of the number of solutions visited to find a (new) local optimum starting with a new solution, called the step length. This approach and the characterization of the neutrality of the χ-GCP are more detailed in [4].

2.3   Experiments

For this work we focus on literature instances of the GCP known to have "difficult upper bound", that is to say that a minimal legal coloring is hard to obtain. Those instances are extracted from the DIMACS Computational Challenge on "Graph Colouring and its Generalisations".[1] There are four classes of instances, according to the type of generation. All the indicators, presented above, are computed from 30 different solutions.

---

[1] http://dimacs.rutgers.edu/Challenges/

Table 1. Average neutral degree and the corresponding ratio for the random solutions and the local optima.

Instances	Data			Neutral Degree			
				Random		Local optima	
	V	χ	| nbh |	nd	ratio %	nd	ratio %
dsjc250.5	250	28	6750	858	12.7	83.8	1.2
dsjc500.1	500	12	5500	800	14.5	144	2.6
dsjc500.5	500	48	23500	2910	12.4	176	0.7
dsjc500.9	500	126	62500	9320	14.9	384	0.6
dsjc1000.1	1000	20	19000	2440	12.8	290	1.5
r250.5	250	65	16000	3470	24.2	1090	9.7
dsjr500.5	500	122	60500	12800	21.1	356	5.9
dsjr500.1c	500	84	41500	4780	11.5	121	0.3
dsjr1000.1c	1000	98	97000	8600	8.9	188	0.2
flat300_28_0	300	28	8100	1010	12.5	90.5	1.1
flat1000_50_0	1000	50	49000	4400	9.0	146	0.3
le450_25c	450	25	10800	1910	17.7	552	5.1
le450_25d	450	25	10800	1900	17.6	496	4.6

The ratio of the neutral degree is the neutral degree over the size of the neighborhood. This measure is also computed for the random solutions and the local optima as it makes the comparison between different instances easier. Table 1 gives the average neutral degree and the corresponding ratio for random solutions and local optima on the GCP instances. For each instance, the number of nodes V, the chromatic number χ and the size of the neighborhood \| nbh \| are also given. This table first shows that, the ratios for random solutions are quite high (up to 24.2 % for the instance r250.5). The neutrality characterises the problem in general. The landscape may have a lot of flat parts. The second observation on the results of Table 1 is that the ratios for local optima are smaller than the ones of random solutions. Hence, depending on the instances, the number of neutral neighbors in the neighborhood of a local optimum may be important or not. Some instances present a high neutral degree (such as 9.7 % for the highest, or around 5 % for others) while some others have a much smaller neutral degree (down to 0.2 % for the instance dsjr1000.1c). These results confirm the *apriori* fact that neutrality is a strong property in the graph coloring problem that may be used in local search strategies to be more efficient. In the following, only the instances where the average neutral degree of the local optima is higher than 1 % are considered.

Table 2 gives the statistics of the number of solutions visited on a plateau (nbS) before finding a portal. The statistics of the step lengths (L) are also given to study the cost/quality trade-off. Clearly, it is very quick to meet a portal even randomly. Indeed, it is necessary to visit only 1 or 2 new solution(s) on the plateau to find a portal to escape. Let us remark that reaching a portal may be quick, but the difficulty is to identify a solution as a portal. However, compared to the steps lengths, it seems to be more interesting to continue the

Table 2. It gives the number of T1, T2 and T3 plateaus. nbS stands for the number of visited solutions before finding a portal and, L for the step length.

Instances	Plateaus			nbS				L			
	T1	T2	T3	Min	Med	Mean	Max	Min	Med	Mean	Max
dsjc250.5	0	0	30	1	1	1.7	6	266	301	301	323
dsjc500.1	0	0	30	1	2	2	5	501	530	532	596
r250.5	0	1	29	1	2	3.3	17	125	148	148	164
flat300_28_0	0	0	30	1	1	1.7	6	344	388	385	406
le_450_25c	0	0	30	1	1	2.3	9	362	396	399	424
le_450_25d	0	0	30	1	2	2.5	11	370	399	400	428

search process by moving on a plateau to find a portal than to restart the search process from a new random solution.

The analysis of the plateaus of the local optima shows that portals, solutions of the plateau with at least one improving neighbor, are quick to reach with a random neutral walk. It assumes that exploiting neutrality in the search process may help to find better solutions. The following section provides insight about the way to exploit neutrality to solve the GCP.

3   Influence of Neutrality on Local Search Performance

3.1   NILS Algorithm

The Neutrality-based Iterated Local Search (NILS) is an algorithm designed to exploit the plateaus of the local optima [5]. It iterates a steepest descent and a perturbation step to escape when the search is blocked on a local optimum. In the perturbation step, from the local optimum, NILS performs neutral moves until finding a portal. If no portal has been found until a maximum number of neutral moves MNS, the solution is kicked. Thus, NILS is a generic local search that benefits from the neutrality of the problem. In the following, NILS is performed on the graph coloring problem in order to emphasize the benefit of exploiting the neutrality of this problem.

3.2   Experiments

Four instances, one of each type, have been selected to perform NILS: dsjc250.5, r250.5, flat_300_28_0 and le_450_25_c. The landscape analysis of theses instances has shown that the plateaus of the local optima get portals that lead to improving solutions. Several MNS values were tested in order to analyze the trade-off between exploiting the plateau and exploring an other part of the search space. MNS values were set to 0, 1, 2 and 5 times the size of the neighborhood. For the value 0, NILS corresponds to a classical ILS that restarts from a new part of the search space. 30 runs were performed for each configuration of NILS. The stopping criterion was set to 2×10^7 evaluations.

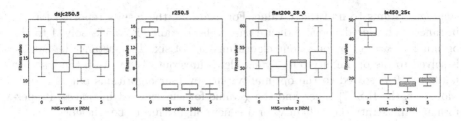

Fig. 1. ILS and NILS performance on 4 instances of the graph coloring problem.

Figure 1 presents the boxplot of the performance of the classical ILS ($MNS = 0$) and the three configurations of NILS ($MNS = \{1, 2, 5\} \times |Nbh|$). This figure shows first, that the performance of NILS are in average better than the ones of the classical ILS. For the instances r250.5 and le_450_25c, the neutral degree ratios were high (respectively 9.7 % and 5.1 %), and the results are very promising as they show a clear improvement over the standard ILS. For the other instances, the neutral degree ratios were lower (1.2 % for dscj250.5, and 1.1 % for flat300_28_0), and the results obtained by NILS are only a little better than the classical ILS. These results lead to the hypothesis that if the neutrality degree ratio of an instance is high, NILS will probably give good results. In other cases, it will not be attractive to use the neutrality, but however the results will not be worse. Thus, exploiting the neutrality in the search process can lead to a better efficiency and should not be discarded.

In these experiments, the performance of NILS is studied under different MNS values for a same total number of evaluations. Results show that for MNS values equal to 1 or 2 times the neighborhood size, performance is fairly similar. However, with a coefficient of 5, results are worse. That implies NILS can be stuck on plateaus on which searching portals is too expensive, and it may be preferable, in these cases, to escape the plateaus not to waste too much time.

4 Discussion

The experimental results have shown that the hard instances of GCP present neutrality where local search algorithms may be blocked on plateaus. Indeed, the classical ILS is not able to find interesting solutions. However, when the neutrality is exploited in the local search, results are improved even if no configuration of NILS gives a legal solution. This may be explained by the fact that these instances are the hardest instances of the literature, and for each, k is set to the χ-value, the best known chromatic number. In 2010, Porumbel et al. [8] made a comparison between their algorithm dedicated to GCP and the 10 best performing algorithms from the literature. Except the Iterated Local Search, all the other algorithms are well-sophisticated and specific to GCP. Indeed, GCP-specific mechanisms are used to improve the search. These mechanisms require a huge knowledge on the GCP to be designed and tuned efficiently. Despite this high level of sophistication, the comparison points out the difficulty for some algorithms to find the χ-value. For example, the results reported for the instances considered above indicate that: The instance r250.5 is solved to the optimality $(k = \chi)$ by only four algorithms out of six. The instance le_450_25c is solved to the optimality only by six algorithms out of ten. And, the instance flat300_28_0 is solved to the optimality only by four algorithms out of eleven. Moreover, the ILS [7] never find the χ-value for the two last instances. Its performance illustrate the difficulty for a generic algorithm to be efficient.

This paper should be considered as a preliminary work on the neutrality of the GCP. Indeed, one points out the neutrality of some hard instances and gives the degree of this neutrality. However, the performance of NILS are not as good as expected, but, it shows the potential of exploiting neutrality to solve the GCP. Since heuristic methods represent the state-of-the-art algorithms [1,9], one wants to investigate how to exploit neutrality in such heuristics.

References

1. Caramia, M., Dell'Olmo, P., Italiano, G.F.: Checkcol: improved local search for graph coloring. J. Discrete Algorithms **4**, 277–298 (2006)
2. Galinier, P., Hertz, A.: A survey of local search methods for graph coloring. Comput. Oper. Res. **33**(9), 2547–2562 (2006)
3. Garey, M.R., Johnson, D.S.: Computers and Intractability; A Guide to the Theory of NP-Completeness. W. H. Freeman and Co., San Francisco (1990)
4. Marmion, M.-E., Blot, A., Jourdan, L., Dhaenens, C.: Neutrality in the graph coloring problem. Technical Report RR-8215, INRIA (2013)
5. Marmion, M.-E., Dhaenens, C., Jourdan, L., Liefooghe, A., Verel, S.: NILS: a neutrality-based iterated local search and its application to flowshop scheduling. In: Merz, P., Hao, J.-K. (eds.) EvoCOP 2011. LNCS, vol. 6622, pp. 191–202. Springer, Heidelberg (2011)
6. Marmion, M.-E., Dhaenens, C., Jourdan, L., Liefooghe, A., Verel, S.: On the neutrality of flowshop scheduling fitness landscapes. In: Coello, C.A.C. (ed.) LION 2011. LNCS, vol. 6683, pp. 238–252. Springer, Heidelberg (2011)
7. Paquete, L., Stützle, T.: An experimental investigation of iterated local search for coloring graphs. In: Cagnoni, S., Gottlieb, J., Hart, E., Middendorf, M., Raidl, G. (eds.) EvoWorkshops 2002. LNCS, vol. 2279, pp. 122–131. Springer, Heidelberg (2002)
8. Porumbel, D.C., Hao, J.K., Kuntz, P.: An evolutionary approach with diversity guarantee and well-informed grouping recombination for graph coloring. Comput. Oper. Res. **37**(10), 1822–1832 (2010)
9. Porumbel, D.C., Hao, J.K., Kuntz, P.: A search space "cartography" for guiding graph coloring heuristics. Comput. Oper. Res. **37**(4), 769–778 (2010)
10. Reidys, C.M., Stadler, P.F.: Neutrality in fitness landscapes. Appl. Math. Comput. **117**, 321–350 (2001)
11. Stadler, P.F.: Landscapes and their correlation functions. J. Math. Chem. **20**, 1–45 (1996)
12. Verel, S., Collard, P., Tomassini, M., Vanneschi, L.: Fitness landscape of the cellular automata majority problem: view from the "Olympus". Theor. Comput. Sci. **378**, 54–77 (2007)
13. Wilke, C.O.: Adaptative evolution on neutral networks. Bull. Math. Biol. **63**, 715–730 (2001)

Kernel Multi Label Vector Optimization (kMLVO): A Unified Multi-Label Classification Formalism

Gilad Liberman[1], Tal Vider-Shalit[2], and Yoram Louzoun[2]([✉])

[1] Gonda Multidisciplinary Brain Research Center,
Bar Ilan University, Ramat Gan, Israel
[2] Department of Mathematics, Bar Ilan University, Ramat Gan, Israel
louzouy@math.biu.ac.il

Abstract. We here propose the kMLVO (kernel Multi-Label Vector Optimization) framework designed to handle the common case in binary classification problems, where the observations, at least in part, are not given as an explicit class label, but rather as several scores which relate to the binary classification. Rather than handling each of the scores and the labeling data as separate problems, the kMLVO framework seeks a classifier which will satisfy all the corresponding constraints simultaneously. The framework can naturally handle problems where each of the scores is related differently to the classifying problem, optimizing both the classification, the regressions and the transformations into the different scores. Results from simulations and a protein docking problem in immunology are discussed, and the suggested method is shown to outperform both the corresponding SVM and SVR.

1 Introduction

Classic supervised learning problems are formulated as a set of (x_i, y_i) pairs, where x_i lies in the problem domain (typically $\mathbb{R}^n$, but may be more complex), and $y_i \in \{0, 1, \ldots, n\}$ for classification problems (with $n = 1$ for decision problems) or $y_i \in \mathbb{R}$ for regression problems. Naturally, not all problems fall into these categories and several generalization have been suggested, where each instance belongs to more than one class or where multiple instances have multiple labels [1,2]. In this study, we keep the assumption that each instance either belongs to some target class or does not; however, the available data might not contain this labeling but rather some indirect measurements. This formulation is related to many real life problems. For example, in the medical domain, the decision whether a subject is ill or not is made not just based on past subjects' data along with their diagnoses, but also on past subjects' data along with their physiological condition scores, appetite and happiness scores, etc.

This notion can be especially helpful in areas where the classification is difficult to obtain, with limited data sets, or where the data suffers from high variation in measurement modality and protocol. The application of a first solution,

G. Nicosia and P. Pardalos (Eds.): LION 7, LNCS 7997, pp. 131–137, 2013.
DOI: 10.1007/978-3-642-44973-4_15, © Springer-Verlag Berlin Heidelberg 2013

the MLVO method, was briefly introduced in a recent publication [3]. Here, we present the extention of this fomalism to a kernel machine - the kMLVO, along with some useful extensions.

The remainder of this manuscript is organized as follows. In Sect. 2, we present the kMLVO framework. Extensions of the kMLVO are shown in Sect. 3. In Sect. 4 we discuss the results on simulated and experimental data sets and compare the performance of different classifiers, and we conclude with a summary in Sect. 5.

2   The kMLVO Framework

We use the SVR formalization for the regression part. This implies linear penalties on regression errors, which improves the robustness to outliers. When using a non-linear kernel, the direct relation to the original problem dimension is lost, and the contribution of w_0 must be indirect (we will return to this issue in the kMLVO extensions).

2.1   Formalism

We note by X the training set, which is now not restricted to be a subset of $\mathbb{R}^n$ but can be of any input space $\mathcal{X}$. Let each instance $x_i \in \mathcal{X}$ of the training set be associated with a class $y_i \in \{-1, 1, \phi\}$, where ϕ represents an uknown value (i.e. we do not know the classification for this point) and with L target values $s_{i,l} \in \mathbb{R} \bigcup \{\phi\}$. The classifier is a function $f : \mathcal{X} \rightarrow \{-1, 1\}$. In a manner analogous to the SVM and SVR primal formulation, the classifier f is a weight vector w in the problem space (or a vector space in which the prolem is transformed into using explicit transformation or a multiplication kernel) and a bias b. The optimality criteria for the classifier is the (soft) maximum margin both for the classification and the regression tasks, with different costs. We then note by C the cost vector (or scalar) for each misclassified sample, and by D the cost matrix for the regression tasks, i.e. $D_{i,l}$ is the cost for the ith sample, for the lth measurement. The problem can be written, ignoring w_0, as:

$$\underset{w,b,\xi,\alpha,\beta,\zeta,\zeta^*}{\text{minimize}} \quad \frac{1}{2}\|w\|^2 + C^T\xi + \sum_{i=1}^{L} D_i^T \zeta_i^+$$

$$\text{subject to} \quad Y(Xw + 1b) \geq 1 - \xi, \ \xi \geq 0 \tag{1}$$

$$(\alpha_i s_i + 1\beta_i) - Xw - 1b \leq 1\epsilon_i + \zeta_i, \ \zeta_i \geq 0, \forall_i, 1 \leq i \leq L$$

$$Xw + 1b - (\alpha_i s_i + 1\beta_i) \leq 1\epsilon_i + \zeta_i^*, \ \zeta_i^* \geq 0, \forall_i, 1 \leq i \leq L$$

Where s_i is the vector of scores for the ith modality, for all of the samples. Note that the units of the continuous scores s_i may differ from the units of the optimal separating hyper-plane of the binary data, thus we added the vectors α, β of length L, with the linear transformations for the corresponding modalities. ϵ is a vector of length L containing the insensitive loss parameters for the different modalities (as in [4]). Note that since for each modality this parameter is applied after the linear transformation, its value is normalized and searching for the optimal value is easier and indicative of the regression fit (which is otherwise hidden).

2.2 Transition to the Dual Problem

We first transform the primal problem into its dual problem and then apply a KKT formalism to it [5, 6]. Then, using the standard SVM technique, the problem can be written in a Lagrangian formalism and using the fact that the appropriate partial derivatives equal 0 at the optimum, we get the final formulation as a quadratic problem:

$$
\underset{\mu, \delta, \delta^*}{\text{maximize}} \quad -\frac{1}{2}(\mu^T Y^T X X^T Y \mu + (\sum_{i=1}^{L} \delta_i^{-T}) X X^T (\sum_{i=1}^{L} \delta_i^{-})) + \mu^T 1 - \sum_{i=1}^{L} \delta_i^{+T} 1 \epsilon_i
$$

$$
- \sum_{i=1}^{l} \delta_i^{-T} X X^T Y \mu
$$

$$
\text{subject to} \quad \mu, \lambda, \delta_i, \delta_i^*, \eta_i, \eta_i^* \geq 0, \forall_i, 1 \leq i \leq L
$$

$$
\mu^T y = 0; 1^T \delta_i^- = 0; s_i^T \delta_i^- = 0, 1 \leq i \leq L \tag{2}
$$

$$
C = \mu + \lambda; D_i = \delta_i^{(*)} + \eta_i^{(*)}, \forall_i, 1 \leq i \leq L
$$

where $\delta_i^- = \delta_i - \delta_i^*$ and $\delta_i^+ = \delta_i + \delta_i^*$, with μ, λ being the Lagrange multipliers corresponding to the classification problem (as in the SVM fromalism) and $\delta_i, \delta_i^*, \eta_i, \eta_i^*$ the Lagrange multipliers corresponding to the i-th modality of the regression problem (as in the SVR formalism, following [4, 7]). Using such a formalism, we enjoy the advantages of a kernel machine, i.e. the optimization is on the support vectors coefficients μ, δ, δ^* and the (possibly high-dimensional) product $X X^T$ can be replaced using any kernel function K. Here the decision function becomes $g(q) = \sum_{j=1}^{n} (\mu_j y_j + \sum_{i=1}^{L} \delta_{i,j}^-) K(x_j, q) + b$, i.e. a weighted sum on the contributions of the kernel function of the support vectors with the classified sample point, where the weight on each support vector considers both the classification and the various regression constraints.

2.3 Handling Missing Values

The method can handle any combination of inputs, by simply setting the corresponding element of the classification cost vector C and/or of the cost matrix D to 0 where a value is missing. This constraints the corresponding support vectors coefficient to be fixed (box constaint of 0) and effectively removes the element from the optimization problem, while keeping all the other information intact.

3   Extending kMLVO

Two possible extensions of the kMLVO framework handle the incorporation of w_0, as in the MLVO, and non-linear regression.

3.1   Incorporation of w_0

Given w_0, we would like to introduce a penalty when diverging from it which is simiar to the one used in the MLVO, i.e. $E_2 = \frac{1}{2} \|w_0 - w\|_2^2$. This however will automatically lead to terms which are not quadratic in the values of x_i. This can be solved by projecting w and w_0 on any basis of the feature space. Given a base $B = \{B_i, \cdots, B_n\}$ of the feature space, $\|w_0 - w\|_2^2 = \sum_{i=1}^{n}(< w_0, B_i >$ $- < w, B_i >)^2 = \sum_{i=1}^{n}(\mu_i - < w, B_i >)^2$ where $< w_0, B_i >= \mu_i, \forall_i$ is the score induced by w_0 for the base vectors. We would like the scores induced by w to be similar, i.e. the problem is transformed into a regression problem. Thus it suffices to add the base vectors $\{B_i, \cdots, B_n\}$ as additional input samples along with their w_0 induced scores. As before, while the MLVO uses squared penalty, the kMLVO uses L_1 penalty.

3.2   Non-linear Regression

Suppose that the scores for the i-th modality are not linear with the optimal (or real) separating hyperplane for the classification problem, but follow $s_{i,k} = f(w^T x_k + b) = f(\sum_{j=1}^{n} \mu_j y_j K(x_j, x_k) + b)$ for some function f. In this case we would like to linearize the scores, i.e. applying f^{-1} before performing the kMLVO. If f (and thus f^{-1}) is unknown, we may let the kMLVO approximate it as a linear combination of score vectors. This can be performed by replacing the term $(\alpha_i s_i + 1\beta_i)$ in the equations with $(\alpha_{i,1} s_{i,1} + \cdots + \alpha_{i,p_i} s_{i,p_i} + 1\beta_i)$. The only additional constraints added to the final dual problem is:

$$s_{i,1}^T \delta_i^- = 0, \ldots, s_{i,p_i}^T \delta_i^- = 0 \tag{3}$$

These different scores can be, but not limited to, the original scores s_i in different powers, etc.

4   Simulations

In order to test whether the proposed formalism outperforms existing methods, we have compared the precision obtained using four different formalisms: SVR, SVM, MLVO and kMLVO on artificial datasets of different dimentionality, noise levels, and sample sizes. Additinaly, 10 % of the points were randomly chosen to be "outliers". For these points, the standard deviation of the added noise was 3 times the sandard deviation of all scores (instead of 0.03 or 0.6).

4.1   kMLVO Results on Simulations

The kMLVO outperforms the other classifiers in the presence of outliers. Such outliers can significantly affect the SVM and SVR formalisms, and we have here tested their effect on the kMLVO formalism. The average performace scores on the different data sets can be seen in Fig. 1. While for weak noise levels the MLVO is dominant, in the stronge noise level, a clear dominancy of the kMLVO can be seen, especially in the region of high number of samples with continous scores and a low number of binary samples. This can be explained by the fact that kMLVO (as SVR) is has L_1 regularization term, while the regression part of the MLVO (as LS-SVR) is regularized with an L_2 term.

In another application, regarding the binding to an immune system molecule, the transporter associated with antigen processing (TAP), the kMLVO outperformed the MLVO and the uni-label classifiers SVM (using binding/non binding data) and SVR (using affinity score), of the commonly used package LibSVM [8] on our data.

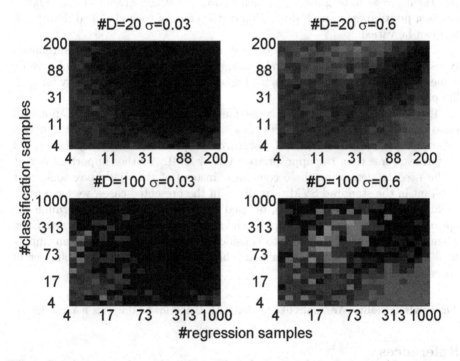

Fig. 1. Simulation results with outliers. The relative number of winners (that is, most accurate estimation of the direction vector) is coded to RGB by red - kMLVO, green - SVM, blue - SVR, and black - MLVO. The titles refer to the dimensionality of the data set and the noise's standard deviation.

5   Discussion

The approach presented can be used as a general supervised learning method when multiple labels of data are available. Such a situation often emerges in biological interactions, such as transcription factor binding or protein-protein interactions. In such cases, observations can either be binary (the presence or absence of an interaction) or continuous (the affinity).

Several extensions of the kMLVO have been proposed. The simplest expansion is the use of multiple continuous scores. Assume samples having continuous scores that are derived from several unit scales (e.g. IC50 and EC50 affinity related measurements). As part of the solution (as described above), we simultaneously fit between the predicted to the continuous score by linear regression. Thus, actually all the available measurements can be merged together, and the problem will be transformed to a set of linear regressions with multiple values of α and β. The algorithm can also be improved if the validity of the different dimensions of the samples in the n dimensional space or the validity of the samples themselves can be guessed. In such a case, the weight given to the similarity to the a priori guess in each dimension or the error of each classified data point (ξ_i) can be varied.

The use of kernels, along with extension for handling multiple measurements types, with different non-linear relations, and inherent consideration of missing values gives the suggested approach a higher flexibility and applicability for real-life problems.

The main limitations of the proposed methodology is that it mainly applies to cases where the number of observations is limited. When the number of observations is very large and biased toward one type of observations, the MLVO performs worse than the appropriate SVM or SVR. Another important caveat is the need to determine three constants, instead of the single box constraint constant in the standrad SVM formalism. In the presented cases, we have predefined the constant to be used, or used an internal test set to determine the optimal constants, and then applied the results to an external test set. Even when these caveats are taken into consideration, the MLVO can be an important methodological approach in many biological cases, where the number of observations is highly limited.

Acknowledgment. We would like to thank M. Beller for editing this manuscript.

References

1. Zhou, Z., Zhang, M., Huang, S., Li, Y.: Multi-instance multi-label learning. Artif. Intell. **176**, 2291–2320 (2011)
2. Boutell, M., Luo, J., Shen, X., Brown, C.: Learning multi-label scene classification. Pattern Recogn. **37**, 1757–1771 (2004)
3. Vider-Shalit, T., Louzoun, Y.: Mhc-i prediction using a combination of t cell epitopes and mhc-i binding peptides. J. Immunol. Methods **374**, 43–46 (2010)

4. Farag, A., Mohamed, R.M: Regression using support vector machines: basic foundations. Technical Report, CVIP Laboratory, University of Louisville (2004)
5. Karush, W.: Minima of functions of several variables with inequalities as side constraints. Master's thesis, Department of Mathematics, University of Chicago (1939)
6. Kuhn, H., Tucker, A.: Nonlinear programming. In: Proceedings of the Second Berkeley Symposium on Mathematical Statistics and Probability, California, vol. 5 (1951)
7. Cortes, C., Vapnik, V.: Support-vector networks. Mach. Learn. **20**, 273–297 (1995)
8. Chang, C.C., Lin, C.J.: LIBSVM: a library for support vector machines. ACM Trans. Intell. Sys. Technol. **2**, 27:1–27:27 (2011)

Robust Benchmark Set Selection
for Boolean Constraint Solvers

Holger H. Hoos[1]([✉]), B. Kaufmann[2], T. Schaub[2], and M. Schneider[2]

[1] Department of Computer Science, University of British Columbia,
Vancouver, BC, Canada
hoos@cs.ubc.ca
[2] Institute of Computer Science, University of Potsdam, Potsdam, Germany
{kaufmann,torsten,manju}@cs.uni-potsdam.de

Abstract. We investigate the composition of representative benchmark
sets for evaluating and improving the performance of robust Boolean constraint solvers in the context of satisfiability testing and answer set programming. Starting from an analysis of current practice, we isolate a set
of desiderata for guiding the development of a parametrized benchmark
selection algorithm. Our algorithm samples a benchmark set from a larger
base set (or distribution) comprising a large variety of instances. This
is done fully automatically, in a way that carefully calibrates instance
hardness and instance similarity. We demonstrate the usefulness of this
approach by means of empirical results showing that optimizing solvers
on the benchmark sets produced by our method leads to better configurations than obtained based on the much larger, original sets.

1 Introduction

The availability of representative sets of benchmark instances is of crucial importance for the successful development of high-performance solvers for computationally challenging problems, such as propositional satisfiability (SAT) and
answer set programming (ASP). Such benchmark sets play a key role for assessing solver performance and thus for measuring the computational impact of
algorithms and/or their vital parameters. On the one hand, this allows a solver
developer to gain insights on the strengths and weaknesses of features of interest. On the other hand, representative benchmark instances are indispensable to
empirically underpin the claims of computational benefit of novel ideas.

A representative benchmark set is composed of benchmark instances stem-
ming from a variety of different benchmark classes. Such benchmark sets have
been assembled (manually) in the context of well-known solver competitions,
such as the SAT and ASP competitions, and then widely used in the research
literature. These sets of competition benchmarks are well-accepted, because
they have been constituted by an independent committee using sensible criteria.
Moreover, these sets evolve over time and thus usually reflect the capabilities
(and limitations) of state-of-the-art solvers; they are also publicly available and
well-documented.

G. Nicosia and P. Pardalos (Eds.): LION 7, LNCS 7997, pp. 138–152, 2013.
DOI: 10.1007/978-3-642-44973-4_16, © Springer-Verlag Berlin Heidelberg 2013

However, instance sets from competitions are not always suitable for benchmarking scenarios where the same runtime cutoff is used for all instances. For example, in the last three ASP competitions, only ≈ 10 % of all instances were non-trivial (runtime over 9 s, i.e., 1 % of the runtime cutoff) for the state-of-the-art ASP solver *clasp*, while all other instances were trivial or unsolvable for *clasp* within the time cutoff used in the competition. While benchmarking, results of benchmarks are (typically) aggregated over all instances. But if the percentage of interesting instances in the benchmark set is too small, the interesting instances have small influence on the aggregated result and the overall result is dominated by uninteresting, i.e., trivial or unsolvable, instances. Hence, a significant change of the runtime behaviour of a new algorithm is harder to identify on such degenerate benchmark sets. In addition, uninteresting instances unnecessarily waste computational resources and thus cause avoidable delays in the benchmarking process.

Moreover, in ASP, competition instances do not necessarily represent real world applications. In the absence of a common modelling language, benchmark instances are often formulated in the most basic common setting and thus bear no resemblance to how real world problems are addressed (e.g., they are usually free of any aggregates).[1] The situation is simpler in SAT, where a wide range of benchmark instances stems from real-world applications and are quite naturally encoded in a low-level format, without the modelling layer present in ASP. Notably, SAT competitions place considerable emphasis on a public and transparent instance selection procedure [1]. However, as we discuss in detail in Sect. 3, competition settings may differ from other benchmarking contexts.

In what follows, we elaborate upon the composition of representative benchmark sets for evaluating and improving the performance of Boolean constraint solvers in the context of ASP and SAT. Starting from an analysis of current practice of benchmark set selection in the context of SAT competitions (Sect. 2), we isolate a set of desiderata for representative benchmark sets (Sect. 3). For instance, sets with a large variety of instances are favourable when developing a default configuration of a solver that is desired to perform well across a wide range of instances. We rely on these desiderata for guiding the development of a parametrized benchmark selection algorithm (Sect. 4).

Overall, our approach makes use of (i) a large base set (or distribution) of benchmark instances; (ii) instance features; and (iii) a representative set of state-of-the-art solvers. Fundamentally, it constructs a benchmark set with desirable properties regarding difficulty and diversity by sampling from the given base set. It achieves diversity of the benchmark set by clustering instances based on their similarity w.r.t a given set of features, while ensuring that no cluster is overrepresented. The difficulty of the resulting set is calibrated based on the given set of solvers. Use of the benchmark sets thus obtained helps save computational resources during solver development, configuration and evaluation, while concentrating on interesting instances.

---

[1] In ASP competitions, this deficit is counterbalanced by a modelling track, in which each participant can use its preferred modelling language.

We empirically demonstrate in Sect. 5 that optimizing solvers on the obtained selection of benchmarks leads to better configurations than obtainable from the vast original set of benchmark instances. We close with a final discussion and some thoughts on future work in Sect. 6.

2   Current Practice

The generation or selection of benchmark sets is an important factor in the empirical analysis of algorithms. Depending on the goals of the empirical study, there are various criteria for benchmark selection. For example, in the field of Boolean constraint solving, regular competitions are used to asses new app-roaches and techniques as well as to identify and recognize state-of-the-art solvers. Over the years, competition organizers came up with sets of rules for selecting subsets of submitted instances to assess solver performance in a fair manner. To begin with, we investigate the rules used in the well-known and widely recognized SAT Competition,[2] which try to achieve (at least) three over-all goals. First, the selection should be broad, i.e., the selected benchmark set should contain a large variety of different kinds of instances to assess the robust-ness of solvers. Second, each selected instance should be significant w.r.t. the ranking obtained from the competition. Third, the selection should be fair, i.e., the selected set should not be dominated by a set of instances from the same source (either a domain or a benchmark submitter).

For the 2009 SAT Competition [2] and the 2012 SAT Challenge [1], instances were classified according to hardness, as assessed based on the runtime of a set of representative solvers. For instance, for the 2012 SAT Challenge, the orga-nizers measured the runtimes of the best five SAT solvers from the Application and Crafted tracks of the last SAT Competition on all available instances and assigned each instance to one of the following classes: *easy* instances are solved by all solvers under 10 % of the runtime cutoff, i.e., 90 CPU seconds; *medium* instances are solved by all solvers under 100 % of the runtime cutoff; *too hard* instances are not solved by any solver within 300 % of the runtime cutoff; and *hard* instances are solved by at least one solver within 300 % of the runtime cutoff but not by all solvers within 100 % of the runtime cutoff. Instances were then selected with the objective to have 50 % *medium* and 50 % *hard* instances in the final instance set and, at the same time, to allow at most 10 % of the final instance set to originate from the same source.

While the *easy* instances are assumed to be solvable by all solvers, the *too hard* instances are presumably not solvable by any solver. Hence, neither class contributes to the solution count ranking used in the competition.[3] On the other hand, *medium* instances help to rank weaker solvers and to detect performance deterioration w.r.t. previous competitions. The *hard* instances are most useful for ranking the top-performing solvers and provide both a challenge and a chance to improve state-of-the-art SAT solving.

---

[2] http://www.satcompetition.org

[3] Solution count ranking assesses solvers based on the number of solved instances.

Although using a large variety of benchmark instances is clearly desirable for robust benchmarking, the rules used in the SAT Competition are not directly applicable to our identified use cases. First, the hardness criteria and distribution used are directly influenced by the use of the solution count ranking system. On the other hand, ranking systems that also consider measured runtimes, like the careful ranking[4] [3], might be better suited for differentiating solver performance. Second, limiting the number of instances from one source to achieve fairness is not needed in our setting. Furthermore, the origin of instances provides only an indirect way of achieving a heterogeneous instance set, as certain instances of different origin may in fact be more similar than other pairs of instances from the same source.

3   Desirable Properties of Benchmark Sets

Before diving into the details of our selection algorithm, let us first explicate the desiderata for a representative benchmark set (cf. [4]).

Large Variety of Instances. As mentioned, a large variety of instances is favourable to assess the robustness of solver performance and to reduce the risk of generalising from results that only apply to a limited class of problems. Such large variety can include different types of problems, i.e., real-world applications, crafted problems, and randomly generated problems; different levels of difficulty, i.e., easy, medium, and hard instances; different instance sizes; or instances with diverse structural properties. While the structure of an instance is hard to assess, a qualitative assessment could be based on visualizing the structure [5], and a quantitative assessment can be performed based on instance features [6,7]. Such instance features have already proven useful in the context of algorithm selection [7,8] and algorithm configuration [9,10].

Adapted Instance Hardness. While easy problem instances are sometimes useful for investigating certain properties of specific solvers, intrinsically hard or difficult to solve problem instances are better suited to demonstrate state-of-the-art solving capabilities through benchmarking. However, in view of the nature of NP-hard problems, it is likely that many hard instances cannot be solved efficiently. Resource limitations, such as runtime cutoffs or memory limits, are commonly applied in benchmarking. Solver runs that terminated prematurely because of violations of resource limits are not helpful in differentiating solver performance. Hence, instances should be carefully selected so that such prematurely terminated runs for the given set of solvers are relatively rare. Therefore, the *distribution of instance hardness* within a given benchmark set should be adjusted based on the given resource limits and solvers under consideration. In particular, instances that are too hard (i.e., for which there is a high probability of a timeout) as well as instances that are too easy, should be avoided, where

---

[4] Careful ranking compares pairs of solvers based on statistically significant performance differences and ranks solvers based on the resulting ranking graph.

hardness is assessed using a representative set of state-of-the-art solvers, as is done, for example, in the instance selection process of SAT competitions [2].

Since computational resources are typically limited, the number of benchmark instances should also be carefully calibrated. While using too few instances can bias the results, using too many instances can cost computational resources without improving the information gained from benchmarking. Therefore, we propose to start with a broad base set of instances, e.g., generated by one or more (possibly parametrized) generators or a collection of previously used competition instance sets, and to select a subset of instances following our desiderata.

Free of Duplicates, Reproducible, and Publicly Available. Benchmark set should be free of duplicates, because using the same instance twice does not provide any additional information about solver performance. Nevertheless, non-trivially transformed instances can be useful for assessing the robustness of solvers [11]. To facilitate reproducibility and comparability, both the problem instances and the process of instance selection should be publicly available. Ideally, problem instances should originate from established benchmark sets and/or public benchmark libraries. To our surprise, these properties are not true for all competition sets. For example, we found duplicates in the SAT Challenge 2012, ASP Competitions 2007 and 2009 (for example, `15-puzzle.init1.gz` and `15puzzle_ins.lp.gz` in the latter).

4   Benchmark Set Selection

Based on our analysis of solver competitions and the resulting desiderata, we developed an instance selection algorithm. Its implementation is open source and freely available at http://potassco.sourceforge.net. In addition, we present a way to assess the relative robustness and quality of an instance set based on the idea of Q-scores [1].

4.1   Benchmark Set Selection Algorithm

Our selection process starts from a given base set of instances I. This set can be a benchmark collection or simply a mix of previously used instances from competitions.

Inspired by past SAT competitions, a representative set of solvers S – e.g., best solvers of the last competition, the state-of-the-art (SOTA) contributors identified in the last competition, or contributors to SOTA portfolios [12] – is used to assess the hardness $h(i) \in \mathbb{R}$ of an instance $i \in I$. Typically, the runtime $t(i, s)$ (measured in CPU seconds) is used to assess the hardness of an instance $i \in I$ for solver $s \in S$. The aggregation of the runtimes of all solvers $s \in S$ on a given instance i can be carried out in several ways, e.g., by considering the minimal ($\min_{s \in S} t(i, s)$) or the average runtime ($\frac{1}{|S|} \cdot \sum_{s \in S} t(i, s)$). The resulting hardness metric is closely related to the intended ranking scheme for solvers. For example, the minimal runtime is a lower bound of the portfolio runtime performance and represents a challenging hardness metric appropriate in the context of solution count ranking. In contrast, the average runtime would be better suited for a careful ranking [3], which uses pairwise comparisons between solvers for each instance, because the pairs of runtimes for two solvers are of limited value if neither of them solved the given instance within the given cutoff time. Since all solvers contribute to the average runtime per instance, this metric will assess instances as hard even if only some solvers time out on time, and selecting instances based on it (as explained in the following) can therefore be expected to result in fewer timeouts overall.

After selecting a hardness metric, we have to choose how the instance hardness should be distributed within the benchmark set. As stated earlier, and under the assumption that the set to be created will not be used primarily in the context of solution count ranking, the performance of solvers can be compared better, if the incidence of timeouts is minimized. This is important, for example, in the context of algorithm configuration (manual or automatic). The incidence of timeouts can be minimized by increasing the runtime cutoff, but this is infeasible or wasteful in many cases. Alternatively, we can ensure that not too many instances on which timeouts occur are selected for inclusion in our benchmark set. At the same time, as motivated previously, it is also undesirable to include too many easy instances, because they incur computational cost and, depending on the hardness metric used, can also distort final performance rankings determined on a given benchmark set.

One way to focus the selection process on the most useful instances w.r.t. hardness, namely those that are neither too easy nor too hard, is to use an appropriately chosen probability distribution to guide sampling from the given base set of instances. For example, the use of a normal (Gaussian) distribution of instance hardness in this context leads to benchmark sets consisting predominantly of instances of medium hardness, but also include some easy and hard instances. Alternatively, one could consider log-normal or exponential distributions, which induce a bias towards harder instances, as can be found in many existing benchmark sets.Compared to the instance selection approach used in SAT competitions [1,2], this method does not require the classification of instances into somewhat arbitrary hardness classes.

The parameters of the distribution chosen for instance sampling, e.g., mean and variance in the case of a normal or log-normal distribution, can be determined based on the hardness metric and runtime limit; e.g., the mean could be chosen as half the cutoff time. By modifying the mean, the sampling distribution can effectively be shifted towards harder or easier benchmark instances.

As argued before, the origin of instances is typically less informative than their structure, as reflected, e.g., in informative sets of instance features. Such informative sets of instance features are available for many combinatorial problems, including SAT [7], ASP [13] and CSP [14], where they have been shown to correlate with the runtime of state-of-the-art solvers and have been used prominently in the context of algorithm selection (see, e.g., [7,8]). To prevent the inclusion of too many similar instances in the benchmark sets, we cluster the

---

Algorithm 1: Benchmark Selection Algorithm

    Input   : instance set I; desired number of instances n; representative set of
                  solvers S; runtimes $t(i, s)$ with $(i, s) \in I \times S$; normalized instance
                  features $f(i)$ for each instance $i \in I$; hardness metric $h : I \to \mathbb{R}$ of
                  instances; desired distribution $\mathcal{D}_h$ regarding h; clustering algorithm
                  ca; cutoff time t_c; threshold e for too easy instances;

    Output : selected instances I^*

1 remove instances from I that are not solved by any $s \in S$ within t_c;
2 remove instances from I that are solved by all $s \in S$ under $e\%$ of t_c ;
3 cluster all instances $i \in I$ in the normalized feature space $f(i)$ into clusters $S(i)$
    using clustering algorithm ca;
4 **while** $|I^*| < n$ and $I \neq \emptyset$ **do**
5   sample $x \in \mathbb{R} \sim \mathcal{D}_h$;
6   select instance $i^* \in I$ with the nearest $h(i^*)$ to x;
7   remove i^* from I;
8   **if** $S(i^*)$ *is not over-represented* **then**
9 | add i^* to I^*;
10   **end**
11 **end**
12 **return** I^*

---

instances based on their similarity in feature space. We then require that a cluster must not be over-represented in the selected instance set; in what follows, roughly reminiscent of the mechanism used in SAT competitions, we say that a cluster is over-represented if it contributes more than 10 % of the instances to the final benchmark set. While other mechanisms are easily conceivable, the experiments we report later demonstrate that this simple criterion works well.

Algorithm 1 implements these ideas with the precondition that the base instance set I is free of duplicates. (This can be easily ensured by means of simple preprocessing.) In Line 1, all instances are removed from the given base set that cannot be solved by all solver from the representative solver set S within the selection runtime cutoff t_c (*rejection of too hard instances*). If solution count ranking is to be used in the benchmarking scenario under consideration, the cutoff in the instance selection process should be larger than the cutoff for benchmarking, as was done in the 2012 SAT Challenge. In Line 2, all instances are removed that are solved by all solvers under e % of the cutoff time (*rejection of too easy instances*). For example, in the 2012 SAT Challenge [1], all instances were removed which were solved by all solvers under 10 % of the cutoff. Line 3 performs clustering of the remaining instances based on their normlized features. To perform this clustering, the well-known k-means algorithm could be used, and the number of clusters could be computed using G-means [10, 15] or by increasing the number of clusters until the clustering optimization does not improve further under a cross validation [16]. In our experiments, we used the latter, I've reworded the following: because the G-means algorithm relies on a normality assumption that is not necessarily satisfied for the instance feature data used here. Beginning with Line 4, instances are sampled within a loop until enough instances are selected or no more instances are left in the base set. To this end, $x \in \mathbb{R}$ is sampled from a distribution $\mathcal{D}_h$ induced by instance hardness metric h, such that for each sample x from hardness distribution $\mathcal{D}_h$, the instance i^* is selected whose hardness $h(i^*)$ is closest to x. Instance i^* is removed from the base instance set I. If the respective cluster $S(i^*)$ is not already over-represented in I^*, instance i^* is added to I^*, the benchmark set under construction.

4.2   Benchmark Set Quality

We would like to ensure that our benchmark selection algorithm produces instance sets that are in some way better than the respective base sets. At the same time, any benchmark set I^* it produces should be representative of the underlying base set I in the sense that if an algorithm performs better than a given baseline (e.g., some prominent solver) on I^* it should also be better on I. However, the converse may not hold, because specific kinds of instances may dominate I but not I^*, and excellent performance on those instances can lead to a situation where an algorithm that performs better on I does not necessarily perform better on I^*.

Bayless et al. [17] proposed a quantitative assessment of instance set utility. Their use case is the performance assessment of (new) algorithms on an instance set I_1 that has practical limitations, e.g., the instances are too large, too hard to solve, or not enough instances are available. Therefore, a second instance set I_2 without these limitations is assessed as to whether it can be regarded as a representative proxy for the instance set I_1 during solver development or configuration. The key idea is that any I_2 that is a representative proxy for I_1 can be used in lieu of I_1 to assess performance of a solver, with the assurance that good performance on I_2 (which is easier to demonstrate or achieve) implies, at least statistically, good performance on I_1.

To assess the utility of an instance set, they use algorithm configuration [9,10,18]. An algorithm configurator is used to find a configuration $c := s(c_I)$ of solver s on instance set I by optimizing, e.g., the runtime of s. If I_2 is a representative proxy for I_1, the algorithm configuration $s(c_{I_2})$ should perform on I_1 as well as a configuration optimized directly on I_1, i.e., $s(c_{I_1})$. The Q-score $Q(I_1, I_2, s, m)$ defined in Eq. (1) is the performance ratio of $s(c_{I_1})$ and $s(c_{I_2})$ on I_1 with respect to a given performance metric m. A large Q-score means I_2 is a good proxy for I_1. The short form of $Q(I_1, I_2, s, m)$ is $Q_{I_1}(I_2)$.

To compare both sets, I_1 and I_2, we want to know whether I_2 is a better proxy for I_1 than vice versa. To this end, we extended the idea in [17] and propose the Q*-score of I_1 and I_2 by computing the ratio of $Q_{I_1}(I_2)$ and $Q_{I_2}(I_1)$ as per Eq. (2). If I_1 is a better proxy for I_2 than vice versa, the Q*-score $Q^*(I_1, I_2)$ is larger than 1.

$$Q(I_1, I_2, s, m) = \frac{m(s(c_{I_1}), I_1)}{m(s(c_{I_2}), I_1)} \tag{1}$$

$$Q^*(I_1, I_2) = \frac{Q_{I_1}(I_2)}{Q_{I_2}(I_1)} \qquad (2)$$

We use the Q^*-score to assess the quality of the sets I^* obtained from our benchmark selection algorithm in comparison to the respective base sets I. Based on this score, we can assess the degree to which our benchmark selection algorithm succeeded in producing a set that is representative of the given base set in the way motivated earlier. Thereby, a Q^*-score ($Q^*(I_1, I_2)$) and a Q-score ($Q_{I_1}(I_2)$) of larger than 1.0 indicates that I_2 is better proxy for I_1 than vice versa and I_2 is a good proxy for I_1.

5   Evaluation

We evaluated our benchmark set selection approach by means of the Q^*-score criterion on widely studied instance sets from SAT and ASP competitions.

Instance Sets. We used three base instance sets to select our benchmark set: SAT-Application includes all instances of the *application* tracks from the 2009 and 2011 SAT Competition and 2012 SAT Challenge; SAT-Crafted includes instances of the *crafted* tracks (resp. hard combinatorial track) of the same competitions; and ASP includes all instances of the 2007 ASP Competition (SLparse track), the 2009 ASP Competition (with the encodings of the Potassco group [19]), the 2011 ASP Competition (decision NP-problems from the system track), and several instances from the ASP benchmark collection platform *asparagus*.[5] Duplicates were removed from all sets, resulting in 649 instances in SAT-Application, 850 instances in SAT-Crafted, and 2,589 instances in ASP.

Solvers. In the context of the two sets of SAT instances, the best two solvers of the application track, i.e., *Glucose* [20] (2.1) and *SINN* [21], and of the hard combinatorial track, i.e., *clasp* [19] (2.0.6) and *Lingeling* [22] (agm), and the best solver of the random track, i.e., *CCASAT* [23], of the 2012 SAT Challenge were chosen as representative state-of-the-art SAT solvers. *clasp* [19] (2.0.6), *cmodels* [24] (3.81) and *smodels* [25] (2.34) were selected as competitive and representative ASP solvers capable of reading the *smodels*-input format [26].

Instance Features. We used efficiently computable, structural features to cluster instances. The 54 *base* features of the feature extractor of *SATzilla* [7] (2012) were utilized for SAT. The seven structural features of *claspfolio* [13] were considered for ASP, namely, tightness (0 or 1), number of atoms, all rules, basic rules, constraint rules, choice rules, and weight rules of the grounded program. For feature computation, a runtime limit of 900 CPU seconds per instance and a z-score normalization was used. Any instance for which the complete set of features could not be computed within 900 s was removed from the set of candidate instances. This led to the removal of 52 instances from the SAT-Application set, 2 from the SAT-Crafted set, and 3 from the ASP set.

---

[5] http://asparagus.cs.uni-potsdam.de/

Execution Environment and Solver Settings. All our experiments were performed on a computer cluster with dual Intel Xeon E5520 quad-core processors (2.26 GHz, 8,192 KB cache) and 48 GB RAM per node, running Scientific Linux (2.6.18-308.4.1.el5). Each solver run was limited to a runtime cutoff of 900 CPU seconds. Furthermore, we set parameter e in our benchmark selection procedure to 10, i.e., instances solved by all solvers within 90 CPU seconds were discarded, and the number of instances to select (n) to 200 for SAT (because of the relatively small base sets) and 300 for ASP. After filtering out *too hard* instances (Line 1 of Algorithm 1), 404 instances remained in `SAT-Application`, 506 instances in `SAT-Crafted` and 2,190 instances in `ASP`; after filtering out *too easy* instances (Line 2), we obtained sets of size 393, 425, and 1,431, respectively.

Clustering. To cluster the instances based on their features (Line 3), we applied k-means 100 times with different randomised initial cluster centroids. To find the optimal number of clusters, we gradually increased the number of clusters (starting with 2) until the quality of the clustering, assessed via 10-fold cross validation and 10 randomised repetitions of k-means for each fold, did not improve any further [16]. This resulted in 13 clusters for each of the two SAT sets, and 25 clusters for the ASP set.

Selection. To measure the hardness of a given problem instance, we used the average runtime over all representative solvers. We considered a cluster to be over-represented (Line 8) if more than 20 % of the final set size (n) were selected for SAT, and more than 5 % in case of ASP; the difference in threshold was motivated by the fact that substantially more clusters were obtained for the ASP set than for `SAT-Application` and `SAT-Crafted`.

Algorithm Configuration. After generating the benchmark sets `SAT-Application*`, `SAT-Crafted*` and `ASP*` using our automated selection procedure, these sets were evaluated by assessing their Q^*-scores. To this end, we used the freely available, state-of-the-art algorithm configurator *ParamILS* [18] to configure the SAT and ASP solver *clasp* (2.0.6). *clasp* is a competitive solver in several areas of Boolean constraint solving[6] that is highly parameterized, exposing 46 performance-relevant parameters for SAT and 51 for ASP. This makes it particularly well suited as a target for automated algorithm configuration methods and hence for evaluating our instance sets. Following standard practice, for each set, we performed 10 independent runs of *ParamILS* of 2 CPU days each and selected from these the configuration with the best training performance as the final result of the configuration process for each instance set.

Sampling Distributions. One of the main input parameters of Algorithm 1 is the sampling distribution. With the help of our Q^*-score criterion, three distributions are assessed: a normal (Gaussian) distribution, a log-normal distribution, and an exponential distribution. The parameters of these distributions were set to the empirical statistics (e.g., empirical mean and variance) of the hardness distribution over the base sets. The log-normal and exponential distributions

---

[6] *clasp* won several first places in previous SAT, PB and ASP competitions.

Table 1. Comparison of set qualities of the base sets I and benchmark sets I^* generated by Algorithm 1; evaluated with Q^*-Scores with $I_1 = I^*$, $I_2 = I$, *clasp* as algorithm A and PAR10-scores as performance metric m

Sampling-distribution	PAR10 on I			PAR10 on I^*			Q^*-score
	c_{def}	c_I	c_{I^*}	c_{def}	c_I	c_{I^*}	
SAT-Application							
Normal	4629	4162	3997	3410	2667	1907	1.46
Log-normal	4629	4162	4683	3875	2601	3487	0.66
Exponential	4629	4162	4192	2969	2380	2188	1.08
SAT-Crafted							
Normal	5226	5120	5056	2429	2155	1752	1.25
Log-normal	5226	5120	5184	3359	3235	3184	1.04
Exponential	5226	5120	5072	1958	1819	1523	1.21
ASP							
Normal	2496	1239	1072	1657	705	557	1.46
Log-normal	2496	1239	1128	3136	1173	678	1.90
Exponential	2496	1239	1324	1648	710	555	1.20

have fat right tails and typically reflect better the runtime behaviour of solvers for NP problems than the normal distribution. However, when using the average runtime as our hardness metric, the instances sampled using a normal distribution are not necessarily atypically easy. For instance, an instance i, on which half of the representative solvers have a timeout while the other half solve the instance in nearly no time, has an average runtime of half of the runtime cutoff. Therefore, the instance is medium hard and will be likely selected by using the normal distribution.

In Table 1, we compare the benchmark sets we obtained from the base sets SAT-Application, SAT-Crafted and ASP when using these three types of distributions, based on their Q^*-scores. On the left of the table, we show the PAR10 performance on the base set I of the default configuration of *clasp* (c_{def}; we use this as a baseline), the configuration c_I found on the base set I, and the configuration c_{I^*} found on the selected set I^*; this is followed by the performance on the benchmark sets I^* generated using our new algorithm. The last column reports the Q^*-score values for the pairs of sets I and I^*.

For all three instance sets, the Q^*-scores obtained via the normal distribution were larger than 1.0, indicating that c_{I^*} performed better than c_I and the set obtained from our benchmark selection algorithm I^* proved to be a good alternative to the entire base set I. Although on the ASP set, by using the log-normal distribution a larger Q^*-score (1.90) was obtained than for the normal distribution (1.46), on the SAT-Application set, using the log-normal distribution did not produce good benchmark sets. When using exponential distributions, Q^*-scores are larger than 1.0 in all three cases, but smaller than those obtained with normal distributions.

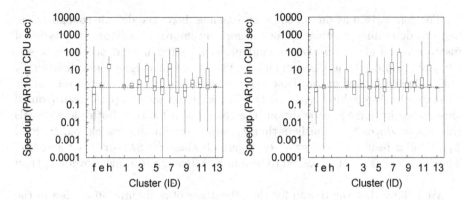

Fig. 1. *Boxplots* indicating the median, quartiles minimum and maximum speedup achieved on the instance clusters within the base set `SAT-Application`; (*left*) compares $c_{default}$ and c_I (high values are favourable for c_I); (*right*) compares $c_{default}$ and c_{I*} (high values are favourable for c_{I*}); special clusters: S_f uncompleted feature computation; S_e too easy, S_h too hard.

When using the normal distribution, configuration c_{I*} performed better than c_I on both sets I and I^* (implying $Q_{I_1}(I_2) > 1.0$). Therefore, configuration on the selected set I^* leads to faster (and more robust) configurations than on the base set I. Furthermore, the benchmark sets produced by our algorithm are smaller and easier than the respective base sets. Hence, less CPU time is necessary to assess the performance of an algorithm on those benchmark sets. For instance, the default configuration of *clasp* needed 215 CPU hours on the base `ASP` set and only 25 CPU hours on the benchmark set `ASP*`. For developing a new algorithm or configuring an algorithm (manually or automatically), fast and informative assessment, as facilitated by our new benchmark set generation algorithm, is very important.

Cluster Assessment. An additional advantage of Algorithm 1 is the fact that it produces a feature-based instance clustering, which can be further used to assess more precisely the performance of algorithms (or configurations). Normally, the performance of an algorithm is assessed over an entire instance set, but with the help of instance clusters, the performance can be assessed on different types of instances. This is useful, for example, in the context of developing a robust solver which should perform equally well across different types of instances. An example for such a solver is the CPLEX solver for mixed integer programming (MIP) problems, which is designed to perform well over a broad range of application contexts, each of which gives rise to different types of MIP instances.

The box plots in Fig. 1 show the speedups (y-axis) of the configurations c_I (left) and c_{I*} (right; while sampling with a normal distribution) against the default configuration c_{def} of *clasp* on each cluster $S_{1..13}$ (x-axis) within the `SAT-Application` base set. Furthermore, three special clusters contain the instances that were discarded in Algorithm 1 because, feature computation could not be completed (S_f), they were too easy (S_e), or too hard (S_h).

The comparison against a common baseline, here: the default configuration, helps to determine whether the new algorithm improved only on some types of instance or on all. For instance, configuration c_I (configured on the base set; left plot) improved the performance by two orders of magnitude on cluster S_8 but is slightly slower on S_9. However, configuration c_{I*} (configured on the set generated by Algorithm 1; right plot) achieved better median performance on all clusters except for S_f. In addition, the comparison between both plots reveals that c_{I*} produces fewer outliers than c_I, especially on clusters S_6, S_9, S_{11} and S_{13}. Similar results (not shown here) were obtained for SAT-Crafted and ASP. Therefore, c_{I*} can be considered to be a more robust improvement over c_{def} than c_I.

We believe that the reason for the robustness of configuration c_{I*} lies in the fact that the (automatic) configuration process tends to be biased by instance types that are highly represented in a given training set. Since Algorithm 1 produces sets I^* that cover instance clusters more evenly than the respective base sets I, the configuration process is naturally guided more towards robust performance improvements across all clusters.

Particular attention should be paid to the special clusters S_f, S_e and S_h for the assessment of c_{I*}, because the instances contained in these clusters are not at all represented in I^*. On none of our experiments with the three types of sampling distributions did we ever observe that the performance of c_{I*} on the *too hard* instances S_h decreased; in fact, it sometimes increased. In contrast, the performance on the *too easy* instances S_e and instances with no features S_f was less consistent, and we observed speedups between 300 and 0.1 in comparison to c_I. Therefore, the threshold for filtering too easy instances e should be set conservatively (below 10 %), to ensure that not too many too easy instances are discarded (we note that this is in contrast to common practice in SAT competitions).

Furthermore, our Algorithm 1 ensures that no cluster is over-represented, but does not ensure a sufficient representation of all clusters in the selected set. For instance, cluster S_4 has 141 instances in the base ASP set but only one instance in ASP* set (with normal distribution). Nevertheless, a low representation of a cluster in the selected set did not necessarily harm the configuration process, and in most observed cases, the configuration c_{I*} performed as well as c_I on the under-represented clusters.

6   Conclusions and Future Work

In this work, we have introduced an algorithm for selecting instances from a base set or distribution to form an effective and efficient benchmark set. We consider a benchmark set to be effective, if a solver configured on it performs at least as well as when configured on the original set, and we consider it to be efficient, if the instances in it are on average easier to solve than those in the base set. By using such benchmark sets, the computational resources required for assessing the performance of a solver can be reduced substantially. Our benchmark selection procedure can use arbitrary sampling distributions; yet, in our experiments, we found that using a normal (Gaussian) distribution is particularly effective. Since our approach filters out instances considered too easy or too hard for the solver under consideration, it can lead to a situation where the performance of a given solver, when configured on the benchmark set, becomes worse on those discarded instances. However, the risk of worsening the performance on too hard instances can be reduced by setting the runtime cutoff of the selection process higher than in the actual benchmark. Then, the selected set contains very challenging instances under the runtime cutoff in the benchmark, which are yet known to be solvable. We have also demonstrated that clustering of instances based on instance features facilitates diagnostic assessments of the degree to which a solver performs well on specific types of instances or across an entire, heterogeneous benchmark set. Our work reported here is primarily motivated by the desire to develop solvers that perform robustly well across a wide range of problem instances, as has been (and continues to be) the focus in developing solvers for many hard combinatorial problems.

In future work, it may be interesting to ensure that semantically different types of instances, such as satisfiable and unsatisfiable instances in the case of SAT, are represented evenly or equivalently as in a given base set. Furthermore, one could consider more sophisticated ways to assess the over-representation of feature-based clusters and to automatically adjust the sampling process based on the number of clusters and their sizes. Finally, we believe that it would be interesting to study criteria for assessing the robustness of solver performance across clusters and to use such criteria for automatic algorithm configuration.

Acknowledgments. B. Kaufmann, T. Schaub and M. Schneider were partially supported by DFG under grants SCHA 550/8-3 and SCHA 550/9-1. H. Hoos was supported by an NSERC Discovery Grant and by the GRAND NCE.

References

1. Balint, A., Belov, A., Järvisalo, M., Sinz, C.: Application and hard combinatorial benchmarks in SAT challenge. In: Proceedings of SAT Challenge 2012: Solver and Benchmark Descriptions. Department of CS Series of Publications B, vol. B-2012-2, pp. 69–71. University of Helsinki (2012)
2. Berre, D., Roussel, O., Simon, L.: http://www.satcompetition.org/2009/ BenchmarksSelection.html (2009). Accessed 09 March 2012
3. Van Gelder, A.: Careful ranking of multiple solvers with timeouts and ties. In: Sakallah, K.A., Simon, L. (eds.) SAT 2011. LNCS, vol. 6695, pp. 317–328. Springer, Heidelberg (2011)
4. Hoos, H., Stützle, T.: Stochastic Local Search: Foundations and Applications. Elsevier/Morgan Kaufmann, San Francisco (2004)
5. Sinz, C.: Visualizing SAT instances and runs of the DPLL algorithm. J. Autom. Reason. **39**, 219–243 (2007)
6. Nudelman, E., Leyton-Brown, K., Hoos, H., Devkar, A., Shoham, Y.: Understanding random SAT: beyond the clauses-to-variables ratio. In: Wallace, M. (ed.) CP 2004. LNCS, vol. 3258, pp. 438–452. Springer, Heidelberg (2004)

7. Xu, L., Hutter, F., Hoos, H., Leyton-Brown, K.: SATzilla: portfolio-based algorithm selection for SAT. J. Artif. Intell. Res. **32**, 565–606 (2008)
8. Kadioglu, S., Malitsky, Y., Sabharwal, A., Samulowitz, H., Sellmann, M.: Algorithm selection and scheduling. In: Lee, J. (ed.) CP 2011. LNCS, vol. 6876, pp. 454–469. Springer, Heidelberg (2011)
9. Hutter, F., Hoos, H.H., Leyton-Brown, K.: Sequential model-based optimization for general algorithm configuration. In: Coello, C.A.C. (ed.) LION 2011. LNCS, vol. 6683, pp. 507–523. Springer, Heidelberg (2011)
10. Kadioglu, S., Malitsky, Y., Sellmann, M., Tierney, K.: ISAC - instance-specific algorithm configuration. In: Proceedings of ECAI'10, pp. 751–756. IOS Press (2010)
11. Brglez, F., Li, X., Stallmann, F.: The role of a skeptic agent in testing and benchmarking of sat algorithms (2002)
12. Xu, L., Hutter, F., Hoos, H., Leyton-Brown, K.: Evaluating component solver contributions to portfolio-based algorithm selectors. In: Cimatti, A., Sebastiani, R. (eds.) SAT 2012. LNCS, vol. 7317, pp. 228–241. Springer, Heidelberg (2012)
13. Gebser, M., Kaminski, R., Kaufmann, B., Schaub, T., Schneider, M.T., Ziller, S.: A portfolio solver for answer set programming: preliminary report. In: Delgrande, J.P., Faber, W. (eds.) LPNMR 2011. LNCS, vol. 6645, pp. 352–357. Springer, Heidelberg (2011)
14. O'Mahony, E., Hebrard, E., Holland, A., Nugent, C., O'Sullivan, B.: Using case-based reasoning in an algorithm portfolio for constraint solving. In: AICS'08 (2008)
15. Hamerly, G., Elkan, C.: Learning the k in k-means. In: Proceedings of NIPS'03. MIT Press (2003)
16. Hill, T., Lewicki, P.: Statistics: Methods and Applications. StatSoft, Tulsa (2005)
17. Bayless, S., Tompkins, D., Hoos, H.: Evaluating instance generators by configuration. Submitted for publication (2012)
18. Hutter, F., Hoos, H., Leyton-Brown, K., Stützle, T.: ParamILS: an automatic algorithm configuration framework. J. Artif. Intell. Res. **36**, 267–306 (2009)
19. Gebser, M., Kaminski, R., Kaufmann, B., Ostrowski, M., Schaub, T., Schneider, M.: Potassco: the potsdam answer set solving collection. AI Commun. **24**(2), 105–124 (2011)
20. Audemard, G., Simon, L.: Glucose 2.1. in the SAT challenge 2012. In: Proceedings of SAT Challenge 2012: Solver and Benchmark Descriptions. Department of CS Series of Publications B, vol. B-2012-2, pp. 23–23. University of Helsinki (2012)
21. Yasumoto, T.: Sinn. In: Proceedings of SAT Challenge 2012: Solver and Benchmark Descriptions. Department of CS Series of Publications B, vol. B-2012-2, pp. 61–61. University of Helsinki (2012)
22. Biere, A.: Lingeling and friends entering the SAT challenge 2012. In: Proceedings of SAT Challenge 2012: Solver and Benchmark Descriptions. Department of CS Series of Publications B, vol. B-2012-2, pp. 33–34. University of Helsinki (2012)
23. Cai, S., Luo, C., Su, K.: CCASAT: solver description. In: Proceedings of SAT Challenge 2012: Solver and Benchmark Descriptions. Department of CS Series of Publications B, vol. B-2012-2, pp. 13–14. University of Helsinki (2012)
24. Giunchiglia, E., Lierler, Y., Maratea, M.: Answer set programming based on propositional satisfiability. J. Autom. Reason. **36**(4), 345–377 (2006)
25. Simons, P., Niemelä, I., Soininen, T.: Extending and implementing the stable model semantics. Artif. Intell. **138**(1–2), 181–234 (2002)
26. Syrjänen, T.: Lparse 1.0 user's manual

Boosting Sequential Solver Portfolios: Knowledge Sharing and Accuracy Prediction

Yuri Malitsky[1]([✉]), Ashish Sabharwal[2], Horst Samulowitz[2],
and Meinolf Sellmann[2]

[1] Department of Computer Science, Brown University, Providence, RI 02912, USA
ynm@cs.brown.edu
[2] IBM Watson Research Center, Yorktown Heights, NY 10598, USA
{ashish.sabharwal,samulowitz,meinolf}@us.ibm.com

Abstract. Sequential algorithm portfolios for satisfiability testing (SAT), such as SATzilla and 3S, have enjoyed much success in the last decade. By leveraging the differing strengths of individual SAT solvers, portfolios employing older solvers have often fared as well or better than newly designed ones, in several categories of the annual SAT Competitions and Races. We propose two simple yet powerful techniques to further boost the performance of sequential portfolios, namely, a generic way of knowledge sharing suitable for sequential SAT solver schedules which is commonly employed in parallel SAT solvers, and a meta-level guardian classifier for judging whether to switch the main solver suggested by the portfolio with a recourse action solver. With these additions, we show that the performance of the sequential portfolio solver 3S, which dominated other sequential categories but was ranked 10th in the application category of the 2011 SAT Competition, can be boosted significantly, bringing it just one instance short of matching the performance of the winning application track solver, while still outperforming all other solvers submitted to the crafted and random categories.

1 Introduction

Significant advances in solution techniques for propositional satisfiability testing, or SAT, in the past two decades have resulted in wide adoption of the SAT technology for solving problems from a variety of fields such as design automation, hardware and software verification, cryptography, electronic commerce, AI planning, and bioinformatics. This has also resulted in a wide array of challenging problem instances that continually keep pushing the design of better and faster SAT solvers to the next level. The annual SAT Competitions and SAT Races have played a key role in this advancement, posing as a challenge a set of so-called "application" category (previously known as the "industrial" category) instances, along with equally, but differently challenging, "crafted" and "random" instances.

Given the large diversity in the characteristics of problems as well as specific instances one would like to solve by translation to SAT, it is no surprise

G. Nicosia and P. Pardalos (Eds.): LION 7, LNCS 7997, pp. 153–167, 2013.
DOI: 10.1007/978-3-642-44973-4_17, © Springer-Verlag Berlin Heidelberg 2013

that different SAT solvers, some of which were designed with a specific set of application domains in mind, work better on different kinds of instances. Algorithm portfolios (cf. [7]) attempt to leverage this diversity by employing several individual solvers and, at runtime, dynamically selecting what appears to be the most promising solver — or a schedule of solvers — for the given instance. This has allowed sequential SAT portfolios such as SATzilla [16] and 3S [8,10] to perform very well in the annual SAT Competitions and Races.

Most of the state-of-the-art sequential algorithm portfolios are based on two main components: (a) a schedule of "short running" solvers to be run first in sequence for some small amount of time (usual some fixed percentage of the total available time such as 10 %) and (b) a "long running" solver to be executed for the remainder of the time which is selected by one or the other Machine Learning technique (e.g., logistic regression, nearest neighbor search, or decision forest). If one of the short running solvers succeeds in solving the instance, then the portfolio terminates successfully. However, all work performed by each short running solver in this execution sequence is completely wasted unless it manages to fully solve the instance. If none of the short running solvers in the schedule succeeds, all faith is put in the one long running solver.

Given this typical sequential portfolio setup, it is natural to consider an extension that attempts to utilize information gained by short running solvers even if they all fail to solve the instance. Further, one may also consider an automated way to carefully revisit the choice of the long running solver whose improper selection may substantially harm the overall portfolio performance. We propose two relatively simple yet powerful techniques towards this end, namely, learnt clause forwarding and accuracy prediction.

We remark that one limitation of current algorithm portfolios is that their performance can never be better than that of the oracle or "virtual best solver" which, for each given instance, (magically) selects an individual solver that will perform best on it. By sharing knowledge, we allow portfolio solvers to, in principle, go beyond VBS performance. Specifically, a distinguishing strength of our proposed clause forwarding scheme is that it enables the portfolio solver to potentially succeed in solving an instance that no constituent SAT solver can.

Learnt clause forwarding focuses on avoiding waste of effort by the short running solvers in the schedule. We propose to share, or "forward," the knowledge gained by the first k solvers in the form of a selection of short learned clauses, which are passed on to the $k + 1$st solver. Conflict-directed clause learning (CDCL) is a very powerful technique in SAT solving, often regarded as the single most important element that allows these solvers to tackle real-life problems with millions of variables and constraints. Forwarding learnt clauses is a cheap but promising way to share knowledge *between solvers* and is commonly employed in parallel SAT solving. We demonstrate that sharing learnt clauses can improve performance in sequential SAT solver portfolios as well.[1]

---

[1] For the specific case of population-based algorithm portfolios, Peng et al. [11] have proposed sharing information through migration of individuals across populations.

Accuracy prediction and recourse aims to use meta-level learning to correct errors made by the portfolio solver when selecting the "primary" or long running solver. Typically, effective schedules allocate a fairly large fraction of the available runtime to one solver, as not doing so would limit the best-case performance of the portfolio to that of an oracle portfolio with a relatively short timeout. This, of course, poses a risk, as a substantial amount of time is wasted if the portfolio selects the "wrong" primary solver. We present a scheme to generate large amounts of training data from existing solver performance data, in order to create a machine learning model that aims to predict the accuracy of the portfolio's primary solver selector. We call this meta-level classifier as a *guardian classifier*. We also use this training data to determine the most promising *recourse action*, i.e., which solver should replace the suggested primary solver.

These techniques are general and may be applied to various portfolio algorithms. However, unlike the development of portfolio solvers that do not share information, experimentation in our setting is much more cumbersome and time consuming. It involves modifying individual SAT solvers and running the designed portfolio solver on each test instance in real time, rather than simply reading off performance numbers from a pre-computed runtime matrix.

We here demonstrate the effectiveness of our techniques using one base portfolio solver, namely 3S, which had shown very good performance in SAT Competition 2011 in the crafted and random categories but was ranked 10th in the application category. Note also that the instances and solvers participating in the 2011 Competition were designed with a 5,000 s time limit in mind, compared to instances and solvers in the 2012 Challenge where the time limit was only 900 s. Our focus is on utilizing algorithm portfolios and our techniques for solving hard instances. Our results, with a time limit roughly equivalent to 5,000 s of the competition machines, show that applying these techniques can boost the performance of 3S on the 2011 competition instances to a point where it is only one instance short of matching the performance of the winning solver, Glucose 2.0 [1], on the 300 application track instances. Moreover, the resulting solver, 3S+fp, continues to dominate all other solvers from the 2011 Competition in the crafted and random categories in which 3S had excelled.

We note that our portfolio solver built using these techniques, called ISS or Industrial SAT Solver, was declared the Best Interacting Multi-Engine SAT Solver in the 2012 SAT Challenge, a category that specifically compared portfolios that share information among multiple SAT engines.

2 Background

We briefly review some essential concepts in constraint satisfaction, SAT, and portfolio research.

Definition 1. *Given a Boolean variable $X \in \{true, false\}$, we call X and $\neg X$ (speak: not X) literals (over X). Given literals $L_1, \ldots, L_k$ over Boolean variables $X_1, \ldots, X_n$, we call $(\bigvee_a L_a)$ a clause (over variables $X_1, \ldots, X_n$). Given clauses*

$C_1, \ldots, C_m$ over variables $X_1, \ldots, X_n$, we call $\bigwedge_a C_a$ a formula in conjunctive normal form (CNF).

Definition 2. *Given Boolean variables $X_1, \ldots, X_n$, a valuation is an assignment of values "true" or "false" to each variable: $\sigma : \{X_1, \ldots, X_n\} \to \{true, false\}$. A literal X evaluates to "true" under σ iff $\sigma(X) = true$ (otherwise it evaluates to "false"). A literal $\neg X$ evaluates to "true" under σ iff $\sigma(X) = false$. A clause C evaluates to true under σ iff at least one of its literals evaluates to "true." A formula evaluates to "true" under σ iff all its clauses evaluate to "true."*

Definition 3. *The Boolean Satisfiability or SAT Problem is to determine whether, for any given formula F in CNF, there exists a valuation σ such that F evaluates to "true."*

The SAT problem has played a prominent role in theoretical computer science where it was the first to be proven to be NP-hard [3]. At the same time, it has driven research in combinatorial problem solving for decades. Moreover, the SAT problem has great practical relevance in a variety of areas, in particular in cryptography and in verification.

2.1 SAT Solvers

While algorithmic approaches for SAT have been developed as early as the beginning of AI research, a boost in SAT solving performance has been achieved since the mid-nineties. Problems with a couple of hundred Boolean variables frequently posed a challenge back then. Today, many problems with hundreds of thousands of variables can be solved as a matter of course. While there exist very different algorithmic approaches to solving SAT problems, the performance of most systematic SAT solvers (i.e., those that can prove unsatisfiability) is frequently attributed to three ingredients:

1. Randomized search decisions and systematically restarting search when it exceeds some dynamic fail limit,
2. Very fast inference engines which only consider clauses which may actually allow us to infer a new Boolean variable for a variable, and
3. Conflict analysis and clause learning.

The last point regards the idea of inferring new clauses during search that are redundant to the given formula but encode, often in a succinct way, the reason why a certain partial truth assignment cannot be extended to any solution. These redundant constraints strengthen our inference algorithm when a different partial valuation cannot be extended to a full valuation that satisfies the given formula for a "similar" reason. One of the ideas that we pursue in this paper is to inform a solver about the clauses learnt by another solver that was invoked previously to try and solve the same CNF formula. This technique is standard in parallel SAT solving but, surprisingly, has not been considered for solver portfolios.

2.2   Solver Portfolios

Another important contribution was the inception of algorithm portfolios [4, 9, 15]. Based on the observation that solvers have complementary strengths and thus exhibit incomparable behavior on different problem instances, the ideas of running multiple solvers in parallel or to select one solver based on the features of a given instance were introduced. Portfolio research has led to a wealth of different approaches and an amazing boost in solver performance in the past decade [8, 16].

Solver Selection: The challenge when devising a solver portfolio is to develop a learning algorithm that, for a given set of training instances, builds a dynamic mechanism that selects a "good" solver for any given SAT instance. To this end, we need a way to characterize a given SAT instance, which is achieved by computing so-called "features." These could be, e.g., the number of clauses or variables, statistics over the number of negated over positive variables per clause, or the clause over variable ratio. Features can also include dynamic properties of the given instance, obtained by running a solver for a very short period of time as a probe and collecting statistics. As the goal of this paper is to devise techniques to improve *existing* portfolios, a full understanding of instance features is unnecessary. We refer the reader to Xu et al. [16] for a comprehensive study of features suitable for SAT.

Solver Scheduling: Recent versions of SATzilla and 3S no longer just choose one among the portfolio's constituent solvers. While still selecting one long running primary solver, they first schedule a sequence of several other solvers for a shorter amount of time. In particular, 3S, our base solver for experimentation, employs a semi-static schedule of solvers, given a test instance F and an total time limit T. It runs a static schedule (independent of F, based solely on prior knowledge from training data) for an internal time limit t (with $t \approx 10\%$ of T) in which several different solvers with different (short) time limits are used. This is followed by a long running solver, scheduled for time $T - t$, based on the features of F computed at runtime.

We will refer to these two components of 3S's scheduling strategy as the *pre-schedule* and the *primary solver*. In this paper, we tackle precisely these two aspects: How can we improve the interplay between the short-running solvers in the pre-schedule while also passing knowledge on to the primary solver, and how can we improve the selection of the long-running primary solver itself.

3   Sharing Knowledge Among Solvers

A motivating factor behind the use of a pre-schedule used in sequential portfolios is *diversity*. By employing very different search strategies, one increases the likelihood of covering instances that may be challenging for some solvers and very easy for others. Diversity has also been an important factor in the design of *parallel* SAT solvers, such as ManySAT [6] and Plingeling [2]. When designing these parallel solvers, it has been observed that the overall performance can be improved by carefully sharing a limited amount of knowledge between the search efforts led by different threads. This knowledge sharing must be carefully done, as it must balance usefulness of the information against the effort of communicating and incorporating it. One effective strategy has been to share information in the form of very short learned clauses, often just unit clauses, i.e., clauses with only one literal (e.g., the winning parallel solver [2] at the 2011 SAT Competition).

3.1   Knowledge Sharing Among Clause-Learning Systematic Solvers

In contrast, current sequential portfolios, while also relying on diversity through the use of a pre-schedule, do not exploit any kind of knowledge sharing. If the first k solvers in the pre-schedule fail to solve the instance, the time they spent is wasted. We propose to avoid this waste by employing the same technique that is used in parallel SAT, namely by *forwarding* a subset of the clauses learned by one solver in the pre-schedule to all solvers that follow it. In our implementation, clause forwarding is parameterized by two positive integer parameters, L and M. Each clause forwarding solver outputs all learned clauses containing up to L literals. Out of this list, the M shortest ones (or fewer, if not enough such clauses are generated) are forwarded to the next solver in the schedule, which then treats these clauses as part of the input formula. While we solely base our choice on what clauses to forward on their lengths, one could also consider more sophisticated measures (e.g., [1]). Note that, unlike clause sharing in today's parallel SAT solvers, in the sequential case clause forwarding incurs a relatively low communication overhead. Nonetheless, it needs to be balanced out with the potential benefits. We implemented clause forwarding in three conflict directed clause learning (CDCL) solvers, henceforth referred to as the *clause forwarding solvers*.

3.2   Impact of Knowledge Sharing on Other Solvers

In addition to CDCL solvers, pre-schedules typically also employ two other kinds of solvers: incomplete local search solvers and "lookahead" based complete solvers. The former usually perform very well on random and some crafted instances, and the latter usually excel in the crafted category and sometimes on unsatisfiable random instances. Since these solvers are not designed to generate or use conflict directed learned clauses, it is not clear *a priori* whether such clauses — which are redundant with respect to the underlying SAT theory — would help these two kinds of solvers as well. In our experiments, we found it best to run these solvers *before* our clause forwarding solvers are used.

The exceptions to this rule were two solvers: march_hi and mxc-sat09, which showed a mixed impact of incorporating forwarded learned clauses. We thus chose to run them both before the forwarding solvers, as in the base portfolio 3S,

Table 1. Gap closed to the virtual best solver (VBS) by using clause forwarding.

	2009	2010	2011	Average
% closed over 3S / VBS gap	12.5	16.67	5.41	11.53

and also after forwarding. Our overall pre-schedule was composed of the original one used by 3S in the 2011 SAT Competition, scaled appropriately to take the difference in machine speeds into account, enhanced with clause forwarding solvers, and reordered to have non-forwarding CDCL solvers appear after the forwarding ones. We note that changing the pre-schedule itself did not significantly alter the performance of 3S. E.g., in the application category, as we will later see in Table 3, the performances of 3S with the original and the updated pre-schedules were very similar.

3.3   Formula Simplification

One other consideration that has a significant impact in practice is, whether to simplify the CNF formula before handing it to the next solver in the schedule, after (up to) M forwarded clauses have been added to it. With some experimentation, we found that minimal simplification of the formula after adding forwarded clauses, performed using SatElite [13] in our case, was the most rewarding. We thus used this clause forwarding setup for the experiments reported in this paper.

3.4   Practical Impact of Clause Forwarding

We will demonstrate in the experiments section that clause forwarding allows 3S to close a significant part of the gap in performance when compared to the best solvers for application instances of the 2011 Competition, along with more information on the choice of training/test splits we consider and the experimental setup we use. We here provide an additional preview of the impact of clause forwarding when using the latest SAT solver available prior to the 2012 Challenge. For this evaluation we consider three train/test splits of instances: the first split uses the 2009 competition instances as test instances and every instance available before 2009 for training; the second and third split are defined similarly but for the 2010 race and 2011 competition, respectively.

Results are presented in Table 1. Here, we consider the *gap in performance* between the portfolio without clause forwarding and the best possible no-knowledge-sharing portfolio, VBS, which uses an oracle to invoke the fastest solver for each given instance. In the table, we show how much of that gap is closed by using clause forwarding. Of course, the portfolio that uses knowledge sharing between solvers is no longer limited in performance by the oracle portfolio, as remarked earlier. However, using the oracle portfolio gives us a good baseline to compare with. As we see, clause forwarding significantly helps on all three competition splits clause. On average, using this technique we are able to close over 10 % of the gap between the pure portfolio and the oracle portfolio.

4   Accuracy Prediction and Recourse

Studying the results of the SAT Competition 2011 one can observe that the best sequential portfolio, 3S only solved 200 out of 300 instances in the application category. However, when analyzing the performance of the solvers the 3S portfolio is composed of, one can also see that the virtual best solver (VBS) based on those solvers can actually solve more than 220 application instances. Hence, the suggestions made by the portfolio are clearly wrong in more than 10 % of all cases. The objective of this section is to lower this performance gap. In the following we first try to determine when the suggestion of a portfolio results in a loss in performance, and second what to do when we believe the portfolio's choice is wrong.

4.1   Accuracy Prediction

One way to potentially improve performance would be to improve the portfolio selector itself (e.g., by multi-class learning). Nonetheless, most classifiers often cannot represent exactly the concept class they are used for. One standard way out in machine learning is to conduct classification in multiple stages, which is what we consider here. Basic classifiers providing a confidence or trust value can function as their own guardian. In Ensemble Learning, more complex recourse classifiers are considered. Our goal here is to design such an approach in the specific context of machine learning methods for SAT solver selection.

We propose a two-stage approach where we augment the existing SAT portfolio classifier by accompanying it with a "guardian" classifier, which aims to predict when the first classifier errs, and a second "selector" classifier that selects an alternative solver whenever the guardian finds that the first selector is probably not right.

To train a guardian and a replacement selector classifier, we first need to capture some characteristics that correlate with the quality of the decision of the portfolio. To that end we propose to create a set of features and label the portfolio's suggestion as "good" or "bad" ($\mathcal{L} = \{\text{good}, \text{bad}\}$). A key question is, how should these two labels be defined. Inspired by the SAT competition context, a "good" decision will be defined as one where an instance can be solved within the given time limit and a "bad" one is when it cannot be.[2]

The definition of a feature vector f to use for a guardian classifier is unfortunately far less straightforward. We, of course, first tried the original features used by 3S but that did not result in an overall improvement in performance. As is typically done in machine learning, we experimented with a few additions and variations, and settled on the following:

---

[2] We also tried labels that identify top performer (e.g., not more than x % slower than the best solver, for various x), but obtained much worse results. The issue here is that it is more ambitious than necessary to predict which solver is best or close to best. Instead, we need to be able to distinguish solvers that are good enough from those that fail. That is, rather than aiming for speed, we optimize for solver robustness.

Table 2. Description of features used by guardian classifier. Solver rank is based on average PAR10 score on neighborhood.

List of employed features	
F1	Distance to closest cluster center
F2	k used for test instance
F3-F7	Min/Max/Average/Median/Variance of distance to closest cluster center
F8	Solver ID selected by k-NN
F9	Solver type: incomplete or complete
F10	Average distance to solved instances by top-2 solvers
F11	VBS time on k-neighborhood of test instance
F12	Number of instances solved by top-5 ranked solvers
F13-F23	PAR10 score/instances solved by top-5 ranked solvers
F24-F34	10 test instance features

We selected 34 features composed of: the first 10 features of the test instance, the Euclidean based distance measures of training instances in the neighborhood to the test instance, and runtime measures of the five best solvers on a restricted neighborhood (see Table 2 for details). These features are inspired by the k-nearest-neighbor classifier that 3S employs.

Consequently, for the guardian we need to learn a classifier function: $f \longmapsto \mathcal{L}$. To this end we require training data. The 3S portfolio is based on data T that is composed of features of and runtimes on 5,467 SAT instances appearing in earlier competitions. We can split T into a training set T_{train} and test set T_{test}. Now, we can run the portfolio restricting its knowledge base to T_{train} and test its performance on T_{test}. For each test instance $i \in T_{test}$ we can compute the corresponding feature vector f_i and obtain the label $\mathcal{L}_i$. Hence, the number of training instances we obtain for the classifier is i. Obviously, one can split T differently over and over by random subsampling, and each time one creates new training data to train the "guardian" classifier.

The question arises whether different splits will not merely regenerate existing knowledge. This depends on the features chosen, but here the feature vector will actually have a high probability to be different for each single split since in each split the neighborhood of a test instance will be different. A thought experiment that makes this more apparent is the following: Assume that, for a single instance i, we sort all other instances according to the distance to i (neighborhood of i). Assume further we select training instances from the neighborhood of i with probability $1/k$ until we have selected k instances (where k is the desired neighborhood size). When $k > 10$ it is obviously very unlikely for an instance to have exactly the same neighbors.

In order to determine an appropriate amount of training data we first randomly split the data set T in a training split $T_{train'}$ and test split $T_{test'}$, before generating the data for the classifier. We then perform the aforementioned splitting to generate training data for the classifier on $T_{train'}$ and test it on the data generated by running k-NN with data $T_{train'}$ on the test set $T_{test'}$. We use 10 different random splits of type $T_{train'}$ and $T_{test'}$ and try to determine the best number of splits for generating training data for the classifier.

While normally one could essentially look at the plain accuracy of the classifier and select the number of splits that result in the highest accuracy, we propose to employ another measure based on the following reasoning. The classifier's "confusion matrix" looks in our context like this (denoting the solver that was selected by the portfolio on instance I with S):

(a) S solves I, and classifier predicts that it can
(b) S solves I, but classifier predicts that it cannot
(c) S can't solve I, but classifier predicts that it can
(d) S can't solve I, and classifier predicts that it cannot

Instances that fall in category (a) reflect a "good" choice by the portfolio (our original selector) and, while correctly detected, there is also nothing for us to gain. In case (c) we cannot exploit the wrong choice of the portfolio since the guardian classifier does not detect it. However, we will also not degenerate the performance of the portfolio. Cases (b) and (d) are the interesting cases. In (b) we collect the false-positives where the classifier predicts that the portfolio's choice was wrong while it was not. Consequently it could be the case that we degrade the performance of the original portfolio selector by altering its decision. All instances falling in category (d) represent the correctly labeled decisions of the primary selector that should be overturned. In (d) lies the *potential* of our method: all instances that fall in this category cannot be solved by solver S that the primary selector chose, and the guardian classifier correctly detected it. Since cases (a) and (c) are somewhat irrelevant to any potential recourse action, we focus on keeping the ratio $\frac{(b)}{(d)}$ as small as possible in order to favorably balance potential losses and wins. Based on this quality measure we determined that roughly 100 splits achieve the most favorable trade off on our data.

4.2   Recourse

When the guardian classifier triggers, we need to select an alternative solver. For this purpose we need to devise a second "recourse" classifier. While we clearly do not want to select the same solver that was suggested by the original portfolio selector, the choices for possible recourse actions is vast and their benefits hardly apparent. We introduce the following recourse strategy:

Since we want to replace the suggested solver S, we assume S is not suitable for the given test instance I. Based on this *conditional probability* we can also infer that the instances solved by S in the neighborhood of size k of I can be removed from its neighborhood. Now, it can be the case that the entire neighborhood of I can be solved by S and therefore we extend the size of the neighborhood by 30 %. If on this extended neighborhood S cannot solve all instances, we choose the solver with the lowest PAR10-score on the instances in the extended neighborhood not solved by S. Otherwise, we choose the solver with the second best ranking by the original portfolio selector. In the context of

3S this is the solver that has the second lowest PAR10-score on the neighborhood of the test instance.

Designing a good recourse strategy poses a challenge. As we will see later in Sect. 5.3, our proposed recourse strategy resulted in solving 209 instances on the 2011 SAT Competition application benchmark, compared to the 204 that 3S solved. We tried a few other simpler strategies as well, which did not fare as well. We briefly mention them here: First, we used the solver that has the second best ranking in terms of the original classifier. For 3S this means choosing the solver with the second lowest PAR10-score on the neighborhood of the test instance. This showed only a marginal improvement, solving 206 instances. We then tried to leverage diversity by mixing-and-matching the two recourse strategies mentioned above, giving each exactly half the remaining time. This resulted in overall performance to drop below 3S without accuracy prediction. Finally, we computed offline a static replacement map that, for each solver S, specifies one fixed solver $f(S)$ that works the best across all training data whenever S is selected by the original classifier but does not solve the instance. This static, feature-independent strategy also resulted in degrading performance. For the rest of this paper, we will not consider these alternative replacement strategies.

5   Empirical Evaluation

In order to evaluate the impact of our two proposed techniques on an existing portfolio solver, we applied them to 3S [8], the best performing sequential portfolio solver at the 2011 SAT Competition.[3] We refer to the resulting enhanced portfolio solver as 3S+f when clause forwarding is used, as 3S+p when accuracy prediction and recourse classifiers are used, and as 3S+fp when both new techniques are applied. We compare their performance to the original 3S, which was the winner in the crafted and random categories of the main sequential track of the 2011 SAT Competition.

As remarked earlier, our techniques are by no means limited to 3S and may be applied to more recent portfolios. However, these techniques are likely to pay off more on harder instances and thus we focus here on the 2011 Competition in which both instance selection and solver design was done with a 5,000 s time limit in mind.

For evaluation, we use the 2011 competition split, i.e., we use the same application (300), crafted (300), and random (600) category instances as the ones used in the main phase of the Competition. The enhanced variants of 3S rely only on the pre-2011 training data that comes with the original 3S. We note that we did conduct experiments using random splits after mixing all instances, but there the performance of the original k-NN classifier of 3S is typically almost perfect, leaving little to no room for improvement. Competition splits exhibit a completely different and perhaps arguably more realistic behavior, as the suboptimal performance of 3S in the application category shows. We thus focused

---

[3]   The source code of 3S can be obtained from http://www.satcompetition.org/

on splits that were neither random nor hand-crafted by us and experimented on competition splits to evaluate the techniques.

All experiments were conducted on 2.3 GHz AMD Opteron 6134 machines with 8 4-core CPUs and 64 GB memory, running Scientific Linux release 6.1. We used a time limit of 6,500 s, which roughly matched the 5,000 s timeout that was used on the 2011 Competition machines. As performance measures we consider the number of instances solved, average runtime, and PAR10 score. PAR10 stands for penalized average runtime, where instances that time out are penalized with 10 times the timeout.

5.1   Implementation Details on Clause Forwarding

We implemented learnt clause forwarding in three CDCL SAT solvers that were used by 3S in the 2011 Competition: CryptoMiniSat 2.9.0 [12], Glucose 1.0 [1], and MiniSat 2.2.0 [14]. The pre-schedule was modified to prolong the time these three clause-learning solvers are run, as discussed earlier. With clause forwarding disabled, 3S with this modified pre-schedule resulted in roughly the same performance on our testbed as 3S with the original pre-schedule used in the Competition (henceforth referred to as 3S-C). In other words, any performance differences we observe can be attributed to clause forwarding and accuracy prediction and recourse, not to the change in the pre-schedule itself.

For clause forwarding, we used parameter values $L = 10$ and $M = 10,000$, i.e., each of the three solvers may share up to 10,000 clauses of size up to 10 for the next solver to be run. The maximum amount of clauses shared is therefore 30,000. We note that these parameters are by no means optimized. Among other variations, we tried sharing an unlimited number of (small) clauses, but this un-surprisingly degraded performance. We expect that these parameters can be tuned better. Nevertheless, the above choices worked well enough to demonstrate the benefits of clause sharing, which is the main purpose of this experimentation.

5.2   Implementation Details on Accuracy Prediction

To predict how good the primary solver suggested by 3S is likely to be, we experimented with several classifiers available in the Weka 3.7.5 data mining and machine learning Java library [5].[4] The results presented here are for the REP-Tree classifier, which is a fast decision tree learner that uses information gain and variance reduction for building the tree, and applies reduced-error pruning. Using training data based on splitting 5,464 pre-2011 instances (the ones 3S is based on) 100 times, as described earlier, we trained a REP-Tree and obtained the following confusion matrix for instances[5] of the 2011 SAT Competition application test data:

---

[4] http://www.cs.waikato.ac.nz/ml/weka/

[5] Note that the numbers do not add up to 300 since, with the classifier, we only consider instances that have not been solved yet by the pre-scheduler and can be solved by at least one of the solvers in our portfolio.

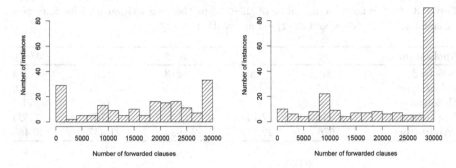

Fig. 1. Histogram showing how often N clauses are forwarded. Left: Crafted instances. Right: Application instances.

(a) 61	(b) 7
(c) 25	(d) 14

Hence, the best possible outcome for a recourse action would be to solve the previously unsolved 14 instances ($\approx 5\%$ of all the 2011 application instances) under (c) and to still be able to solve the 7 instances ($\approx 2\%$) under (b). While in the best case we could gain 14 instances and lose none, it is obviously not clear whether one would achieve *any* gain at all, or even solve at least the 7 instances that originally used to be solved. Fortunately, with our recourse strategy, we witness a significant gain in overall performance. We integrated the classifier in 3S in the following way: When 3S suggests the primary solver, if indicated by our guardian REP-Tree model, we intercept its decision and alter it as proposed by our recourse strategy.

5.3   Results on 2011 SAT Competition Data

Since our base portfolio solver, 3S, already works best on random and crafted instances considered, the objective is to close the large gap between the best sequential portfolios and the best individual solvers in the application track, while not degrading the performance of the portfolio on crafted and random categories.

To this end, let us first note that adding the methods proposed in this paper have no significant impact on 3S performance on random and crafted instances. On random instances, knowledge sharing hardly takes place since CDCL based complete solvers are barely able to learn any short clauses on these instances.

For crafted instances, a limited amount of clause forwarding does happen, but much less than in application instances. In Fig. 1 we show how many instances in our test set share how many clauses. On the left we see that, on crafted instances, we mostly share a modest amount of clauses between solvers, if any. The plot on the right shows the situation for application instances. Here it is usually the case that the solvers share the fully allowed 30,000 clauses.

Table 3. Performance comparison of 3S-C from the competition and its four new variants: 3S, 3S+f, 3S+p, and 3S+fp on application.

Application	3S-C	3S	3S+f	3S+p	3S+fp
# Solved	205	204	213	209	214
# Unsolved	95	96	87	91	86
% Solved	68.3	68.0	71.0	69.7	71.3
Avg runtime	2,764	2,744	2,537	2,707	2,524
PAR10 score	22,676	22,311	20,693	21,437	20,485

Interestingly, the clause sharing in crafted instance causes a slight decrease in performance, but this is outweighed by the positive impact of our prediction and recourse classifiers which actually improve the performance of the solver presented here over 3S on crafted instances. In summary, the solver presented here works as well as 3S on random instances, and insignificantly better than 3S on crafted instances.

It remains to test if the methods proposed here can boost 3S performance to a point where it is competitive on application instances as well.

Table 3, column 1, shows the performance of the 3S version available from the competition website (3S-C). The number of instances it solves here differs slightly from the competition results due to hardware and experimental differences. Comparing 3S-C with 3S (where we changed the pre-schedule to allow more clauses to be learned), we observe that the difference is very small. 3S-C solves just 1 instance more than 3S, letting us conclude that the subsequently reported performance gains are not due to differences in the pre-schedule itself.

Both 3S+f (3S with clause forwarding) and 3S+p (3S with prediction and recourse) are able to improve on 3S in a significant fashion: 3S+f is able to solve 9 more instances, and 3S+p solves 5 more. Note that in the application category it is usually the case that the winning solver only solves a couple of more instances than the second-ranked solver. Indeed, the difference between the 10th ranked 3S in the competition and the winner was only 15 instances. That is to say, prediction and recourse closes 33 % of the gap to the winning solver, and clause forwarding even over 60 %.

The combination of clause forwarding and prediction and recourse in 3S+fp is able to solve 214 instances. This is just one instance shy of the best sequential solver Glucose 2.0 at the 2011 SAT Competition which we re-ran on our hardware using the same experimental settings. Note that 3S uses only pre-2011 solvers. Furthermore, we found that the average runtime of 3S+fp is close to Glucose 2.0 as well, which also indicates that in contexts where objectives other than the number of solved instances are of interest, 3S+fp is very competitive.

6   Conclusion

We presented two novel generic techniques for boosting the performance of SAT portfolios. The first approach shares the knowledge discovered by SAT solvers that run in sequence, while the second improves solver selection accuracy by detecting when a selection is likely to be inferior and proposing a more promising recourse selection. Applying these generic techniques to the SAT portfolio 3S resulted in significantly better performance on application instances while not reducing performance on crafted and random categories, making the resulting solver, 3S+fp excel on all categories in our evaluation using the 2011 SAT Competition data and solvers.

References

1. Audemard, G., Simon, L.: Predicting learnt clauses quality in modern SAT solvers. In: 21st IJCAI, Pasadena, CA, pp. 399–404, July 2009
2. Biere, A.: Plingeling: solver description. SAT Race (2010)
3. Cook, S.A.: The complexity of theorem-proving procedures. In: STOC, pp. 151–158. ACM (1971)
4. Gomes, C.P., Selman, B.: Algorithm portfolios. AI J. **126**(1–2), 43–62 (2001)
5. Hall, M., Frank, E., Holmes, G., Pfahringer, B., Reutemann, P., Witten, I.H.: The WEKA data mining software: an update. SIGKDD Explor. **11**(1), 10–18 (2009)
6. Hamadi, Y., Sais, L.: ManySAT: a parallel SAT solver. JSAT **6**, 245–262 (2009)
7. Rice, J.R.: The algorithm selection problem. Adv. Comput. **15**, 65–118 (1976)
8. Kadioglu, S., Malitsky, Y., Sabharwal, A., Samulowitz, H., Sellmann, M.: Algorithm selection and scheduling. In: Lee, J. (ed.) CP 2011. LNCS, vol. 6876, pp. 454–469. Springer, Heidelberg (2011)
9. Leyton-Brown, K., Nudelman, E., Andrew, G., McFadden, J., Shoham, Y.: A portfolio approach to algorithm selection. In: IJCAI, pp. 1542–1543 (2003)
10. Malitsky, Y., Sabharwal, A., Samulowitz, H., Sellmann, M.: Non-model-based algorithm portfolios for SAT. In: Sakallah, K.A., Simon, L. (eds.) SAT 2011. LNCS, vol. 6695, pp. 369–370. Springer, Heidelberg (2011)
11. Peng, F., Tang, K., Chen, G., Yao, X.: Population-based algorithm portfolios for numerical optimization. IEEE Trans. Evol. Comput. **14**(5), 782–800 (2010)
12. Soos, M.: CryptoMiniSat 2.9.0. http://www.msoos.org/cryptominisat2 (2010)
13. Sorensson, N., Een, N.: SatELite 1.0. http://minisat.se (2005)
14. Sorensson, N., Een, N.: MiniSAT 2.2.0. http://minisat.se (2010)
15. Xu, L., Hutter, F., Hoos, H.H., Leyton-Brown, K.: SATzilla-07: the design and analysis of an algorithm portfolio for SAT. In: Bessière, C. (ed.) CP 2007. LNCS, vol. 4741, pp. 712–727. Springer, Heidelberg (2007)
16. Xu, L., Hutter, F., Hoos, H.H., Leyton-Brown, K.: SATzilla: portfolio-based algorithm selection for SAT. JAIR **32**(1), 565–606 (2008)

A Fast and Adaptive Local Search Algorithm for Multi-Objective Optimization

Duy Tin Truong[✉]

University of Trento, Trento, Italy
truong@disi.unitn.it

Abstract. Although population-based algorithms are robust in solving Multi-objective Optimization Problems (MOP), they often require a large number of function evaluations. In contrast, individual-solution based algorithms are fast but can be stuck in local minima. To solve these problems, we introduce a fast and adaptive local search algorithm for MOP. Our algorithm is an individual-solution algorithm with a flexible mechanism for switching between the exploration and exploitation phase to escape from local minima. The experimental results on the *DTLZ* benchmark show that our algorithm significantly outperforms the popular evolutionary algorithm **NSGAII** and three other simulated annealing algorithms for MOP.

1 Introduction

Multi-objective Optimization Problems (MOP) appear in many practical applications in engineering, finance, transportation, etc. They are characterized by multiple objective functions which need to be jointly optimized. The Pareto-optimal solutions of MOP are the solutions where no single objective can be improved without worsening at least another objective. The set of all Pareto-optimal solutions in the objective space is called the Pareto front. Several algorithms have been proposed to approximate the Pareto front with a single run. They can be split into two groups: population method and individual-solution method. The first group contains evolutionary algorithms (EAs) [1,2]. In each generation, EAs modify a population of solutions by genetic operators such that the next population contains high-quality *and* diverse solutions. Although EAs are robust for many problems, they often require a large number of function calls to evaluate the quality of solutions. In contrast, the individual-solution algorithms [5,6] try improving only one solution in each iteration, by replacing it with a better one in its neighborhood. When that solution is Pareto-optimal, it is saved as a representative solution on the Pareto front. As only one solution is evaluated in each iteration, the number of function evaluations can be significantly reduced. However, without a good trade-off between exploration and exploitation, these algorithms can be stuck in local minima.

To solve the above difficulties, we introduce a **F**ast and **A**daptive Local **S**earch **A**lgorithm for Multi-objective **O**ptimization (abbreviated as **FASAMO**).

G. Nicosia and P. Pardalos (Eds.): LION 7, LNCS 7997, pp. 168–173, 2013.
DOI: 10.1007/978-3-642-44973-4_18, © Springer-Verlag Berlin Heidelberg 2013

It is an individual-solution algorithm with an automatic mechanism for switching between the exploration and exploitation phase to escape from local minima.

The rest of the paper is organized as follows. We conclude this section by summarizing some related works. Then, we describe our algorithm in Sect. 2 and compare it with four other algorithms in Sect. 3.

Related Work. One of the first individual-solution algorithms for MOP is the multi-objective simulated annealing algorithm (MOSA) proposed by Serafini [5] (denoted as **SMOSA** in this paper). In each iteration, **SMOSA** optimizes a weighted sum of the objectives. To diversify the optimal solution set, the algorithm modifies slightly the weight vector in each search step. Ulungu et al. [7] later introduces a population-based MOSA algorithm (denoted as **UMOSA**). **UMOSA** first generates a set of random solutions and associates each initial solution with a random weight vector. Then, in each generation, **UMOSA** optimizes each solution by the scalarizing function with the weight vector of that solution. Smith et al. also proposes another individual-solution MOSA algorithm in [6] (denoted as **DMOSA**) which minimizes an energy function based on the dominance count to approximate the Pareto front. The energy function is defined such that solutions near the Pareto front and in sparse regions have small energy. Instead of improving individual solutions, the popular algorithm **NSGAII** proposed by Deb et al. [2] tries to maintain populations of diverse and high-quality solutions. In each generation, it selects the best solutions in the current population to form the parent population. Then, it applies genetic operators on these parents to generate a next offspring population.

2   Fast and Adaptive Local Search Algorithm for MOP

Our algorithm **FASAMO** uses the dominance-based energy function as in [6]. Let $nonDomSet$ be the set of non-dominated solutions obtained so far by the algorithm. The energy $energy(\mathbf{x}_i)$ of a solution $\mathbf{x}_i$ w.r.t $nonDomSet$ is defined as the portion of the solutions in $nonDomSet$ that dominate $\mathbf{x}_i$: $|\{\mathbf{x}_j|\mathbf{x}_j \in nonDomSet$ & $\mathbf{x}_j$ dominates $\mathbf{x}_i\}|$ / $|nonDomSet|$. An important property of this energy function is that solutions near the Pareto front and in spare regions have small energy. Besides, if the size of $nonDomSet$ is less than 100, we increase the energy resolution by sampling 100 points on the attainment surface of the approximate front (formed by $nonDomSet$) as in [6].

FASAMO consists of two main phases: exploration and exploitation. In the exploration phase, the perturbation level is set to a large value to help **FASAMO** identify quickly local fronts. In each iteration, the algorithm improves the current solution by replacing it with one of its neighbor solutions having smaller energy. When the current solution is on a local optimal front, the algorithm switches to the exploitation phase for exploiting that local front. In this phase, the perturbation level is first decreased to help the algorithm determine the best solutions on that front. Then, it is increased, thus the algorithm can discover other optimal solutions around these best solutions and move gradually to a neighbor region. This process is repeated periodically by using a cosine function to adjust the perturbation level. Besides, two phases of the algorithms can be switched according to the approximate front and the current solution. This gives the algorithm chances to correct the mistake of switching to the exploitation phase too soon. In addition, to increase the convergence speed, if a perturbation step is successful then it will be performed again in the next step.

Algorithm 1: Fast and Adaptive Local Search Algorithm for MOP

$\mathbf{x}_{current}$ = a random solution; $nonDomSet = \{\mathbf{x}_{current}\}$
$searchDirection = random$; $perturbationLevel = maxPerturbation$
for $i = 1$ *to* $maxEvaluations$ **do**
    $\mathbf{x}_{new} = \mathbf{x}_{current}$
    if $searchDirection = random$ **then**
        | Perturb $\mathbf{x}_{new}$ randomly with the scale of $perturbationLevel$.
    else
        └ Perturb $\mathbf{x}_{new}$ as the previous perturbation step does.
    Add $\mathbf{x}_{new}$ to $nonDomSet$.
    if $energy(\mathbf{x}_{new}) \leq energy(\mathbf{x}_{current})$ *or* $\mathbf{x}_{new}$ *improves* $nonDomSet$ **then**
        $\mathbf{x}_{current} = \mathbf{x}_{new}$
        | $searchDirection = goOn$
    else
        └ $searchDirection = random$
    // Switch phases
    if $phase = exploration$ **then**
        if $\mathbf{x}_{current}$ *is unimproved sequentially for* $maxUnimproved$ *times* **then**
            └ $phase = exploitation$
    else
        $cosFactor = 0.5 \times (1 + cos(2\pi\frac{(i \mod cosFreq)}{cosFreq}))$ // $cosFactor \in [0,1]$
        $perturbationLevel = maxPerturbation \times cosFactor$
        if $\mathbf{x}_{current}$ *is improved for* $maxImproved$ *times in this phase* **then**
            $phase = exploration$
            └ $perturbationLevel = maxPerturbation$
return $nonDomSet$

The pseudo-code of **FASAMO** is shown as in Algorithm 1. **FASAMO** is initialized by a random solution. In this paper, each solution $\mathbf{x}_i$ is presented as a vector of D real variables $\{\mathbf{x}_i^k\}_{k=1}^D$. In each iteration, a new solution $\mathbf{x}_{new}$ is generated by first copying from the current solution $\mathbf{x}_{current}$. Then, this new solution $\mathbf{x}_{new}$ is modified based on the search direction. If the search direction is $random$, then the algorithm picks randomly a variable $\mathbf{x}_{new}^k$ and adds to it a random value Δ^k generated from a Laplacian distribution with the mean of 0 and the scale of $perturbationLevel$. Otherwise, the previous perturbation step is performed again with a doubled length by adding $2\Delta^l$ to $\mathbf{x}_{new}^l$ where l was the index of the perturbed variable in the previous iteration and Δ^l was the perturbation value added on that variable. The reason for doubling the perturbation length is to increase the convergence speed and avoid executing a large number of small perturbation steps. Besides, we use the Laplacian distribution since it has a fatter tail than the normal distribution. This gives the algorithm a high probability of exploring the regions that are far from the current solution. Next, the new solution $\mathbf{x}_{new}$ is added to the non-dominated solution set *nonDomSet*. If $\mathbf{x}_{new}$ dominates at least one solution in *nonDomSet*, then $\mathbf{x}_{new}$ is considered as it has improved *nonDomSet*. If the new solution improves the non-dominated solution set or its energy is smaller than or equal to the current solution energy, then the algorithm moves to the new solution and sets the search direction to *goOn*. Otherwise, the search direction is set to *random*. After that, the algorithm considers the improvement status to switch between two phases and adjust the perturbation level. In detail, if the algorithm is in the exploration phase and the number of times that the algorithm sequentially cannot improve the current solution reaches a maximum value *maxUnimproved*, then the algorithm switches to the exploitation phase as the current solution is now on a local front. When the algorithm is in the exploitation phase, the perturbation level *perturbationLevel* is adjusted by a cosine function with a frequency of *cosFreq*. If the number of times that the current solution is improved equals to a maximum value *maxImproved*, then the algorithm switches again to the exploration phase. The main reason of this step is that the improvement of the current solution implies that the current solution is now in a new and more potential region, thus the large search steps can be performed to quickly reach the local minima of that region or identify other potential neighbor regions.

3   Experiments

This section presents the experiments to compare our algorithm and four other algorithms: **SMOSA**, **UMOSA**, **DMOSA** and **NSGAII** on the *DTLZ* benchmark [3] (with the number of variables and objectives as suggested in [3]). All algorithms are implemented on the *jmetalcpp* framework [4]. The parameters of **FASAMO** are set as follows: $maxUnimproved = 100$, $maxImproved = 10$, $maxPerturbation = 0.3$ and $cosFreq = 1 + 10 * numberOfVariables$. The odd value of *cosFreq* is used to eliminate the zero value of the perturbation level. The parameters of **DMOSA** is set as suggested in [6]. The default parameters of **NSGAII** in the *jmetalcpp* framework are used in the experiments. The maximum temperature T_{max} of **SMOSA** and **UMOSA** is set to 1.0. For measuring the distance to the true Pareto front and the coverage of the solution set, we use the generational distance metric [8] and the hyper volume metric [9] (see the implementation for the minimization problem in the *jmetalcpp* framework), respectively. Besides, the Pareto front samples on the website of *jmetalcpp* are used when computing the metrics.

Experimental Results on the Number of Function Evaluations: In this experiment, we measure the number of evaluations required by each algorithm to obtain the approximate front with the generational distance of 0.01 (to the true Pareto front). We run each algorithm 10 times on each problem. In a run, for every 100 evaluations, the algorithms check whether the generational distance of the approximate front is less than or equal to 0.01. If this condition is hold or the number of evaluations equals to 100000, the algorithms stop and report the number of evaluations.

Table 1. Average number (± Standard Deviation) of function evaluations. NA (Not Available) means that a number of runs (written in brackets) did not reach the required target: generational distance of 0.01 within 100000 iterations.

	DTLZ1	DTLZ2	DTLZ3	DTLZ4	DTLZ5	DTLZ6	DTLZ7
FASAMO	4180 ± 120	410 ± 123	8760 ± 113	2580 ± 1207	340 ± 12	280 ± 25	370 ± 22
SMOSA	NA (10)	9500 ± 189	NA (7)	NA (9)	11050 ± 3272	8050 ± 1596	6330 ± 768
DMOSA	6850 ± 173	3290 ± 98	12590 ± 154	NA (3)	3140 ± 113	3140 ± 240	2880 ± 69
UMOSA	NA (10)	14370 ± 262	NA (10)	NA (10)	22010 ± 2280	24410 ± 540	27230 ± 53
NSGAII	26090 ± 1356	2490 ± 60	53650 ± 2355	2790 ± 218	2060 ± 18	43530 ± 1021	4920 ± 56

Table 1 shows the average number of evaluations required by five algorithms on seven problems. We only compute the average evaluation number for the algorithms which can finish in all 10 run times. As can be seen, on all problems, **FASAMO** requires a much smaller number of evaluations compared to that one of the other algorithms. Besides, on the difficult problems like *DTLZ1*, *DTLZ3*, *DTLZ4*, **SMOSA**, **DMOSA** and **UMOSA** cannot approach the generational distance of 0.01 within 100000 iterations.

Experimental Results on Generational Distance and Hyper Volume: In this experiment, we fix the number of evaluations to compare the performance on the generational distance and hyper volume. We set the number of evaluations of five algorithms on each problem as the average number of evaluations required by **FASAMO** to reach the generational distance of 0.01 (presented in Table 1). However, as the population size of **UMOSA** and **NSGAII** is 100, we round up these values to the nearest multipliers of 100. Table 2a presents the generational distance of five algorithms on seven problems. Except for problem *DTLZ4*, on the other six problems, **FASAMO** obtains a much smaller generational distance compared to that one of the other algorithms. In more detail, the second smallest generational distance is at least 10 times larger than that one of **FASAMO**. As for problem *DTLZ4*, **FASAMO** is slightly outperformed by **NSGAII**. In this problem, a very large portion of decision space is mapped to a single point in the objective space [3,6]. Thus, when **FASAMO** samples in this special region of decision space, a large

Table 2. Performance comparison

	DTLZ1	DTLZ2	DTLZ3	DTLZ4	DTLZ5	DTLZ6	DTLZ7
(a) Generational distance with (Mean ± Standard Deviation) ×10^{-2}							
FASAMO	1.41 ± 0.41	0.59 ± 0.09	2.04 ± 0.67	1.81 ± 0.31	1.15 ± 0.13	7.24 ± 2.25	1.01 ± 0.03
SMOSA	71.27 ± 18.86	14.32 ± 1.66	90.80 ± 18.19	22.06 ± 2.13	23.31 ± 2.16	220.67 ± 17.38	79.72 ± 2.51
DMOSA	117.11 ± 25.61	19.16 ± 0.92	823.03 ± 74.59	16.68 ± 0.93	28.86 ± 1.85	245.17 ± 14.92	85.71 ± 4.65
UMOSA	891.00 ± 60.62	9.90 ± 0.12	2722.03 ± 145.93	12.34 ± 0.00	16.13 ± 0.10	133.53 ± 1.57	70.82 ± 1.12
NSGAII	1450.47 ± 62.91	6.71 ± 0.04	2082.86 ± 4.59	**1.04 ± 0.04**	12.31 ± 0.10	115.75 ± 0.04	51.25 ± 1.51
(b) Hyper volume with (Mean ± Standard Deviation) ×10^{-2}							
FASAMO	56.61 ± 6.72	27.38 ± 1.25	24.58 ± 7.77	12.70 ± 1.31	5.33 ± 0.13	5.68 ± 0.34	21.55 ± 0.06
SMOSA	14.82 ± 4.69	1.17 ± 1.14	6.19 ± 0.26	0.20 ± 0.06	0.00 ± 0.00	0.00 ± 0.00	0.00 ± 0.00
DMOSA	0.01 ± 0.00	0.01 ± 0.00	0.00 ± 0.00	0.00 ± 0.00	0.00 ± 0.00	0.00 ± 0.00	0.00 ± 0.00
UMOSA	0.00 ± 0.00	0.70 ± 0.22	0.00 ± 0.00	0.00 ± 0.00	0.01 ± 0.02	0.00 ± 0.00	0.00 ± 0.00
NSGAII	0.00 ± 0.00	3.03 ± 0.44	0.00 ± 0.00	**29.15 ± 0.03**	0.09 ± 0.02	0.00 ± 0.00	0.00 ± 0.00

number of evaluations can be wasted as identical solutions in the objective space are produced. In contrast, **NSGAII** may overcome this issue better as it uses crossover operators and has an explicit diversifying mechanism.

Table 2b shows the hyper volume of five algorithms on seven problems. Similarly to the case of generational distance, **FASAMO** is slightly worse than **NSGAII** on the *DTLZ4* problem. However, on the other six problems, **FASAMO** outperforms significantly the other four algorithms. Especially, on two problems *DTLZ6* and *DTLZ7*, except for **FASAMO**, the other algorithms have no solution in the coverage volume of the true Pareto front.

4   Conclusion

In this paper, we propose a fast and adaptive local search algorithm for MOP, called **FASAMO**. Our algorithm is an individual-solution algorithm with an automatic mechanism for switching between two phases of exploration and exploitation. The experiments on seven problems of the *DTLZ* benchmark show that our algorithm significantly outperforms the popular evolutionary algorithm **NSGAII** and three other multi-objective simulated annealing algorithms.

References

1. Coello, C.A.C., Lamont, G.B., Van Veldhuizen, D.A.: Evolutionary algorithms for solving multi-objective problems, vol. 5. Springer, Heidelberg (2007)
2. Deb, K., Pratap, A., Agarwal, S., Meyarivan, T.: A fast and elitist multiobjective genetic algorithm: Nsga-ii. IEEE Trans. Evol. Comput. **6**(2), 182–197 (2002)
3. Deb, K., Thiele, L., Laumanns, M., Zitzler, E.: Scalable multi-objective optimization test problems. In: Proceedings of the Congress on Evolutionary Computation (CEC-2002), Honolulu, USA, pp. 825–830 (2002)
4. Durillo, Juan J., Nebro, Antonio J.: jmetal: a java framework for multi-objective optimization. Adv. Eng. Softw. **42**, 760–771 (2011)
5. Serafini, P.: Simulated annealing for multiple objective optimization problems. In: Tzeng, G., Wang, H., Wen, U., Yu, P. (eds.) Multiple Criteria Decision Making. Expand and Enrich the Domains of Thinking and Application, pp. 283–292. Springer, Heidelberg (1994)
6. Smith, K.I., Everson, R.M., Fieldsend, J.E., Murphy, C., Misra, R.: Dominance-based multiobjective simulated annealing. IEEE Trans. Evol. Comput. **12**(3), 323–342 (2008)
7. Ulungu, E.L., Teghem, J., Fortemps, P.H., Tuyttens, D.: Mosa method: a tool for solving multiobjective combinatorial optimization problems. J. Multi-Criteria Decis. Anal. **8**(4), 221–236 (1999)
8. Van Veldhuizen, D.A., Lamont, G.B.: Evolutionary computation and convergence to a pareto front. In: Late Breaking Papers at the Genetic Programming 1998 Conference, pp. 221–228 (1998)
9. Zitzler, E., Thiele, L.: Multiobjective evolutionary algorithms: a comparative case study and the strength pareto approach. IEEE Trans. Evol. Comput. **3**(4), 257–271 (1999)

An Analysis of Hall-of-Fame Strategies in Competitive Coevolutionary Algorithms for Self-Learning in RTS Games

Mariela Nogueira[1], Carlos Cotta[2], and Antonio J. Fernández-Leiva[2(✉)]

[1] University of Computers Science, Havana, Cuba
[2] University of Málaga, Málaga, Spain
mnogueira@uci.cu, {ccottap,afdez}@lcc.uma.es

Abstract. This paper explores the use of Hall-of-Fame (HoF) in the application of competitive coevolution for finding winning strategies in *RobotWars*, a two-player real time strategy (RTS) game developed in the University of Malaga for research purposes. The main goal is testing different approaches in order to implement the concept of HoF as part of the self learning mechanism in competitive coevolutionary algorithms. Five approaches were designed and tested, the difference between them being based on the implementation of HoF as a long or short-term memory mechanism. Specifically they differ on the police followed to keep the members in the champions' memory during an updating process which deletes the weakest individuals, in order to consider only the robust members in the evaluation phase. It is shown how strategies based on periodical update of the HoF set on the basis of quality and diversity provide globally better results.

Keywords: Coevolution · RTS game · Game's strategy · Evaluation process · Memory mechanism

1 Introduction

Artificial Intelligence (AI) implementation represents a challenge in game development: a deficient game AI surely decreases the satisfaction of the player. Game AI has been traditionally coded manually, causing well-known problems such as loss of reality, sensation of artificial stupidity, and predictable behaviors, among others; to overcome them a number of advanced AI techniques have recently been proposed in the literature. Coevolution, a biologically inspired technique based on the interaction between different species, represents one of the most interesting techniques to this end.

Coevolutionary systems are usually based on two kinds of interactions: one in which different species interact symbiotically (i.e., the cooperative approach) and other in which species compete among them (i.e., the competitive approach). In cooperation-based approaches, an individual is typically decomposed in different

G. Nicosia and P. Pardalos (Eds.): LION 7, LNCS 7997, pp. 174–188, 2013.
DOI: 10.1007/978-3-642-44973-4_19, © Springer-Verlag Berlin Heidelberg 2013

components that evolve simultaneously and the fitness depends on the interaction between these components; in competition-based approaches, an individual competes with other individuals for the fitness value and, if appropriate, will increase its fitness at the expense of its counterparts, whose fitnesses decrease. This latter approach resembles an army race in which the improvement of some individuals causes the improvement in others, and vice versa.

This paper deals with the application of competitive coevolution (CC) as a self learning mechanism in RTS games. As a first contribution, the paper analyzes the performance of different approaches in order to apply the concept of Hall-of-Fame (HoF) defined by Rosin and Belew in [1] as a long-term memory in competitive coevolutionary algorithms; the analysis is conducted in the context of the real-time strategy (RTS) game *RobotWars*. The goal is to produce automatically game strategies to govern the behavior of an army in the game that can also beat its opponent counterpart. As a second contribution this work proposes alternatives for optimizing two key aspects in the implementation of the HoF, which are, the diversity of the solutions, and the growth of the champions' memory.

This paper is organized as follows. Next, –and given that we focus in this work on exploring different variants of HoF as an evaluation and memory mechanism in competitive coevolutionary settings– we present an overview of competitive coevolution in games. In Sect. 3 we explain the game which is our arena for competitive coevolution. Section 4 describes ours variants for implementing a Hall-of-Fame based competitive coevolutionary algorithm. In Sect. 5 we analyze the results obtained by each variants in many experiments. And finally in Sect. 6 we closure this investigation.

2   Background on Competitive Coevolution in Games

Coevolution has been shown to be successful on a number of applications but it also has a number of drawbacks [2]. The primary remedy to cope with the inherent pathologies of coevolution consists of proposing new forms of evaluating individuals during coevolution [1], and memorization of a number of successful solutions to guide the search is one of the most employed. Following this idea, Rosin and Belew [3] already proposed the use of a *Hall-of-Fame* (HoF) based mechanism as an archive method and, since then, there have been similar proposals such as those described in [4–6]. According to [7] the question of how to actually use the memory in the coevolution tends to fall into two areas: inserting individuals from memory into the coevolution, or evaluating individuals from the populations against the memory. Precisely, our investigation fits to this latter area, we have implemented different variants of Hall-of-Fame for controlling evaluation process in a CC algorithm, which tries to find wining strategies for the game *RobotWars*.

Several experiments have showed significants results in the application of coevolutionary models in competitive environments; for example the study described in [8] on competitive fitness functions in the Tic Tac Toe game, the application of simple competitive models for evolving strategies in a pursuit-evasion game [9], or the evolution of both morphology and behaviors of artificial creatures through competition in a predator-prey environment [10]. Competitive coevolution continues to be useful nowadays, and has been used heavily in complex scenarios like those that emerge in strategy games; so, Smith et al. [11] coevolved artificial intelligent opponents with the objective of training human players in the context of a game of type capture-the-flag. Also, Avery and Louis [12] analyzed the employment of coevolution for creating a tactical controller for small groups of game entities in a real-time capture-the-flag game; a representation for generating adaptive tactics using coevolved Influence Maps was proposed, and the result was the attainment of an autonomous entity that plays in coordination with the rest of the team to achieve the team objectives. More recently, Dziuk and Miikkulainen [13] explores several methods for automatically shaping the coevolutionary process, and this is done by modifying the fitness function as well as the environment during evolution.

Other research that addresses the use of the HoF concept as an evaluation mechanism in a competitive-coevolutive environment was given by Pawel Lichocki in [14]. He implemented the HoF with three useful extensions included: uniqueness, manual teachers, and Competitive Fitness Sharing [3]. The results of this work showed that HoF works better than SET (Single Elimination Tournament) [8], but this method was not sufficient to prevent the lack of diversity in the population. In our variants of HoF the probability of a repeated member being inserted in the memory is minimal, because we do coevolutions by turns of two independent populations, and each coevolutive turn begins with a new population which evolves until finding an unique champion, or reaching the maximum number of continuous coevolution without success; and this champion must defeat all the members of the opponents' HoF. The use of manual teachers is also possible in our system: the first coevolutive iteration may be started with *random* or *offensive* –manually defined– strategies.

Another interesting perspective was presented in [15] where the authors using the game of Tempo as a test space, tried to ease the selection of optimal strategies by clustering the solutions in the population of a coevolutionary system through the concept of similarity. This cluster system integrated a long-term memory that valued the changes produced in the environment to trigger appropriate coevolution. The game of Tempo has also been used with the aim of improving the creation of smart agents in [16,17].

3   Game Description

This section is devoted to *RobotWars*,[1] our arena for competitive coevolution which will be presented for first time in [18]. *RobotWars* is a test environment that allows two virtual players (i.e., game AIs) to compete in a 3 dimensional scenario of a RTS war game, and thus it is not a standard game itself in the sense that no human players intervene interactively during the game; however it

---

[1] http://www.lcc.uma.es/~afdez/robotWars

is a perfect scenario to test the goodness and efficacy of (possibly hand-coded) strategies to control the game AI and where the human player can influence the game by setting its initial conditions.

In *RobotWars*, two different armies (each of them controlled by a virtual player - i.e., a game AI) fight in a map (i.e., the scenario) that contains multiple obstacles and has limited dimensions. Each army consists of a number of different units (including one general) and the army that first wipes out the enemy general is considered the winner. Virtual players are internally coded as a 4 dimension matrix where the first dimension has 7 different values corresponding to the *type of unit* (i.e., general, infantry, chariot, air force, antiaircraft, artillery, cavalry), the second dimension to *objective closeness* (i.e., a binary value: 0 if the unit is closer to the enemy general than to its friendly general, and 1 otherwise), the third dimension to *numeric advantage* (i.e., are there, in a nearby space, more friendly units than enemy units? A binary answer:yes/no), and the fourth dimension to *health* (i.e., an amount that indicates the health level as high, medium or low). Each position of the matrix acts as a gen and stores one of the following 5 actions: *attack, advance, recede, crowd together*, or *no operation*.

The whole matrix represents a strategy that controls, deterministically, the behavior of an army during the game. For a specific type of unit there are 12 possible different states (i.e., $2 \times 2 \times 3$, all the possible value combinations considering the last three dimensions of the matrix), and basically, in a specific turn of the game each unit of the army will execute the action stored in the state in which the unit perceives that it is. Note that all the units (irrespective of their type) are managed by the same controller, and in any instant of the game, the team behavior will be the result of summing up all the action taken by each of its constituent units. Note however that this does not mean all the units execute the same action because the action to be executed by a unit will depend on its particular situation in the match and its specific category.

4   Hall-of-Fame Based Competitive Coevolutionary Algorithm and Variants

Our objective is to apply competitive coevolution techniques to automate the generation of victorious strategies for the game described above. According to the strategies encoding shown in the previous section, the search space is $5^{7 \times 2 \times 2 \times 3} = 5^{84}$, which is really huge, and cannot be efficiently explored using implicit enumeration techniques due to the inherently complex and non-linear behavior of game simulations. Thus, the use of metaheuristic techniques is approached.

Using our RTS game we test five variants of a competitive coevolutionary algorithm that uses the HoF as a memory mechanism to keep the winning strategies found in each coevolutionary step, to this end the best individual from each generation is retained for future testing. In our approach, each army (i.e., player) maintains its own HoF, in which its own winning strategies (with respect to the set of winning strategies of its opponent) found in each coevolutionary step will be saved.

Regarding the use and implementation of HoF some aspects must be defined. The first is the criteria for inserting a new member in the memory. Also we have considered different policies for maintaining the champions in the set, regarding this issue one has to take into account the contribution of the individual (i.e., the champion) to the search process as, for instance, it might be the case that some opponents that belong to very old generations do not show a valuable performance in comparison with opponents generated in recent generations and thus they might be easily beaten; it is therefore crucial to remove those champions not contributing to the solution what, in other words, represents a mechanism to control the size of the champions' memory. Another relevant aspect concerns to the selection of those strategies from HoF that will be employed in the evaluation process; considering all the champions might produce more consistent solutions at the expense of a very high computational cost (note that a simulation of the match must be executed for each champion involved in the evaluation; we will provide more details on this further on). Next we present our HoF-based competitive coevolutionary algorithm (HofCC) and five variants that precisely differ in the policy of establishing the aspects mentioned previously.

4.1   Basic HofCC

Algorithm 1 shows the schema of our basic algorithm HofCC. A specific strategy is considered *winning* if it achieves a certain score (see below) when it deals with each of the strategies belonging to the set of winning strategies of its opponent (i.e., the enemy Hall-of-Fame). The initial objective is to find a winning strategy of player 1 with respect to player 2 (i.e., the initial opponent) so that the HoF of player 2 is initially loaded with some strategies (randomly or manually initialized: line 2). Then a standard evolutionary process tries to find a strategy for player 1 that can be considered as victorious (lines 7–13). A strategy is considered winning if its fitness value is above a certain threshold value ϕ (line 14) that enables the tuning of the selective pressure of the search process by considering higher/lower quality strategies; in case of success (line 14), this strategy is added to the HoF of player 1 (line 16) and the process is initiated again but with the players' roles interchanged (line 17); otherwise (i.e., no winning strategy is found) the search process is restarted again. If after a number of coevolutionary steps no winning strategy is found the search is considered to have stagnated and the coevolution finishes (see while condition in line 4). At the end of the whole process we obtain as a result two sets of winning strategies associated respectively to each of the players.

Regarding to the evaluation of candidates for a specific player p (where $p \in$ {*player 1, player 2*}), the fitness of an specific strategy is computed by facing it against a selected subset of the (winning) strategies in the Hall-of-Fame of its opponent player (that we call the *selected opponent set*). Given a specific strategy s its fitness is computed as follows:

$$fitness(s) = \frac{\sum_{j=1}^{k} \left(p_j^s + extras_s(j)\right)}{k} \tag{1}$$

Algorithm 1: HofCC()

```
1  nCoev ← 0; A ← player₁; B ← player₂; φ ← thresholdvalue;
2  HoF_A ← ∅; HoF_B ← INITIALOPPONENT();
3  pop ← EVALUATE(HoF_B); // Evaluate initial population
4  while nCoev < MaxCoevolutions ∧ NOT(timeout) do
5      pop ←RANDOMSOLUTIONS(); // pop randomly initialized
6      i ← 0;
7      while (i < MaxGenerations) ∧ (fitness(best(pop)) < φ) do
8          parents ←SELECT (pop);
9          childs ← RECOMBINE (parents, p_X);
10         childs ← MUTATE (childs, p_M);
11         pop ← REPLACE(childs);
12         pop ← EVALUATE(HoF_B);
13         i ← i + 1;
14     end while
15     if fitness(best(pop)) ≥ φ then //winner found!
16         nCoev ← 0; // start new search
17         HoF_A ← HoF_A ∪ {best(pop)}
18         temp ← A; A ← B; B ← temp; // interchange players' roles
19     else
20         nCoev ← nCoev + 1; // continue search
21     end if
22 end while
```

---

where $k \in \mathbb{N}$ is the cardinality of the selected opponent set, $p_j^s \in \Re$ returns ϕ points if strategy s beats strategy h_j belonging to the selected opponent set (i.e., victorious case), $\frac{\phi}{2}$ in case of a draw, and 0 if h_j wins to strategy s; Also:

$$extras_s(j) = c - nturn_{sj} + c * \Delta health_{sj} + c * \Delta Alive_{sj} \qquad (2)$$

where $c \in \mathbb{N}$ is a constant, $nTurn_{sj} \in \mathbb{N}$ is the number of turns spent on the game to achieve a victory of s over its opponent h_j (0 in case of draw or defeat), $\Delta health_{sj} \in \mathbb{N}$ is the difference between the final health of the army (i.e., sum of the health of all its living units) controlled by strategy s at the end of the match and the corresponding health of the enemy army, and $\Delta Alive_{sj} \in \mathbb{N}$ is its equivalent with respect to the number of living units at the end of the combat. This fitness definition was formulated based on our game experience, and it values the victory above any other result.

4.2 HofCC Variants

This section is devoted to describe five variants of our HofCC algorithms; basically these variants differ in the nature (i.e., in this case, size) of the HoF when it is used as a memory mechanism, ranging from short-term memory versions to long-term memory instances.

4.2.1 HofCC-Complete

In this variant the Hall-of-Fame acts as a long-term memory by keeping all the winners found in previous coevolutions, and all of them are also used in the evaluation process. So, in the coevolutionary step n each possible solution of army A fight again each solution in $\{B_1, B_2, ...B_{n-1}\}$, where B_i is the champion found by army B in the i (for $1 \leq i \leq n-1$) coevolutionary step.

Note that the cardinality of the selected opponent set and the cardinality of the HoF in the coevolutionary step n are equal (i.e., $k = n$ in Eq. 1).

4.2.2   HofCC-Reduced

Here the HoF acts as the minimum short-term memory mechanism by minimizing the number of battles required for evaluating an individual; that is to say, in the n co-evolutionary step, each individual in army A only faces the latest champion inserted in the HoF of army B (i.e., B_{n-1}). Note that this means $k = 1$ in Eq. 1.

4.2.3   HofCC-Diversity

In this proposal the HoF acts as a long-term memory mechanism, but the content of the HoF is updated by removing those members that provide less diversity. The value of *diversity* that an individual in the HoF provides is calculated by the genotypic distance as follows: we manage the memory of champions as a matrix in which each row represents a solution and each column a gen (i.e., an action in the strategy). Then, we compute the entropy value for a specific column j as follows:

$$H_j = - \sum_{i=1}^{k} (p_{ij} \log p_{ij}) \tag{3}$$

where p_{ij} is the probability of action i in column j, and k is the length of the memory. Finally the entropy of the whole set is defined by the following formula:

$$H = \sum_{j=1}^{n} H_j \tag{4}$$

The higher the value of H the greater the diversity of the set. For determining the diversity's contribution of a specific solution, we calculate the value of entropy with this solution inside the set, and the corresponding value with this solution out of the set, and finally, the difference of these two values represents the contribution of diversity.

The number of individuals to be deleted from the memory should be set by the programmer as a percentage value (α) representing the portion of the HoF to be removed; in other words, the HoF (with cardinality $\#HoF$) is ordered according to the diversity value in a decreased order and the last $\lceil \frac{\#HoF \times n}{\alpha} \rceil$ individuals in this ordered sequence are removed. The frequency of updating (λ) is also a parameter of this version (i.e., the HoF is updated every λ coevolutions)

The motivation of this proposal is to maintain certain diversity among the members of the HoF, and at the same time to reduce (or maintain an acceptable value for) the size of the memory. With this idea, we assume that the deleted individuals will not affect the quality of the found solutions.

Here, the cardinality of the selected opponent set k in the evaluation phase (see Eq. 1) is the cardinality of the opponent HoF after executing the updating of the memory (i.e., after removing the individuals).

4.2.4   HofCC-Quality

In this version, we follow a similar approach to that applied in HofCC-Diversity but now the HoF is ordered with respect to a measure of quality that is defined as the number of defeats that an individual obtained in the previous coevolutionary step; in other words, a simple counter variable associated with each member of the HoF stores the number of defeats that were computed for the corresponding member during the evaluation process of the opponent army in the previous coevolutive turn.

Based on our game experience, we assume that this metric is representative of the strength of a solution, and the aim is to keep only the robust individuals in the champions' memory by removing the weak strategies.

As in the HofCC-Diversity, the parameters α and λ have to be set, and the cardinality of the selected opponent set k is exactly the same.

4.2.5   HofCC-U

This variant of HofCC follows the idea of optimizing the memory of champions, but in this case we propose a multiobjective approach where each solution has a diversity value and also a quality value as described previously associated with it. Then, a percentage value (α) from the set of dominated solutions according to the multiobjective values is removed; if the set of dominate solutions is empty then HoF is ordered according to the measure of quality and the solutions with worst quality will be removed.

As in the previous algorithms (HofCC-Quality and HofCC-Diversity) the frequency of updating the HoF is an important parameter that must be defined.

This proposal uses a different fitness function (to that shown in Eq. 1) whose definition was inspired by the Competitive Fitness Sharing (CFS) [3]. The main idea is that a defeat against opponent X has more importance if there are other individuals that defeated X. So, a penalization value N for each individual i (for $1 \leq i \leq k$) in the population is then calculated as follows:

$$N_i = 1 - \frac{\sum_{j=1}^{k} \frac{v_{ij}}{V(j)}}{k} \qquad (5)$$

where $v_{ij} = 1$ is the individual i of the population defeats the strategy (or champion) j in the HoF (whose cardinality is k) and 0 otherwise; and

$$V(j) = \sum_{i=1}^{n} v_{ij}$$

is the number of individuals in the population which defeat opponent j of the HoF. As a consequence, $N_i \approx 0$ if the candidate i defeats all opponents of HoF and the solution i itself is one of the few candidates to do so; $N_i = 1$ if it defeats no opponent; and $0 < N_i < 1$ depending on how many times it wins and how common is to beat certain opponents. The fitness of a candidate i is then computed as follows:

$$F_i = P_i - cN_i \tag{6}$$

where P_i is the result obtained in the battles by Eq. 1, and $c \in \mathbb{N}$ is a coefficient that scales N_i in order to make it meaningful with respect to value P.

5   Experiments and Analysis

Due to the high computational cost to execute the experiments we have considered a unique battle scenario without obstacles and in which the formation (i.e., morphology) of armies is the type of *tortoise vs. tortoise*, and the initial predefined enemy strategy is *random*.

The five variants of the HofCC described in previous section has been considered for the experiments.

5.1   Configuration of the Experiments

Eleven instances of our algorithms were used: one for HofCC-Complete (HofC); one for HofCC-Reduce (HofR); and three for each of the HofCC-Diversity, HofCC-Quality, and HofCC-U varying according to the values of $\alpha = 10\%$ (i.e., *(HofDiv-10, HofQua-10, HofU-10)*, $\alpha = 30\%$ *((HofDiv-30, HofQua-30, HofU-30))*, and $\alpha = 50\%$ *((HofDiv-50, HofQua-50, HofU-50))*. Also we set $\lambda = 3$ for all the versions of HofCC-Diversity, HofCC-Quality, and HofCC-U.

Ten runs per each algorithm instance were executed, and in all of them we have used a steady-sate genetic algorithm (GA - note that this corresponds to Lines 7–13 in Algorithm 1) with the aim of finding a winning strategy with respect to a set of strategies (stored in the HoF of the opponent) that were considered winning in previous stages of the coevolutionary algorithm; this GA employed binary tournament for selection, Bernoulli crossover, bit-flip mutation, elitist replacement, $MaxCoevolution = 15$ (it represents the numbers of continuous coevolutions in the same army without finding a champion solution), $Maxgenerations = 300$, population size was set to 100, the limit of evaluation was 230000 (i.e., the timeout value), and standard values for crossover and mutation probabilities were used ($p_X = .9$ and $p_M = 1/nb$ respectively where nb is the number of genes); and $\phi = 2000$ so that a strategy that defeats all the strategies in the HoF of its opponent is surely considered as victorious, although others can also be. The choice of these values is due to a previous analysis of the mean fitness reached by individuals in the competitions. We also set the constant c in Eq. (6) to 200, a representative value of the range of scores that were obtained from the battle simulator after executing a number of games.

Our analysis has been guided by the following indicators which are applied for all runs of each algorithm: *Best fitness*: shows the best value of fitness found by the search process; *Average fitness*: shows the average fitness value reached in the coevolutive cycle; *Number of evaluations*: indicates the total number of battles which are realized during the evaluation process; *Number of defeats*: indicates the total number of defeats obtained in an *All(vs)All* fighting among the best solutions found by each version of algorithm.

In what follows, we analyze the results obtained in ten independent executions for the eleven versions of HofCC, and focus on the mentioned indicators; we have used the well-known non-parametric statistical tests to compare ranks namely Kruskal-Wallis test [19] with a significance of 95 %. When this test detects significant differences in the distributions, we have performed multiple tests using the Dunn–Sidak method [20] in order to determine which pairs of means are significantly different, and which are not. Next, the results obtained in the experiments for each indicator are presented.

5.2   Results of Average Fitness

Figure 1 shows the behavior of the average fitness for each algorithm instance. In this figure the algorithms are sorted according to the median of the distribution (i.e., the vertical red line). The Kruskal-Wallis test confirms that the differences between values are statistically significant (see the first row in Table 1). The HofR algorithm reaches the worst results for this indicator, such results may be a sign that this algorithm does not exploit the search space sufficiently because in this version the HoF acts as a short memory mechanism, whereby it is easier to find a champion than for the rest of the algorithms. Moreover, note that the best results are obtained by algorithms which optimize the use of HoF (in terms of diversity, quality, or both), and at the same time do not reduce too significantly the memory size; note also that the average fitness value decreases in those cases where the HoF suffered a reduction of 50 % during the updating process. In the results of multiple tests for the value of average fitness, the HofR distribution has significant differences respect to the majority of the algorithms

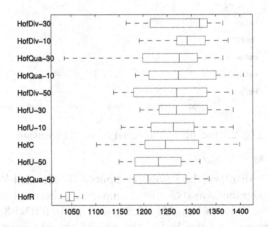

Fig. 1. Average fitness obtained in each algorithm

Table 1. Results of Kruskal-Wallis test for all the indicators ($p$value < 0.05)

Indicator	$p$value
Average fitness	$2.6205E - 004$
Best fitness	0.0175
Number of evaluations	0,2214
Numbers of defeats	$5.6909E - 007$

(except HofU-50, and HofQua-50), and the rest of the versions have a similar behavior.

5.3   Results of Best Fitness

Figure 2 shows the results of best fitness for each algorithm in the executions. Note that HofR finds the higher values. This situation can be generated by the fact that in HofR the individuals compete only against one opponent during the evaluation process, therefore it is easier to obtain higher scores. For the rest of the algorithms a high value in this indicator may be given by a further intensification in the search. As another interesting point, note that in many cases the algorithms with poor results in terms of average fitness, have good results here; except in the case of the HofDiv-10 and HofU-30 which maintain good results in both analysis.

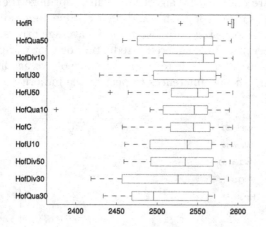

Fig. 2. Best fitness obtained by each algorithm

Table 1 (row 2) displays the results computed by Kruskal-Wallis' test confirming there are significative differences among the distributions. The results of multicompare test shows that the HofQua-30 and HofDvi-30 have relevant differences with respect to the HofR; the rest of the algorithms have similar values.

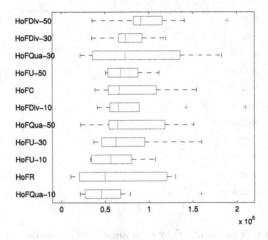

Fig. 3. Numbers of evaluations employed by each algorithm

5.4   Results of Number of Evaluations

For the case of the number of evaluations the results are shown in Fig. 3 and according to the statistical tests performed (see Table 1), there is no significant statistical difference in the distribution data. In this indicator we noted that the increasing in the number of evaluations is in consonance to the length of the coevolutive cycle; except in the case of HofR which presents a very long cycle and has no influence because during the evaluation process of this algorithm the individuals face a single opponent, and this decreases the number of evaluations significantly. Consider that the coevolutive cycle's length is determined by the number of coevolutions that use the algorithm to find an undefeated champion (i.e a member of the HoF which can not be defeated by the opponent side), and it helps to identify whether the problem difficulty increases as best solutions are obtained, or if it remains stable. In all algorithms (except HofR) the rigor of the competition increases until it reaches the point at which the algorithm can not exceed the level of specialization achieved. However, in HofR the cycles were very long, because the quality of the solutions was stagnated; and it was necessary to limit the length of cycles up to 500 iterations.

5.5   Results of Number of Defeats

For this test, the last champions (i.e., the last member added to the HoF) found by each algorithm instance (in each execution) fought in an *All versus All* tournament. The results with respect to the amount of defeats are shown in Fig. 4; and Table 1 (row 4). The main differences are in the values of HofR, and HofC which still maintain the same poor results as the previous indicators. On the other hand, HofDiv-10 and HofDiv-30 again obtains the best values. Curiously, the HofQua-30 which had the worst results in the analysis of best fitness has a

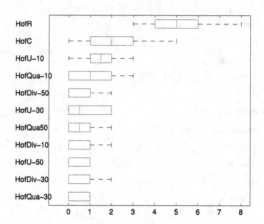

Fig. 4. Numbers of defeats obtained by each algorithm in an *All (vs) All* tournament

low ranking of defeats here, this is certainly an indicator that the fitness measure used is insufficient. The instances of HofQua-10 and HofU-10 have a similar behavior with high numbers of defeats. Another detail that attracts attention is that variants that reduce the HoF by 50 % in the previous indicators have not shown encouraging results, except for the HofDiv-50, however in this analysis we can see that they are in the middle top of the ranking.

5.6   Summary of the Results

We can conclude that, for all the experiments, the versions that incorporate updating the HoF were more efficient than HofR and HofC. In the case of the best fitness analysis the HofR shows higher values, however in the fighting tournaments the strategies generated by this algorithm were the weakest, so we can say that the fitness function is not sufficiently representative of the individuals' strength, and undoubtedly the loss of transitivity in this algorithm causes a total disengagement of the search.

Regarding to the numbers of evaluations there are not significant differences, and here let us underline that the game used does not allow very long evolutionary cycles due to performance limitations. This means the HoF obtained is not large, so it is not possible to demonstrate the benefits of those algorithms which optimize the size of the HoF and in turn reduce the number of necessary evaluations.

By analyzing the families of algorithms we can conclude that the HofDiv's variants shown the best performance for all the indicators. In the fitness analysis HofDiv shown the best results, while the HofU and HofQua reached similar behavior. And as general rule, for these indicators, the variants which update HoF by 10 and 30 % obtain a better performance; however in the tournaments, the algorithms which use a high percentage showed some slight advantages, although the differences between the versions that update the HoF are not statistically significant for any indicator.

We have also shown that the fitness values are not related with the solution robustness by executing fighting tournaments; this is a particular result which can be interpreted as follows. During the coevolutive cycle the search space is explored, but the self-learning mechanism falls into a local optimum and gets trapped there, so that the solutions found improve their fitness values without a global improvement in a more general context.

6 Conclusions and Future Work

The analyzed results allow us to conclude that whenever we update the memory of champions using a selection criteria the search process provides better solutions. In our experiments the updating of HoF using the *diversity* as the selection criteria showed the best performance. We also detected that affecting the transitivity among the solutions has a direct influence on the quality of the search result, and therefore, the removing of members in the champions' memory should be done carefully. However, it is important to decrease to avoid performance problems; in our experiments we did not suffer this, but there are cases in which the exhaustive exploration of the search space requires to significantly enlarging the size of the champions' memory what can affect the performance of the algorithm considerably.

As for future work we propose testing these algorithms in other games. Trying at that time to achieve a more complete representation of individuals by genetic encoding, and a better adjustment of the evaluation function. The games that we will choose should allow us to experiment with long coevolutionary cycles, to carry out a performance analysis of the algorithms.

Acknowledgements. This work is partially supported by Spanish MICINN under project ANYSELF (TIN2011-28627-C04-01), and by Junta de Andalucía under project P10-TIC-6083 (DNEMESIS).

References

1. Rosin, C.D., Belew, R.K.: Methods for competitive co-evolution: finding opponents worth beating. In: ICGA, pp. 373–381 (1995)
2. Ficici, S.G., Bucci, A.: Advanced tutorial on coevolution. In: Proceedings of the: GECCO Conference Companion on Genetic and Evolutionary Computation, pp. 3172–3204. ACM, New York (2007)
3. Rosin, C., Belew, R.: New methods for competitive coevolution. Evol. Comput. **5**(1), 1–29 (1997)
4. de Jong, E.: Towards a bounded pareto-coevolution archive. In: Congress on Evolutionary Computation, CEC2004, vol. 2, pp. 2341–2348. IEEE, New York (2004)
5. Jaskowski, W., Krawiec, K.: Coordinate system archive for coevolution. [21], pp. 1–10
6. Yang, L., Huang, H., Yang, X.: A simple coevolution archive based on bidirectional dimension extraction. In: International Conference on Artificial Intelligence and Computational Intelligence: AICI'09, vol. 1, pp. 596–600. IEEE, Washington (2009)

7. Avery, P.M., Greenwood, G.W., Michalewicz, Z.: Coevolving strategic intelligence. In: IEEE Congress on Evolutionary Computation, pp. 3523–3530. IEEE (2008)

8. Angeline, P.J., Pollack, J.B.: Competitive environments evolve better solutions for complex tasks. In: ICGA, pp. 264–270 (1993)

9. Reynolds, C.: Competition, coevolution and the game of tag. In: Brooks, R., Maes, P. (eds.) Proceedings of Artificial Life IV, pp. 59–69. MIT Press, Cambridge (1994)

10. Sims, K.: Evolving 3D morphology and behavior by competition. Artif. Life **1**(4), 353–372 (1994)

11. Smith, G., Avery, P., Houmanfar, R., Louis, S.J.: Using co-evolved RTS opponents to teach spatial tactics. In: Yannakakis, G.N., Togelius, J. (eds.) CIG, pp. 146–153. IEEE, New York (2010)

12. Avery, P., Louis, S.J.: Coevolving team tactics for a real-time strategy game. [21], pp. 1–8

13. Dziuk, A., Miikkulainen, R.: Creating intelligent agents through shaping of coevolution. In: IEEE Congress on Evolutionary Computation, pp. 1077–1083 (2011)

14. Lichocki, P.: Evolving players for a real-time strategy game using gene expression programming. Master thesis, Poznan Universtity of Technology (2008)

15. Avery, P.M., Michalewicz, Z.: Static experts and dynamic enemies in coevolutionary games. In: IEEE Congress on Evolutionary Computation, pp. 4035–4042 (2007)

16. Johnson, R., Melich, M., Michalewicz, Z., Schmidt, M.: Coevolutionary tempo game. In: Congress on Evolutionary Computation, CEC'04, vol. 2, pp. 1610–1617 (2004)

17. Avery, P., et al.: Coevolving a computer player for resource allocation games: using the game of Tempo as a test space. Ph.D. thesis, School of Computer Science University of Adelaide (2008)

18. Nogueira, M., Gálvez, J., Cotta, C., Fernández-Leiva, A.: Hall of Fame based competitive coevolutionary algorithms for optimizing opponent strategies in a new RTS game. In: Fernández-Leiva, A., et al., (eds.) 13th Annual European Conference on Simulation and AI in Computer Games (GAMEON'2012), Málaga, Spain, Eurosis, pp. 71–78, November 2012

19. Kruskal, W., Wallis, W.: Use of ranks in one-criterion variance analysis. J. Am. Stat. Assoc. **47**(260), 583–621 (1952)

20. Sokal Robert, R., Rohlf James, F.: Biometry: The Principles and Practice of Statistics in Biological Reseach. W.H. Freeman and Company, New York (1995)

21. Proceedings of the IEEE Congress on Evolutionary Computation, CEC 2010, Barcelona, Spain, 18–23 July 2010. In: IEEE Congress on Evolutionary Computation, IEEE (2010)

Resources Optimization in (Video) Games: A Novel Approach to Teach Applied Mathematics?

Dario Maggiorini[✉], Simone Previti, Laura Anna Ripamonti, and Marco Trubian

Dip. di Informatica, Università di Milano, Via Comelico 39 20135 Milan, Italy
{maggiorini, ripamonti, trubian}@di.unimi.it, rekotc@gmail.com

Abstract. In spite of the efficacy of Operations Research (OR), its tools are still underused, due to the difficulties that people experience when describing a problem through a mathematical model. For this reason, teaching how to approach and model complex problems is still an open issue. A strong relation exists between (video) games and learning: for this reason we explore to which extent (real time) simulation video games could be envisaged to be an innovative, stimulating and compelling approach to teach OR techniques.

Keywords: Simulation · Video games · Applied mathematic · Operations research · Simulation games · Teaching applied mathematics · Player experience in game

1 Introduction

The adoption of effective mathematical tools is of paramount importance when facing complex problems involving a wise resources allocation. This means achieving a deep understanding of each problem in terms of goals and constraints to its solution. This approach belongs to the domain of Operations Research (OR), which, beside modelling and solving complex decisional problems, allows to produce formal justifications for the choices among which the decision maker has to choose. In spite of the proven efficacy of OR in specific contexts (e.g. business management, complex systems management, etc.), its tools are still underused, mainly due to the difficulties that many people experience when facing the job of describing a problem through a rigorous mathematical model. For this reason, teaching how to approach and model complex problems is still an open issue in the field of applied mathematics [1]. Typically, OR is applied to optimization problems in the following fields: services distribution in a closed area, goods production, productive structures organization and management, territory optimal exploitation. All this issues do not belong only to the "real life" domain, but very often appear also in simulation or strategy (video) games [2]. Our goal, in the present work, is to explore to which extent (real time) simulation video games could be exploited as an innovative, stimulating and compelling approach to the teaching of OR techniques, hence helping students to overcome part of the difficulties that they often experience when modeling complex problems.

G. Nicosia and P. Pardalos (Eds.): LION 7, LNCS 7997, pp. 189–195, 2013.
DOI: 10.1007/978-3-642-44973-4_20, © Springer-Verlag Berlin Heidelberg 2013

We will not dig into how the game should be presented to a classroom or how the modelling phase should be taught, but on a preliminary aspect: (dis)proving that simulation video games contain enough combinatorial structure to imply interesting optimization problems and enough numerical data to allow building reasonable instances.

The paper is organized as follows: Sect. 2 briefly recalls the grounding upon which we have built our work. The following Sect. 3 describes how it is possible to formalize quite complex OR problems starting from the video game Caesar IV. Section 4 summarizes several perspective results, draws conclusions and describes future developments.

2   (Serious) Video Games and Teaching Applied Math

The explicit use of mathematics in games is no news in popular scientific literature, (see, e.g., the prolific work of Gardner [3]). Moreover, at the very core of every game there is a limited formal system (see e.g. [4–6]). In spite of this, a game-based approach in teaching OR is not frequent, except for few excellent examples [7].

Beside the "traditional" way to convey teaching, learning patterns have changed radically [8] and further developments are easily foreseeable: younger are "native speakers" in the language of digital media. New generations are experiencing new forms of computer and video game entertainment and, as Prensky underlines, this has shaped their preferences and abilities, while offering an enormous potential for their learning, both as children and as adults [9]. The idea of exploiting video games as a teaching media has been "formalized" in 2002: in that year, the Woodrow Wilson Center for International Scholar in Washington, D.C. founded the Serious Games Initiative. Although we instinctively tend to associate "serious games" to video games, this term was already in use long before computers and electronic devices became a way to convey entertainment [10]. Anyway, no generally accepted definition of what a Serious Game (SG) is exists yet. The one provided by Zyda seems good enough for our goals: *Serious Game: a mental contest, played with a computer in accordance with specific rules, that uses entertainment to further government or corporate training, education, health, public policy, and strategic communication objectives* ([8], p. 26). It is important to enlighten that SGs are not only a way to *instruct* people about something, but also way to convey knowledge within a context that is *motivationally rich*. The usefulness of games as learning tools is a well known phenomenon, that becomes of overwhelming relevance in the first years of our life [11–13], and it is demonstrated that games are able to guarantee high learning effectiveness in quite short time [5, 14, 15]; e.g., they are adopted as training tools in medical, military, and business fields [16]. Moreover, as [14] point out, games are able to train cognitive functions – such as memory –, and to teach how to exert control over emotions [17–20]. It might be worth spending also a few words to recall what scientists have proved about the relation between games and our brain. Although we instinctively recognize than *games* and *fun* are tightly related, both concept and their interrelation are quite slippery to define [21–26]. The investigation of these issues has led neuroscientists and cognitive psychologists to examine how *playing* a game and

Table 1. Video Games and Simulation Problems

Video game	Simulation problem
Caesar IV [29]	Maximum profit
	Optimal configuration of services
	Facility location
Age of Empires II	Creation of an army
	Time to create an army
Domino	Optimal configuration
Armor Picross	Optimal configuration
World of Warcraft	Knapsack model

learning are connected [27, 28]. The basic observation is that humans have always used games as playgrounds for learning and exercising safely specific skills (see, e.g., the relation that, in the past, linked archery tournaments to the ability to catch food). During this process, human brain secretes endorphins (which makes a game an enthralling and *fun* activity), is highly focused on recognizing recurring *patterns* in problems, and on creating appropriate neural routines to deal with them. Once the pattern is fully caught by the player, the game becomes boring, but the *skill* has been accurately acquired. To exploit this phenomenon, we have envisaged a set of par-ticularly "difficult to approach" problems, typical of the OR domain, and for each problem we have identified at least one popular video game which can be exploited as a case study to help students in the process of refining their modelling skills (Table 1). For each game we have developed and tested the related OR problems, but for reasons of paper length, in the following we will focus mainly on the *maximum profit problem* for the game Caesar IV.

3 Caesar IV: Maximizing the City Profit

Caesar IV [29] is a *managerial strategic* video game in *real time*, set in the ancient Rome period, developed by Tilted Mill Entertainment. The game aims at simulating the management of a city, and the player ultimate goal is to develop the "ideal" city, endowed with whichever good or service its population may need, while at the same time, maximizing its income. It is as an ideal playground to verify how OR techniques can be proficiently applied to video games. We have designed, developed and tested models based on Caesar (namely: profit maximization, optimal service configuration, facility location) able to allow the maximization of the player performances. This means exploiting the "rewarding" effect implicit in the accomplishment of in-game goals that characterizes each (video) game as tool for motivating learners to approach the modeling problem. As an example, we briefly describe the approach to the *maximization of the city profit*. From the point of view of OR, the goal of the game maps into the optimization of the starting budget of the player, that is to say the optimal allocation of starting resources in order to acquire the maximization of the city income deriving from trading goods with other cities. Since goods can be produced either from scratch (e.g. olives) or from raw materials (e.g. olive oil), a first set of decisions focuses on the selection of the best (in terms of trade-off between production costs and income provided) mix of goods to produce. This decision is influenced also by the types of goods produced by or tradable (in defined quantities) with neighboring cities. Once the mix has been chosen, the necessity of other decisions arises: each good can be produced only by a specific plant (characterized by a construction cost), which, in turn, is operated by workers characterized by specific sets of needs. As a consequence, the player must face the necessity to preserve enough of each good to satisfy domestic consumption. The problem is further complicated by the fact that goods must be distributed among the population according to the needs of different classes. Also, different social classes will need different types of houses (with different cost, space occupation, etc.).

In other words, the player aims at maximizing the difference between total revenues (R) and total costs (C). R is the sum of the revenues produced by selling abroad certain amounts of goods at their *fixed* prices. Costs are, as usual, the sum between total fixed costs and total variable costs, constrained by the total starting budget available to the player. Total fixed costs are the sum of the costs sustained to build production plants, houses, warehouses and to open commercial routes, and total variable costs are produced by the total amount of goods imported. To model the problem, it is necessary to define sets, constraints and several variables, whose values indicates if a certain activity is in place, the quantity of specific plants, goods, etc. Once the objective function, the relevant sets and variables are in place, several constraints mirroring the restriction in the gameplay are needed for the following in-game objects:

– *production plants*: many of them need raw materials to produce goods. Raw materials can be produced or imported. A production plant should be built only if the raw materials it requires can be acquired;
– *goods for import/export*: if no neighbouring city needs a specific good, there is no reason to set up its production for export. The production of a certain good should be constrained by the fact that a trading route with possible customers is active and by the total quantity of that good that can be absorbed by the market;
– *production plants*: to define the number of each production plant to build, we need to constraint their number to the actual production (i.e. if the player does not produce wine, she does not need wineries), and then to model the fact that a certain good can be both a final product and a raw material. Moreover, the same good may be imported from abroad suppliers. Finally, we must take into account the production necessary to satisfy domestic consumption and export, plus the import;
– *workforce*: the production plants should be operated by the appropriate workforce. We need two constraints, the first one models the number of houses needed to host the workforce, the second one the correct number of workers for operating all the plants (and warehouses) set up by the player.

4 Conclusions and Further Developments

Our goal was to verify to what extent several problems presented in an appropriate selection of video games can be described "ex-post" through formal systems, aimed at supporting teaching activities in the field of applied mathematics (i.e. simulation and OR). Obviously, this approach would arise a greater interest in the student/player if the effort put in the modelling process actually produces better in-game results (since players are always struggling to enhance their in-game performances). For these reasons we have developed and tested (using AMPL - www.ampl.com) several among the above mentioned models (namely those based on Caesar IV, Armor Picross and Domino). In the case of Armor Picross, for example, the simulation allowed the player to find very quickly an optimal configuration that respects all the constraints (and, in several cases, the game map can produce more than one optimal solution). Anyway, the most interesting results emerged from the simulations run on Caesar IV. In this case, we have run many simulations, gradually reducing the starting budget, in order to test which are the most suitable starting choices when the budget is very limited. The solutions supplied by the simulations have then been adopted by a human player in the game: as a result, the profit deriving from trading was higher then the profit that a skilled player can typically obtain. As we showed, the models that can be built starting from a real-time simulation video game are not trivial and require a non-trivial effort to be fully understood by undergraduate students. Same drawbacks are also present: for example, while numerical data for several problems (e.g. maximizing the city profit) are presented by the game in tabular form, gathering data for several other problems can be a truly boring task. In this latter case, the simulated environment could be used mainly as a support tool to introduce real-life problems and to build their related models. As our preliminary computational campaign also shows, the non trivial impact on in-game performance of right decisions can be used to encourage student's interest. Actually, video games could be exploited as tools to convey the teaching of applied mathematics, provided that they offer problems matching also with the skills required by each specific teaching unit and student. This is, by the way, coherent with one of the fundamental requirements of game design: a "fun" game provides a challenge that perfectly matches the player skills, drawing her into the "flow" [30] (hence, also the difficulty of the game should grow at the same pace of the player ability [5, 30]).

The next steps in our research will focus on further developing the corpus of problems derived from video games that can be exploited for OR teaching purposes and in testing their use with undergraduate students in Computer Science.

References

1. Kaiser, G., Blum, W., Borromeo Ferri, R., Stillman, G. (eds.): Trends in Teaching and Learning of Mathematical Modelling, vol. 1. Springer, Netherlands (2011)
2. Adams, E.: Fundamentals of Game Design. New Riders, Indianapolis (2009)
3. Gardner, M.: My Best Mathematical and Logic Puzzles. Dover Recreational Math. Dover, Mineola (1994)

4. Crawford, C.: On Game Design. New Riders Publishing, Thousand Oaks (2003)
5. Koster, R.: A Theory of Fun for Game Design. Paraglyph Press, Scottsdale (2005)
6. Fullerton, T.: Game Design Workshop: A Playcentric Approach to Creating Innovative Games. Morgan Kaufmann, Burlington (2008)
7. DePuy, G.W., Taylor, D.: Using board puzzles to teach operations research. INFORMS Trans. Educ. 7(2), 160–171 (2007)
8. Zyda, M.: From visual simulation to virtual reality to games. Computer 38(9), 25–32 (2005) (IEEE Computer Society Press, Los Alamitos)
9. Prensky, M.: Don't Bother Me Mom, I'm Learning!. Paragon House Publisher, St. Paul (2005)
10. Abt, C.: Serious Games. The Viking Press, New York (1970)
11. Din, F.S., Calao, J.: The effects of playing educational videogames on kindergarten achievement. Child Study J. 2, 95–102 (2001)
12. Durik, A.M., Harackiewicz, J.M.: Achievement goals and intrinsic motivation: coherence, concordance, and achievement orientation. J. Exp. Soc. Psychol. 39, 378–385 (2003)
13. Ritterfeld, U., Weber, R.: Video games for entertainment and education. In: Vorderer, P., Bryant, J. (eds.) Playing Video Games – Motives, Responses, and Consequences. Lawrence Erlbaum, Mahwah (2006)
14. Squire, K., Jenkins, H.: Harnessing the Power of Games in Education. Insight 3(1), 5–33 (2003)
15. Susi, T., Johannesson, M., Backlund, P.: Serious games – an overview. Technical Report HS-IKI-TR-07-001, University of Skovde, Sweden (2007)
16. Ripamonti, L.A., Peraboni, C.: Managing the design-manufacturing interface in VEs through MUVEs: a perspective approach. Int. J. Comput. Integr. Manuf. 23(8–9), 758–776 (2010) (Taylor & Francis)
17. Michael, D., Chen, S.: Serious Games: Games that Educate, Train, and Inform. Thomson Course Technology, Boston (2006)
18. Mitchell, A., Savill-Smith, C.: The Use of Computer and Video Games for Learning: A Review of the Literature. Learning and Skills Development Agency, London (2004)
19. Wong, W. L., Shen, C., Nocera, L., Carriazo, E., Tang, F., Bugga, S., Narayanan, H., Wang, H., Ritterfeld, U.: Serious Video Game Effectiveness. In: ACE'07, Salzburg, Austria (2007)
20. Ripamonti, L.A., Maggiorini, D.: Learning in virtual worlds: a new path for supporting cognitive impaired children. In: Schmorrow, D.D., Fidopiastis, C.M. (eds.) FAC 2011. LNCS, vol. 6780, pp. 462–471. Springer, Heidelberg (2011)
21. Caillois, R.: Man, Play, and Games. The Free Press, Glencoe (1961)
22. Huizinga, J.: Homo Ludens. The Beacon Press, Boston (1950)
23. Juul, J.: The game, the player, the world: looking for a heart of gameness. In: Copier, M., Raessens, J. (eds.) Level Up: Digital Games Research Conference Proceedings, pp. 30–45. Utrecht University, Utrecht (2003)
24. Crawford, C.: The Art of Computer Game Design. McGraw-Hill/Osborne Media Berkeley, California (1984)
25. Rollings, A., Adams, E.: A Rollings and E. Adams on Game Design. New Riders, California (2003)
26. Salen, K., Zimmerman, E.: Game design and meaningful play. In: Raessens, J., Goldstein, J. (eds.) Handbook of Computer Game Studies, pp. 59–79. MIT Press, Cambridge (2005)
27. Miller, G.A.: The magical number seven, plus or minus two: some limits on our capacity for processing information. Psychol. Rev. 63, 81–97 (1956)

28. Johnson, S.: Mind Wide Open: Your Brain and the Neuroscience of Everyday Life. Scribner, New York (2004)
29. Caesar IV. http://caesar4.heavengames.com/
30. Csikszentmihalyi, M.: Flow: The Psychology of Optimal Experience. Harper & Row, New York (1990)

CMF: A Combinatorial Tool to Find Composite Motifs

Mauro Leoncini[1,3](✉), Manuela Montangero[1,3], Marco Pellegrini[3], and Karina Panucia Tillán[2]

[1] Dipartimento di Scienze Fisiche, Informatiche e Matematiche, University of Modena and Reggio Emilia, Modena, Italy
{manuela.montangero,leoncini}@unimore.it
[2] Dip. di Scienze e Metodi dell'Ingegneria, Univ. di Modena e Reggio Emilia, Modena, Italy
83672@unimore.it
[3] Istituto di Informatica e Telematica, CNR, Pisa, Italy
marco.pellegrini@iit.cnr.it

Abstract. Controlling the differential expression of many thousands genes at any given time is a fundamental task of metazoan organisms and this complex orchestration is controlled by the so-called *regulatory genome* encoding complex regulatory networks. *Cis-Regulatory Modules* are fundamental units of such networks. To detect Cis-Regulatory Modules "in-silico" a key step is the discovery of recurrent clusters of DNA binding sites for sets of cooperating Transcription Factors. *Composite motif* is the term often adopted to refer to these clusters of sites. In this paper we describe CMF, a new efficient combinatorial method for the problem of detecting composite motifs, given in input a description of the binding affinities for a set of transcription factors. Testing with known benchmark data, we attain statistically significant better performance against nine state-of-the-art competing methods.

1 Introduction

Transcription Factors (or simply *factor*) are particular proteins that bind to short specific stretches of DNA (called *TFBS - Transcription Factor Binding Sites*) in the proximity of genes and participate in regulating the expression of those genes [1]. The "language" of gene regulation is a complex one since a single factor regulates multiple genes, and a gene is usually regulated over time by a cohort of cooperating factors. This network of interactions is still far from being completely uncovered and understood even for well studied model species. Groups of factors that concur in regulating the expression of groups of genes form functional elements of such complex network and are likely to have TFBS in the proximity of the regulated genes. TFBSs are often described by means of *Position Weight Matrices* (*PWMs*) (see Sect. 2 for a quick recap).

Over the last two decades more than a hundred computational methods have been proposed for the de-novo prediction "in silico" of single functional TFBSs

G. Nicosia and P. Pardalos (Eds.): LION 7, LNCS 7997, pp. 196–208, 2013.
DOI: 10.1007/978-3-642-44973-4_21, © Springer-Verlag Berlin Heidelberg 2013

(often called *single motifs*, or simply motifs) [2–5]. Moreover, several hundreds of validated PWMs for identifying TFBS are available in databases such as TRANSFAC [6] and JASPAR [7]. Observe that, although these PWM have been subject to validation in some form, the highly degenerate nature of the TFBS implies that, when scanning sequences for PWM matches, false positive non-functional matches are quite likely.

In this paper we address the problem of discovering groups of TFBSs that are functional for a set of cooperating factors, given in input a set of PWMs that describe the binding affinities of the single factors. This is known as the *Composite Motif Discovery* problem in the literature [8]. For us, a composite motif will be simply a set of TFBSs that are close by in a stretch of DNA, i.e., we do not pose any constraints on the order or the spacing between the participating TFBSs (but see [9] for other possible models).

The composite motif discovery problem has been the subject of a number of studies, and we refer to [10] for a survey. In addition we observe that the phenomenon of *clustering* of TFBS is used also by tools that try to predict the location and composition of *Cis-Regulatory Modules* (see, e.g., [11]), which then address composite motif discovery problems of some sort.

In this paper we present a new tool (*CMF*) for composite motifs discovery that adopts a two stage approach: first it looks for candidate single TFBSs in the given sequences, and then uses them to devise the prospective composite motifs by using mainly combinatoric techniques. CMF borrows the idea of the two stage approach from a previous tool we developed for the related problem of structured motif detection [12]. Using the data set and the published results in [8,13] we can readily compare CMF's performance against the eight state of the art methods listed in [8] and other three more recent methods [13–15], showing that our tool is highly competitive with the others.

The detection of TFBS and composite motifs is a complex challenging problem (witness the wide spectrum of approaches) which is far from having a satisfactory solution [16], thus there is ample scope for improvements from both the modeling and the algorithmic one points of view. *CMF* introduces several new key ideas within a combinatorial approach, which, on one hand, have been shown empirically to be valid on challenging benchmarks, and on the other hand may prove useful in developing future more advanced solutions.

The rest of the paper is organized as follows: Sect. 2 introduces preliminary notions and definitions, Sect. 3 describes the algorithm adopted by CMF and, finally, Sect. 4 reports experimental results.

2   Preliminary Notions

In this Section we introduce the fundamental notions used in the description of the algorithm that forms the computational core of CMF.

Given the DNA alphabet $\mathcal{D} = \{A, C, G, T\}$, a short word $w \in \mathcal{D}^*$ is called an *oligonucleotide*, or simply *oligo* (typically $|w| \leq 20$), and we say that w *occurs* in $S \in \mathcal{D}^*$ if and only if w is a substring of S.

From a computational point of view, a *DNA motif* (or simply *motif*) is a representation of a set of oligos, that are meant to describe possible factor binding loci. The representation can be made according to one of a number of models presented in the literature. Here we adopt the well-known *Position Weight Matrices (PWMs)*. A PWM $M = (m_{b,j})$, $b \in \mathcal{D}, j = 1, \ldots, k$, is a $4 \times k$ real matrix. The element $m_{b,j}$ gives a score for nucleotide b being found at position j in the subset of length-k oligos that M is intended to represent. Scores are typically computed from frequency values.

Among the different ways in which oligos can be associated to PWMs (e.g., [17–19]), here we adopt perhaps the simplest one. Consider a word $w = w_1 w_2 \ldots w_k$ over $\mathcal{D}^k$, and define the *score* of w, according to M, simply as the sum of the scores of all nucleotides: $S_M(w) = \sum_{j=1}^{k} m_{w_j,j}$. The maximum possible score given by M to any word in $\mathcal{D}^k$ is clearly $S_M = \sum_{j=1}^{k} \max_{b \in D} m_{b,j}$. Then we say the M *represents* word w iff $\frac{S_M(w)}{S_M} \geq \tau$, for some threshold value $\tau \in (0, 1]$. In the following, we will identify motifs with their matrix representation.

Let $\mathcal{S} \subseteq \mathcal{D}^*$ denote the set of N input sequences. A motif M has a *match* in $S \in \mathcal{S}$ if and only if there is a substring of S that is represented by M. As in [13], we call *discretization* the process of determining the matches of a motif in a set of DNA sequences.

A *motif class* is a set of motifs. Ideally, in CMF all the motifs in a class describe potential binding sites for the same factor. For this reason, we often freely speak of *factors* to refer to motif classes. A *factor match* in a DNA sequence is thus a match of any of the motifs in the class associated to that factor. Note that motif classes have the ability to represent oligos of different lengths, since different matrices usually exist for the same factor that have a different number of columns. Let $\mathcal{F}$ be the set of factors having matches in $\mathcal{S}$. We consider a one-to-one mapping between $\mathcal{F}$ and an arbitrary alphabet $\mathcal{R}$ of $|\mathcal{F}|$ symbols, which we refer to $\mathcal{R}$ as the *mapping alphabet*.

A *combinatorial group* (or just *group*) is a collection of not necessarily distinct factors that have close-by matches in a sufficiently large fraction of the input sequences.[1] The minimum fraction allowed for a collection of factors to be considered a combinatorial group is termed *quorum*. The *width* or *span* of a collection of factor matches in a sequence S is the "distance" (measured in bps) between the first bps of first and last factor match of the group in S.

Finally, a *Composite Motif* is simply a collection of close-by factor matches in some input sequence, representing CMF's *best guess* for functional factor binding regions. Note that no quorum constraint is imposed to composite motifs. Indeed, as collection of factor matches, composite motifs are clearly unique objects. As we shall see in Sect. 3, CMF builds composite motifs by extending the matchings of some combinatorial group.

---

[1] Assuming the number N of sequences is clearly understood, we silently equate the fraction $q \in (0, 1]$ and the absolute number of sequences $\lceil q \cdot N \rceil$.

In set-theoretic terms, groups are multisets. In CMF they are represented as character sorted strings over the mapping alphabet $\mathcal{R}$. In the algorithm of Sect. 3.3 we will make use of some operations than involve multisets. We first recall, using two simple examples, the customary definitions of intersection and symmetric difference:

$$xxyyyz \cap xyyw = xyy$$
$$xxyyyz \setminus xyyw = xyz$$

Note that we have adopted the string representation for multisets. We next consider pairs $\langle M, n \rangle$, where M is a multiset and n is a positive integer, and sets P of such pairs which only include maximal pairs. That is, if $\langle M, n \rangle \in P$ then there is no other pair $\langle \bar{M}, \bar{n} \rangle$ in P such that $\bar{M} \supseteq M$ and $\bar{n} \geq n$, where inclusion takes multiplicity into account.

We define special union and intersection operations, denoted by $\vee$ and $\wedge$, over sets of maximal pairs. The definition of $\vee$ is easy:

$$P \vee Q = \{p : p \text{ is maximal in } P \cup Q\}$$

We first define $\wedge$ for singleton sets:

$$\{\langle M_1, n_1 \rangle\} \wedge \{\langle M_2, n_2 \rangle\} = \{\langle M_1 \cap M_2, n_1 + n_2 \rangle\}_{\text{if } M_1 \cap M_2 \neq \emptyset} \cup$$
$$\{\langle M_1, n_1 \rangle\}_{\text{if } M_1 \setminus M_2 \neq \emptyset} \cup \{\langle M_2, n_2 \rangle\}_{\text{if } M_2 \setminus M_1 \neq \emptyset}$$

Then, for arbitrary sets $P_1 = \{p_i^{(1)}\}_{i=1,\dots,h}$ and $P_2 = \{p_j^{(2)}\}_{j=1,\dots,k}$:

$$P_1 \wedge P_2 = \vee_{i,j} \left(\left\{ p_i^{(1)} \right\} \wedge \left\{ p_j^{(2)} \right\} \right).$$

3 Algorithm

CMF main operation mode is composite motifs discovery in a set $\mathcal{S} = \{S_1, \dots, S_N\}$ of DNA sequences, using a collection of PWMs.[2]

PWMs can be passed to CMF in either a single or multiple files. In the latter case, CMF assumes that each file contains PWMs for only one given factor. Actually, when the input set is prepared using matrices taken from an annotated repository (e.g., the TRANSFAC database [20]), assuming the knowledge of the corresponding factors is not an artificial scenario. However, here we describe the main steps implementing CMF's operation mode on input a single PWM file, namely:

1. (Optional) *PWM clustering*, to organize the matrices in classes believed to belong to different factors;
2. *Discretization*, to detect PWM matches in the input sequences;
3. *Group and composite motif finding.*

---

[2] Even if not taken into consideration in this paper, CMF is also able to run a number of third-party motif discovery tools to "synthesize" PWMs.

3.1 PWM Clustering

By default, CMF assumes that the PWMs in the input file correspond to different factors, and hence it does not perform any clustering. However, in many cases the number of matrices available, which describe the binding affinities of the factors involved in the upstream experimental protocol, is much larger than the number of such factors. If the latter information is available to the user, then clustering may be highly useful both to improve the accuracy and to reduce the group finding complexity. Another circumstance in which clustering is advisable (not discussed here) is when the input matrices are produced by third-party motif discovery tools.

To perform the clustering, CMF first builds a weighted adjacency graph whose nodes are the matrices and edges the pairs (M_1, M_2) such that the similarity[3] between M_1 and M_2 is above a given threshold. Then, CMF executes a single-linkage partitioning step of the graph vertices; finally, it identifies the dense cores in each set of the partition, via pseudo-cliques enumeration [22], returning them as the computed clusters.

The experiments described in Sect. 4 suggest that, when the PWM file mainly includes good matrices corresponding to possibly different factors, then even a simple clustering algorithm like the one mentioned above is able to correctly separate them into the "right" groups (factors). In general, however, performing a good partitioning of the input matrices when the fraction of "noisy" PWM increases (as is the case when CMF is used downstream de-novo motif discovery software tools) is one of the major issues left open for further work.

3.2 Discretization

Even with the most accurate PWM description of a motif, the problem of determining the "true" motif matches in the input sequences is all but a trivial task. Whatever the algorithm adopted, there is always the problem of setting some thresholds τ to distinguish matches from non-matches, a choice that may have a dramatic impact on the tool's performance.

In general, low thresholds improve sensitivity while high thresholds may improve the rate of positive predicted values (PPVs). A reasonable strategy is to moderately privilege sensitivity during discretization, with the hope to increase the positive predicted rate thanks to the combinatorial effect of close-by matches. Indeed, keeping initial low thresholds may give the benefit of not filtering out low-score matches.[4] On the other hand, complexity issues demand that the number of possible combinations of motif matches, which the composite motifs should emerge from, will not explode. Now, for factors with many matrices, low thresholds may incur in a very high number of matches and these in turn affect the number of potential composite motifs.

---

[3] Currently, CMF invokes RSAT's utility `compare-matrices` for this purpose [21], which uses pairwise normalized correlation

[4] Sometimes referred to as *weak signals* in the literature.

In light of the above arguments, we formulate the following general and simple qualitative criterion: assign factors (motif classes) with many/few matrices a high/low threshold. All the experiments of Sect. 4 were performed with fixed threshold values. Although these can be varied (in the configuration file, hence in a completely transparent way to the typical user), the overall good results suggests that the above criterion may have some merits, to be further investigated.

3.3   Composite Motif Finding

The previous two steps result in a set of factors (motif classes) and a set of factors matches, which are the "input" to the Composite Motif Finding step. This is in turn divided into two main sub-processes:

(a) *Finding combinatorial groups.* CMF uses a simple search strategy, with the aim of trading computation time for accuracy. Let $\{W_1, \ldots, W_r\}$ be a set of (internal) window sizes and let $\{q_1, \ldots, q_s\}$ be a set of (internal) quorum values, with $W_1 < W_2 < \ldots < W_r$ and $1 \geq q_1 > q_2 > \ldots > q_s > 0$. For a given window size value W and sequence S_i, we say that a multiset m over $\mathcal{R}$ is *feasible* iff each letter/factor of m corresponds to a match in S_i and the span of all the matches in S_i is bounded by W.

The algorithm that computes the combinatorial groups can be described as follows.

1. Set $W = W_1$ and $q = q_1$.
2. For $i = 1, \ldots, N$, compute the maximal multisets $M_1^{(i)}, \ldots, M_{n_i}^{(i)}$ that are feasible for W and S_i, and form the set of pairs

$$P_i = \{\langle M_1^{(i)}, 1 \rangle, \ldots, \langle M_{n_i}^{(i)}, 1 \rangle\}$$

3. Set $G_1 = P_1$
4. For $i = 2, \ldots, N$ compute

$$G_i = G_{i-1} \wedge P_i$$

5. Discard from G_N all the pairs $\langle M, n \rangle$ such that $n < \lceil q \cdot N \rceil$.
6. If the (remaining) multisets in G_N include all the letters of $\mathcal{R}$ or $W = W_r$ and $q = q_s$, then set $G = \{M : \langle M, n \rangle \in G_N\}$ and return G.
7. In alternate order (whenever possible) advance W or q to the next value and jump to step 2.

The above general description has only explanatory purposes, since a direct implementation would be highly inefficient. For instance, when relaxing the quorum value, step 2 can be avoided, since the multisets $M_j^{(i)}$ have already been computed. On the other hand, the pairwise intersections of step 4 can be performed quite efficiently thanks to the character sorted string representation of multisets of factors.

By the properties of the $\vee$ and $\wedge$ operators, the pairs $\langle M, n \rangle$ included in G_N are maximal, with n satisfying the last fixed quorum value. Note, however, that even with the weakest parameter values (i.e., widest window and smallest quorum), some factors may not be represented in G. This is not necessarily a problem, since the user may have provided PWMs for irrelevant factors.

(*b*) *Computing the composite motifs.* For any combinatorial group g in G, CMF first retrieves its actual matches from the input sequences; then tries to extend each group of matches by possibly including other strong factors matches that do not make the extended group unfeasible with respect to the window constraint. This is done independently for each sequence. All these extended group matches form the composite motifs that CMF gives in output under the ANR (Any NumbeR) model. Under the ZOOPS (Zero Or One Per Sequence) model, groups and composite motifs are further filtered basing on the most recurrent span width (details not reported here for lack of space).

3.4   Computational Cost

The cost of the bare CMF algorithm is dominated by the Composite Motif finding step or, more precisely, by the combinatorial group finding subprocess. This is easily seen to be exponential in the length of the longest group g (regarded as a string over $\mathcal{R}$) in any of the initial sets M_i's, simply because g may have an exponential number of maximal subgroups that satisfy also the quorum constraint. In turn, the length of g may be of the order of composite motif width and hence of sequence length. At the other extreme, there is the situation where we only have two (of few) factors and look for sites where both factors bind (as for the TRANSCompel datasets of Sect. 4). In this case the cost of the subprocess is linear in the number of sequences.

When combinatorial group finding is fast (as in all the experiments we have performed) the computational bottlenecks move to other parts of the code, i.e., outside of the software module that implements the core CMF algorithm. In particular, the computation of PWM pairwise similarities takes quadratic time in the number of PWMs, which can be pretty high in a number of scenarios.

4   Experiments

In this section we present the results obtained from a number of experiments performed on the twelve benchmark datasets presented in [8].[5]

We compare CMF against the eight tools considered in the assessment paper (CisModule [23], Cister [24], Cluster-Buster (CB) [25], Composite Module Analyst (CMA) [26], MCAST [27], ModuleSearcher (MS) [28], MSCAN [29] and Stubb [30]). We also consider two other (more recent) tools, named COMPO, developed by the same research group that performed the assessment [13], and MOPAT [14]. We based our choice on tools whose code was available or for which we could find reported results for all datasets taken into consideration in this paper. We also compare CMF against CORECLUST [15] on just one dataset, the only one for which data are available.

---

[5] In the following, we refer to [8] as to the *assessment paper*.

4.1 Datasets

We use the TRANSCompel as well as the liver and muscle datasets presented in [8]. The TRANSCompel benchmark includes ten datasets corresponding to as many composite motifs, each consisting of two binding sites for different factors.

In [8], all the matrices corresponding to a same factor were grouped to form an "equivalence set", and treated as they were one. These matrices form what is called, in the assessment paper, the *noise_0 benchmark*. To simulate conditions in which input data are fuzzier, we also consider the so-called *noise_50 benchmark* presented in [8], in which each dataset is composed of an equal number of good and random (i.e., taken at random from TRANSFAC) matrices.

Two additional benchmarks are discussed in [8], namely *liver* and *muscle*, having very different characteristics from the previous ones. Liver includes sequences with up to nine binding sites from four different factors, while muscle includes sequences with up to eight sites from five factors.

Statistics for tools evaluated in [8] were downloaded from the site http://tare.medisin.ntnu.no/composite/composite.php. Regarding CO-MPO, we computed the statistics for liver and muscle datasets starting from the prediction files made available by the authors at the address http://tare.medisin.ntnu.no/compo/. For the TRANSCompel datasets (noise_0 and noise_50), we directly used the statistic results provided at the same address.

4.2 Scoring Predictions

We compare CMF against all the other eleven tools using the *correlation coefficient (CC)*. We also compare CMF and COMPO (the best performing tool among CMF's competitors) using other popular statistics, namely: *Sensitivity (Sn)*, *Positive Predicted Values (PPV)*, *Performance Coefficient (PC)*, and *Average Site Performance (ASP)* (see [31] for definitions). All the mentioned statistics are computed at the *nucleotide-level*. CMF and COMPO are also compared using *motif level* statistics.

4.3 Results

In all the experiments, CMF was run with fixed configuration file, with $W = \{50, 75, 100, 125, 150\}$, $q = \{0.9, 0.8, 0.7, 0.6., 0.5., 0.4, 0.3, 0.2, 0.1\}$ (see Sect. 3.3).

Nucleotide level analysis. Table 1 shows the results obtained by CMF compared to eleven competitor algorithms on the whole collection of twelve datasets (noise_0, liver, and muscle). The results suggest that CMF is indeed competitive with other state of the art tools. In the attempt to assess the significance of the results of Table 1, we first performed a Friedman non-parametric test (see, e.g., [32]) that involved eleven tools (all but CORECLUST, because of the limited availability of homogeneous data with which to perform the comparisons). As it can be easily argued, here the null hypothesis (i.e., that all the considered algorithms behave similarly, and hence that the average ranks over the all datasets are essentially the same), can be safely rejected, with a P-value around $2.2 \cdot 10^{-9}$.

Table 1. CC results for noise_0, liver, and muscle data, with best figures in bold-face. CB = Cluster-Buster, MS = ModuleSearcher, CMA = Composite Module Analyst, CM = CisModule, C = CORECLUST.

Dataset/tool	CMF	COMPO	CB	Cister	MSCAN	MS	MCAST	Stubb	CMA	CM	MOPAT	C
AP1-Ets	**0.52**	0.19	0.24	0.00	0.11	0.30	0.20	0.15	0.22	−0.0	0.37	
AP1-NFAT	0.11	0.06	0.04	0.00	0.00	0.05	0.14	−0.01	**0.15**	−0.02	0.14	
AP1-NFkB	**0.76**	0.59	0.49	0.19	0.36	0.29	0.26	0.35	0.55	0.05	0.18	
CEBP-NFkB	**0.74**	0.70	0.72	0.45	0.56	0.56	0.60	0.36	0.60	−0.03	0.38	
Ebox-Ets	**0.59**	0.55	0.16	0.26	0.44	0.20	0.23	0.14	0.18	0.05	0.15	
Ets-AML	**0.49**	0.42	0.30	0.07	0.31	0.38	0.26	0.23	0.33	0.03	0.27	
IRF-NFkB	**0.92**	0.73	0.77	0.62	0.91	0.85	0.41	0.41	0.69	0.04	0.57	
NFkB-HMGIY	0.26	0.31	0.35	0.10	0.30	**0.40**	0.23	0.07	0.15	−0.03	0.13	
PU1-IRF	**0.92**	0.28	0.16	0.27	0.00	0.43	0.16	0.17	0.24	−0.01	0.21	
Sp1-Ets	**0.20**	0.05	0.09	**0.20**	0.00	0.00	0.13	0.19	0.15	0.02	0.09	
Liver	0.49	0.57	**0.59**	0.31	0.51	0.42	0.50	0.48	0.36	−0.01	0.33	
Muscle	**0.56**	0.52	0.41	0.36	0.50	0.46	0.30	0.24	0.46	0.29	0.37	**0.56**

We then performed the post hoc tests associated to the Friedman statistics, by considering CMF as the new proposed methods to be compared against the other ten tools. Table 2 shows the P-values of the ten comparisons, adjusted (according to the Hochberg step-up procedure [32]) to take into account possible type-I errors in the whole set of comparisons. For nine competing algorithms we obtained figures below the critical 0.05 threshold; only in case of COMPO we cannot reject with high confidence the null hypothesis, namely that the observed average ranks of the two algorithms (CMF and COMPO) are different by chance only.

Table 2. Adjusted P-values for post hoc comparisons of CFM against other 10 tools: MS = ModuleSearcher, CMA = Composite Module Analyst, CM = CisModule

COMPO	ClusterBuster	MS	CMA	MSCAN	MCAST	MOPAT	Cister	Stubb	CM
0.1961	0.0455	0.0455	0.0455	0.0308	0.0102	0.0012	$3.4e^{-4}$	$1.9e^{-5}$	$6.9e^{-10}$

We next concentrate on the comparison between CMF and COMPO, which is the best performing tool among the CMF competitors considered here. Table 3 compares CMF and COMPO on a wider sets of statistics. For the noise_0 benchmark, the results shown combine the results of the corresponding ten datasets (i.e., counting the total numbers of positive, positive predicted, negative, and negative predicted nucleotides). For the noise_50 benchmark, the combined results are the average of the figures obtained on ten runs on each datasets. In each run, the "good" matrices were mixed with different sets of decoy PWMs (see [8]).

Motif level analysis. We compared CMF and COMPO to understand their ability to correctly tell the matrices (possibly within an equivalence set) whose matches

Table 3. Further nucleotide level comparisons between CMF and COMPO. We report here the results most favorable to COMPO, as the authors provide three different files with predictions for each datasets.

Statistics	Noise_0	Noise_50	Liver	Muscle	Tool
PPV	**0.67**	**0.45**	0.67	**0.60**	CMF
	0.40	0.37	**0.85**	0.52	COMPO
Sn	**0.54**	**0.49**	**0.429**	0.65	CMF
	0.47	0.48	0.425	**0.69**	COMPO
PC	**0.42**	**0.31**	0.35	**0.45**	CMF
	0.28	0.26	**0.40**	0.42	COMPO
ASP	**0.60**	**0.47**	0.55	**0.62**	CMF
	0.44	0.42	**0.64**	0.60	COMPO
CC	**0.58**	**0.45**	0.49	**0.56**	CMF
	0.41	0.39	**0.57**	0.52	COMPO

belong to a predicted composite motif. In contrast to [8], we do not consider here true negative predictions, as we regard the concept of *true negative* not well defined at the motif level.[6] Hence we only computed Sensitivity, Positive Predicted Value, and Performance Coefficient as motif level statistics.

Table 4. Motif level results for CMF and COMPO on the noise_0 dataset

Statistics	AP1-Ets	AP1-NFAT	AP1-NFkB	CEBP-NFkB	Ebox-Ets	Ets-AML	Tool
Sn	**0.647**	0.045	**0.75**	0.625	0.417	**0.9**	CMF
	0.47	**0.364**	0.625	**0.75**	**0.5**	0.8	COMPO
PPV	0.733	0.5	1	1	0.833	0.9	CMF
	1	1	1	1	1	1	COMPO
PC	**0.524**	0.043	**0.75**	0.625	0.385	**0.818**	CMF
	0.47	**0.364**	0.625	**0.75**	**0.5**	0.8	COMPO

Statistics	IRF-NFkB	NFkB-HMGIY	PU1-IRF	Sp1-Ets	Combined	Tool
Sn	1	0.357	1	**0.438**	**0.574**	CMF
	0.833	**0.429**	0.6	0	0.506	COMPO
PPV	1	0.625	1	**0.875**	0.861	CMF
	1	1	1	0	**0.932**	COMPO
PC	1	0.294	1	**0.412**	**0.525**	CMF
	0.833	**0.429**	0.6	0	0.488	COMPO

---

[6] Note that in the already cited paper by Tompa et al. [31], true negative predictions at the motif level are not considered.

Table 5. CMF motif level statistics for the muscle and liver datasets

Liver dataset			Muscle dataset		
Sn	PVV	PC	Sn	PVV	PC
0.5	0.728	0.42	0.74	0.66	0.54

Table 4 reports the performances at motif level obtained using our computed CMF predictions and the predictions made by COMPO on the TRANSCompel datasets. The results are essentially similar, with a slightly better Performance Coefficient (the sole comprehensive measure computed at motif level) exhibited by CMF. The results suggest once more that our software is competitive with current state of the art tools.

Finally, Table 5 reports CMF statistics on the muscle and liver datasets. We do not include a comparison against COMPO here since the way to correctly and fairly interpret COMPO's prediction is not completely clearly to us. First of all, the authors present three different prediction sets, obtained under different configuration runs. Secondly, all the prediction files contain multiple identical predictions, which negatively influences the PPV counts.

5   Conclusions

In this paper we have presented CMF, a novel tool for Composite Motif detection, a computational problem which is well-known to be very difficult. Indeed, to date, no available software for (simple or composite) motif discovery can be clearly identified as the "best one" under all application settings. Knowing this, we are also aware that more comparisons are required, in different experimental frameworks, for general conclusions to be drawn about the competitiveness of CMF.

However, we think that some interesting findings have emerged from this work, all related to the power of simple motif combinations. First of all, that the good results exhibited by CMF have been obtained without using any sophisticated statistic filtering criteria; the combination of "right" simple sites were often strong enough to emerge from a huge set of potentially active motif clusters. Secondly, that the conceptually simple CMF architecture, based on a two-stage approach to composite motif finding (i.e., first detect simple motifs, then combine them to form clusters of prospective functional motifs) proved to be competitive against other, more sophisticated approaches (see also [12]). In the third place, that lowering the thresholds that "define" (in silico) the DNA occupancy by a transcription factor, can be appropriate a strategy that can be kept hidden to the user.

On the other hand, the same issues outlined in the preceding paragraph suggest possible directions to improve CMF performance. For instance, incorporating a statistical filtering may enhance the PPV rate of the prospective composite motifs devised by simple site combinations. However, we think that the most delicate aspect has to do with thresholding and discretization. It is a growing popular belief among biologists that the DNA occupancy is determined mostly by chromatin accessibility (rather than DNA-factor affinities), with the occupancy scale being a continuum of thermodynamics levels. Turning this knowledge into a computable property of the potential binding site seems indeed a hard challenge.

Acknowledgments. The present work is partially supported by the Flagship project *InterOmics* (PB.P05), funded by the Italian MIUR and CNR organizations, and by the joint IIT-IFC Lab for Integrative System Medicine (LISM).

References

1. Davidson, E.H.: The Regulatory Genome: Gene Regulatory Networks in Development and Evolution, 1st edn. Academic Press, San Diego (2006)
2. Pavesi, G., Mauri, G., Pesole, G.: In silico representation and discovery of transcription factor binding sites. Brief. Bioinform. **5**, 217–236 (2004)
3. Sandve, G.K., Drabløs, F.: A survey of motif discovery methods in an integrated framework. Biol. Direct. **1**, 11 (2006)
4. Häußler, M., Nicolas, J.: Motif discovery on promotor sequences. Research report RR-5714, INRIA (2005)
5. Zambelli, F., Pesole, G., Pavesi, G.: Motif discovery and transcription factor binding sites before and after the next-generation sequencing era. Brief. Bioinf. (2012)
6. Wingender, E., et al.: Transfac: a database on transcription factors and their DNA binding sites. Nucl. Acids Res. **24**, 238–241 (1996)
7. Sandelin, A., Alkema, W., Engström, P.G., Wasserman, W.W., Lenhard, B.: Jaspar: an open-access database for eukaryotic transcription factor binding profiles. Nucl. Acids Res. **32**, 91–94 (2004)
8. Klepper, K., Sandve, G., Abul, O., Johansen, J., Drabløs, F.: Assessment of composite motif discovery methods. BMC Bioinform. **9**, 123 (2008)
9. Sinha, S.: Finding regulatory elements in genomic sequences. Ph.D. thesis, University of Washington (2002)
10. Van Loo, P., Marynen, P.: Computational methods for the detection of cis-regulatory modules. Brief. Bioinform. **10**, 509–524 (2009)
11. Ivan, A., Halfon, M., Sinha, S.: Computational discovery of cis-regulatory modules in drosophila without prior knowledge of motifs. Genome Biol. **9**, R22 (2008)
12. Federico, M., Leoncini, M., Montangero, M., Valente, P.: Direct vs 2-stage approaches to structured motif finding. Algorithms Mol. Biol. **7**, 20 (2012)
13. Sandve, G., Abul, O., Drablos, F.: Compo: composite motif discovery using discrete models. BMC Bioinform. **9**, 527 (2008)
14. Hu, J., Hu, H., Li, X.: Mopat: a graph-based method to predict recurrent cis-regulatory modules from known motifs. Nucl. Acids Res. **36**, 4488–4497 (2008)
15. Nikulova, A.A., Favorov, A.V., Sutormin, R.A., Makeev, V.J., Mironov, A.A.: Coreclust: identification of the conserved CRM grammar together with prediction of gene regulation. Nucl. Acids Res. **40**, e93 (2012). doi:**10.1093/nar/gks235**
16. Vavouri, T., Elgar, G.: Prediction of cis-regulatory elements using binding site matrices - the successes, the failures and the reasons for both. Curr. Opin. Genet. Develop. **15**, 395–402 (2005)

17. Kel, A., Gößling, E., Reuter, I., Cheremushkin, E., Kel-Margoulis, O., Wingender, E.: Matchtm: a tool for searching transcription factor binding sites in DNA sequences. Nucl. Acids Res. **31**, 3576–3579 (2003)
18. Chen, Q.K., Hertz, G.Z., Stormo, G.D.: Matrix search 1.0: a computer program that scans DNA sequences for transcriptional elements using a database of weight matrices. Comp. Appl. Biosci.: CABIOS **11**, 563–566 (1995)
19. Prestridge, D.S.: Signal scan: a computer program that scans DNA sequences for eukaryotic transcriptional elements. Comp. Appl. Biosci.: CABIOS **7**, 203–206 (1991)
20. Matys, V., et al.: TRANSFAC and its module TRANSCompel: transcriptional gene regulation in eukaryotes. Nucl. Acids Res. **34**, D108–D110 (2006)
21. Thomas-Chollier, M., et al.: RSAT: regulatory sequence analysis tools. Nucl. Acids Res. **36**, W119–W127 (2008)
22. Uno, T.: Pce: Pseudo clique enumerator, ver. 1.0 (2006)
23. Zhou, Q., Wong, W.H.: Cismodule: De novo discovery of cis-regulatory modules by hierarchical mixture modeling. Proc. Natl. Acad. Sci. **101**, 12114–12119 (2004)
24. Frith, M.C., Hansen, U., Weng, Z.: Detection of cis -element clusters in higher eukaryotic dna. Bioinformatics **17**, 878–889 (2001)
25. Frith, M.C., Li, M.C., Weng, Z.: Cluster-Buster: finding dense clusters of motifs in DNA sequences. Nucl. Acids Res. **31**, 3666–3668 (2003)
26. Kel, A., Konovalova, T., Waleev, T., Cheremushkin, E., Kel-Margoulis, O., Wingender, E.: Composite module analyst: a fitness-based tool for identification of transcription factor binding site combinations. Bioinformatics **22**, 1190–1197 (2006)
27. Bailey, T.L., Noble, W.S.: Searching for statistically significant regulatory modules. Bioinformatics **19**, ii16–ii25 (2003)
28. Aerts, S., Van Loo, P., Thijs, G., Moreau, Y., De Moor, B.: Computational detection of cis -regulatory modules. Bioinformatics **19**, ii5–ii14 (2003)
29. Johansson, Ö., Alkema, W., Wasserman, W.W., Lagergren, J.: Identification of functional clusters of transcription factor binding motifs in genome sequences: the mscan algorithm. Bioinformatics **19**, i169–i176 (2003)
30. Sinha, S., van Nimwegen, E., Siggia, E.D.: A probabilistic method to detect regulatory modules. Bioinformatics **19**, i292–i301 (2003)
31. Tompa, M., et al.: Assessing computational tools for the discovery of transcription factor binding sites. Nat. Biotechnol. **23**, 137–144 (2005)
32. García, S., Fernández, A., Luengo, J., Herrera, F.: Advanced nonparametric tests for multiple comparisons in the design of experiments in computational intelligence and data mining: experimental analysis of power. Inf. Sci. **180**, 2044–2064 (2010)

Hill-Climbing Behavior on Quantized NK-Landscapes

Matthieu Basseur[✉] and Adrien Goëffon

LERIA, University of Angers, Angers, France
{matthieu.basseur, adrien.goeffon}@univ-angers.fr

Abstract. This paper provides guidelines to design climbers considering a landscape shape under study. In particular, we aim at competing best improvement and first improvement strategies, as well as evaluating the behavior of different neutral move policies. Some conclusions are assessed by an empirical analysis on non-neutral (NK-) and neutral (quantized NK-) landscapes. Experiments show the ability of first improvement to explore rugged landscapes, as well as the interest of accepting neutral moves at each step of the search.

1 Introduction

Basic iterative improvement methods like climbers are generally used as components of more sophisticated local search techniques or metaheuristics. A climber consists in reaching a local optimum by iteratively improving a single solution with local modifications. Although most of metaheuristics use climbers or variants as intensification mechanism, they mainly focus on determining how to escape local optima. Nevertheless, several important questions have to be considered while designing a climber. Usually, the conception effort of any local search algorithm focus on the design of the neighborhood structure as well as how to build the solution initiating the search. However there are several questions which are regularly considered during the conception process, but not really empirically or theoretically investigated. Among them, one can identify two main issues. First, the choice of pivoting rule: are different pivoting rules leading to similar local optima qualities in comparable computational effort? To the best of our knowledge, there is no real consensus on the benefit of using a best-improvement strategy rather than a first-improvement one, or vice versa. Second, the neutral moves policy: should we restrict the use of neutral moves for escaping local optima? The use of neutral moves during the climbing should be experimentally analyzed. In particular, it is contradictory that traditional climbers only allow strictly improving moves, while a derived search strategy, e.g. simulated annealing, systematically accept neutral moves.

Most of local search and evolutionary computation contributions are focusing on the design and evaluation of advanced and original search mechanisms. However, those aforementioned elementary components are rarely discussed in the experimental analysis.

G. Nicosia and P. Pardalos (Eds.): LION 7, LNCS 7997, pp. 209–214, 2013.
DOI: 10.1007/978-3-642-44973-4_22, © Springer-Verlag Berlin Heidelberg 2013

Since the efficiency of advanced search methods is usually dependent to the problem under consideration, it should be interesting to determine if their elementary components are themselves dependent to the considered search space topology. In this study, we aim to evaluate the behavior and efficiency of basic search methods on search spaces of different size, rugosity and neutrality. To this end, we will use NK-landscapes [3] model to simulate problems with different structures and sizes. Since the basic NK model does not induce landscapes with neutrality, we will also focus on *quantized NK-lanscapes* (NKq [5]).

Recently, Ochoa et al. [6] investigated the behavior of first and best improvement algorithms on NK-landscapes, by exhaustive exploration of small search spaces. In this paper, we extend the study by evaluating both pivoting rules and neutral moves effects on large size problems, which are those considered while using metaheuristics. To achieve this, we will evaluate empirically the relative efficiency of climber variants. The aim of this study is to compare basic ways to navigate through a search space, rather than to propose an efficient sophisticated algorithm to solve NK instances.

In Sect. 2, we will discuss on climbers key components and NK(q)-landscapes models. In Sect. 3, we compare the ability of climber variants a large scale of landscape shapes. Finally, the last section reports the main contributions and point out future investigations.

2   Climbers and NK-Landscapes

We assume that the reader is familiar with the notions of combinatorial optimization and local search algorithms. For more detailed definitions and comprehensive review of local search principles, we refer the reader to [2].

A *hill-climbing* algorithm (or *climber*) is a basic local search strategy which navigates through the search space in allowing only non-deteriorating moves. Given an initial configuration called *starting point*, a traditional climber iteratively moves to better neighbors, until it reaches a local optimum. Such a search mechanism, also known as *iterative improvement*, allows to distinguish several variants which are discussed hereafter.

2.1   Climber Components

The design of a climber implies several choices, whose effects are not clearly established. Let us point out two mains conception issues that need to be discussed.

Pivoting Rule. The *best improvement* strategy (or *greedy hill-climbing*) consists in selecting, at each iteration, a neighbor which achieves the best fitness. This implies to generate the whole neighborhood at each step of the search, unless an incremental evaluation of all neighbors can be performed. On the contrary, the *first improvement* strategy accepts the first evaluated neighbor which satisfies the moving condition. This avoids the systematic generation of the entire neighborhood and allows more conceptual options.

Neutral Move Policy. A *basic hill-climbing* algorithm does not allow neutral moves (i.e. moves to neutral neighbors) during the search, and only performs improving moves until reaching a local optimum. Question of neutral moves can be considered to escape local optima (*neutral perturbation*, NP) when the fitness landscape contains a substantial proportion of neutral transitions (on smooth landscapes). Another variant, called *stochastic hill-climbing*, can accept indifferently neutral or improving neighbors throughout the search, even before reaching a local optimum. It is not that obvious to determine the influence of the neutral move policy on the quality of the configurations reached. However, it is interesting to note that the more advanced simulated annealing algorithm, which allows some deteriorating moves during the search, systematically accepts neutral moves under consideration.

There are other aspects which could be discussed. For instance the neighborhood evaluation can be made with or without replacement, and its generation order can be done deterministically or randomly. Nevertheless, these choices are greatly dependent on the problem under study and are not discussed here.

2.2   NK-Landscapes and Neutrality

NK-Landscapes. The NK family of landscapes [3] is a problem-independent model for constructing multimodal landscapes. NK-landscapes use a basic search space, with binary strings as configurations and bit-flip as neighborhood (two configurations are neighbors iff their Hamming distance is 1). Characteristics of an NK-landscape are determined by two parameters N and K. N refers to the size of binary string configurations, which defines the search space size ($|\mathcal{X}| = 2^N$). K specifies the rugosity level of the landscape; indeed, the fitness value of a configuration is given by the sum of N terms, each one depending on $K + 1$ bits of the configuration. Thus, by increasing the value of K from 0 to $N - 1$, NK-landscapes can be tuned from smooth to rugged.

In NK-landscapes, the fitness function $f : \{0,1\}^N \to [0,1)$ to be maximized is defined as follows.

$$f(x) = \frac{1}{N} \sum_{i=1}^{N} c_i(x_i, x_{i_1}, \ldots, x_{i_K}).\tag{1}$$

where $c_i : \{0,1\}^{K+1} \to [0,1)$ defines the component function associated with each variable x_i, $i \in \{1, \ldots, N\}$, and where $K < N$.

NK-landscapes instances are both determined by the $(K + 1)$-uples $(x_i, x_{i_1}, \ldots, x_{i_K})$ and the $2^N.(K + 1)$ c_i result values corresponding to a fitness contribution matrix C whose values are randomly generated in $[0,1)$. The usual precision of random values imply that plateaus are almost absent on NK-landscapes.

NKq-Landscapes. To add neutrality to NK-landscapes, Newman et al. introduced quantised NK-landscapes [5], by fluctuating the discretization level of c_i

result values. Indeed, limiting their possible values increase the number of neutral neighbors. Thus, NKq implies a third parameter $q \geqslant 2$ which specifies the c_i. functions codomain size. The maximal degree of neutrality is reached when $q = 2$ (C is then a binary matrix), and decreases while q increases.

3   Comparison of Hill-Climbing Strategies

In this section, we aim at evaluating the ability of climbers to explore various landscapes. Thus, different climbing strategies introduced Sect. 2 will be applied on landscapes defined in Sect. 2.2.

The experimental analysis will compare five climbers variants combining pivoting rule (PR) alternatives and the neutral move policy (NMP). All climbers start from a random configuration, and stop after $10.N^2$ configuration evaluations (unless for basic move policy which stops when a local optimum is reached). Let us notice that this maximal number of evaluations has been set to allow a convergence of the search, after observing no significant improvements for longer searches.

Each climber will be executed 10,000 times on a benchmark set of 48 instances: 16 basic NK-landscapes parametrizations, as well as 32 instances which corresponds to the 32 NKq landscapes parametrizations s.t. $N \in \{256, 1024\}$ and $K \in \{1, 2, 4, 8\}$. For each instance, the five climbers start their searches from a single set of 10,000 starting points, in order to cancel the initialization bias.

Empirical Analysis

Experiment results are given on Tables 1 and 2 which focus respectively on the NK and NKq instances. For each couple climber/instance, we report the average fitness of the 10,000 resulting configurations. For each instance, the best average value appears in bold. Moreover, we indicate in grey methods which are not statistically outperformed by any other method (w.r.t. the Mann-Whitney test).

Table 1. Climbers results on NK-landscapes. Only two variants, with no neutral moves, are outputed.

PR	First	Best	PR	First	Best
256_1	.7021	**.7079**	1024_1	.6969	**.7039**
256_2	.7066	**.7094**	1024_2	.7146	**.7197**
256_4	**.7235**	.7204	1024_4	**.7246**	.7242
256_8	**.7166**	.7122	1024_8	**.7216**	.7174

Results obtained on the basic NK-landscapes are given in Table 1. In this table, results include only two variants which correspond to the pivoting rule alternatives. Indeed, the basic NK-landscapes do not contains a significant number of neutral neighbors. Then, experiments show equivalent results whatever the neutral move policy being adopted. Anyway, this table provides us a significant piece of information while comparing the best improvement and the first improvement pivoting rules. Best improvement statistically outperforms first improvement when $K \in \{1, 2\}$, and first improvement appears more efficient while K increases. In other words, best improvement is well-suited to explore smooth landscapes, whereas first improvement seems more adapted to explore a rugged one.

Table 2. Climbers results on NKq landscapes (q = $\{10, 5, 3, 2\}$).

NMP PR	Basic First	Basic Best	Stoch. First	Basic+NP First	Basic+NP Best	NMP PR	Basic First	Basic Best	Stoch. First	Basic+NP First	Basic+NP Best
256_1_q10	.7206	.7275	.7271	.7267	.7323	1024_1_q10	.7144	.7209	.7203	.7195	.7256
256_1_q5	.7377	.7434	.7518	.7516	.7557	1024_1_q5	.7336	.7418	.7514	.7509	.7550
256_1_q3	.7626	.7682	.7938	.7938	.7948	1024_1_q3	.7611	.7691	.7955	.7953	.7971
256_1_q2	.8095	.8143	.8672	.8670	.8668	1024_1_q2	.7883	.7920	.8416	.8411	.8414
256_2_q10	.7245	.7278	.7318	.7307	.7340	1024_2_q10	.7360	.7388	.7434	.7422	.7462
256_2_q5	.7390	.7463	.7610	.7608	.7640	1024_2_q5	.7532	.7581	.7768	.7763	.7783
256_2_q3	.7713	.7784	.8183	.8159	.8193	1024_2_q3	.7788	.7836	.8259	.8253	.8260
256_2_q2	.8095	.8144	.8907	.8879	.8886	1024_2_q2	.8216	.8268	.9067	.9036	.9045
256_4_q10	.7441	.7438	.7503	.7498	.7490	1024_4_q10	.7464	.7447	.7540	.7511	.7501
256_4_q5	.7582	.7588	.7865	.7838	.7819	1024_4_q5	.7619	.7637	.7897	.7850	.7858
256_4_q3	.7894	.7879	.8436	.8407	.8429	1024_4_q3	.7924	.7928	.8469	.8424	.8432
256_4_q2	.8323	.8343	.9363	.9297	.9288	1024_4_q2	.8334	.8348	.9348	.9286	.9294
256_8_q10	.7368	.7343	.7425	.7402	.7369	1024_8_q10	.7440	.7372	.7479	.7467	.7428
256_8_q5	.7575	.7527	.7762	.7741	.7710	1024_8_q5	.7606	.7580	.7830	.7789	.7772
256_8_q3	.7871	.7835	.8255	.8196	.8200	1024_8_q3	.7918	.7879	.8350	.8299	.8292
256_8_q2	.8285	.8277	.9098	.9039	.9018	1024_8_q2	.8380	.8340	.9191	.9125	.9121

NKq instances experiments lead to relevant outcomes. One can see in Table 2 that neutral moves are necessary to climb landscapes containing even a small level of neutrality. Indeed, basic climbers are always statistically outperformed by others. Moreover, this table emphasizes significant differences between the three strategies allowing neutral moves. First, stochastic climbers reach bests results on most instances, especially on more rugged and/or neutral landscapes (high K, low q). This is particularly interesting since, to our knowledge, basic policies – with or without neutral perturbations – are more traditionally used while designing metaheuristics. However, a best improvement strategy combined with neutral perturbations remains suitable in smooth landscapes, especially with lowest levels of neutrality. Globally, one observe that the search space size given by parameter N does not influence the overall tendency of the results; although efficiency differences between policies tends to be more significant for larger search spaces.

Let us precise that in our original experimental analysis, two other models of neutrality have been experimented: *probabilistic* NK-landscapes [1] as well as *rounded* NK-landscapes, designed by simply rounding fitnesses. These experiments also lead to significant outcomes, which are not outputted here due to a lack of space.

4 Conclusion

Climbers are, often considered as basic components of advanced search methods. However, influence of their conception choices are rarely discussed through advanced studies. In this paper we have focused on the capacity of different hill-climbings versions to reach good configurations in various landscapes. In particular, we compared the first and best improvement strategies as well as three different neutral move policies. In order to provide an empirical analysis on a large panel of representative instances, we used NK-landscapes with different sizes and rugosity levels. On landscapes with no neutrality, we show that best improvement performs better on smooth landscapes, while first improvement is well-suited on more rugged ones. To evaluate the impact of neutral move policies, we used quantized NK-lanscapes (NKq) as model of neutrality. First, one observes that stochastic hill-climbings globally reach better configurations than other variants. In other words, at each step of the search, it makes sense to perform the first non-deteriorating move instead of extending the neighborhood evaluation.

Perspectives of this work mainly includes the extension of this analysis to Iterative Local Search methods [4]. Indeed, several questions arise while considering iterated versions. First, we have to determine to what extent efficient climbers can improve iterated searches. Last, a similar study performed in an iterated context will determine if the overall influence of structural choices remain unchanged.

References

1. Barnett, L.: Ruggedness and neutrality - the NKp family of fitness landscapes. In: Alive VI: Sixth International Conference on Artificial Life, pp. 18–27. MIT Press (1998)
2. Hoos, H., Stützle, T.: Stochastic Local Search: Foundations & Applications. Morgan Kaufmann Publishers Inc., San Francisco (2004)
3. Kauffman, S.A.: The Origins of Order: Self-Organization and Selection in Evolution, 1st edn. Oxford University Press, USA (1993)
4. Lourenço, H.R., Martin, O., Stützle, T.: Iterated local search. In: Glover, F., Kochenberger, G. (eds.) Handbook of Metaheuristics. International Series in Operations Research and Management Science, vol. 57, pp. 321–353. Kluwer Academic, Norwell (2002)
5. Newman, M.E.J., Engelhardt, R.: Effects of selective neutrality on the evolution of molecular species. Proc. Roy. Soc. B **265**(1403), 1333–1338 (1998)
6. Ochoa, G., Verel, S., Tomassini, M.: First-improvement vs. best-improvement local optima networks of NK landscapes. In: Proceedings of the 11th International Conference on Parallel Problem Solving From Nature, Krakow Pologne, pp. 104–113, September 2010

Neighborhood Specification for Game Strategy Evolution in a Spatial Iterated Prisoner's Dilemma Game

Hisao Ishibuchi[✉], Koichiro Hoshino, and Yusuke Nojima

Department of Computer Science and Intelligent Systems, Graduate School of Engineering,
Osaka Prefecture University, 1-1 Gakuen-cho, Naka-ku, Sakai, Osaka 599-8531, Japan
{hisaoi,nojima}@cs.osakafu-u.ac.jp,
kouichirou.hoshino@ci.cs.osakafu-u.ac.jp

Abstract. The prisoner's dilemma is a two-player non-zero-sum game. Its iterated version has been frequently used to examine game strategy evolution in the literature. In this paper, we discuss the setting of neighborhood structures in its spatial iterated version. The main characteristic feature of our spatial iterated prisoner's dilemma game model is that each cell has a different scheme to represent game strategies. In our computational experiments, one of four representation schemes is randomly assigned to each cell in a two-dimensional grid-world. An agent at each cell has a game strategy encoded by the assigned representation scheme. In this situation, an agent may have no neighbors with the same representation scheme as the agent's scheme. The existence of such an agent has a negative effect on the evolution of cooperative behavior. This is because strategies with different representation schemes cannot be recombined. When no neighbors have the same representation scheme as the agent's scheme, no recombination can be used for generating a new strategy for the agent. In our former study, we used a larger neighborhood structure for such an agent. As a result, each agent has a different neighborhood structure and a different number of neighbors. This makes it difficult to discuss the effect of the neighborhood size on the evolution of cooperative behavior. In this paper, we propose the use of the following setting: Each agent has the same number of neighbors with the same representation scheme as the agent's scheme. This means that each agent has the same number of qualified neighbors as its mates. We also examine a different spatial model where the location of each agent is randomly specified as a point in a two-dimensional continuous space instead of a grid-world.

Keywords: Evolutionary games · Prisoner's dilemma game · Iterated prisoner's dilemma (IPD) · Spatial IPD games · Neighborhood structures · Representation

1 Introduction

In general, an appropriate choice of a representation scheme for solution encoding has been an important research topic in evolutionary computation. This is also the case in evolutionary games. Ashlock et al. [1–3] showed that the evolution of cooperative

G. Nicosia and P. Pardalos (Eds.): LION 7, LNCS 7997, pp. 215–230, 2013.
DOI: 10.1007/978-3-642-44973-4_23, © Springer-Verlag Berlin Heidelberg 2013

behavior in the IPD (Iterated Prisoner's Dilemma) game strongly depended on the choice of a representation scheme for game strategy encoding. They examined a number of different representation schemes, and obtained totally different results from different settings. In each run in their computational experiments, a single representation scheme was assigned to all agents to examine the relation between the choice of a representation scheme and the evolution of cooperative behavior.

Simultaneous use of two representation schemes was examined in Ishibuchi et al. [4]. One of the two schemes was randomly assigned to each agent. When the IPD game was not played between agents with different schemes, totally different results were obtained from each scheme. However, similar results were obtained from each scheme when the IPD game was played between agents with different schemes. The simultaneous use of different representation schemes was also examined in other areas of evolutionary computation (e.g., see [5] for the use in island models).

It has been demonstrated in many studies on the IPD game that spatial structures of agents have a large effect on the evolution of cooperative behavior (e.g., [6–9]). A two-dimensional grid-world with a neighborhood structure is often used in such a spatial IPD game. An agent in each cell plays the IPD game against its neighbors (i.e., local interaction through the IPD game). A new strategy for each agent is generated from current strategies of the agent and its neighbors (i.e., local mating). Local interaction and local mating usually use the same neighborhood structure. The use of different neighborhood structures was examined in Ishibuchi and Namikawa [10].

Motivated by the above-mentioned studies on representation schemes [1–5] and spatial structures [6–10], we examined a spatial IPD game model with a number of different representation schemes in our former study [11]. One of four representation schemes was randomly assigned to each cell in a two-dimensional grid-world. An example of such a random assignment is shown for a two-dimensional 11×11 grid-world in Fig. 1 where the von Neumann neighborhood with four neighbors is also illustrated. In Fig. 1, the center cell of the illustrated von Neumann neighborhood has a representation scheme D. However, none of its four neighbors has the representation scheme D. Thus no strategies of the four neighbors can be used to generate a new strategy for the center cell. As a result, only the current strategy of the center cell is used to generate its new strategy (i.e., new strategies are generated by mutation).

In our former study [11], we used a larger neighborhood when a cell had no neighbors with the same representation scheme as the cell's scheme. That is, each cell had a different neighborhood and a different number of neighbors. This makes it very difficult to discuss the effect of the neighborhood size on the strategy evolution.

In this paper, we examine the following three settings of neighborhood for local mating in our spatial IPD game model in a two-dimensional grid-world:

(i) Basic setting: A given neighborhood structure for local mating is always used for all cells. If a cell has no neighbors with the same representation scheme as the cell's scheme, a new strategy is always generated from the current one by mutation. Otherwise, a new strategy is generated by local mating, crossover and mutation.

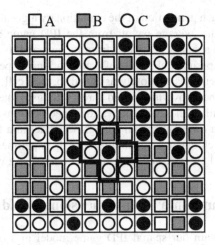

Fig. 1. An example of random assignment of four representation schemes (A, B, C and D) over a two-dimensional 11 × 11 grid-world.

(ii) Setting in our former study [11]: When a cell has no neighbors with the same representation scheme as the cell's scheme, a larger neighborhood structure is used so that the cell has at least one neighbor with the same scheme. In this setting, a new strategy at each cell is generated by local mating, crossover and mutation.

(iii) Setting with the same number of neighbors for local mating: In this setting, a pre-specified number of cells with the same representation scheme are defined for each cell as neighbors for local mating. That is, each cell has the same number of neighbors for local mating. A new strategy at each cell is generated by local mating, crossover and mutation.

For comparison, we also examine a different spatial IPD game model with a different spatial structure where each agent is randomly assigned to a point in a two-dimensional continuous space. We do not use a two-dimensional grid-world in this spatial IPD game model. For each agent, a pre-specified number of the nearest agents with the same representation scheme are defined as its neighbors for local mating.

One advantage of this model is that the number of neighbors for local mating can be arbitrarily specified in each computer simulation. Another advantage is that the definition of neighbors is clear and easy to implement. One disadvantage is that the neighborhood relation is not always symmetrical. That is, "X is in the nearest neighbors of Y" does not always mean "Y is in the nearest neighbors of X". The above-mentioned settings (ii) and (iii) of neighbors for local mating in a two-dimensional grid-world also have such an asymmetric property of neighborhood.

This paper is organized as follows. In Sect. 2, we explain our spatial IPD game model in a two-dimensional grid-world with different representation schemes. We use two neighborhood structures in our spatial IPD game model: One is for local mating and the other is for local interaction. Game strategies and representation schemes are also explained in Sect. 2. In Sect. 3, we explain our cellular genetic algorithm for strategy evolution. In each generation of the evolutionary algorithm, the fitness of each agent is evaluated as the average payoff from the IPD game against its neighbors. A new strategy of each agent is generated from its own and its neighbors' strategies. In Sect. 4, we report experimental results of computational experiments where one of four representation schemes is randomly assigned to each cell in a two-dimensional 11×11 grid-world. The above-mentioned three settings of neighborhood for local mating are compared with each other using various specifications of neighborhood size. In Sect. 5, we explain a different spatial IPD game model in a two-dimensional continuous space where each agent is randomly located. Experimental results using this spatial IPD game model are reported in Sect. 6. In Sect. 7, we conclude this paper.

2   Spatial IPD Game in a Two-Dimensional Grid-World

In this section, we explain our spatial IPD game model in a two-dimensional grid-world with two neighborhood structures and different representation schemes. As in many other studies on the spatial IPD game, we assume the torus structure of the two-dimensional grid-world. Our spatial IPD game model in this section is the same as in our previous study [11] except for the specification of local mating neighborhood.

Payoff Matrix: The PD (Prisoner's Dilemma) game is a two-player non-zero sum game with two actions: cooperation and defection. We use a frequently-used standard payoff matrix in Table 1. When both the agent and the opponent cooperate in Table 1, each of them receives the payoff 3. When both of them defect, each of them receives the payoff 1. The agent receives the maximum payoff 5 by defecting when the opponent cooperates. In this case, the opponent receives the minimum payoff 0. In Table 1, the defection is a rational action because the agent always receives the larger payoff by defecting than cooperating in each of the two cases of the opponent action. The defection is also a rational action for the opponent. However, the payoff 1 by mutual defection is smaller than the payoff 3 by mutual cooperation. That is, the rational actions of the agent and the opponent lead to the smaller payoff 1 than the payoff 3 by their irrational actions. This is the dilemma in Table 1.

IPD Game Strategies: The IPD game is an iterated version of the PD game. The agent plays the PD game with the payoff matrix in Table 1 against the same opponent for a pre-specified number of rounds. In the first round, no information is available with respect to the previous actions of the opponent. When the agent and the opponent choose their actions for the second round, they can use the information about

Table 1. Payoff matrix in our spatial IPD game.

Agent's action	Opponent's action	
	C: Cooperation	D: Defection
C: Cooperation	Agent payoff: 3	Agent payoff: 0
	Opponent payoff: 3	Opponent payoff: 5
D: Defection	Agent payoff: 5	Agent payoff: 1
	Opponent payoff: 0	Opponent payoff: 1

their actions in the first round. In the third round, the information about the actions in the first two rounds is available. A game strategy for the IPD game determines the next action based on a finite memory about previous actions. In this paper, we use the following four representation schemes to encode IPD game strategies:

(1) Binary strings of length 3,
(2) Real number strings of length 3,
(3) Binary strings of length 7,
(4) Real number strings of length 7.

Each value in these strings shows the probability of cooperation in the next action in a different situation. Strings of length 3 determine the next action based on the opponent's previous action whereas strings of length 7 use the opponent's actions in the previous two rounds. As an example, we show a binary string strategy "101" called TFT (tit for tat) in Table 2. This string chooses the cooperation at the first round, which corresponds to the first value "1" of the string "101". When the opponent cooperated in the previous round, the cooperation is selected using the third value "1". Only when the opponent defected in the previous round, the defection is selected using the second value "0". Table 3 shows an example of a binary string strategy of length 7 ("1110111"). The first value "1" is used for the first round. The next two values "11" are used for the second round. The choice of an action in each of the other rounds is specified by the last four values "0111". This strategy, which is called TF2T (tit for two tats), defects only when the opponent defected in the previous two rounds. In the case of real number strings, the cooperation is probabilistically chosen using the corresponding real number as the cooperation probability.

Local Interaction Neighborhood in the Grid-World: We use the 11×11 grid-world with the torus structure in Fig. 1. Each cell has an agent, a game strategy, and a representation scheme. Each agent plays the IPD game against its local interaction neighbors. Let us denote the set of local interaction neighbors of Agent i by $N_{IPD}(i)$. If $N_{IPD}(i)$ includes five or less neighbors, Player i plays the IPD game against all neighbors in $N_{IPD}(i)$. If $N_{IPD}(i)$ includes more than five neighbors, five opponents are randomly selected from $N_{IPD}(i)$. It should be noted that any agent is not allowed to play the IPD game against itself. It was demonstrated that the IPD game of each agent against itself had a large positive effect on the evolution of cooperative behavior [12].

We examine six specifications of $N_{IPD}(i)$ in Fig. 2 with 4, 8, 12, 24, 40 and 48 neighbors. We also examine an extreme specification of $N_{IPD}(i)$ for Agent i where all the other 120 agents are included in $N_{IPD}(i)$.

Local Mating Neighborhood in the Grid-World: A new strategy for each agent is generated from its own and its neighbors' strategies. Let $N_{GA}(i)$ be a set of neighbors

Table 2. Binary string strategy of length 3 called TFT ("101").

Round	Opponent's previous action	Cooperation probability
1st Round	–	1
Other Rounds	Defect	0
Other Rounds	Cooperate	1

Table 3. Binary string strategy of length 7 called TF2T ("1110111").

Round	Opponent's previous actions	Cooperation probability
1st Round	–	1
2nd Round	Defect	1
2nd Round	Cooperate	1
Other Rounds	Defect and defect	0
Other Rounds	Defect and cooperate	1
Other Rounds	Cooperate and defect	1
Other Rounds	Cooperate and cooperate	1

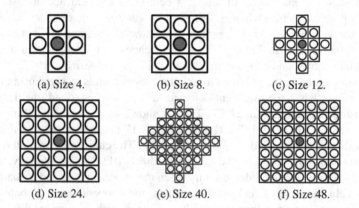

(a) Size 4.   (b) Size 8.   (c) Size 12.

(d) Size 24.   (e) Size 40.   (f) Size 48.

Fig. 2. Six neighborhood structures examined in this paper.

of Agent i for local mating. As in the case of local interaction neighborhood $N_{\text{IPD}}(i)$, we examine as local mating neighborhood $N_{\text{GA}}(i)$ the six specifications in Fig. 2 and the extreme specification including all the other agents. Only neighbors with the same representation scheme as Agent i are qualified as mates of Agent i. If no neighbors are qualified, the current strategy of Agent i is selected to generate its new strategy. We consider the following three settings to handle this undesirable case:

(i) **Basic Setting:** We do not modify local mating neighborhood even in this case.

(ii) **Use of a Larger Neighborhood:** In our former study [11], we used a larger neighborhood structure for Agent i if Agent i had no qualified neighbors. More specifically, a larger neighborhood structure with at least one qualified neighbor was searched in the order (a) → (b) → (c) → (d) → (e) → (f) in Fig. 2. If no qualified neighbor is included even in Fig. 2(f), all agents were handled as neighbors.

(iii) **The Same Number of Qualified Neighbors:** In this new setting, all agents have the same number of qualified neighbors. First all neighbors are systematically sorted (e.g., see Fig. 3). Then qualified neighbors are added to $N_{\text{IPD}}(i)$ up to the pre-specified number. We can use any kind of order of neighbors.

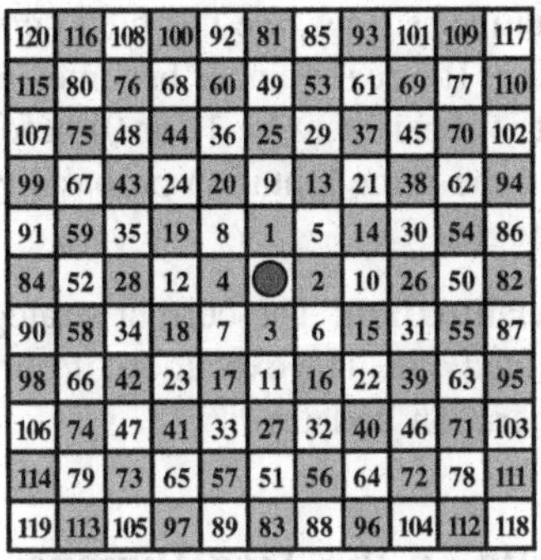

120	116	108	100	92	81	85	93	101	109	117
115	80	76	68	60	49	53	61	69	77	110
107	75	48	44	36	25	29	37	45	70	102
99	67	43	24	20	9	13	21	38	62	94
91	59	35	19	8	1	5	14	30	54	86
84	52	28	12	4	●	2	10	26	50	82
90	58	34	18	7	3	6	15	31	55	87
98	66	42	23	17	11	16	22	39	63	95
106	74	47	41	33	27	32	40	46	71	103
114	79	73	65	57	51	56	64	72	78	111
119	113	105	97	89	83	88	96	104	112	118

Fig. 3. Order of neighbors.

3 Cellular Genetic Algorithm for Game Strategy Evolution

In this section, we explain our cellular genetic algorithm for game strategy evolution. In our algorithm, each cell has a representation scheme and an agent with a game strategy. The fitness value of each strategy is evaluated through the IPD game against its neighbors. A new strategy is generated from its own and its neighbors' strategies with the same representation scheme. The current strategy is always replaced with a newly generated one. The following are more detailed explanations.

Representation Scheme Assignment: One of the four representation schemes (i.e., binary strings of length 3, real number strings of length 3, binary strings of length 7, and real number strings of length 7) is randomly assigned to each agent. More specifically, the 121 agents in the 11×11 grid-world are randomly divided into four subsets with 30, 30, 30 and 31 agents. Each representation scheme is randomly assigned to a different subset (i.e., to all agents in that subset). Each agent uses the assigned representation scheme throughout the current execution of our cellular genetic algorithm. The random assignment of a representation scheme to each agent is updated in each run of our cellular genetic algorithm.

Initial Strategies: Each agent randomly generates an initial strategy using the assigned representation scheme. For binary strings, each bit is randomly specified as 0 or 1 with the same probability. Real numbers in real number strings are randomly specified using the uniform distribution over the unit interval [0, 1].

Fitness Evaluation: The fitness value of each agent is calculated as the average payoff obtained from the IPD game with 100 rounds against up to five opponents in its local interaction neighborhood $N_{IPD}(i)$. If an agent has five or less neighbors, the average payoff is calculated over the IPD game against all neighbors. Otherwise, five neighbors are randomly chosen as opponents to calculate the average payoff.

Parent Selection, Crossover and Mutation: Two parent strategies are selected for each agent from its own and its qualified neighbors' strategies in $N_{GA}(i)$ using binary tournament selection with replacement. A new strategy is generated from the selected pair by crossover and mutation. For binary strings, we use one-point crossover and bit-flip mutation. For real number strings, we use blend crossover (BLX-α [13]) with $\alpha = 0.25$ and uniform mutation. If a real number becomes more than 1 (or less than 0) by the crossover operator, it is repaired to be 1 (or 0) before the mutation. The same crossover probability 1.0 and the same mutation probability $1/(5 \times 121)$ are used for binary and real number strategies.

Generation Update and Termination: The current strategy of each agent is always replaced with a newly generated one. The execution of our cellular genetic algorithm is terminated after 1000 generation updates.

4 Experimental Results on the Two-Dimensional Grid-World

In this section, we report experimental results with four representation schemes. The reported results are average results over 500 runs of our cellular genetic algorithm for each setting of local interaction and mating neighborhood structures. For comparison, we also report experimental results using a single representation scheme.

Experimental Results using a Single Representation Scheme: For comparison, we first report experimental results with a single representation scheme. One of the four representation schemes was assigned to all the 121 agents. We examined 7×7 combinations of the seven neighborhood specifications for local interaction $N_{\mathrm{IPD}}(i)$ and local mating $N_{\mathrm{GA}}(i)$. Experimental results are summarized in Fig. 4 where each bar shows the average payoff over 1000 generations in each of 500 runs.

Experimental Results with the Basic Setting: As explained in Sect. 2, some agents have no qualified neighbors as mates when the four representation schemes are randomly assigned over the 121 agents. Table 4 summarizes the average percentage of those agents with no qualified neighbors as mates. In the case of the smallest neighborhood with four neighbors, many agents (i.e., 32.3 % of the 121 agents) have no qualified neighbors as mates. A new strategy for each of those agents is generated from its current strategy by mutation. No selection pressure towards good strategies with higher average payoff is applied to those agents.

In Fig. 5, we show experimental results with the basic setting where the specified local mating neighborhood structure is never modified. As shown in Table 4, many agents have no qualified neighbors as mates in this setting. In Fig. 5, lower average payoff was obtained from the smallest two neighborhood structures for local mating (i.e., $N_{\mathrm{GA}}(i)$ with four and eight neighbors) than the other larger structures. We can also observe in Fig. 5 that similar results were obtained from the four representation schemes when they were randomly assigned over the 121 agents (whereas totally different results were obtained from each representation scheme in Fig. 4).

Experimental Results with the Use of a Larger Neighborhood: As in our former study, we used a larger neighborhood structure for each agent when the agent had no qualified neighbors as its mate. In this setting, each agent has at least one qualified neighbor as its mate. This means that each agent has at least two candidates for its parent strategies. As a result, some selection pressure was applied to strategies of all

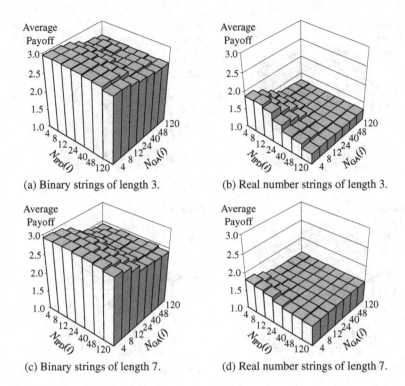

(a) Binary strings of length 3.   (b) Real number strings of length 3.

(c) Binary strings of length 7.   (d) Real number strings of length 7.

Fig. 4. Experimental results with a single representation scheme for all the 121 agents.

Table 4. Average percentage of agents with no qualified neighbors as mates.

Neighborhood size	4	8	12	24	40	48
Average percentage	32.3 %	10.0 %	2.8 %	0.1 %	0.0 %	0.0 %

agents. Experimental results are summarized in Fig. 6. There is no large difference between Figs. 5 and 6. Closer comparison between these two figures shows that the average payoff from $|N_{GA}(i)| = 4$ in Fig. 6(b) and (d) was decreased from Fig. 5. This is because uncooperative strategies are more likely to be evolved in Fig. 6(b) and (d) with $|N_{GA}(i)| = 4$ than in the corresponding situations in Fig. 5 including 32.3 % agents with no qualified mates.

In Table 5, we show the average number of qualified neighbors in computational experiments for Figs. 5 and 6. In Fig. 6, a larger neighborhood structure was used for agents with no qualified neighbors under the originally specified neighborhood structure. Thus the average number of qualified neighbors is larger in Fig. 6 than Fig. 5. For example, when we used the von Neumann neighborhood with four neighbors, 32.3 % of agents had no qualified neighbors on average. A larger neighborhood structure was assigned to each of those agents. As a result, the average number of qualified neighbors was increased from 0.97 in Fig. 5 to 1.49 in Fig. 6 (see Table 5).

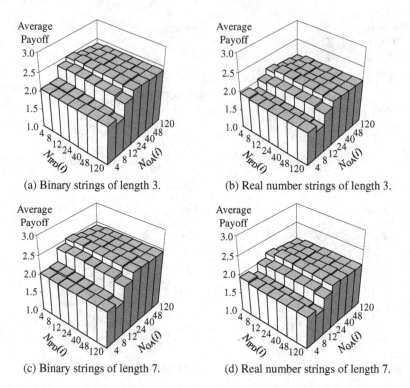

(a) Binary strings of length 3.   (b) Real number strings of length 3.

(c) Binary strings of length 7.   (d) Real number strings of length 7.

Fig. 5. Experimental results with the randomly assigned four representation schemes. The specified neighborhood structure for local mating was never modified (i.e., the basic setting).

Simulation Results with the Same Number of Qualified Neighbors: In the third setting of local mating neighborhood $N_{GA}(i)$, all neighbors of each agent are sorted in a systematic manner. Then qualified neighbors are added to $N_{GA}(i)$ up to a pre-specified number. In our computational experiments, we used the order of neighbors in Fig. 3. Since our two-dimensional 11×11 grid-world has the torus structure, we can use Fig. 3 for all agents (not only for the agent at the center of the grid-world). As the number of qualified neighbors, we examined seven specifications: 1, 2, 3, 6, 10, 12 and 30. These values are 1/4 of the size of the seven neighborhood structures examined in the previous computational experiments. When 30 was used as the number of qualified neighbors in $N_{GA}(i)$, the actual number of qualified neighbor was 29 or 30 since 121 agents were divided into four subsets with 30, 30, 30 and 31 agents.

Experimental results are summarized in Fig. 7. The axis with $N_{GA}(i)$ in Fig. 7 is the number of qualified neighbors whereas it is the size of $N_{GA}(i)$ including unqualified neighbors in Figs. 5 and 6. Experimental results in Figs. 5–7 are similar to one another. Closer examination on experimental results for $|N_{GA}(i)| = 4$ and $|N_{GA}(i)| = 8$ in Figs. 5–7 may suggest that Fig. 7 looks like something between Figs. 5 and 6 (see Table 5 for the average number of qualified neighbors in Figs. 5 and 6).

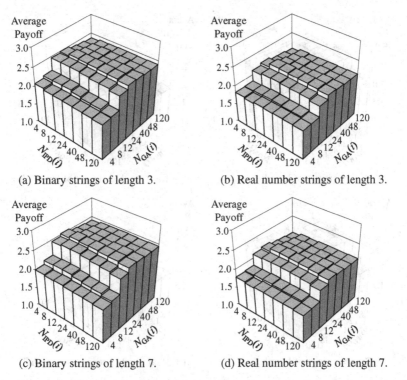

(a) Binary strings of length 3.

(b) Real number strings of length 3.

(c) Binary strings of length 7.

(d) Real number strings of length 7.

Fig. 6. Experimental results with the randomly assigned four representation schemes. A larger neighborhood structure was used as the local mating neighborhood for each agent when the agent has no qualified neighbor as its mate.

Table 5. The average number of qualified neighbors in Figs. 5 and 6.

Neighborhood size	4	8	12	24	40	48
Qualified neighbors in Fig. 5	0.97	1.93	2.91	5.84	9.74	11.68
Qualified neighbors in Fig. 6	1.49	2.13	3.00	5.84	9.74	11.68

5   Spatial IPD Game in a Two-Dimensional Continuous Space

In this section, we explain a different spatial IPD game model in a two-dimensional continuous space. In our model in this section, the location of each of 121 agents is specified as a point in a two-dimensional unit square $[0, 1] \times [0, 1]$ with the torus structure. Of course, we can use any continuous space in our model. The number of agents can also be arbitrarily specified.

As in the previous section, we assume that one of the four representation schemes is randomly assigned to each agent. An example of the location of each agent and its representation scheme is shown in Fig. 8. The neighborhood size can be arbitrary specified for each agent because we can choose an arbitrarily specified number of the

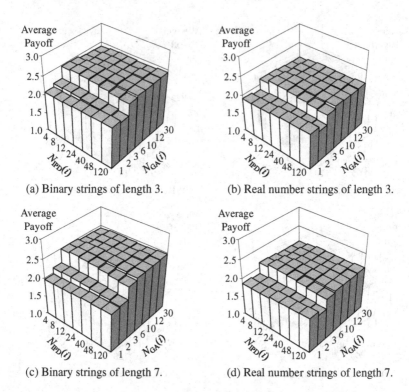

(a) Binary strings of length 3.   (b) Real number strings of length 3.

(c) Binary strings of length 7.   (d) Real number strings of length 7.

Fig. 7. Experimental results with the randomly assigned four representation schemes. Each agent had the pre-specified number of qualified neighbors as its mate.

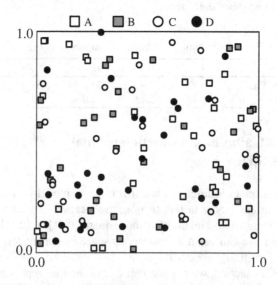

Fig. 8. An example of 121 agents with a different representation scheme in the unit square. The four representation schemes are shown by A, B, C and D (i.e., A: binary strings of length 3, B: real number strings of length 3, C: binary strings of length 7, and D: real number strings of length 7). Those representation schemes are randomly assigned to 30, 30, 30 and 31 agents.

nearest neighbors for each agent. We use the Euclidean distance between agents when we choose the nearest neighbors. The neighborhood structure $N_{IPD}(i)$ for local interaction includes the nearest agents independent of the representation scheme of each agent. The neighborhood structure $N_{GA}(i)$ for local mating includes the nearest agents with the same representation scheme as Agent i. Except for the specifications of these two neighborhood structures, we can use the same specifications for the IPD game and the cellular genetic algorithm for our spatial IPD game in the continuous space $[0, 1] \times [0, 1]$ in this section as those in the 11×11 grid-world in Sect. 4.

6 Experimental Results on the Continuous Space

In each run of our cellular genetic algorithm, the locations of the 121 agents were randomly specified in the unit square as in Fig. 8. One of the four representation schemes was randomly specified to each agent in the same manner as in Sect. 4 (i.e., the 121 agents were divided into four subsets with 30, 30, 30 and 31 agents in each run). As the number of neighbors in $N_{IPD}(i)$ for local interaction and $N_{GA}(i)$ for local

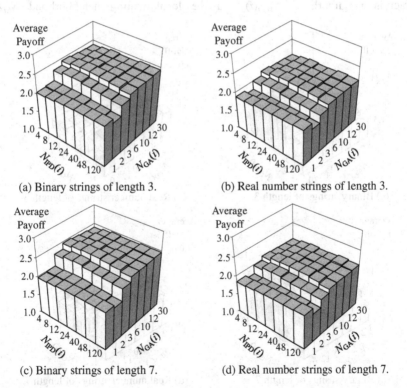

(a) Binary strings of length 3.

(b) Real number strings of length 3.

(c) Binary strings of length 7.

(d) Real number strings of length 7.

Fig. 9. Experimental results of our different spatial IPD game in the two-dimensional unit square with the randomly assigned four representation schemes over 121 agents. Each agent had the same number of qualified neighbors as its mates.

mating, we examined the same 7×7 combinations as in Fig. 7: $|N_{IPD}(i)| = 4, 8, 12,$ 24, 40, 48, 120 and $|N_{GA}(i)| = 1, 2, 3, 6, 10, 12, 30$. When $|N_{GA}(i)|$ was specified as 30, the actual number of neighbors in $N_{GA}(i)$ was 29 or 30.

Experimental results are summarized in Fig. 9. We can see that similar results were obtained in Figs. 5–7 with the two-dimensional grid-world and Fig. 9 with the two-dimensional unit square. This is because each specification of the local mating neighborhood $N_{GA}(i)$ included almost the same number of qualified neighbors in Figs. 5–7 and Fig. 9. One interesting observation is that Fig. 9 is more similar to Fig. 5 than Fig. 7 whereas $N_{GA}(i)$ includes exactly the same number of qualified neighbors in Figs. 7 and 9. It may need further examinations and more computational experiments to explain why this observation was obtained.

In order to examine the effect of local interaction through the IPD game between agents with different representation schemes on the evolution of cooperative behavior, we performed additional computational experiments by excluding neighbors with different representation schemes from the interaction neighborhood $N_{IPD}(i)$. That is, we performed computational experiments under the following condition: each agent was allowed to play the IPD game against its neighbors with the same representation scheme. Under this condition, we examined the same seven specifications for the local interaction neighborhood $N_{IPD}(i)$ and the local mating neighborhood $N_{GA}(i)$.

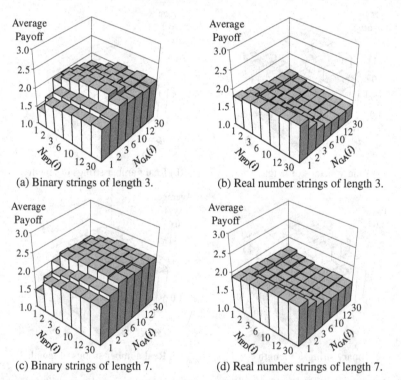

Fig. 10. Experimental results with no local interaction through the IPD game between agents with different representation schemes. All the other settings of computational experiments are the same as Fig. 9.

Experimental results over all of the 49 combinations of those seven specifications are summarized in Fig. 10. Experimental results in Fig. 10 were totally different from those in Fig. 9. In Fig. 9, similar results were obtained from each of the four representation schemes. That is, the four plots in Fig. 9 are similar to one another. However, experimental results from each representation scheme were totally different in Fig. 10. This is because there was no interaction through the IPD game between agents with different representation schemes in Fig. 10.

7 Conclusions

In this paper, we discussed the handling of agents with no qualified neighbors for local mating. Those agents generated new strategies from their current strategies by mutation. No selection pressure towards better strategies with higher average payoff was applied to those agents. Thus those agents might have negative effects on the evolution of cooperative behavior. We examined the three settings with respect to the handling of those agents with no qualified neighbors through computational experiments. However, the effect of those agents was not so clear. This is because the existence of those agents prevents the population from converging not only to cooperative strategies but also to uncooperative strategies. We also proposed the use of a different spatial IPD game in a two-dimensional continuous space. Since we do not use any discrete structure in the continuous space, we can arbitrarily specify the number of agents and the size of neighborhood structures. The location of each agent can also be arbitrarily specified in the continuous space. That is, the use of the spatial IPD game in the continuous space makes it possible to perform various computational experiments in a more flexible manner than the case of the grid-world. Those computational experiments are left for future research.

References

1. Ashlock, D., Kim, E.Y., Leahy, N.: Understanding representational sensitivity in the iterated prisoner's dilemma with fingerprints. IEEE Trans. Syst. Man Cybern.: Part C **36**, 464–475 (2006)
2. Ashlock, D., Kim, E.Y.: Fingerprinting: visualization and automatic analysis of prisoner's dilemma strategies. IEEE Trans. Evol. Comput. **12**, 647–659 (2008)
3. Ashlock, D., Kim, E.Y., Ashlock, W.: Fingerprint analysis of the noisy prisoner's dilemma using a finite-state representation. IEEE Trans. Comput. Intell. AI Games **1**, 154–167 (2009)
4. Ishibuchi, H., Ohyanagi, H., Nojima, Y.: Evolution of strategies with different representation schemes in a spatial iterated prisoner's dilemma game. IEEE Trans. Comput. Intell. AI Games **3**, 67–82 (2011)
5. Skolicki, Z., De Jong, K.A.: Improving evolutionary algorithms with multi-representation island models. In: Yao, X., Burke, E.K., Lozano, J., Smith, J., Merelo-Guervós, J., Bullinaria, J.A., Rowe, J.E., Tiño, P., Kabán, A., Schwefel, H.-P. (eds.) PPSN 2004. LNCS, vol. 3242, pp. 420–429. Springer, Heidelberg (2004)

6. Oliphant, M.: Evolving cooperation in the non-iterated prisoner's dilemma: the importance of spatial organization. In: Brooks, R.A., Maes, P. (eds.) Artificial Life IV, pp. 349–352 (1994)
7. Grim, P.: Spatialization and greater generosity in the stochastic prisoner's dilemma. BioSystems **37**, 3–17 (1996)
8. Brauchli, K., Killingback, T., Doebeli, M.: Evolution of cooperation in spatially structured populations. J. Theor. Biol. **200**, 405–417 (1999)
9. Seo, Y.G., Cho, S.B., Yao, X.: The impact of payoff function and local interaction on the N-player iterated prisoner's dilemma. Knowl. Inf. Syst. **2**, 461–478 (2000)
10. Ishibuchi, H., Namikawa, N.: Evolution of iterated prisoner's dilemma game strategies in structured demes under random pairing in game playing. IEEE Trans. Evol. Comput. **9**, 552–561 (2005)
11. Ishibuchi, H., Hoshino, K., Nojima, Y.: Evolution of strategies in a spatial IPD game with a number of different representation schemes. In: Proceedings of 2012 IEEE Congress on Evolutionary Computation, pp. 808–815 (2012)
12. Ishibuchi, H., Hoshino, K., Nojima, Y.: Strategy evolution in a spatial IPD game where each agent is not allowed to play against itself. In: Proceedings of 2012 IEEE Congress on Evolutionary Computation, pp. 688–695 (2012)
13. Eshelman, L.J., Schaffer, J.D.: Real-coded genetic algorithms and interval-schemata. Foundations of Genetic Algorithms 2, pp. 187–202. Morgan Kaufman, San Mateo (1993)

A Study on the Specification of a Scalarizing Function in MOEA/D for Many-Objective Knapsack Problems

Hisao Ishibuchi[✉], Naoya Akedo, and Yusuke Nojima

Department of Computer Science and Intelligent Systems, Graduate School of Engineering,
Osaka Prefecture University, 1-1 Gakuen-cho, Naka-ku, Sakai, Osaka 599-8531, Japan
{hisaoi,nojima}@cs.osakafu-u.ac.jp,
naoya.akedo@ci.cs.osakafu-u.ac.jp

Abstract. In recent studies on evolutionary multiobjective optimization, MOEA/D has been frequently used due to its simplicity, high computational efficiency, and high search ability. A multiobjective problem in MOEA/D is decomposed into a number of single-objective problems, which are defined by a single scalarizing function with evenly specified weight vectors. The number of the single-objective problems is the same as the number of weight vectors. The population size is also the same as the number of weight vectors. Multi-objective search for a variety of Pareto optimal solutions is realized by single-objective optimization of a scalarizing function in various directions. In this paper, we examine the dependency of the performance of MOEA/D on the specification of a scalarizing function. MOEA/D is applied to knapsack problems with 2-10 objectives. As a scalarizing function, we examine the weighted sum, the weighted Tchebycheff, and the PBI (penalty-based boundary intersection) function with a wide range of penalty parameter values. Experimental results show that the weighted Tchebycheff and the PBI function with an appropriate penalty parameter value outperformed the weighted sum and the PBI function with no penalty parameter in computational experiments on two-objective problems. However, better results were obtained from the weighted sum and the PBI function with no penalty parameter for many-objective problems with 6-10 objectives. We discuss the reason for these observations using the contour line of each scalarizing function. We also suggest potential usefulness of the PBI function with a negative penalty parameter value for many-objective problems.

Keywords: Evolutionary multiobjective optimization · Many-objective problems · MOEA/D · Scalarizing functions

1 Introduction

Recently, high search ability of MOEA/D (multiobjective evolutionary algorithm based on decomposition [1]) has been reported in many studies [2–8]. MOEA/D is a simple and efficient scalarizing function-based EMO (evolutionary multiobjective optimization) algorithm. MOEA/D has clear advantages over Pareto dominance-based

G. Nicosia and P. Pardalos (Eds.): LION 7, LNCS 7997, pp. 231–246, 2013.
DOI: 10.1007/978-3-642-44973-4_24, © Springer-Verlag Berlin Heidelberg 2013

EMO algorithms such as NSGA-II [9] and SPEA2 [10]. For example, its scalarizing function-based fitness evaluation is very fast even for many-objective problems. Its hybridization with local search and other heuristics is often very easy.

One important issue in the implementation of MOEA/D is the choice of an appropriate scalarizing function. In our former study [11], we proposed an idea of automatically switching between the weighted sum and the weighted Tchebycheff during the execution of MOEA/D. We also proposed an idea of simultaneously using multiple scalarizing functions [12]. In the original study on MOEA/D [1], the weighted sum and the weighted Tchebycheff were used for multiobjective knapsack problems while the weighted Tchebycheff and the PBI (penalty-based boundary intersection) function were used for multiobjective continuous optimization. Good results were obtained from a different scalarizing function for a different test problem.

In this paper, we compare the weighted sum, the weighted Tchebycheff and the PBI function with each other, which were used in the original study on MOEA/D [1]. We examine a wide range of penalty parameter values in the PBI function. As test problems, we use multiobjective knapsack problems with 2-10 objectives. One of our test problems is the 2-500 (two-objective 500-item) knapsack problem in Zitzler and Thiele [13]. Many-objective test problems with 4-10 objectives are generated by adding new objectives to the 2-500 problem. Objectives in some test problems are strongly correlated [14] while objectives in other problems are not. MOEA/D with a different scalarizing function shows a different search behavior on each test problem.

This paper is organized as follows. In Sect. 2, we explain our implementation of MOEA/D, which is its basic version with no archive population [1]. In Sect. 2, we also explain the three scalarizing functions used in [1]: the weighted sum, the weighted Tchebycheff, and the PBI function. In Sect. 3, we compare the three scalarizing functions through computational experiments on the 2-500 problem. Experimental results by each scalarizing function are explained using its contour lines. It is also shown that the PBI function with different penalty parameter values has totally different contour lines. In Sect. 4, we examine the performance of each scalarizing function through computational experiments on many-objective problems. In Sect. 5, we suggest a potential usefulness of the PBI function with a negative penalty parameter value. Finally we conclude this paper in Sect. 6.

2 MOEA/D Algorithm

We explain the basic version of MOEA/D [1] with no archive population for the following m-objective maximization problem:

$$\text{Maximize} \quad f(x) = (f_1(x), f_2(x), \ldots, f_m(x)), \tag{1}$$

where $f(x)$ is an m-dimensional objective vector, $f_i(x)$ is the ith objective to be maximized, and x is a decision vector. The use of the basic version is to concentrate on a choice of a scalarizing function. The maintenance of an archive population needs additional special care and large computation load in many-objective optimization.

The multiobjective problem in (1) is decomposed into a number of single-objective problems defined by a scalarizing function with different weight vectors. We examine the weighted sum, the weighted Tchebycheff and the PBI function used in [1]. The weighted sum is written using the weight vector $\lambda = (\lambda_1, \lambda_2, ..., \lambda_m)$ as

$$g^{WS}(x|\lambda) = \lambda_1 \cdot f_1(x) + \lambda_2 \cdot f_2(x) + \cdots + \lambda_m \cdot f_m(x). \tag{2}$$

The weighted sum is to be maximized in its application to our multiobjective maximization problem. The weighted Tchebycheff in [1] is written using the weight vector λ and a reference point $z^* = (z_1^*, z_2^*, ..., z_m^*)$ as

$$g^{TE}(x|\lambda, z^*) = \max_{i=1,2,...,m} \left\{ \lambda_i \cdot |z_i^* - f_i(x)| \right\}. \tag{3}$$

The weighted Tchebycheff is to be minimized. As in computational experiments in [1], we use the following specification of the reference point z^*:

$$z_i^* = 1.1 \cdot \max\{f_i(x)|x \in \Omega(t)\}, \quad i = 1, 2, ..., m, \tag{4}$$

where $\Omega(t)$ is the population at the tth generation.

Zhang and Li [1] also used the PBI (penalty-based boundary intersection) function:

$$g^{PBI}(x|\lambda, z^*) = d_1 + \theta d_2, \tag{5}$$

$$d_1 = \frac{\|(z^* - f(x))^T \lambda\|}{\|\lambda\|}, \tag{6}$$

$$d_2 = \|f(x) - (z^* - d_1\lambda)\|. \tag{7}$$

In this formulation, z^* is the reference point in (4) and θ is a user-definable penalty parameter. In [1], θ was specified as $\theta = 5$. The PBI function is to be minimized. For detailed explanations on scalarizing functions and their characteristic features, see textbooks on multiobjective optimization (e.g., see Miettinen [15])

In each of the three scalarizing functions, all weight vectors $\lambda = (\lambda_1, \lambda_2, ..., \lambda_m)$ satisfying the following two conditions are used for the m-objective problem:

$$\lambda_1 + \lambda_2 + \cdots + \lambda_m = 1, \tag{8}$$

$$\lambda_i \in \left\{ 0, \frac{1}{H}, \frac{2}{H}, ..., \frac{H}{H} \right\}, \quad i = 1, 2, ..., m, \tag{9}$$

where H is a user-definable positive integer. For example, we have a set of evenly specified two-dimensional 101 vectors for $H = 100$ and $m = 2$: $\lambda = (0, 1), (0.01, 0.99), (0.02, 0.98), ..., (0.99, 0.01), (1, 0)$. The number of weight vectors can be calculated as $N = {}_{H+m-1}C_{m-1}$. The number of weight vectors is the same as the population size.

Let N be the number of weight vectors (i.e., N is the population size). We denote N weight vectors as λ^k, $k = 1, 2, ..., N$. Each weight vector λ^k has the nearest T weight vectors as its neighbors where T is a user-definable positive integer. In MOEA/D, first

N solutions are randomly generated as an initial population in the same manner as in other evolutionary algorithms. Each solution is assigned to a different weight vector.

Let x^k be the solution assigned to the kth weight vector λ^k. For each weight vector λ^k with the current solution x^k, a pair of parent solutions is randomly selected from its neighbors to generate a new solution y^k by crossover and mutation. The newly generated solution y^k is compared with the current solution x^k using a scalarizing function with the kth weight vector λ^k. If the newly generated solution y^k is better, the current solution x^k is replaced with y^k. The newly generated solution y^k is also compared with each of its neighbors. Let λ^h be a neighbor of the kth weight vector λ^k. The newly generated solution y^k for the kth weight vector λ^k is compared with the current solution x^h of each neighbor λ^h using the scalarizing function with the neighbor's weight vector λ^h. If y^k is better, the current solution x^h is replaced with y^k. In this manner, the newly generated solution for each weight vector is locally compared with its own and its neighbors' current solutions.

In MOEA/D, local and global parent selection can be probabilistically used. In our implementation, we always use the above-mentioned local parent solution. The upper bound on the number of replaced current solutions with a newly generated one can be specified in MOEA/D. We do not use any upper bound on the number of replaced current solutions in our implementation. MOEA/D also has an option of using an archive population to store non-dominated solutions. We do not use any archive population. All of these settings are to clearly examine the effect of the choice of a scalarizing function on the performance of MOEA/D under a simple situation.

3   Experimental Results on a Two-Objective Knapsack Problem

In this section, we report experimental results on the 2-500 (two-objective 500-item) knapsack problem in Zitzler and Thiele [13]. This problem is written as follows:

$$\text{Maximize} \quad f_i(x) = \sum_{j=1}^{n} p_{ij} x_j, \quad i = 1, 2, \tag{10}$$

$$\text{subject to} \quad \sum_{j=1}^{n} w_{ij} x_j \le c_i, \quad i = 1, 2, \tag{11}$$

$$x_j = 0 \text{ or } 1, \quad j = 1, 2, \ldots, n, \tag{12}$$

where n is the number of items (i.e., $n = 500$), x is a 500-bit binary string, p_{ij} is the profit of item j according to knapsack i, w_{ij} is the weight of item j according to knapsack i, and c_i is the capacity of knapsack i [13]. The values of each profit p_{ij} and each weight w_{ij} were randomly specified integers in [10, 100], and the constant value ci was set as a half of the total weight value (i.e., $c_i = (w_{i1} + w_{i2} + \cdots + w_{in})/2$) in [13]. In the next section, we generate many-objective knapsack problems by adding new objectives to the 2-500 problem.

We applied MOEA/D with the weighted sum, the weighted Tchebycheff, and the PBI function ($\theta = 5$) to the 2-500 problem using the following specifications:

Coding: 500-bit binary string, Population size: 200,
Termination condition: 200×2000 solution evaluations,
Parent selection: Random selection from the neighborhood,
Crossover: Uniform crossover (Probability: 0.8),
Mutation: Bit-flip mutation (Probability: 1/500 for each bit),
Constraint Handling: Maximum profit/weight ratio-based repair in [13],
Number of runs of MOEA/D with each scalarizing function: 100 runs.

Solutions in the final generation of a single run of MOEA/D with each scalarizing function are shown in Fig. 1a–c. Figure 1d shows the 50 % attainment surface over 100 runs of MOEA/D with each scalarizing function. In Fig. 1, better distributions of solutions were obtained by the weighted Tchebycheff and the PBI function ($\theta = 5$).

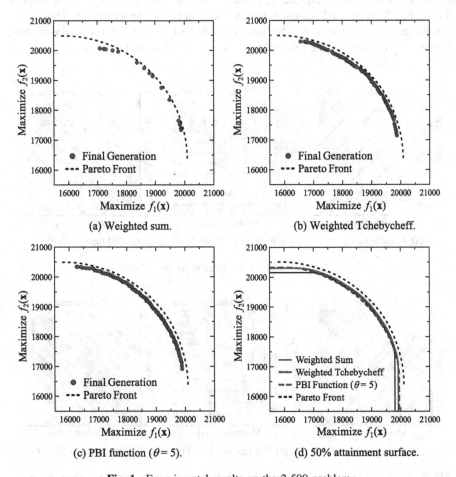

(a) Weighted sum.

(b) Weighted Tchebycheff.

(c) PBI function ($\theta = 5$).

(d) 50% attainment surface.

Fig. 1. Experimental results on the 2-500 problem.

In Fig. 2, we show the contour lines of each scalarizing function for the weighted vector $\lambda = (0.5, 0.5)$. Contour lines have been used to explain scalarizing functions in the literature [15]. When we used the weighted sum in Fig. 2a, the same solution was often obtained from different weight vectors. As a result, many solutions were not obtained in Fig. 1a. In the case of the weighted Tchebycheff in Fig. 2b, the same solution was not often obtained from different weight vectors. Thus more solutions were obtained in Fig. 1b than Fig. 1a. The best results with respect to the number of solutions were obtained in Fig. 1 from the PBI function with $\theta = 5$ in Fig. 1c due to the sharp edge of the contour lines in Fig. 2c. In Fig. 1d, the PBI function slightly outperformed the weighted Tchebycheff with respect to the width of the obtained solution sets along the Pareto front.

The shape of the contour lines of the PBI function totally depends on the value of θ. Figure 3 shows the contour lines of the PBI function with some other values of θ. When $\theta = 0$, the contour lines in Fig. 3a are similar to those of the weighted sum in Fig. 2a. The contour lines of the PBI function with $\theta = 1$ in Fig. 3b look somewhat similar to those of the weighted Tchebycheff in Fig. 2b. The valley of the contour lines in Fig. 3c with $\theta = 50$ is much longer than that in Fig. 2c with $\theta = 5$. This means that the contour lines with $\theta = 50$ have a sharper edge, which leads to slower convergence speed towards the reference point (i.e., towards the Pareto front).

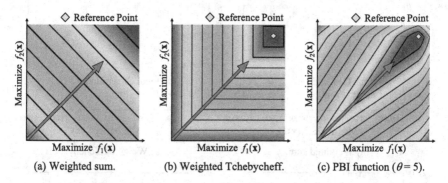

Fig. 2. Contour lines of the three scalarizing functions.

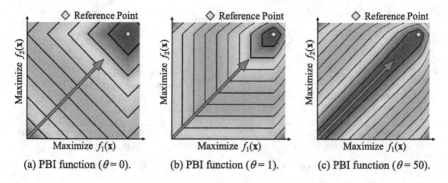

Fig. 3. Contour lines of the PBI function with different values of the penalty parameter θ.

Figure 4 shows experimental results using each value of the penalty parameter θ in Fig. 3. As expected, the use of $\theta = 50$ in Fig. 4c deteriorated the convergence ability of MOEA/D from Fig. 1c with $\theta = 5$. Figure 4a with $\theta = 0$ is similar to Fig. 1a with the weighted sum. However, Fig. 4b with $\theta = 1$ is not similar to Fig. 1b with the weighted Tchebycheff. This is because the contour lines of the two scalarizing functions are not similar to each other for other weight vectors (see Fig. 5).

Figure 5 shows the contour lines of each scalarizing function for the five weight vectors: $\lambda = (1, 0)$, $(0.75, 0.25)$, $(0.5, 0.5)$, $(0.25, 0.75)$, $(0, 1)$. The contour lines of the weighted sum are similar to those of the PBI function with $\theta = 0$ in Fig. 5. Thus similar results were obtained from these two scalarizing functions in Figs. 1 and 4. Except for the PBI function with $\theta = 1$, the contour lines are (almost) vertical or (almost) horizontal when $\lambda = (1, 0)$ and $\lambda = (0, 1)$. Thus a wide variety of solutions were obtained in Figs. 1 and 4 except for the case of the PBI function with $\theta = 1$.

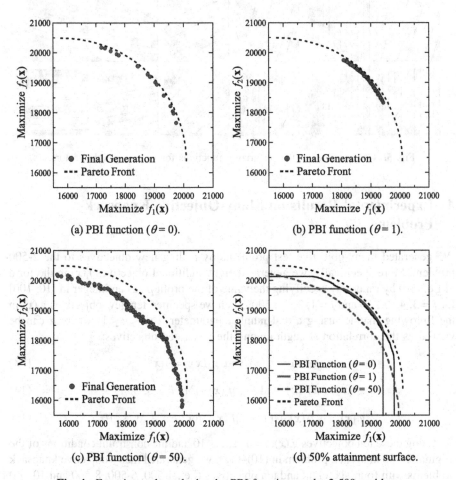

Fig. 4. Experimental results by the PBI function on the 2-500 problem.

Weighted Sum   Tchebycheff   PBI ($\theta=0$)   PBI ($\theta=1$)   PBI ($\theta=5$)   PBI ($\theta=50$)

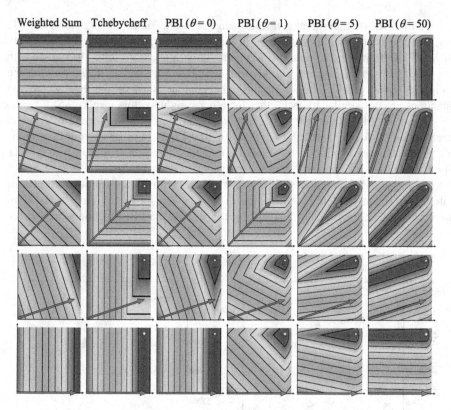

Fig. 5. Contour lines of the scalarizing functions for various weight vectors.

4   Experimental Results on Many-Objective Knapsack Problems

We generated many-objective test problems by adding new objectives to the 2-500 problem. More specifically, we generated eight additional objectives $f_i(x)$ in the form of Eq. (10) by randomly specifying the value of the profit p_{ij} as an integer in [10, 100] for $i = 3, 4, ..., 10$ and $j = 1, 2, ..., 500$. Then we specified ten new objectives $g_i(x)$ in the following manner using a real number parameter α ($0 \leq \alpha \leq 1$) where α can be viewed as the correlation strength among the generated objectives:

$$g_1(x) = f_1(x) \text{ and } g_2(x) = f_2(x), \tag{13}$$

$$g_i(x) = \alpha f_1(x) + (1 - \alpha)f_i(x) \text{ for } i = 3, 5, 7, 9, \tag{14}$$

$$g_i(x) = \alpha f_2(x) + (1 - \alpha)f_i(x) \text{ for } i = 4, 6, 8, 10. \tag{15}$$

Using these ten objectives $g_i(x)$, $i = 1, 2, ..., 10$ and the constraint conditions of the original 2-500 knapsack problem in (10)–(12), we generated many-objective knapsack problems with four, six, eight and ten objectives (i.e., 4-500, 6-500, 8-500 and 10-500 problems). It should be noted that all of those test problems have the same constraint condition as in the 2-500 problem. This means that all of our test problems have the same feasible solution set as in the 2-500 problem.

In one extreme case with $\alpha = 1$ in (13)–(15), our many-objective test problems have only two different objectives $g_1(x)$ and $g_2(x)$ (i.e., $f_1(x)$ and $f_2(x)$ of the 2-500 problem). In the other extreme case with $\alpha = 0$, all objectives $g_i(x)$ are totally different from each other and have no strong correlation since profit values were randomly specified in each objective. We examined six values of α: $\alpha = 0, 0.2, 0.4, 0.6, 0.8, 1$.

We applied MOEA/D with the weighted sum, the weighed Tchebycheff and the PBI function ($\theta = 0, 0.1, 0.5, 1, 5, 10, 50$) to our many-objective 4-500, 6-500, 8-500 and 10-500 problems with $\alpha = 0, 0.2, 0.4, 0.6, 0.8, 1$. For comparison, we also applied NSGA-II and SPEA2 to the same test problems. Computational experiments were performed using the same parameter specifications as in Sect. 3 except for the population size in MOEA/D due to the combinatorial nature of the number of weight vectors. The population size in MOEA/D was specified as 220 (4-500), 252 (6-500), 120 (8-500), and 220 (10-500). In NSGA-II and SPEA2, the population size was always specified as 200. We evaluated the performance of each algorithm on each test problem using the average hypervolume over 100 runs. The origin of the objective space of each test problem (i.e., $(0, 0, \ldots, 0)$) was used as a reference point for hypervolume calculation.

Experimental results are summarized in Tables 1, 2, 3 and 4 where all experimental results are normalized using the results of MOEA/D with the weighted sum for each test problem. The average hypervolume value by MOEA/D with the weighted sum is normalized as 1.00. In Table 1, experimental results on the 2-500 problem are also included for comparison. The largest normalized average hypervolume value for each test problem is highlighted by bold face in Tables 1, 2, 3 and 4. Good results were obtained for all test problems by MOEA/D with the weighted sum and the PBI function with $\theta = 0$. The weighted Tchebycheff did not work well on many-objective problems with no or small correlation among objectives (e.g., 10-500 with $\alpha = 0.2$) while it worked well on many-objective problems with highly correlated objectives (e.g., 10-500 with $\alpha = 0.8$). It also worked well on the 2-500 and 4-500 problems. For the 2-500 problem, the best results were obtained by the PBI function with $\theta = 5$ and $\theta = 10$.

Table 1. Normalized average hypervolume on the 4-500 problems.

Problem (4-500)	NSGA-II	SPEA2	Weighted sum	Tchebycheff	PBI (θ) 0	0.1	0.5	1	5	10	50
$\alpha = 0.0$	0.86	0.85	**1.00**	**1.00**	**1.00**	0.96	0.82	0.78	0.92	0.94	0.95
$\alpha = 0.2$	0.90	0.89	1.00	**1.01**	1.00	0.97	0.85	0.83	0.94	0.95	0.96
$\alpha = 0.4$	0.95	0.92	1.00	**1.01**	1.00	0.97	0.88	0.87	0.96	0.97	0.98
$\alpha = 0.6$	0.96	0.94	1.00	**1.01**	1.00	0.97	0.89	0.88	0.97	0.99	1.00
$\alpha = 0.8$	0.95	0.94	1.00	**1.01**	1.00	0.97	0.89	0.88	0.97	0.99	1.00
$\alpha = 1.0$	0.94	0.94	1.00	**1.02**	1.00	0.98	0.88	0.87	0.97	0.99	1.00
2-500	0.96	0.96	1.00	1.01	1.00	0.99	0.93	0.97	**1.02**	**1.02**	1.01

Table 2. Normalized average hypervolume on the 6-500 problems.

Problem (6-500)	NSGA-II	SPEA2	Weighted sum	Tchebycheff	PBI (θ)						
					0	0.1	0.5	1	5	10	50
$\alpha = 0.0$	0.78	0.74	**1.00**	0.94	**1.00**	0.96	0.79	0.67	0.76	0.77	0.78
$\alpha = 0.2$	0.86	0.81	**1.00**	0.97	**1.00**	0.96	0.82	0.74	0.81	0.82	0.83
$\alpha = 0.4$	0.93	0.89	**1.00**	0.99	**1.00**	0.97	0.86	0.80	0.87	0.89	0.90
$\alpha = 0.6$	0.97	0.92	**1.00**	**1.00**	**1.00**	0.97	0.87	0.82	0.89	0.91	0.93
$\alpha = 0.8$	0.96	0.92	1.00	**1.02**	1.00	0.97	0.86	0.82	0.89	0.90	0.92
$\alpha = 1.0$	0.94	0.93	1.00	**1.01**	1.00	0.98	0.86	0.82	0.89	0.91	0.93

Table 3. Normalized average hypervolume on the 8-500 problems.

Problem (8-500)	NSGA-II	SPEA2	Weighted sum	Tchebycheff	PBI (θ)						
					0	0.1	0.5	1	5	10	50
$\alpha = 0.0$	0.73	0.68	**1.00**	0.90	**1.00**	0.96	0.78	0.66	0.68	0.70	0.71
$\alpha = 0.2$	0.83	0.77	**1.00**	0.91	**1.00**	0.96	0.82	0.72	0.74	0.76	0.77
$\alpha = 0.4$	0.92	0.86	**1.00**	0.94	**1.00**	0.97	0.85	0.78	0.81	0.82	0.83
$\alpha = 0.6$	0.97	0.91	**1.00**	0.99	**1.00**	0.97	0.86	0.80	0.83	0.85	0.86
$\alpha = 0.8$	0.99	0.94	1.00	**1.02**	1.00	0.98	0.87	0.81	0.84	0.86	0.87
$\alpha = 1.0$	0.98	0.97	1.00	**1.01**	1.00	1.00	0.88	0.82	0.85	0.87	0.89

Table 4. Normalized average hypervolume on the 10-500 problems.

Problem (10-500)	NSGA-II	SPEA2	Weighted sum	Tchebycheff	PBI (θ)						
					0	0.1	0.5	1	5	10	50
$\alpha = 0.0$	0.66	0.60	**1.00**	0.87	**1.00**	0.95	0.76	0.63	0.62	0.64	0.65
$\alpha = 0.2$	0.77	0.71	**1.00**	0.87	**1.00**	0.95	0.79	0.69	0.68	0.70	0.72
$\alpha = 0.4$	0.88	0.82	**1.00**	0.91	**1.00**	0.96	0.83	0.74	0.75	0.77	0.78
$\alpha = 0.6$	0.96	0.90	**1.00**	0.98	**1.00**	0.97	0.84	0.78	0.78	0.80	0.82
$\alpha = 0.8$	1.00	0.94	1.00	**1.02**	1.00	0.98	0.85	0.79	0.80	0.81	0.83
$\alpha = 1.0$	1.00	0.99	1.00	**1.01**	1.00	**1.01**	0.88	0.81	0.82	0.84	0.86

As pointed out by many studies [16–21], Pareto dominance-based EMO algorithms do not work well on many-objective problems. In Tables 1, 2, 3 and 4, we can observe the performance deterioration of NSGA-II and SPEA2 by the increase in the number of objectives. In the case of a two-objective maximization problem, a solution is dominated by other solutions in its upper right region of the objective space. The relative size of this region can be viewed as being 1/4 of the objective space. The relative size of this region exponentially decreases as the number of objectives increases (e.g., 1/1024 in the case of ten objectives). This is the reason why Pareto dominance-based EMO algorithms do not work well on many-objective problems.

The same reason can be used to explain why the performance of MOEA/D with the weighted Tchebycheff was deteriorated by the increase in the number of objectives. Let us assume a solution of the 2-500 problem at the corner of a contour line of the weighted Tchebycheff in Fig. 2b. This solution is outperformed by other solutions in its upper right region. The relative size of this region can be viewed as being 1/4 of the objective space. The relative size of this region exponentially decreases as the number of objective increases. As a result, it becomes very difficult for MOEA/D to find a better solution with respect to the weighted Tchebycheff. That is, the convergence of solutions towards the reference point using the weighted Tchebycheff is slowed down by the increase in the number of objectives. The performance deterioration of MOEA/D due to the increase in the number of objectives was more severe in Tables 1, 2, 3 and 4 for the PBI function than the weighted Tchebycheff except for the case of zero or small penalty values. This is because the contour lines of the PBI function have a sharper edge than those of the weighted Tchebycheff. It is more difficult for MOEA/D to find a better solution with respect to the PBI function with a large penalty parameter value than the case of the weighted Tchebycheff.

In Tables 1, 2, 3 and 4, we also observe that NSGA-II worked well on many-objective problems with highly correlated objectives (e.g., 10-500 with $\alpha = 0.8$) as pointed out in [14]. This is because the relative size of the above-mentioned dominating region in the objective space does not exponentially decrease with the number of highly correlated objectives (i.e., many solutions are not non-dominated with each other in the case of highly correlated objectives). As a result, the performance of NSGA-II and SPEA2 was not severely deteriorated by the increase in the number of objectives in Tables 1, 2, 3 and 4. For the same reason, MOEA/D with the weighted Tchebycheff worked well on many-objective problems with highly correlated objectives.

5 Use of a Negative Penalty Parameter Value

In Tables 1, 2, 3 and 4, the performance of MOEA/D with the PBI function on many-objective problems was improved by using a smaller value for the penalty parameter θ (e.g., see the results on the 10-500 problem with $\alpha = 0.0$). From this observation, one may think that better results would be obtained from a negative value of θ.

To examine this issue, we further performed computational experiments using three negative values of θ: $\theta = -1, -0.5, -0.1$. Experimental results are summarized in Tables 5, 6, 7 and 8 where the normalized average hypervolume is calculated in the same manner as in Tables 1, 2, 3 and 4. The origin of the objective space of each test problem was used in Tables 5, 6, 7 and 8 for hypervolume calculation as in Tables 1, 2, 3 and 4. We can see from Tables 5, 6, 7 and 8 that the best results were obtained

Table 5. Normalized average hypervolume on 4-500 (reference point: origin).

Problem (4-500)	Weighted sum	Tchebycheff	PBI (θ)									
			-1	-0.5	-0.1	0	0.1	0.5	1	5	10	50
$\alpha = 0.0$	1.00	1.00	0.78	0.95	**1.03**	1.00	0.96	0.82	0.78	0.92	0.94	0.95
$\alpha = 0.2$	1.00	1.01	0.80	0.98	**1.03**	1.00	0.97	0.85	0.83	0.94	0.95	0.96
$\alpha = 0.4$	1.00	1.01	0.80	0.99	**1.02**	1.00	0.97	0.88	0.87	0.96	0.97	0.98
$\alpha = 0.6$	1.00	1.01	0.79	0.97	**1.02**	1.00	0.97	0.89	0.88	0.97	0.99	1.00
$\alpha = 0.8$	1.00	1.01	0.75	0.94	**1.02**	1.00	0.97	0.89	0.88	0.97	0.99	1.00
$\alpha = 1.0$	1.00	**1.02**	0.71	0.92	1.01	1.00	0.98	0.88	0.87	0.97	0.99	1.00
2-500	1.00	1.01	0.93	0.99	1.01	1.00	0.99	0.93	0.97	**1.02**	**1.02**	1.01

Table 6. Normalized average hypervolume on 6-500 (reference point: origin).

Problem (6-500)	Weighted sum	Tchebycheff	PBI (θ)									
			-1	-0.5	-0.1	0	0.1	0.5	1	5	10	50
$\alpha = 0.0$	1.00	0.94	0.53	0.83	**1.03**	1.00	0.96	0.79	0.67	0.76	0.77	0.78
$\alpha = 0.2$	1.00	0.97	0.55	0.86	**1.04**	1.00	0.96	0.82	0.74	0.81	0.82	0.83
$\alpha = 0.4$	1.00	0.99	0.56	0.88	**1.03**	1.00	0.97	0.86	0.80	0.87	0.89	0.90
$\alpha = 0.6$	1.00	1.00	0.53	0.86	**1.02**	1.00	0.97	0.87	0.82	0.89	0.91	0.93
$\alpha = 0.8$	1.00	**1.02**	0.46	0.81	1.01	1.00	0.97	0.86	0.82	0.89	0.90	0.92
$\alpha = 1.0$	1.00	**1.01**	0.42	0.74	1.00	1.00	0.98	0.86	0.82	0.89	0.91	0.93

Table 7. Normalized average hypervolume on 8-500 (reference point: origin).

Problem (8-500)	Weighted sum	Tchebycheff	PBI (θ)									
			-1	-0.5	-0.1	0	0.1	0.5	1	5	10	50
$\alpha = 0.0$	1.00	0.90	0.00	0.56	**1.02**	1.00	0.96	0.78	0.66	0.68	0.70	0.71
$\alpha = 0.2$	1.00	0.91	0.00	0.57	**1.03**	1.00	0.96	0.82	0.72	0.74	0.76	0.77
$\alpha = 0.4$	1.00	0.94	0.00	0.58	**1.03**	1.00	0.97	0.85	0.78	0.81	0.82	0.83
$\alpha = 0.6$	**1.00**	0.99	0.00	0.59	**1.00**	**1.00**	0.97	0.86	0.80	0.83	0.85	0.86
$\alpha = 0.8$	1.00	**1.02**	0.00	0.48	0.99	1.00	0.98	0.87	0.81	0.84	0.86	0.87
$\alpha = 1.0$	1.00	**1.01**	0.00	0.41	0.97	1.00	1.00	0.88	0.82	0.85	0.87	0.89

Table 8. Normalized average hypervolume on 10-500 (reference point: origin).

Problem (10-500)	Weighted sum	Tchebycheff	PBI (θ)									
			-1	-0.5	-0.1	0	0.1	0.5	1	5	10	50
$\alpha = 0.0$	1.00	0.87	0.00	0.36	**1.01**	1.00	0.95	0.76	0.63	0.62	0.64	0.65
$\alpha = 0.2$	1.00	0.87	0.00	0.38	**1.04**	1.00	0.95	0.79	0.69	0.68	0.70	0.72
$\alpha = 0.4$	1.00	0.91	0.00	0.43	**1.03**	1.00	0.96	0.83	0.74	0.75	0.77	0.78
$\alpha = 0.6$	**1.00**	0.98	0.00	0.41	**1.00**	**1.00**	0.97	0.84	0.78	0.78	0.80	0.82
$\alpha = 0.8$	1.00	**1.02**	0.00	0.30	0.97	1.00	0.98	0.85	0.79	0.80	0.81	0.83
$\alpha = 1.0$	1.00	**1.01**	0.00	0.23	0.94	1.00	**1.01**	0.88	0.81	0.82	0.84	0.86

from the PBI function with $\theta = -0.1$ for almost all test problems except for those with highly correlated objectives.

These experimental results in Tables 5, 6, 7 and 8, however, are somewhat misleading. In Fig. 6a, we show all solutions in the final generation of a single run of MOEA/D on the 2-500 problem for the case of the PBI function with $\theta = -0.1$. We can see that no solutions were obtained around the center of the Pareto front. Such a distribution of the obtained solutions can be explained by the concave shape of the contour lines of the PBI function with $\theta = -0.1$, which are shown in Fig. 6b. Due to the concave contour lines, MOEA/D with the PBI function tends to search for solutions around the edges of the Pareto front even when the weight vector λ is specified as (0.5, 0.5). As a result, no solutions were obtained around the center of the Pareto front in Fig. 6a.

In Tables 1, 2, 3, 4, 5, 6, 7 and 8 we used the origin of the objective space as the reference point for hypervolume calculation. Since the origin is far from the Pareto front of each test problem, extreme solutions around the edge of the Pareto front have

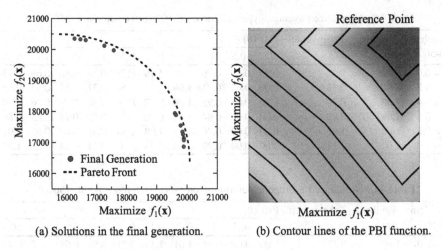

(a) Solutions in the final generation.   (b) Contour lines of the PBI function.

Fig. 6. Experimental results by the PBI function with $\theta = -0.1$ on the 2-500 problem.

Table 9. Normalized average hypervolume on 4-500 (reference point: 15000).

Problem (4-500)	Weighted sum	Tchebycheff	PBI (θ)									
			-1	-0.5	-0.1	0	0.1	0.5	1	5	10	50
$\alpha = 0.0$	**1.00**	0.77	0.00	0.07	0.92	**1.00**	0.98	0.63	0.52	0.92	0.91	0.85
$\alpha = 0.2$	**1.00**	0.84	0.00	0.26	**1.00**	**1.00**	0.94	0.61	0.53	0.88	0.91	0.90
$\alpha = 0.4$	1.00	0.91	0.00	0.37	**1.02**	1.00	0.94	0.65	0.60	0.91	0.94	0.95
$\alpha = 0.6$	1.00	0.93	0.00	0.28	**1.02**	1.00	0.95	0.68	0.65	0.95	0.99	1.01
$\alpha = 0.8$	1.00	0.97	0.00	0.18	1.00	1.00	0.95	0.70	0.67	0.96	1.00	**1.03**
$\alpha = 1.0$	1.00	1.06	0.00	0.09	0.86	1.00	1.03	0.79	0.75	1.06	1.11	**1.14**
2-500	1.00	**1.04**	0.03	0.48	0.95	1.00	0.97	0.79	0.91	**1.04**	1.02	0.94

large effects on hypervolume calculation [22]. This is why the best results were obtained from the PBI function with $\theta = -0.1$ in Tables 5, 6, 7 and 8. Good results from $\theta = -0.1$ also suggest the importance of diversity improvement even for many-objective problems [23].

In Tables 9, 10, 11 and 12, we recalculated the normalized average hypervolume using another reference point closer to the Pareto front for each test problem. For each test problem, we used (15000, 15000, ..., 15000) as the reference point for hypervolume calculation. This setting of hypervolume calculation increases the importance of the convergence of solutions around the center of the Pareto front. Thus the normalized average hypervolume value by the PBI function with $\theta = -0.1$ was deteriorated for almost all test problems in Tables 9, 10, 11 and 12 (e.g., see the last row in Table 9 on the 2-500 problem).

Table 10. Normalized average hypervolume on 6-500 (reference point: 15000).

Problem (6-500)	Weighted sum	Tchebycheff	PBI (θ)									
			-1	-0.5	-0.1	0	0.1	0.5	1	5	10	50
$\alpha = 0.0$	1.00	0.42	0.00	0.00	0.76	1.00	**1.10**	0.78	0.38	0.66	0.69	0.71
$\alpha = 0.2$	**1.00**	0.53	0.00	0.02	0.91	**1.00**	0.99	0.64	0.37	0.59	0.63	0.65
$\alpha = 0.4$	**1.00**	0.70	0.00	0.08	0.95	**1.00**	0.97	0.66	0.46	0.70	0.76	0.79
$\alpha = 0.6$	**1.00**	0.82	0.00	0.06	0.94	**1.00**	0.98	0.71	0.53	0.78	0.83	0.88
$\alpha = 0.8$	1.00	1.02	0.00	0.02	0.76	1.01	**1.09**	0.81	0.65	0.89	0.95	1.01
$\alpha = 1.0$	1.00	1.13	0.00	0.00	0.72	1.00	1.25	1.03	0.84	1.14	1.23	**1.31**

Table 11. Normalized average hypervolume on 8-500 (reference point: 15000).

Problem (8-500)	Weighted sum	Tchebycheff	PBI (θ)									
			-1	-0.5	-0.1	0	0.1	0.5	1	5	10	50
$\alpha = 0.0$	1.00	0.67	0.00	0.00	0.51	1.01	**1.35**	1.21	0.58	0.66	0.77	0.83
$\alpha = 0.2$	1.00	0.58	0.00	0.00	0.72	1.00	**1.11**	0.81	0.43	0.51	0.57	0.62
$\alpha = 0.4$	1.00	0.70	0.00	0.00	0.80	1.00	**1.05**	0.76	0.47	0.57	0.63	0.68
$\alpha = 0.6$	1.00	0.97	0.00	0.00	0.73	0.99	**1.17**	0.94	0.65	0.77	0.86	0.92
$\alpha = 0.8$	1.00	**1.37**	0.00	0.00	0.63	1.00	1.32	1.29	0.95	1.13	1.24	1.31
$\alpha = 1.0$	1.00	1.37	0.00	0.00	0.45	0.99	1.55	1.80	1.43	1.66	1.81	**1.90**

Table 12. Normalized average hypervolume on 10-500 (reference point: 15000).

Problem (10-500)	Weighted sum	Tchebycheff	PBI (θ)									
			-1	-0.5	-0.1	0	0.1	0.5	1	5	10	50
$\alpha = 0.0$	1.00	0.69	0.00	0.00	0.39	1.01	**1.47**	1.39	0.59	0.55	0.68	0.78
$\alpha = 0.2$	1.00	0.57	0.00	0.00	0.64	1.00	**1.15**	0.82	0.38	0.37	0.44	0.51
$\alpha = 0.4$	1.00	0.65	0.00	0.00	0.73	0.98	**1.06**	0.72	0.41	0.43	0.49	0.55
$\alpha = 0.6$	1.00	0.93	0.00	0.00	0.68	0.99	**1.19**	0.94	0.62	0.66	0.74	0.81
$\alpha = 0.8$	1.00	1.37	0.00	0.00	0.51	0.98	**1.38**	**1.38**	0.99	1.05	1.16	1.25
$\alpha = 1.0$	1.00	1.46	0.00	0.00	0.33	1.00	1.77	**2.31**	1.75	1.89	2.02	2.19

6 Conclusions

In this paper, we examined the choice of a scalarizing function in MOEA/D for many-objective knapsack problems. Good results were obtained from the weighted sum over various settings of test problems. With respect to the specification of the penalty parameter in the PBI function, we obtained the following interesting observations:

(i) The best hypervolume values were obtained from a small negative parameter value (i.e., $\theta = -0.1$) when a reference point was far from the Pareto front. In this case, the search of MOEA/D was biased towards the edges of the Pareto front.

(ii) When a reference point was close to the Pareto front, the best hypervolume values were obtained from a small positive parameter value (i.e., $\theta = 0.1$).

(iii) Almost the same results were obtained from $\theta = 0$ and the weighed sum.

References

1. Zhang, Q., Li, H.: MOEA/D: a multiobjective evolutionary algorithm based on decomposition. IEEE Trans. Evol. Comput. **11**, 712–731 (2007)
2. Chang, P.C., Chen, S.H., Zhang, Q., Lin, J.L.: MOEA/D for flowshop scheduling problems. In: Proceedings of IEEE Congress on Evolutionary Computation, pp. 1433–1438 (2008)
3. Zhang, Q., Liu, W., Li, H.: The performance of a new version of MOEA/D on CEC09 unconstrained MOP test instances. In: Proceedings of IEEE Congress on Evolutionary Computation, pp. 203–208 (2009)
4. Li, H., Zhang, Q.: Multiobjective optimization problems with complicated pareto sets, MOEA/D and NSGA-II. IEEE Trans. Evol. Comput. **13**, 284–302 (2009)
5. Ishibuchi, H., Sakane, Y., Tsukamoto, N., Nojima, Y.: Evolutionary many-objective optimization by NSGA-II and MOEA/D with large populations. In: Proceedings of IEEE International Conference on Systems, Man, and Cybernetics, pp. 1820–1825 (2009)
6. Zhang, Q., Liu, W., Tsang, E., Virginas, B.: Expensive multiobjective optimization by MOEA/D with Gaussian process model. IEEE Trans. Evol. Comput. **14**, 456–474 (2010)
7. Tan, Y.Y., Jiao, Y.C., Li, H., Wang, X.K.: A modification to MOEA/D-DE for multiobjective optimization problems with complicated pareto sets. Inf. Sci. **213**, 14–38 (2012)
8. Zhao, S.Z., Suganthan, P.N., Zhang, Q.: Decomposition-based multiobjective evolutionary algorithm with an ensemble of neighborhood sizes. IEEE Trans. Evol. Comput. **16**, 442–446 (2012)
9. Deb, K., Pratap, A., Agarwal, S., Meyarivan, T.: A fast and elitist multiobjective genetic algorithm: NSGA-II. IEEE Trans. Evol. Comput. **6**, 182–197 (2002)
10. Zitzler, E., Laumanns, M., Thiele, L.: SPEA2: Improving the Strength Pareto Evolutionary Algorithm. TIK-Report 103, Department of Electrical Engineering, ETH, Zurich (2001)
11. Ishibuchi, H., Hitotsuyanagi, Y., Ohyanagi, H., Nojima, Y.: Effects of the existence of highly correlated objectives on the behavior of MOEA/D. In: Takahashi, R.H.C., Deb, K., Wanner, E.F., Greco, S. (eds.) EMO 2011. LNCS, vol. 6576, pp. 166–181. Springer, Heidelberg (2011)
12. Ishibuchi, H., Sakane, Y., Tsukamoto, N., Nojima, Y.: Simultaneous use of different scalarizing functions in MOEA/D. In: Proceedings of Genetic and Evolutionary Computation Conference, pp. 519–526 (2010)
13. Zitzler, E., Thiele, L.: Multiobjective evolutionary algorithms: a comparative case study and the strength pareto approach. IEEE Trans. Evol. Comput. **3**, 257–271 (1999)
14. Ishibuchi, H., Hitotsuyanagi, Y., Ohyanagi, H., Nojima, Y.: Effects of the existence of highly correlated objectives on the behavior of MOEA/D. In: Takahashi, R.H.C., Deb, K., Wanner, E.F., Greco, S. (eds.) EMO 2011. LNCS, vol. 6576, pp. 166–181. Springer, Heidelberg (2011)
15. Miettinen, K.: Nonlinear Multiobjective Optimization. Kluwer, Dordrecht (1999)
16. Hughes, E. J.: Evolutionary many-objective optimisation: many once or one many? In: Procdings of IEEE Congress on Evolutionary Computation, pp. 222–227 (2005)
17. Purshouse, R.C., Fleming, P.J.: On the evolutionary optimization of many conflicting objectives. IEEE Trans. Evol. Comput. **11**, 770–784 (2007)
18. Sato, H., Aguirre, H.E., Tanaka, K.: Local dominance and local recombination in MOEAs on 0/1 multiobjective knapsack problems. Eur. J. Oper. Res. **181**, 1708–1723 (2007)
19. Ishibuchi, H., Tsukamoto, N., Hitotsuyanagi, Y., Nojima, Y.: Effectiveness of scalability improvement attempts on the performance of NSGA-II for many-objective problems. In: Proceedings of Genetic and Evolutionary Computation Conference, pp. 649–656 (2008)

20. Ishibuchi, H., Tsukamoto, N., Nojima, Y.: Evolutionary many-objective optimization: a short review. In: Proceedings of IEEE Congress on Evolutionary Computation, pp. 2424–2431 (2008)
21. Kowatari, N., Oyama, A., Aguirre, H., Tanaka, K.: Analysis on population size and neighborhood recombination on many-objective optimization. In: Coello, C.A.C., Cutello, V., Deb, K., Forrest, S., Nicosia, G., Pavone, M. (eds.) PPSN 2012, Part II.LNCS, vol. 7492, pp. 22–31. Springer, Heidelberg (2012)
22. Ishibuchi, H., Nojima, Y., Doi, T.: Comparison between single-objective and multi-objective genetic algorithms: performance comparison and performance measures. In: Proceedings of IEEE Congress on Evolutionary Computation, pp. 3959–3966 (2006)
23. Ishibuchi, H., Tsukamoto, N., Nojima, Y.: Diversity improvement by non-geometric binary crossover in evolutionary multiobjective optimization. IEEE Trans. Evol. Comput. **14**, 985–998 (2010)

Portfolio with Block Branching
for Parallel SAT Solvers

Tomohiro Sonobe[(✉)] and Mary Inaba

Graduate School of Information Science and Technology,
University of Tokyo, Tokyo, Japan
tominlab@gmail.com

Abstract. A portfolio approach has become a widespread method for parallelizing SAT solvers. In comparison with a divide-and-conquer approach, an important feature of the portfolio approach is that there is no need to conduct load-balancing for workers. Instead of load-balancing, the portfolio makes a diversification of workers by differentiating their search parameters. However, it is difficult to achieve effective diversification in a massively parallel environment because the number of combinations of the search parameters is limited. Thus, many overlaps of the search spaces between the workers can occur in such an environment. In order to prevent these overlaps, we propose a novel diversification method, called *block branching*, for the portfolio approach. Preliminary experimental results show that our approach works well, even in a small parallel setting (sixteen processes), and shows potential for a massively parallel environment.

Keywords: SAT solver · Portfolio · Diversification

1 Introduction

The Boolean satisfiability (SAT) problem asks whether an assignment of variables exists that can evaluate a given formula as true. A formula is given in Conjunctive Normal Form (CNF), which is a conjunction of clauses. A clause is a disjunction of literals, where a literal is a positive or a negative form of a Boolean variable. The solvers for this problem are called SAT solvers. Today, there are many real applications [6] of SAT solvers, such as AI planning, circuit design, and software verification. Many state-of-the-art SAT solvers are based on the Davis-Putnam-Logeman-Loveland (DPLL) algorithm. In recent decades, conflict-driven clause learning and non-chronological backtracking, Variable State Independent Decaying Sum (VSIDS) decision heuristic [7], and restart [3] were added to DPLL, which improved the performance of DPLL SAT solvers significantly. Today, these solvers are called Conflict Driven Clause Learning (CDCL) solvers [1].

State-of-the-art parallel SAT solvers are also built upon CDCL solvers. The mainstream approach to parallelizing SAT solvers is portfolio [5]. In this approach, all workers conduct the search competitively and cooperatively without

G. Nicosia and P. Pardalos (Eds.): LION 7, LNCS 7997, pp. 247–252, 2013.
DOI: 10.1007/978-3-642-44973-4_25, © Springer-Verlag Berlin Heidelberg 2013

load-balancing of the search spaces. By contrast, a divide-and-conquer approach often has difficulty with load-balancing because the selection of splitting variables can generate unbalanced sub problems. We believe that the portfolio approach is effective in environments that consist of a small number of processes. However, we also believe that this approach has limitations in conducting effective searches in massively parallel environments because there can be many overlaps of the search spaces between the workers. In order to prevent them, diversification [4] has to be implemented by differentiating the search parameters, such as decision heuristic, restart span, and clause-sharing strategy between the workers.

However, it is apparent that existing methods for creating diversification cannot work well in an environment with, for example, over 100 processes, since the number of combinations of parameter settings is limited and similar types of workers are then searching similar search spaces. To address this issue, we consider the differentiation of the search spaces between the workers, rather than only considering their search activities. In a portfolio approach, the workers conduct the search independently, except for learnt clause sharing. However, we can force each worker to search intensively for specific variables. In this way, each worker focuses on specific and different search spaces, and effective diversification can be achieved.

In this paper, we propose a novel diversification method, called *block branching*, which divides variables in a given instance into blocks and assigns them to the workers. With this method, we can achieve diversification of the search-space, in contrast to existing methods, which focus on differentiation of the search activities of the workers. Preliminary experimental results indicate that our approach is effective, even in a small parallel setting (sixteen processes), and shows potential for massively parallel environments.

In Sect. 2, we explain the details of block branching. We show the experimental results in Sect. 3 and conclude the paper in Sect. 4.

2   Block Branching

Our method is quite simple. Firstly, we divide the variables in a given CNF into blocks, in such a way that the variables in each block have close relationships. Then, we assign them to each worker. Each worker conducts a search, focusing on the variables in the assigned block. In this manner, we can easily reduce the overlaps of the search spaces between the workers. In order to focus on the specific variables in the search, the workers must periodically increase their VSIDS scores, in order to change the decision order. The method used to increase the VSIDS scores is based on Counter Implication Restart (CIR) [10]. For every several restarts, we increase the scores of the variables vigorously, immediately after the restart, in order to force the solver to select them as decision variables. In [2], each worker fixes three variables that are randomly chosen as the root of the search tree. Our method does not always assign values to the target variables at the top of the search tree.

We utilize binary clauses in a given CNF for dividing the variables (literals) into blocks. The basic concept comes from our previous work [9] and we explain the procedure in Fig. 1. Assuming that there are four variables (eight literals), State #1 indicates that each literal belongs to its own block. Next, the first binary clause, $(a \lor b)$, is evaluated in State #2. This clause logically stands for $(\neg a \Rightarrow b)$ and $(\neg b \Rightarrow a)$. In the case of the former clause, the literal b is assigned to True, immediately after the assignment of $a = False$, and the same assignment is made for the latter clause. In other words, the literal b is dominated by the literal $\neg a$. Thus, we can make a block represented by the literal $\neg a$, which consists of the literal $\neg a$ and the literal b. In this manner, we can identify the two blocks in State #4 by analyzing three binary clauses. We believe that variables in a certain block have close relationships, and an intensive search for these variables can achieve effective diversification for the whole search. Note that although there can be illogical relationships between some literals in a block, we permit such irrelevancies since we want to divide the variables into blocks in the simplest possible way. We use the Union-Find algorithm to detect the representative literals and merge the two blocks effectively. For the merging process, we set a threshold in order to change the number of blocks because we have to adjust it to the number of working processes. When the new block-size after the merging exceeds the threshold, the process is cancelled. We use a binary search algorithm for identifying the suitable threshold.

3 Experimental Results

We implemented our proposal on MiniSAT 2.2, using a Message Passing Interface (MPI) in order to run it in a distributed environment. In this implementation, we create a master process that manages the worker processes and a learnt clause database. Each worker sends a solution if found, or learnt clauses whose length is less than, or equal to four. These clauses are exported to other workers through the database in the master process. The master sends the blocks to the workers before the search, and the workers increase the VSIDS scores of all (or thirty, at most) of the assigned variables for every five (or ten) restarts (we call this "INTERVAL"). The workers also use Counter Implication Restart and differentiated parameter settings (e.g., the span of the restart) for diversification.

We used 300 instances from the application category of the SAT Competition 2011. The experiments were conducted on a Linux machine with two Intel Xeon six-core CPUs, running at 3.33 GHz and 144 GB of RAM. Timeout was set to 5000 s for each instance. The number of running processes was set to sixteen, and thus, these results are preliminary. We used four types of solvers: block branching with INTERVAL = 5 (bb_INT5_p16), block branching with INTERVAL = 10 and variable selection for a maximum of thirty variables from the block (bb_INT10_var30_p16), no block branching (no_bb_p16), and ParaCIRMiniSAT [8] with eight threads (ParaCIRMiniSAT_p8). Note that ParaCIRMiniSAT is almost same as our new solver, except our solver uses block bra nching, while ParaCIRMiniSAT is parallelized by OpenMP, to be run in

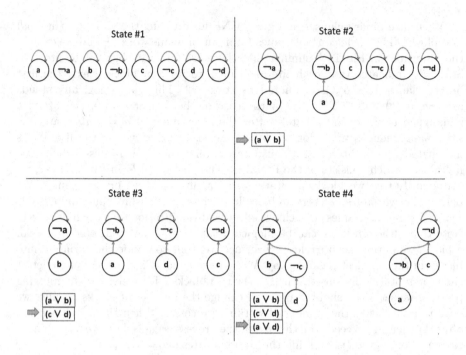

Fig. 1. An illustration of the process of Union-Find for some binary clauses

a shared-memory machine. The cactus plot of the results of the four solvers is shown in Fig. 2 and the details are shown in Table 1. Our proposed solver achieved better performance than ParaCIRMiniSAT within the time limit. In total, it could solve eight more instances than ParaCIRMiniSAT and three more instances than the solver with no block branching (no_bb_p16). Even in the small parallel settings (sixteen processes), it was proven that block branching can improve the performance of the base solver. We are sure that block branching can achieve stronger diversification in massively parallel environments.

Table 1. The details of the results: 252 (SAT: 111, UNSAT: 141) instances could be solved at least one solver. The instances that could not be solved within 5000 s are calculated as 5000.

	SAT (111)	UNSAT (141)	Total	Total time for 252 instances
bb_INT5_p16	107	132	239	190040
bb_INT10_var30_p16	108	131	239	176250
no_bb_p16	106	130	236	188790
ParaCIRMiniSAT_p8	105	126	231	200070

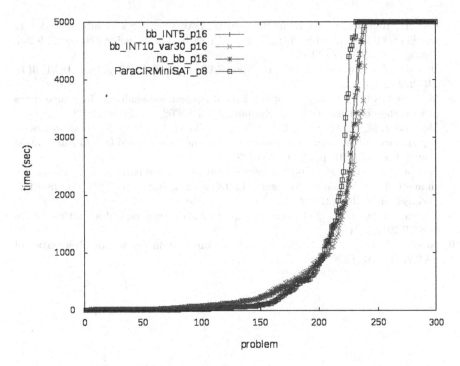

Fig. 2. The experimental results of four solvers using 300 instances from the SAT Competition 2011

4   Conclusion

We have proposed a novel diversification method, block branching, for parallel SAT solvers in massively parallel environments. In this method, variables are divided into blocks and assigned to each worker. The workers conduct an intensive search for the given variables, and this resulted in search-space diversification. Preliminary experiments indicate that the proposed method works well, even in a small parallel environment.

References

1. Biere, A., Heule, M., van Maaren, H., Walsh, T. (eds.): Handbook of Satisfiability. Frontiers in Artificial Intelligence and Applications, vol. 185. IOS Press, Amsterdam (2009)
2. Bordeaux, L., Hamadi, Y., Samulowitz, H.: Experiments with massively parallel constraint solving. In: Proceedings of the 21st International Jont Conference on Artifical Intelligence, IJCAI'09, pp. 443–448 (2009)
3. Gomes, C.P., Selman, B., Kautz, H.: Boosting combinatorial search through randomization. In: Proceedings of the 15th National/10th Conference on Artificial Intelligence, AAAI'98/IAAI'98, pp. 431–437 (1998)

4. Guo, L., Hamadi, Y., Jabbour, S., Sais, L.: Diversification and intensification in parallel SAT solving. In: Cohen, D. (ed.) CP 2010. LNCS, vol. 6308, pp. 252–265. Springer, Heidelberg (2010)
5. Hamadi, Y., Jabbour, S., Sais, L.: ManySAT: a parallel SAT solver. JSAT **6**(4), 245–262 (2009)
6. Marques-Silva, J.: Practical applications of boolean satisfiability. In: Proceedings of Workshop on Discrete Event Systems, WODES'08, pp. 74–80 (2008)
7. Moskewicz, M.W., Madigan, C.F., Zhao, Y., Zhang, L., Malik, S.: Chaff: engineering an efficient SAT solver. In: Proceedings of the 38th Annual Design Automation Conference, DAC'01, pp. 530–535 (2001)
8. Sonobe, T., Inaba, M.: Counter implication restart for parallel SAT solvers. In: Hamadi, Y., Schoenauer, M. (eds.) LION 2012. LNCS, vol. 7219, pp. 485–490. Springer, Heidelberg (2012)
9. Sonobe, T., Inaba M.: Division and alternation of decision variables. In: Pragmatics of SAT 2012 (2012)
10. Sonobe, T., Inaba, M., Nagai, A.: Counter implication restart. In: Pragmatics of SAT 2011 (2011)

Parameter Setting with Dynamic Island Models

Caner Candan[(✉)], Adrien Goëffon, Frédéric Lardeux, and Frédéric Saubion

LERIA, University of Angers, Angers, France
caner.candan@univ-angers.fr

Abstract. In this paper we proposed the use of a dynamic island model which aim at adapting parameter settings dynamically. Since each island corresponds to a specific parameter setting, measuring the evolution of islands populations sheds light on the optimal parameter settings efficiency throughout the search. This model can be viewed as an alternative adaptive operator selection technique for classic steady state genetic algorithms. Empirical studies provide competitive results with respect to other methods like automatic tuning tools. Moreover, this model could ease the parallelization of evolutionary algorithms and can be used in a synchronous or asynchronous way.

1 Introduction

Island Models [7] have been introduced in order to better manage diversity in population-based algorithms. A well-known drawback of evolutionary algorithms (EA) is indeed the premature convergence of their population.

Island models provide a natural abstraction for dividing the population into several subsets, distributed on the islands. An independent EA is run on each of these islands. Individuals are allowed to migrate from one island to another in order to insure information sharing during the resolution and to maintain some diversity on each island thanks to incoming individuals. Another immediate advantage of such models is that they facilitate the parallelization of EAs. The basic model can be refined, especially concerning the following aspects:

- **Migration policies:** Generally, migrations between the different islands are performed according to predefined rules [5]. Individuals may be chosen in order to reinforce the islands'population characteristics [6]. Recently, dynamic policies have been proposed [1]. A transition matrix is updated during the search process in order to dynamically regulate the diversity and the size of the different islands' populations, according to their quality.
- **Search algorithms:** In classic island models, each island uses the same EA. It may be interesting to consider on each island different algorithms or different possible configurations of an algorithm, by changing its parameters.

We propose to use island models as a new method for improving parameter management in EAs. Although the efficiency of EAs is well-established on numerous optimization problems, their performance and robustness depend on the correct setting of its components by means of parameters. Different type of parameters

G. Nicosia and P. Pardalos (Eds.): LION 7, LNCS 7997, pp. 253–258, 2013.
DOI: 10.1007/978-3-642-44973-4_26, © Springer-Verlag Berlin Heidelberg 2013

can be considered from the solving components (e.g., the selection process or the variation operators) until numerical parameters that modify the behavior of a given EA (e.g., the population size or the application rates of the variation operators). Most of the time, these parameters have strong interactions and it is very difficult to forecast how they will affect the performance of the EA. We focus here on the use of variation operators in population based algorithms: i.e., choosing at each iteration which variation operator should be applied.

Automatic tuning (off-line setting before solving) tools [3] have been developed to search for good parameters values in the parameters' space, which is defined as the crossproduct of the parameters values and the possible instances of the problem. Specific heuristics are used in order to sample efficiently the possible values of parameters that are expected to have a significant impact on the performance. Another possible approach for parameter setting is to control the value of the parameters during the search. Adaptive operator selection (AOS) [3] consists in providing an adaptive mechanism for selecting the suitable variation operator to apply on individuals at each iteration of the EA process.

Dynamic island models can be used as an AOS method for classic steady state GA. The main principle is to distribute the operators of the GA on the different islands. We use a dynamic migration policy in order to achieve a suitable distribution of the individuals on the most promising island. The purpose of this paper is to show that our approach is able to identify the most efficient operators of a given EA. This result can be used to improve the performance of an algorithm whose behavior is difficult to handle by non specialist users but also to help algorithm's designers to improve their EAs.

2   Dynamic Islands for Managing Operators

An optimization problem is defined by a search space S, and an objective function $f : S \rightarrow R$. Solving the problem consists thus in finding an element of S that has an optimal value with regards to f. An island model can be defined by the following elements:

- Dimensions: a size n, a set of islands $\mathcal{I} = \{i_1, \cdots, i_n\}$, a set of algorithms $\mathcal{A} = \{A_1, \cdots, A_n\}$, a set of populations $\mathcal{P} = \{P_1, \cdots, P_n\}$, each P_i is a subset of S. Each population P_k is assigned to island I_k. Each A_k assigned to island i_k is indeed an EA that apply variation operators to the individuals of its population (see [2] for more details on EAs).
- A topology: an undirected graph $(\mathcal{I}, V)$ where $V \subseteq \mathcal{I}^2$ is a set of edges between islands of $\mathcal{I}$.
- A migration policy: a squared matrix M of size n, such that $M(i,j) \in [0,1]$ represents the probability for an individual to migrate from island i to island j. This matrix is supposed to be coherent with the topology, i.e., if $\nexists(i,j) \in V$ then $T(i,j) = 0$.

Different choices can be made for the topology: complete graph, ring... A node (i,i) in the graph allows individuals to have the possibility of staying on the

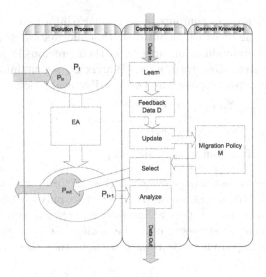

Fig. 1. General mechanism for an island in a dynamic model

same island. The size of the global population is fixed along the process but the size of each P_k changes constantly according to the migrations. Dynamic islands models uses a migration policy that evolves during the search. Each island is equipped with two main processes that define how its population evolves and how the migration policy is controlled.

Figure 1 highlights that the migration policy is modified during the run according to incoming feedback received from other islands. This feedback allows the island to know about the improvement that its individuals have obtained on other islands. The basic learning principle consists in sending more individuals to islands that have been able to improve them significantly and less to islands that are currently less efficient for these individuals. To avoid brutal changes, these effects are evaluated on a time window. The selection component uses the migration matrix M to select the individuals that are sent to other islands. Of course, some individuals may stay on the same island. The analyze component aims at evaluating the improvements of the individuals after the EA has been applied and send this feedback information to the island these individuals were originated from. According to the previous notations, we define the following basic processes and data structures:

– A data matrix D is a square matrix of size n. $D(i,j)$ is the improvement obtained by individuals of island i when they are processed on island j.
– The algorithm A_i applies the operator o_i assigned to this island on every individuals of the population $P_i : A_i(P_i) = \{o_i(s)|s \in P_i\}$.
– The learn process is very simple here since we just keep the last performance (time window of size 1). Therefore we have $[D \oplus D_{in}](i,j) = D(i,j)$ if $D_{in}(i,j) = 0$ and $[D \oplus D_{in}](i,j) = D_{in}(i,j)$ otherwise.

– The *analyze*(*P*) process computes the feedback information and sends it to each island (including its own). In our case, this information will be the average improvement of all individuals in function of their previous localization, during the last evolution step. *Analyze* is a data matrix D_{out} containing only 0 except for the *i*-th column. We propose here two possible evaluations:

- **Mean** $D_{out}(k, i) = \frac{\Sigma_{s \in P_i[k]} f(s)}{card(P_i[k])}$.
- **Max** $D_{out}(k, i) = \max_{s \in P_i[k]}(f(s))$

where $P_i[k] = \{s \in P_i | s$ comes from island $k\}$

- The *update* process uses the data matrix D (i.e., the feedback from other island) into order to modify the migration matrix. Of course, only the line corresponding to island i is modified. We compute an intermediate reward vector R. We propose to use an intensification strategy: only the island where individuals of i have obtained the best improvement is rewarded (note that there could be several such best islands).

$$R(k) = \begin{cases} \frac{1}{card(B)} & \text{if } k \in B, \\ 0 & \text{otherwise,} \end{cases}$$
$$\text{with } B = \underset{k}{\operatorname{argmax}} D(i, k)$$

then $M(i, k) = (1 - \beta)(\alpha.M(i, k) + (1 - \alpha)R(k)).\beta N(k))$, where N is a stochastic vector such that $||N|| = 1$. The parameter α represents the importance of the knowledge accumulated during the last migrations (inertia or exploitation) and β is the amount of noise, which is necessary to explore alternative search space areas by means of individuals (see [1]).

3   Experimental Results

The NK family of landscapes is a problem-independent model for constructing multimodal landscapes. An NK-landscape is commonly defined by triplet $(\mathcal{X}, \mathcal{N}, f)$, with the set of N-length binary strings as search space $\mathcal{X}$ and *1-flip* as neighborhood relation $\mathcal{N}$ (two configurations are neighbors iff their Hamming distance is 1). The fitness function $f : \{0,1\}^N \to [0,1)$, to be maximized, is defined as follows:

$$f(x) = \frac{1}{N} \sum_{i=1}^{N} c_i(x_i, x_{i_1}, \ldots, x_{i_K}) \tag{1}$$

where $c_i : \{0,1\}^{K+1} \to [0,1)$ defines the component function associated with each variable x_i, $i \in \{1, \ldots, N\}$, and where $K < N$.

Parameters N and K define the landscape characteristics. The configurations length N is naturally determining the size of the search space, while K specifies the rugosity level of the landscape. Indeed, the fitness value of a configuration is given by the sum of N terms, each one depending on $K + 1$ bits of the configuration. Thus, by increasing the value of K from 0 to $N - 1$, NK-landscapes

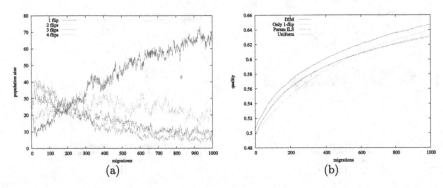

Fig. 2. DIM behavior for the AOS problem on an NK landscape instance (a) Comparisons among DIM, ParamILS, Uniform and 1-flip (b)

can be tuned from smooth to rugged. In particular, if $K = 0$, then the landscape contains only one local (global) optimum; on the contrary, setting K to $N - 1$ leads to a random fitness assignment.

The set of operators used in the experiments is {1-flip, 2-flip, 3-flip, 4-flip}. Number of individuals is set to 100, while 1000 migrations are allowed. Between each migration process, only one operator is applied on each individual. Finally, we specify the *mean* strategy as analyze process. Figure 2(a) presents results for an instance with $N = 1024$ and $K = 4$.

Larger neighborhoods are more intensively used at the beginning of the search. After 200 migrations, the application rates are reversed and 1-flip is detected to be the most effective operator. This behavior seems coherent since the first stage of the search consists in performing more important changes on configurations in order to reach rapidly high fitnesses, while final stages involve intensifying the search to explore close solutions.

Previous experiment shows that the Dynamic Island Model allows to tune the application rates of operators during the search. In order to assess if the efficiency of dynamic adaptation, Fig. 2(b) represents the evolution of the average configurations fitnesses for DIM and three others methods: *1-flip* strategy, which uses only the 1-flip operator; *Uniform selection* strategy with the 4 operators (at each step each a random operator is applied with a uniform probability); *ParamILS strategy* which applies one of the 4 operators at each step with a probability found by paramILS [4]. For this NK-landscape instance, application rates found are 31 % for 1-flip, 3 % for 2-flip, 53 % for 3-flip and 13 % 4-flip.

DIM provides the best result for a 1000 migrations search (Fig. 2(b)). It is interesting to see the difference between DIM and 1-flip whereas DIM mainly uses the 1-flip operator (more than 70 %). As expected, the use of additional operators at the beginning of the search gives a important benefit comparing to the basic 1-flip strategy. The paramILS and uniform strategies provide similar results and are systematically outperformed by DIM.

DIM provides competitive results and does not require many parameters. In order to study the influence of the learning process, we propose to use the multi-armed bandit problem.

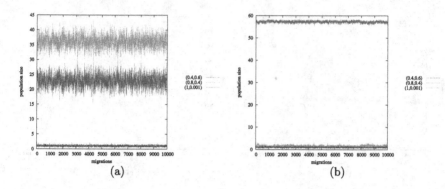

Fig. 3. Multi-armed bandit problem with two learning strategies: (a) for *mean* and (b) for *best*.

An *armed bandit* is a machine which rewards the gambler with a fixed probability. It can be represented by a couple (p, r) where p is the probability to gain r. While playing, the gambler does know neither p nor r. The *multi-armed bandit* is a set of armed bandits where each arm has its own probability and its own reward. Figure 3 shows the Dynamic Island Model behavior with the *mean* (a) and the *max* (b) strategy for the multi-armed bandit problem. In this experiment, 3 operators have been used, corresponding to a 3-armed bandit with the following rewards: (0.4,0.6), (0.8,0.4) and (1,0.001).

The *mean* strategy leads to mainly use the operator with the higher expected value (operator (0.8,0.4)) whereas the *best* strategy leads to only use the operator with the higher reward (operator (0.4,0.6)).

References

1. Candan, C., Goëffon, A., Lardeux, F., Saubion, F.: A dynamic island model for adaptive operator selection. In: Proceedings of Genetic and Evolutionary Computation Conference (GECCO'12), pp. 1253–1260 (2012)
2. Eiben, A., Smith, J.: Introduction to Evolutionary Computing. Natural Computing Series. Springer, Heidelberg (2003)
3. Hamadi, Y., Monfroy, E., Saubion, F. (eds.): Autonomous Search. Springer, Heidelberg (2012)
4. Hutter, F., Hoos, H.H., Leyton-Brown, K., Stützle, T.: ParamILS: an automatic algorithm configuration framework. J. Artif. Int. Res. **36**(1), 267–306 (2009)
5. Rucinski, M., Izzo, D., Biscani, F.: On the impact of the migration topology on the island model. CoRR, abs/1004.4541 (2010)
6. Skolicki, Z., Jong, K.A.D.: The influence of migration sizes and intervals on island models. In: Proceedings of Genetic and Evolutionary Computation Conference (GECCO'05), pp. 1295–1302 (2005)
7. Whitley, D., Rana, S., Heckendorn, R.B.: The island model genetic algorithm: on separability, population size and convergence. J. Comput. Inf. Tech. **7**, 33–47 (1998)

A Simulated Annealing Algorithm for the Vehicle Routing Problem with Time Windows and Synchronization Constraints

Sohaib Afifi[1]([✉]), Duc-Cuong Dang[1,2], and Aziz Moukrim[1]

[1] Université de Technologie de Compiègne,
Laboratoire Heudiasyc, UMR 7253 CNRS, Compiègne 60205, France
[2] School of Computer Science, ASAP Research Group, University of Nottingham,
Jubilee Campus, Wollaton Road, Nottingham NG8 1BB, UK
{sohaib.afifi,duc-cuong.dang,aziz.moukrim}@hds.utc.fr

Abstract. This paper focuses on solving a variant of the vehicle routing problem (VRP) in which a time window is associated with each customer service and some services require simultaneous visits from different vehicles to be accomplished. The problem is therefore called the VRP with time windows and synchronization constraints (VRPTWSyn). We present a simulated annealing algorithm (SA) that incorporates several local search techniques to deal with this problem. Experiments on the instances from the literature show that our SA is fast and outperforms the existing approaches. To the best of our knowledge, this is the first time that dedicated local search methods have been proposed and evaluated on this variant of VRP.

Keywords: Vehicle routing · Synchronization · Destruction/repair · Simulated annealing

1 Introduction

The vehicle routing problem (VRP) [9] is a widely studied combinatorial optimization problem in which the aim is to design optimal tours for a set of vehicles serving a set of customers geographically distributed and respecting some side constraints. We are interested in a particular variant of VRP, the so-called VRP with time windows and synchronization constraints (VRPTWSyn). In such a problem, each customer is associated with a time window that represents the interval of time when the customer is available to receive the vehicle service. This means that if the vehicle arrives too soon, it should wait until the opening of the time window to serve the customer while too late arrival is not allowed. Additionally, for some customers, more than one visit, e.g. two visits from two

This work is partially supported by the Regional Council of Picardie and the European Regional Development Fund (ERDF), under PRIMA project.

G. Nicosia and P. Pardalos (Eds.): LION 7, LNCS 7997, pp. 259–265, 2013.
DOI: 10.1007/978-3-642-44973-4_27, © Springer-Verlag Berlin Heidelberg 2013

different vehicles, are required to complete the service. Visits associated to a particular customer need to be synchronized, e.g. having the same start time.

VRPTWSyn was first studied in [3] with an application in health care services for elders. In such services, the timing and coordination are crucial and therefore the temporal constraints. The readers are referred to [4] for a complete review of those constraints involved in vehicle routing. As an extension of VRP, VRPTWSyn is clearly NP-Hard. There are only a few attempts to solve this problem in the literature [2,3]. In those works, even heuristic ones, integer linear programming is the key ingredient and the methods often require much computational time to deal with large instances. Motivated by the potential applications and by the challenge of computational time, in this work we propose a Simulated Annealing algorithm (SA) for solving VRPTWSyn. Our SA incorporates several local search methods dedicated to the problem. It produces high quality solutions in a very short computational time compared to the other methods of the literature. New best solutions are also detected.

2   Simulated Annealing Algorithm

The main idea of a Simulated Annealing algorithm (SA) [6] is to occasionally accept degraded solutions in the hope of escaping the current local optimum. The probability of accepting a newly created solution is computed as $e^{-\frac{\Delta}{T}}$, where Δ is the difference of fitness between the new solution and the current one and T is a parameter called the current temperature. This parameter is evolved during the search by imitating the cooling process in metallurgy.

Our SA is summarized in Algorithm 1. The algorithm is implemented with a reheating mechanism, due to lines 6 and 25. The simulated annealing routine is from line 9 to line 23. In the algorithm, we use n to denote the number of visits. The other functions are described as follows.

2.1   Constructive Heuristic

The procedure $BestInsertion(X)$ contains a constructive heuristic to build a solution from scratch $(X = \emptyset)$ or from a partial solution. At each iteration of the heuristic, a visit with the less insertion cost is chosen to be inserted in the associated route. Extra routes will be added when it is impossible to insert the remained visits to the existing routes. The heuristic is actually terminated with all visits being routed.

In order to evaluate the insertion cost in constant time $O(1)$, some additional computations for each visit are archived and updated during the process. When an insertion is applied, the update is propagated through different routes because of the related synchronization constraints. The propagation may loop infinitely if the cross synchronizations are not prohibited, e.g. visiting u then v by the first vehicle, visiting p then q by the second one, and finally realizing that u and q are the same customer as well as v and p (see Fig. 1). In our implementation, such issues are avoided by carefully computing beforehand for each visit the set

---

Algorithm 1: Simulated annealing algorithm for VRPTWSyn.

Output: X_{best}, the best solution found so far by the algorithm;

1   $X \leftarrow BestInsertion(\emptyset)$;
2   $X \leftarrow LocalSearch(X)$;
3   $X_{best} \leftarrow X$;
4   $reheat \leftarrow 0$;
5   **while** $(reheat < rhmax)$ **do**
6     $T \leftarrow T_0$;
7     $iter \leftarrow 0$;
8     $X_{lbest} \leftarrow X$;
9   **while** $(iter < itermax)$ **do**
10       $X' \leftarrow Diversification(X, 1, d)$;
11       $X' \leftarrow LocalSearch(X')$;
12       $\Delta \leftarrow Fitness(X') - Fitness(X)$;
13       $iter \leftarrow iter + 1$;
14       $r \sim U(0, 1)$;
15   **if** $(r < e^{-\frac{\Delta}{T}})$ **then**
16         $X \leftarrow X'$;
17         $T \leftarrow \alpha \times T$;
18   **if** $(Fitness(X) < Fitness(X_{lbest}))$ **then**
19           $iter \leftarrow 0$;
20           $X_{lbest} \leftarrow X$;
21   **if** $(Fitness(X) < Fitness(X_{best}))$ **then**
22             $X_{best} \leftarrow X$;
23             $reheat \leftarrow 0$;

24     $X \leftarrow Diversification(X, \frac{n}{2}, n)$;
25     $reheat \leftarrow reheat + 1$;

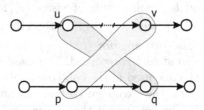

Fig. 1. A cross synchronization

of valid positions (for insertion) from the existing routes. This process is known as the computation of transitive closures.

2.2 Diversification Process

The function $Diversification(X, d_{min}, d_{max})$ first removes a number (randomly generated between d_{min} and d_{max}) of visits from the current solution, then rebuilds it using the above constructive heuristic. This function is actually an implement of the *destruction/repair* operator [1]. The aim is to obtain a new solution from the current one without losing much of the quality, thanks to the constructive heuristic.

In addition, a dynamic priority management is also administered to identify critical visits. Each visit is associated with a priority number initialized to 0. This number is increased by 1 unit whenever the insertion of the visit causes the creation of an extra route. Visits having the highest priority, i.e. frequently causing extra routes, are in fact critical. Therefore, they need to be inserted during the early stages of the constructive heuristic. With this dynamic management, the search is guided back to the feasible space whenever it hits the infeasible one. In general, we remarked that the portion of explored infeasible solutions over feasible ones is varied from one instance to another. This solely depends on the size of the time windows, e.g. the algorithm hits infeasible solutions more frequently with an instance having small time windows.

2.3   Local Search Procedure

The two following neighborhoods were adapted to the synchronization constraints and used in our local search procedure:

2-opt* (exchanges of paths between different routes [7]): in a 2-opt operator, we look for the possibility of exchanging two links with two others in the same route in order to find a local improvement. For the case of multiple vehicles, we use 2-opt* to denote the same principle of exchange but related to two distinct routes. This operator consequently implies the exchanges of paths between the two routes. It is particularly suitable for our case because it is hardly possible for the classical 2-opt to find an improvement due to preserved order of visits from the time windows. Our 2-opt* is implemented as follows: a subset of visits is randomly selected and for each couple of visits $\{r_i^k, r_j^{k'}\}$, we consider the arcs (r_i^k, r_{i+1}^k) and $(r_j^{k'}, r_{j+1}^{k'})$ (where r_i^k denotes the visit at position i in route k). If the exchange of these two arcs for $(r_i^k, r_{j+1}^{k'})$ and $(r_j^{k'}, r_{i+1}^k)$ ensures the feasibility then the associated cost is recorded. The feasibility check is handled by the same process as the one used in the constructive heuristic to avoid cross synchronizations. Therefore, the exchange cost is evaluated in constant time for each couple $\{r_i^k, r_j^{k'}\}$. The best one is then memorized and the exchange is applied.

or-opt (relocation of visits in the same route [8]): in this operator, we look for the possibility of relocating a sequence of (1, 2 or 3) visits from its original place to another one in the same route. The implementation of this operator is similar to 2-opt* operator: a random selection at the beginning then a feasibility check.

Our *LocalSearch(X)* function is then the following: at each iteration, a random neighborhood w is chosen from the current set W, initialized to {2-opt*, or-opt}. Neighborhood w is then removed from W and repeatedly applied to the current solution until no improvement is found. If at least an improvement was detected by w, then the other neighborhood will be put back to W (in case it was removed). The procedure is terminated when W is empty.

3 Results

We tested our algorithm on the instances introduced by [3]. The benchmark comprises 10 sets grouped in 3 categories based on the number of customers. Each set has 5 varieties of instances, those are named after the width of the time windows. Our algorithm is coded in C++ and all experiments were conducted on an Intel Core i7-2620M 2.70 GHz. This configuration is comparable to the computational environment employed by Bredström and Rönnqvist [2,3] (a 2.67 GHz Intel Xeon processor). According to the protocol proposed in [2], all the methods were tested with the varieties of S (small), M (medium) and L (large) time windows. After several experiments on a subset of small instances, we decided to fix the parameters as follows: $T_0 = 20$, $\alpha = 0.99$, $d = 3$, $itermax = 10 \times n$ and $rhmax = 10$.

Table 1 shows our results compared to the literature. Instances, in which all methods report the same results, are discarded from this table. Columns n, m, s and $Best$ show the number of visits, the number of vehicles, the number of synchronizations and the best known solution from all methods (including ours) respectively for each instance. A star symbol (*) is used in $Best$ to indicate that

Table 1. Comparison of results and CPU times

Data	n	m	s	Best	MIP		H		BP1		BP2		SA	
					Sol	CPU	Sol	CPU	Sol	CPU	Sol	CPU	Sol	CPU
1L	20	4	2	3.39*	3.44	3600.00	**3.39**	120.00	**3.39**	107.41	**3.39**	11.91	**3.39**	0.29
2L	20	4	2	3.42*	3.58	3600.00	**3.42**	120.00	**3.42**	2.72	**3.42**	7.41	**3.42**	0.64
3M	20	4	2	3.33*	3.41	3600.00	**3.33**	120.00	**3.33**	17.57	**3.33**	4.31	**3.33**	0.92
4M	20	4	2	5.67*	5.91	3600.00	5.75	120.00	**5.67**	27.53	**5.67**	2.55	**5.67**	0.72
4L	20	4	2	5.13*	5.83	3600.00	5.30	120.00	**5.13**	9.74	**5.13**	7.69	**5.13**	4.66
6S	50	10	5	8.14*	-	-	-	-	**8.14**	3600.00	**8.14**	197.92	**8.14**	93.78
6M	50	10	5	7.70	-	-	-	-	7.71	3600.00	**7.70**	3600.00	**7.70**	3358.60
6L	50	10	5	7.14*	-	-	-	-	**7.14**	3279.48	**7.14**	3600.00	**7.14**	2440.95
7S	50	10	5	8.39*	-	-	-	-	**8.39**	1472.39	**8.39**	169.30	**8.39**	163.03
7M	50	10	5	7.49	-	-	-	-	7.67	3600.00	7.56	3600.00	**7.49**	199.23
7L	50	10	5	6.86	-	-	-	-	6.88	3600.00	6.88	3600.00	**6.86**	144.94
8S	50	10	5	9.54*	-	-	-	-	**9.54**	931.95	**9.54**	850.52	**9.54**	149.95
8M	50	10	5	8.54*	-	-	-	-	**8.54**	3600.00	**8.54**	3490.57	**8.54**	276.46
8L	50	10	5	8.07	-	-	-	-	8.62	3600.00	8.11	3600.00	**8.07**	335.72
9S	80	16	8	12.13	-	-	-	-	-	3600.00	12.21	3600.00	**12.13**	397.876
9M	80	16	8	10.94	-	-	-	-	11.74	3600.00	11.04	3600.00	**10.94**	641.838
9L	80	16	8	10.67	-	-	-	-	11.11	3600.00	10.89	3600.00	**10.67**	376.24
10S	80	16	8	8.82	-	-	-	-	-	3600.00	9.13	3600.00	**8.82**	3099.28
10M	80	16	8	8.01	-	-	-	-	8.54	3600.00	8.10	3600.00	**8.01**	757.87
10L	80	16	8	7.75	-	-	-	-	-	3600.00	-	3600.00	**7.75**	3247.71

the solution is proved to be optimal. The other column headers are: MIP for the results of the default CPLEX solver reported in [3]; H for the heuristic proposed in [3] which is based on the local-branching technique [5]; BP1 and BP2 for the results of the two branch-and-price algorithms presented in [2] and finally SA for our simulated annealing algorithm. Columns Sol and CPU correspond to the best solution found by each method and the associated total computational time. Bold numbers in Sol indicate that the solution quality reaches *Best*.

From these results, we remark that SA finds all known optimal solutions (20 of 30) in very short computational times compared to the other methods. Quality of the other solutions is also better than the one found in the literature. The algorithm strictly improved the best known solutions for 9 instances of the data sets. Those instances are $7M$, $7L$, $8L$, $9S$, $9M$, $9L$, $10S$, $10M$ and $10L$. To summarize, our SA is clearly fast and efficient.

4   Conclusion

The paper presented a simulated annealing based heuristic for VRPTWSyn. Numerical results on the benchmark proposed by [3] demonstrate the competitiveness of the algorithm for such a problem. They also demonstrate that destruction/repair operator and local search methods can be efficiently adapted to the case of the synchronization constraints. As future work, we intend to investigate the performance of the SA on other variants of VRPTWSyn, such as the one with customer-driver preferences and the one with route balance constraints [3]. We also plan to investigate the use of the obtained solutions as a warm start for exact methods, such as mixed integer programming, to solve the open instances of VRPTWSyn to the optimality.

References

1. Bouly, H., Moukrim, A., Chanteur, D., Simon, L.: An iterative destruction/construction heuristic for solving a specific vehicle routing problem. In: MOSIM'08 (2008) (in French)
2. Bredström, D., Rönnqvist, M.: A branch and price algorithm for the combined vehicle routing and scheduling problem with synchronization constraints (Feb 2007)
3. Bredström, D., Rönnqvist, M.: Combined vehicle routing and scheduling with temporal precedence and synchronization constraints. Eur. J. Oper. Res. **191**(1), 19–31 (2008)
4. Drexl, M.: Synchronization in vehicle routing a survey of vrps with multiple synchronization constraints. Transp. Sci. **46**(3), 297–316 (2012)
5. Fischetti, M., Lodi, A.: Local branching. Math. Program. **98**(1–3), 23–47 (2003)
6. Kirkpatrick, S., Gelatt, C.D., Vecchi, M.P.: Optimization by simulated annealing. Science **220**, 671–680 (1983)
7. Potvin, J.Y., Kervahut, T., Garcia, B.L., Rousseau, J.M.: The vehicle routing problem with time windows part I: tabu search. INFORMS J. Comput. **8**(2), 158–164 (1996)

8. Solomon, M.M., Desrosiers, J.: Time window constrained routing and scheduling problems. Transp. Sci. **22**, 1–13 (1988)
9. Toth, P., Vigo, D.: The Vehicle Routing Problem. Monographs on Discrete Mathematics and Applications. Society for Industrial and Applied Mathematics, Philadelphia (2002)

Solution of the Maximum k-Balanced Subgraph Problem

Rosa Figueiredo[1]([✉]), Yuri Frota[2], and Martine Labbé[3]

[1] CIDMA, Department of Mathematics, University of Aveiro,
3810-193 Aveiro, Portugal
rosa.figueiredo@ua.pt
[2] Department of Computer Science, Fluminense Federal University,
Niterói–RJ 24210-240, Brazil
yuri@ic.uff.br
[3] Département d'Informatique, Université Libre de Bruxelles,
CP 210/01 B-1050 Brussels, Belgium
mlabbe@ulb.ac.be

Abstract. A signed graph $G = (V, E, s)$ is k-balanced if V can be partitioned into at most k sets in such a way that positive edges are found only within the sets and negative edges go between sets. We study the problem of finding a subgraph of G that is k-balanced and maximum according to the number of vertices. This problem has applications in clustering problems appearing in collaborative × conflicting environments. We describe a 0-1 linear programming formulation for the problem and implement a first version of a branch-and-cut algorithm based on it. GRASP metaheuristics are used to implement the separation routines in the branch-and-cut. We also propose GRASP and ILS-VND procedures to solve heuristically the problem.

Keywords: Combinatorial optimization · Balanced signed graph · Branch-and-cut · Metaheuristics · Portfolio analysis

1 Introduction

Let $G = (V, E)$ be an undirected graph where $V = \{1, 2, \ldots, n\}$ is a set of vertices and E is a set of m edges connecting pairs of vertices. Consider a function $s : E \rightarrow \{+, -\}$ that assigns a sign to each edge in E. An undirected graph G together with a function s is called a *signed graph*. For a vertex set $S \subseteq V$, let $E[S] = \{(i, j) \in E \mid i, j \in S\}$ denote the subset of edges induced by S. Let $G = (V, E, s)$ denote a signed graph. We assume here that a signed graph has no parallel edges. An edge $e \in E$ is called *negative* if $s(e) = -$ and *positive* if $s(e) = +$. Let E^- and E^+ denote, respectively, the sets of negative and positive edges in G. Let k be a given parameter satisfying $1 \le k \le n$. A signed graph G is *k-balanced* if its vertex set V can be partitioned into sets $N_1, N_2, \ldots, N_l$, with $l \le k$, in such a way that $\cup_{1 \le i \le l} E[N_i] = E^+$.

G. Nicosia and P. Pardalos (Eds.): LION 7, LNCS 7997, pp. 266–271, 2013.
DOI: 10.1007/978-3-642-44973-4_28, © Springer-Verlag Berlin Heidelberg 2013

The *Maximum k-balanced subgraph problem* (k-MBS problem) is the problem of finding a subgraph $H = (V', E', s)$ of G such that H is k-balanced and maximizes the cardinality of V'.

Signed graphs were introduced in 1946 [9] with the purpose of describing sentiment relations between people pertaining to the same social group and to provide a systematic statement of social balance theory. Since then, signed graphs have shown to be a very attractive discrete structure for social networks researchers. The k-MBS problem has applications in clustering problems appearing in collaborative $\times$ conflicting environments. The particular case $k = 2$ yields the problem of finding a maximum balanced subgraph in a signed graph [3,4]. Applications of the 2-MBS in different research areas were discussed recently in [3]. These applications are: the detection of embedded structures, portfolio analysis in risk management and community structure. These applications can be extended to the general case, the k-MBS problem.

Solution approaches have been proposed for the particular case defined for $k = 2$. In [7], a simple linear procedure to detect whether a signed graph is balanced was presented. Descriptions of the polytope associated with the 2-MBS problem were given in [1,4]. A greedy heuristic was proposed in [6] for this particular case and it is able to find an optimal solution whenever G is a balanced signed graph. Figueiredo et al. [4] proposed a branch-and-cut algorithm to the exact solution of the 2-MBS problem. Recently, a GRASP heuristic and an improved version of the greedy heuristic were proposed in [3]. To the best of our knowledge, the general case of the k-MBS problem has never been treated in the literature before.

Our intention is the efficient solution of the k-MBS problem and the identification of difficult instances for which heuristic approaches are required. For that purpose, we describe an integer linear programming formulation to the problem and implement a branch-and-cut algorithm based on it. Metaheuristic procedures were used to improve the branch-and-cut algorithm via efficient separation routines. We also present and compare two heuristic approaches to solve the k-MBS: a GRASP [14] and an ILS-VND [12] procedure.

2 Integer Linear Programming Formulation

A representative formulation [2,5] is proposed here for the k-MBS problem. We define $A = \{(i,j) \mid i \in V, j \in \bar{N}^-(i), i < j\}$ and $A^0 = A \cup V^2$. An arc $(i,j) \in A^0$ indicates that vertex i can represent vertex j. Let us define $D(i) = \{j \in V \mid (i,j) \in A^0\}$ and $O(j) = \{i \in V \mid (i,j) \in A^0\}$. We use binary decision variables $x \in \{0,1\}^{|V|+|A|}$ to define a partition of G in $l \leq k$ clusters. For each vertex $i \in V$, $x_i^i = 1$ if i is a representative vertex and $x_i^i = 0$ otherwise. For each arc $(i,j) \in A$, we define $x_j^i = 1$ if vertex j is represented by vertex i and $x_j^i = 0$ otherwise.

Constraints (1) establish that vertex j must be represented by at most one vertex. The total number of representative vertices is limited to k by constraint (2). Constraints (3) forbid vertex j to be represented by vertex i unless i is a representative vertex. Consider a negative edge $(i, j) \in E^-$. Constraints (4), written for (i, j), ensure that vertices i and j cannot be represented by a same vertex. Consider a positive edge $(i, j) \in E^+$. Constraints (5), written for (i, j), ensure that vertices i and j are represented by the same vertex whenever both i and j belong to the feasible solution. Constraints (6) impose binary restrictions to the variables. Finally, the objective function looks for a maximum subgraph. The formulation follows.

$$\text{maximize} \quad \sum_{(i,j)\in A^0} x_j^i$$

$$\text{subject to} \quad \sum_{i\in O(j)} x_j^i \leq 1, \qquad\qquad\qquad\qquad \forall\, j \in V, \quad (1)$$

$$\sum_{i\in V} x_i^i \leq k, \qquad\qquad\qquad\qquad\qquad\qquad\qquad (2)$$

$$x_j^i \leq x_i^i, \qquad\qquad\qquad\qquad\qquad \forall\, (i,j) \in A, \quad (3)$$

$$x_i^p + x_j^p \leq x_p^p, \qquad\qquad \forall\, (i,j) \in E^-, \forall\, p \in O(i) \cap O(j), \quad (4)$$

$$\sum_{p\in S} x_i^p + \sum_{p\in O(j)\setminus S} x_j^p \leq 1, \qquad \forall\, (i,j) \in E^+, \forall\, S \subseteq O(i), \quad (5)$$

$$x_j^i \in \{0,1\}, \qquad\qquad\qquad\qquad \forall\, (i,j) \in A^0. \quad (6)$$

3 Branch-and-Cut Algorithm

Due to the space limitations, we only scratch the principal components of this algorithm. The branch-and-cut algorithm implemented has two basic components: the initial formulation and the cut generation. The initial formulation is composed by inequalities (1), (2), (3), (4), all the trivial inequalities and by a subset of inequalities (5) (with $|S| = 1$).

A partial description of the polytope associated with the formulation describe in the last section was done. We described families of valid inequalities associated with substructures of the graph (negative cliques, positive cliques and holes) and based on a related problem (the stable set problem). The cut generation component consists, basically, in detecting cliques and holes in G. A GRASP heuristic was used for finding clique cuts and a modification of Hoffman and Padberg's heuristic [10] for finding odd holes cuts.

4 Primal Heuristics

We implemented an ILS-VND and a GRASP procedure for the solution of the k-MBS problem. The ILS-VND heuristic works as follows. An initial solution is reached by a simple random procedure and a local optimum solution is found by a local search VND algorithm [13]. The VND algorithm uses five different

Table 1. Random instances not solved by the branch-and-cut algorithm within the time limit.

Instance				Branch-and-cut			
$|V|$	d	$|E^-|/|E^+|$	k	LR root	LB	UB	Nodes
50	.25	.5	2	36.92	25	29	1591
50	.50	.5	2	27.74	17	27	273
50	.75	.5	2	22.91	12	22	299
50	.25	1	10	37.77	27	37	243
50	.25	.5	10	38.09	26	37	97
50	.50	1	10	29.48	18	28	202
50	.50	2	10	29.72	22	25	2164
50	.50	.5	10	29.75	17	29	150
50	.75	1	10	22.77	14	22	258
50	.75	.5	10	24.07	12	23	106

neighborhoods. Instead of restarting the same procedure from a completely new solution, the ILS heuristic [12] applies the local search (VND) repeatedly to the solutions achieved by perturbing the local optimum solutions previously visited. The GRASP heuristic is an iterative procedure that has two phases associated with each iteration: a construction phase and a local search phase. In the construction phase an initial solution is reached by the same simple random procedure used in the ILS-VND heuristic. The local search phase is also implemented by the VND algorithm used in the ILS-VND heuristic.

Notice that, the GRASP heuristic generates a new random solution at each iteration while the ILS-VND heuristic uses a perturbation scheme with the aim to escape a local optimum.

5   Preliminary Computational Results

The branch-and-cut algorithm and the heuristic procedures are coded in C++ running on a Intel(R) Pentium(R) 4 CPU 3.06 GHz, equipped with 3 GB of RAM. We use Xpress-Optimizer 20.00.21 to implement the components of the enumerative algorithm. The CPU time limit is set to 1h for the branch-and-cut and to 5 min for the heuristics.

We run our experiments with the branch-and-cut algorithm on a set of 74 random instances. The random instances were generated by varying $|V|$, graph density d, rate $|E^-|/|E^+|$ and parameter k, respectively, in sets $\{20, 30, 40, 50\}$, $\{0.25,\ 0.50,\ 0.75\}$, $\{0.5,\ 1.0,\ 2.0\}$ and $\{2, |V|/5\}$. All random instances with up to 40 vertices were solved to optimality; those with 20 and 30 vertices were solved in few seconds. For each unsolved instance, Table 1 exhibits the value of the linear relaxation (LR) at the root of the branch-and-cut tree; upper (UB) and lower (LB) bounds obtained and the total number of nodes in the branch-and-cut tree. The metaheuristics used in the separation procedures showed to be quite efficient: effective cuts were found and the time spent in separation

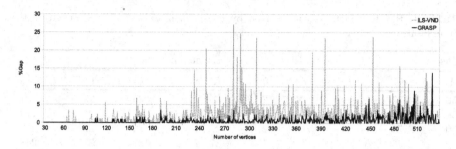

Fig. 1. GRASP and ILS-VND results on portfolio instances with $k = 2$

was around 20 % of CPU time. However, for half of the instances in Table 1, the upper bound after 1h of computation is very close to the value of the LR at the root. This seems to indicate that stronger inequalities are needed to optimally solve large instances of this problem.

We run our experiments with the GRASP and the ILS-VND procedures on a set of portfolio instances described in [3]. This test set is composed by instances with the number of vertices varying in the set $\{30, 60, 90, \ldots, 510\}$ and a threshold value t (used to define the set of positive and negative edges) varying in the set $\{0.300, 0.325, 0.350, 0.375, 0.400\}$. For each combination of these values, 10 different signed graphs were randomly chosen, which means that each signed graph represents a different subset of stocks and totalize 850 instances. For more details on the definition of these instances, we refer the reader to [8, 11]. Figure 1 presents the results obtained for portfolio instances when $k = 2$. The x-axis exhibits instances ordered primarily by number of vertices and secondly by the threshold value while the y-axis exhibits the percentage gaps. Clearly, the GRASP metaheuristic has found the best heuristic solution for almost all portfolio instances when $k = 2$. Similar results were obtained for portfolio instances with $k = |V|/5$.

6   Future Research

In this work, we described a representatives formulation for the k-MBS problem. With the purpose to develop an exact approach to its solution, we investigated some classes of valid inequalities for the associated polytope. Based on this study we implemented a first version of a branch-and-cut algorithm to the problem. Some computational experiments were carried out over a set of random instances. They suggest us some directions on the development of an efficient exact solution approach such as: the investigation of strengthening families of valid inequalities, the development of more efficient separation routines and the development of a tailored branching rule. Primal heuristics were also implemented to the problem: a GRASP heuristic and an ILS-VND heuristic. Computational experiments were executed on a set of portfolio instances. The GRASP procedure achieved better results for almost all the instances. We need to investigate if a similar behavior is observed for other sets of instances in the literature (e.g. the community structure instances defined in [3]). For each test set, we also need to investigate which heuristic is the best option when a quick solution (in some seconds) is needed.

Acknowledgements. Rosa Figueiredo is supported by FEDER founds through COMPETE-Operational Programme Factors of Competitiveness and by Portuguese founds through the CIDMA (University of Aveiro) and FCT, within project PEst-C/MAT/UI4106/2011 with COMPETE number FCOMP-01-0124-FEDER-022690.

References

1. Barahona, F., Mahjoub, A.R.: Facets of the balanced (acyclic) induced subgraph polytope. Math. Program. **45**, 21–33 (1989)
2. Campelo, M., Correa, R.C., Frota, Y.: Cliques, holes and the vertex coloring polytope. Inf. Process.Lett. **89**, 1097–1111 (2004)
3. Figueiredo, R., Frota, Y.: The maximum balanced subgraph of a signed graph: applications and solution approaches. Paper submitted (2012)
4. Figueiredo, R., Labbé, M., de Souza, C.C.: An exact approach to the problem of extracting an embedded network matrix. Comput. Oper. Res. **38**, 1483–1492 (2011)
5. Frota, Y., Maculan, N., Noronha, T.F., Ribeiro, C.C.: A branch-and-cut algorithm for partition coloring. Networks **55**, 194–204 (2010)
6. Gülpinar, N., Gutin, G., Mitra, G., Zverovitch, A.: Extracting pure network submatrices in linear programs using signed graphs. Discrete. Appl. Math. **137**, 359–372 (2004)
7. Harary, F., Kabell, J.A.: A simple algorithm to detect balance in signed graphs. Math. Soc. Sci. **1**, 131–136 (1980)
8. Harary, F., Lim, M., Wunsch, D.C.: Signed graphs for portfolio analysis in risk management. IMA J. Manag. Math. **13**, 1–10 (2003)
9. Heider, F.: Attitudes and cognitive organization. J. Psychol. **21**, 107–112 (1946)
10. Padberg, M., Hoffman, K.L.: Solving airline crew scheduling problems. Manag. Sci. **39**, 657–682 (1993)
11. Huffner, F., Betzler, N., Niedermeier, R.: Separator-based data reduction for signed graph balancing. J. Comb. Optim. **20**, 335–360 (2010)
12. Martin, Q.C., Stutzle, T., Lourenço, H.R.: Iterated Local Search. In: Handbook of Metaheuristics, pp. 1355–1377. Kluwer Academic Publishers, Norwell (2003)
13. Hansen, P., Mladenović, N.: Variable neighborhood search. Comput. Oper. Res. **24**(11), 1097–1100 (1997)
14. Resende, M.G.C., Ribeiro, C.C.: GRASP: greedy randomized adaptive search procedures. In: Burke, E.K., Kendall, G. (eds.) Search Methodologies, 2nd edn, pp. 285–310. Springer, New York (2013)

Racing with a Fixed Budget
and a Self-Adaptive Significance Level

Juergen Branke[(⊠)] and Jawad Elomari

Warwick Business School, Coventry CV4 7AL, UK
Juergen.Branke@wbs.ac.uk, J.Elomari@warwick.ac.uk

Abstract. F-Race is an offline parameter tuning method which efficiently allocates samples in order to identify the parameter setting with the best expected performance, out of a given set of parameter settings. Using non parametric statistical tests, F-Race discards parameter settings which perform significantly worse than the current best, allowing the surviving parameter settings to be tested on more instances and hence obtaining better estimates for their performance. The statistical tests require setting significance levels which directly affect the algorithm's ability of detecting the best parameter setting, and the total runtime. In this paper, we show that it is not straightforward to set the significance level and propose a simple modification to automatically adapt the significance level such that the failure rate is minimized. This is tested empirically using data drawn from probability distributions with pre-defined characteristics. Results indicate that, under a strict computational budget, F-Race with online adaptation performs significantly better than its counterpart with even the best fixed value.

1 Introduction

Racing algorithms were first introduced in [1] to solve the model selection problem in Machine Learning. Within the parameter tuning context, a Racing algorithm takes as an input a number of parameter settings k, a computational budget N, and a problem instance generator I. Running each parameter setting i on an instance j consumes a portion of N and returns a performance value for that parameter setting on that instance U_{ij}. These performance values are used to identify inferior parameter settings and systematically discard them from the race. When only one parameter setting remains, or when N is consumed, the race terminates. If N is consumed with more than one surviving parameter setting, the one with the best aggregated performance value is selected as the winner. In both cases, only one parameter setting is returned by the algorithm to be used to solve other instances of the problem (i.e. test instances). The assumption made here is that the training set is representative and the parameter setting which performed best on the training set will also perform best on the test set.

Racing initially runs all parameter settings on $n_0 \ll N$ instances to get an estimate of their performance, then a two-stage statistical test is carried out to determine which parameter setting(s) to keep and which to discard. The first stage only detects if at least one parameter setting is significantly different from the rest, tests such as

G. Nicosia and P. Pardalos (Eds.): LION 7, LNCS 7997, pp. 272–280, 2013.
DOI: 10.1007/978-3-642-44973-4_29, © Springer-Verlag Berlin Heidelberg 2013

Analysis of Variance (ANOVA), cf. [2], for normally distributed data or ranked-based ANOVA, such as Friedman's F-test or the Kruskal-Wallis, for non-normal data are usually used. If there is at least one parameter setting that is significantly different from the rest, the second stage test is carried out to identify inferior parameter settings using tests such as the paired t-test for normally distributed data, or any non-parametric post hoc test for non-normal data, see [3, 4] for various suggestions. Following this filtering stage, the surviving parameter settings are tested on a new instance and the tests are applied again. This continues until only one parameter setting remains, or until N is consumed. See Fig. 1 for an illustration.

F-Race [5] is a Racing algorithms which uses the F-test followed by a non-parametric pair-wise comparison with the current best to discard inferior parameter settings. It requires setting a significance level α for both tests. If α is set to a high value, parameter settings, including the true best, are likely to be dropped out based on only a few samples and the algorithm terminates before consuming the entire budget. On the other hand, if it is set to a small value, F-Race will be very conservative, rarely discarding any parameter setting, and will allocate the computational budget almost equally between the competing parameter settings. In both cases the performance of F-Race is sub-optimal, which is reflected in the U-shaped curve in Fig. 2, where F-Race is to find the best out of 10 parameter settings with a limited budget of 1000 function evaluations maximum. Another observation from Fig. 2 is that the chosen α does not correspond to the actual failure rate (percentage of runs F-Race fails to select the true best parameter setting), for example an α of 0.08 achieves a 0.12 failure rate, and an α of 0.001 achieves a 0.097 failure rate. This demonstrates that setting an appropriate value for the parameter α is not straightforward.

In this paper, we propose a simple modification to F-Race which automatically adapts α such that the failure rate f, after having used up a pre-determined N, is minimized. Using a fixed N reflects a typical real-world scenario where a decision has to be made within a limited time. To calculate f we use data drawn from probability distributions with pre-defined means, variances, and covariances and assume that the best parameter setting is the one with the lowest mean. The rest of the paper is organized as follows: Sect. 2 reviews related work on Racing algorithms, Sect. 3 describes how to adapt α and the experimental setup, Sect. 4 presents the empirical results, and Sect. 5 concludes the paper.

Iterations / Instances			Parameter settings				
			PS1	PS2	PS3	PS4	
n_0	Iter 1	Instance 1	U_{11}	U_{12}	U_{13}	U_{14}	
	Iter 2	Instance 2	U_{21}	U_{22}	U_{23}	U_{24}	
	Iter 3	Instance 3	U_{31}	U_{32}	U_{33}	U_{34}	Run the first test and drop PS3 and PS4
Iteration 4		Instance 4	U_{41}	U_{42}			Run the second test and don't drop anything
Iteration 5		Instance 5	U_{51}	U_{52}			Run the third test and drop PS1
...		...					
Iteration M		Instance M					Winner identified after 5 iterations

Fig. 1. An example run of a Racing algorithm

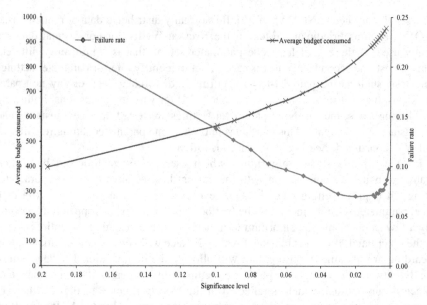

Fig. 2. An example of the performance of F-Race at different significance levels. F-Race chooses from a set of 10 parameter settings and a maximum budget of 1000 function evaluations.

2 Literature Review

Racing algorithms have been applied in a number of papers to tune algorithm parameters. In [6] the authors compared the performance of five different metaheuristics on the University Course Timetabling Problem, aiming to find the best algorithm for that specific problem domain, if possible. To allow for a fair comparison, all algorithms were tested under the same conditions and had their parameters tuned with F-Race. Results showed that even for a very specific problem domain, no one algorithm was able to outperform the rest on all instances. A similar application of F-Race can be found in [7] where the authors tuned the parameters of five metaheuristics for the Quadratic Assignment Problem and compared the results to those obtained by the same algorithms with default parameter settings, results favour the tuned algorithms in almost all cases. Balaprakash et al. [8] improved the performance of an estimation-based local search algorithm by combining heuristically two variance reduction techniques: Importance Sampling and Adaptive Sampling. The authors applied Iterated F-Race to tune the parameters of several Importance Sampling variants. In the same fashion Racing algorithms have been used to tune different algorithms applied in various fields, examples include: [9] in Graph colouring, [10, 11] in bioinformatics, [12] in portfolio selection, and [13] in neural network training.

Racing has indirectly been used in parameter tuning, specifically for selection within a metaheuristic. In [14] the authors used a Racing algorithm to reduce the computational cost of running a $(1 + \lambda)$ Evolutionary Strategy tuning a number of numerical and categorical parameters of a Genetic Algorithm; instead of evaluating all

λ individuals equally to find the best one, Racing was used to efficiently allocate the computational budget on the most promising individuals. In a similar fashion Racing was combined with Ant Colony Optimization in [15], Mesh Adaptive Direct Search [16], Bound Optimization By Quadratic Approximation, Covariance Matrix Adaptation Evolution Strategy, and Uniform Random and Iterated Random Sampling in [17].

F-Race can only select from the initial set of parameter settings provided to it. If a better parameter setting exists for that algorithm and it was not included in the initial set, it will never be discovered. An interesting modification to F-Race to overcome this issue can be found in [18] where the authors insert a new parameter settings into the race after each iteration, this parameter setting is first tested on as many instances as the others, then the statistical tests are carried out. Yet another exploration mechanism added to F-Race can be found in [19] where the authors created an iterative version of F-Race (I/F-Race); each iteration is a single race, and with each iteration the initial set of parameter settings is biased towards the best. In their implementation biasing was done analogous to an Estimation of Distribution Algorithm working at a higher level than F-Race.

In all of these applications, and many others, the significance levels of the statistical tests were set by the user to a fixed value all throughout the race. In the following section we propose a simple modification to automatically adapt α such that f is minimized given a fixed N. We restrict the application of this modification to F-Race.

3 Methodology and Experimental Setup

In the following, we assume that F-Race is run under a fixed budget constraint N, finishing at a time $t < N$ has no benefit. If more than one parameter setting remains in the race after N is consumed, the race is aborted and the parameter setting with the best estimated performance is selected. The basic idea of the proposed approach is to allow F-Race to consume the entire budget even if the race terminates beforehand. Starting with k, N, I, and α (the latter set to a relatively high default value and shown to be irrelevant) F-Race is run until a single parameter setting remains. Because α has been chosen large, it is likely that the race terminates before N is consumed. If this happens, we roll back to the point/iteration where the first parameter setting was discarded, lower α by a factor of ρ, and then repeat the statistical tests on all the parameter settings. Because of the smaller α, F-Race is now more conservative, and less likely to discard parameter settings. Obviously, all the samples already collected are maintained in memory and used again in subsequent iterations if needed, only those parameter settings which had been previously discarded, but are now in the race due to the lower α, are sampled. This "reset" is done as many times as needed until N is consumed, each time the previous α is discounted by ρ. See Fig. 3.

Assessment of the proposed method is based primarily on f calculated over many replications r. This requires prior knowledge of which is the best parameter setting; therefore, we draw numbers from multivariate normal distributions with pre-defined means, variances, and covariances. Using such data will help better understand how the method works and where it fails. In addition to f, the dropout rate d (portion of

Iteration	Reset 1					Reset 2			Reset 3			
	PS0	PS1	PS2	PS3	PS4	PS0	PS1	PS3	PS0	PS1	PS3	PS4
0	1	1	1	1	1							
1	1	1	1	1	1							
2	1	1	1	1	1							
3	1	1	1	1	1							
4	1	1	1	1	1							
5	1	1	1	1	1							
6	1	1				1	1	1	1	1	1	1
7	1	1				1	1	1	1	1	1	1
8	1	1				1	1	1	1	1	1	1
9	1	1				1	1	1	1	1	1	1
10	1	1				1	1	1	1	1	1	1
11						1		1	1	1	1	1
12	1	1				1		1	1	1	1	1
13	1	1				1		1	1	1	1	
14	1	(1)				1		1	1	1	1	
15						1		1	1	1	1	
16						(1)		1	1	1	1	
17									1	1	1	
18									1	1	1	
19									1	(1)	1	

Left-side annotations: n_0 spans iterations 0–5; "First drop out for Reset 1" at iteration 6; "First drop out for Reset 2" at iteration 11 (dashed arrow); "Winner" points to the circled entries.

Fig. 3. An example of Racing with reset

times the best parameter setting was discarded from the race over all replications) is reported. Clearly $f \geq d$ as it accounts for failures due to dropout, and failures due to the variation in the data which could lead to selecting a sub-optimal parameter setting appearing to be better than the true best.

The proposed method F-Race_R is compared to the standard F-Race, which uses a fixed α, and to Equal Allocation. Comparisons are based on f and d plotted against different α values, in specific: an α value is set for both F-Race and F-Race_R (an initial one) and they are run until termination. If the returned parameter settings does not correspond to the one with the true best mean, that method is considered to have failed on that replication. The same process is repeated for all r to find f and d. This represents one point on the plot, other values are obtained by changing α. F-Race_R introduces its own parameter ρ, which determines the new α with each reset. It is shown that F-Race_R is robust to different values of ρ. We expect that the performance of F-Race_R is insensitive to ρ, but that it should not be set too low, as decreasing α by a large amount may result in the optimal value of α being skipped.

The specific settings of the experiments are:

- $k = 10$, $N = 1000$ FEs, $n_0 = 10$ instances, $r = 3000$, $\rho = 0.5, 0.2$
- α values used to construct the f and d plots: 0.2, 0.1:0.01 decreasing by 0.01, and 0.009:0.001 decreasing by 0.001. These represent initial values for F-Race_R
- Sampling distribution of the performance values: $N(i, 10^2) \forall i = 1, \ldots, k$ or $N(U(1, 10), U(24, 48)) \forall i = 1, \ldots, k$

- Correlation values used were: 0, 0.1, 0.3, and 0.6. Although the 0-correlation case does not apply to F-Race for tuning parameter settings, it was included to observe the performance of the algorithm under such conditions.

In the original implementation of F-Race [19], the F-test is replaced with the Wilcoxon matched-pairs signed-ranks test if only two parameter settings remain in the race. The Wilcoxon test statistic T follows a Wilcoxon distribution, for which tabulated critical values are available for a number of significance levels and sample sizes; however, the α values which F-Race_R may require with every reset are not available and can only be calculated by enumeration, which is infeasible for the application of F-Race_R. A normal approximation of the Wilcoxon T statistic is possible if there are many ties, which is not the case since utility values are from probability distributions, and the sample size is large, although there is no general agreement on what is a large enough sample size [4]. Given these difficulties, we chose not to replace the F-test with the Wilcoxon test if two parameter settings remain. This applies for both F-Race and F-Race_R.

4   Results

We first examine the performance of F-Race in Figs. 4 and 5. As expected, f drops as α decreases because a lower α reduces the probability of early termination, and then f increases again as it behaves more like Equal Allocation which does not make use of information gathered during the run and is thus less efficient. Having the f and d lines close to each other, especially for high values of α, means that F-Race is making most of its incorrect selections because the true best parameter setting has been dropped out from the race. This is reduced for low values of α as F-Race becomes more cautious.

The performance of F-Race is greatly improved using the reset idea, this is evident from the f and d curves of F-Race_R. First, the F-Race_R-Fail curve is almost steady for any value of α between 0.2 and 0.009, which indicates that F-Race_R is able to adapt α regardless of its initial value. Obviously, this does not apply for very small initial values of α, at which only few resets, if any, are possible and the behaviour of the algorithm eventually has to approach that of Equal Allocation. It is safe to start with a high α as there does not seem to be any penalty for doing so. Second, the F-Race_R-Drop is at a very low level, which means that F-Race_R hardly ever drops out the best parameter settings.

The most interesting outcome is that F-Race_R was able to make correct selection more often than even the best setting of α for F-Race. To see if the difference in performance is significant, we compared the best f achieved by F-Race with any f achieved by F-Race_R at a relatively high α (0.2 and 0.1 were selected even though they are not the best setting for α for F-Race_R). A 1-sample sign test with a significance level of 0.05 was conducted over 30 replications. The null hypothesis that the differences between observations are equal to zero vs. that they are less than zero (indicating the F-Race_R has a lower f than F-Race). This non-parametric test was chosen because the f data does not follow a normal distribution and it is not symmetric. As seen from Table 1 f is significantly better for F-Race_R at all correlation levels.

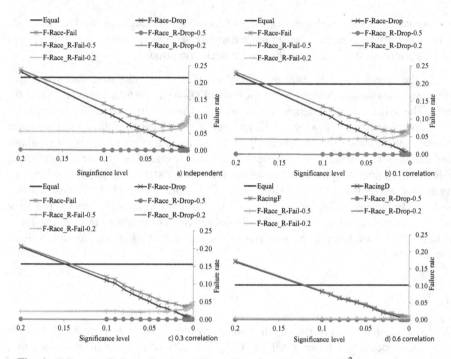

Fig. 4. F-Race vs. F-Race_R using utility values drawn from $N(i, 10^2)$ *for* $i = 1, ..., k$

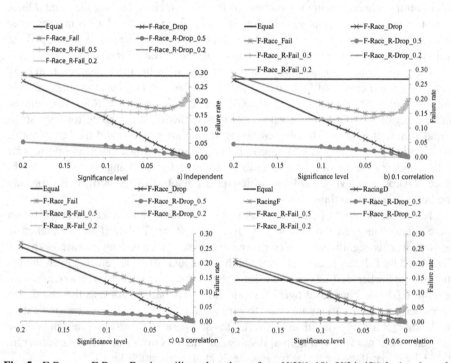

Fig. 5. F-Race vs. F-Race_R using utility values drawn from $N(U(1, 10), U(24, 48))$ *for* $i = 1, ..., k$

Table 1. Comparing the best f obtained by F-Race with any f obtained by F-Race_R obtained at a single high default value of α

Sampling distribution of utility values	Correlation	F-Race_R @ 0.2 p-value	F-Race_R @ 0.1 p-value
$N(i, 10^2)\forall i = 1, \ldots, k$	0	0.000	0.000
	0.1	0.000	0.000
	0.3	0.001	0.000
	0.6	0.000	0.000
$N(U(0, 10), U(24, 48))\forall i = 1, \ldots, k$	0	0.001	0.000
	0.1	0.001	0.001
	0.3	0.000	0.001
	0.6	0.000	0.001

5 Conclusion

A new method to automatically adjust the significance level of the F-Race algorithm for best performance given a fixed budget was presented. It was shown that a chosen significance level does not correspond to the actual failure rate the user observes, and choosing an appropriate significance level is not straightforward. The proposed method, F-Race_R, allows the user to set a computational budget and it will adapt the significance level accordingly such that the failure rate is minimized. This is achieved by systematically lowering the initial significance level each time the race terminates until the entire budget is consumed.

Experiments were carried out using performance values drawn from normal distributions with known means, variances, and covariances. Results show that F-Race_R is quite robust to the initial significance level chosen, as long as it is not too low, demonstrating its ability to adapt it online. Finally, and perhaps most importantly, F-Race_R is able to find significance levels which achieve lower failure rates than any fixed significance level used in F-Race. The 1-sample sign test indicates that the improvement in the failure rate is significant. F-Race_R comes with its own new parameter, the reduction factor. However, as shown from the experiments, F-Race_R is also robust to different values of the reduction factor.

More experiments are still needed to better understand the full potential, and limitations, of the proposed method. Utility values drawn from other normal and non-normal distributions need to be tested, in addition to using actual utility values from algorithms solving real optimization problems.

References

1. Maron, O., Moore, A.: Hoeffding races: accelerating model selection search for classification and function approximation. Adv. Neural Inf. Process. Syst. **6**, 59–66 (1994)
2. Schaffer, J.D., Caruana, R.A., Eshelman, L.J., Das, R.: A study of control parameters affecting online performance of genetic algorithms for function optimization. In: International Conference on Genetic Algorithms, pp. 51–60 (1989)

3. Conover, W.J.: Practical Nonparametric Statistics. Wiley, New York (1999)
4. Sheskin, D.: Handbook of Parametric and Nonparametric Statistical Procedures, 5th edn. Chapman and Hall/CRC, New York (2011)
5. Birattari, M., Stutzle, T., Paquete, L., Varrentrapp, K.: A Racing algorithm for configuring metaheuristics. In: Genetic and Evolutionary Computation Conference, pp. 11–18 (2002)
6. Rossi-Doria, O., et al.: A comparison of the performance of different metaheuristics on the timetabling problem. In: Burke, E.K., De Causmaecker, P. (eds.) PATAT 2003. LNCS, vol. 2740, pp. 329–351. Springer, Heidelberg (2003)
7. Paquete, L., Stutzle, T.: A study of stochastic local search algorithms for the biobjective quadratic assignment problem with correlated flow matrices. Eur. J. Oper. Res. **169**(3), 943–959 (2006)
8. Balaprakash, P., Birattari, M., Stützle, T., Dorigo, M.: Adaptive sample size and importance sampling in estimation-based local search for the probabilistic traveling salesman problem. Eur. J. Oper. Res. **199**(1), 98–110 (2009)
9. Chiarandini, M., Stutzle, T.: Stochastic local search algorithms for graph set T-colouring and frequency assignment. Constraints **12**(3), 371–403 (2007)
10. Di Gaspero, L., Roli, A.: Stochastic local search for large-scale instances of the haplotype inference problem by pure parsimony. J. Algorithms **63**(3), 55–69 (2008)
11. Lenne, R., Solnon, C., Stutzle, T., Tannier, E., Birattari, M.: Reactive stochastic local search algorithms for the genomic median problem. In: European Conference on Evolutionary Computation in Combinatorial Optimization, pp. 266–276 (2008)
12. Di Gaspero, L., di Tollo, G., Roli, A., Schaerf, A.: Hybrid local search for constrained financial portfolio selection problems. In: Van Hentenryck, P., Wolsey, L.A. (eds.) CPAIOR 2007. LNCS, vol. 4510, pp. 44–58. Springer, Heidelberg (2007)
13. Blum, C., Socha, K.: Training feed-forward neural networks with ant colony optimization: an application to pattern classification, p. 6
14. Yuan, B., Gallagher, M.: Combining meta-EAs and Racing for difficult EA parameter tuning tasks. In: Lobo, F., Lima, C., Michalewicz, Z. (eds.) Parameter Setting in Evolutionary Algorithms, pp. 121–142. Studies in Computational IntelligenceSpringer, Berlin (2007)
15. Birattari, M., Balaprakash, P., Dorigo, M.: The ACO/F-Race algorithm for combinatorial optimization under uncertainty. In: Doerner, K., Gendreau, M., Greistorfer, P., Gutjahr, W., Hartl, R., Reimann, M. (eds.) Metaheuristics, Operations Research/Computer Science Interfaces Series, pp. 189–203. Springer, Heidelberg (2007)
16. Yuan, Z., Stützle, T., Birattari, M.: MADS/F-Race: mesh adaptive direct search meets F-Race. In: García-Pedrajas, N., Herrera, F., Fyfe, C., Benítez, J., Ali, M. (eds.) IEA/AIE 2010, Part I. LNCS, vol. 6096, pp. 41–50. Springer, Heidelberg (2010)
17. Yuan, Z., Montes de Oca, M., Birattari, M., Stützle, T.: Continuous optimization algorithms for tuning real and integer parameters of swarm intelligence algorithms. Swarm Intell. **6**(1), 49–75 (2012)
18. Chiarandini, M., Birattari, M., Socha, K., Rossi-Doria, O.: An effective hybrid algorithm for university course timetabling. J. Sched. **9**(5), 403–432 (2006)
19. Birattari, M., Yuan, Z., Balaprakash, P., Stützle, T.: F-Race and iterated F-Race: an overview. In: Bartz-Beielstein, T., Chiarandini, M., Paquete, L., Preuss, M. (eds.) Experimental Methods for the Analysis of Optimization Algorithms, pp. 311–336. Springer, Berlin (2010)

An Efficient Best Response Heuristic for a Non-preemptive Strictly Periodic Scheduling Problem

Clément Pira[1,2]([⊠]) and Christian Artigues[1,2]

[1] CNRS, LAAS, 7 avenue du colonel Roche, F-31400 Toulouse, France
[2] Université de Toulouse, LAAS, F-31400 Toulouse, France
{pira,artigues}@laas.fr

Abstract. We propose an enhanced version of a original heuristic first proposed in [1,2] to solve a NP-hard strictly periodic scheduling problem. Inspired by game theory, the heuristic reaches an equilibrium by iteratively solving best response problems. Our contribution is to greatly improve its efficiency, taking advantage of the two-dimensionality of the best response problem. The results show that the new heuristic compares favorably with MILP solutions.

Keywords: Periodic scheduling · Equilibrium · Two dimensional optimization

1 Problem and Method

We consider a periodic scheduling problem, arising for example in the avionic field, where a set of N periodic tasks (measure of a sensor, etc.) has to be scheduled on P processors distributed on the plane [1–4]. In this problem, each task i has a fixed period T_i which cannot be modified. A solution is given by an assignment of the tasks to the processors and, for each task, by the start time t_i of one of its occurrences. Each task has a processing time p_i and no two tasks assigned to the same processor can overlap during any time period. In fact, we adopt a slightly more general model in which processing times p_i are generalized by positive latency delays $l_{i,j} \geq 0$. The former case is the particular case where $l_{i,j} = p_i$ for all other tasks j (Fig. 1).

In this paper, we only consider the case where the offsets t_i are integers, which is important to prove convergence of the heuristic.

A more complete exposition of the material presented in this article can be found in the technical report [6].

This work was funded by the French Midi-Pyrenee region (allocation de recherche post-doctorant n°11050523) and the LAAS-CNRS OSEC project (Scheduling in Critical Embedded Systems).

G. Nicosia and P. Pardalos (Eds.): LION 7, LNCS 7997, pp. 281–287, 2013.
DOI: 10.1007/978-3-642-44973-4_30, © Springer-Verlag Berlin Heidelberg 2013

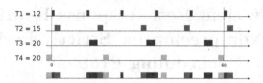

Fig. 1. $N = 4$ non-overlapping periodic tasks on $P = 1$ processor

1.1 Problem Definition

Non-overlapping Constraints. We first focus on the monoprocessor problem. Given two tasks i and j, we want a latency delay $l_{i,j} \geq 0$ to be respected whenever an occurrence of j starts after an occurrence of i. Said differently, we want the smallest positive difference between an occurrence of j and an occurrence of i to be greater than $l_{i,j}$. The set of occurrences of i is $t_i + T_i\mathbb{Z}$, while the set of occurrences of j is $t_j + T_j\mathbb{Z}$. When doing the difference, and using Bézout identity, we obtain that the set of possible differences is $(t_j - t_i) + g_{i,j}\mathbb{Z}$ where $g_{i,j} = \gcd(T_i, T_j)$. The smallest positive representative of this set is $(t_j - t_i)$ mod $g_{i,j}$. Therefore, the constraint we want to pose is simply:

$$(t_j - t_i) \bmod g_{i,j} \geq l_{i,j}, \qquad \forall(i,j) \in \mathcal{G} \tag{1}$$

The graph $\mathcal{G}$ involved in constraint (1) contains the arcs (i,j) for which $l_{i,j} > 0$ (since otherwise the equation is trivially satisfied). Since some complications are introduced when only one of the delays $l_{i,j}$ or $l_{j,i}$ is strictly positive, we will suppose in the following that $\mathcal{G}$ is symetric (which can always be enforced [6]). Seeing it as an undirected graph, we will write $\mathcal{G}(i)$ for the set of neighbors of i, i.e. the set of tasks with which i is constrained.

Objective to Maximize. In a context of robustness, it could be natural to maximize the feasibility of the system. More concretely, we want an execution of a task to be as far as possible from every other executions of another task which precedes or follows it. Indeed, due to uncertainties, we could imagine that a task lasts longer than expected. This increase in the duration is naturally proportional to the original processing time. Hence, we make all the delays $l_{i,j}$ proportional to a common factor $\alpha \geq 0$ that we try to optimize (see [1,2]).

$$
\begin{aligned}
\max \quad & \alpha & & \text{(2)} \\
\text{s.t.} \quad & (t_j - t_i) \bmod g_{i,j} \geq l_{i,j}\alpha & \forall(i,j) \in \mathcal{G} & \text{(3)} \\
& t_i \in \mathbb{Z} & \forall i & \text{(4)} \\
& \alpha \geq 0 & & \text{(5)}
\end{aligned}
$$

We easily check that a schedule is feasible for the feasibility problem *iff* the optimization problem has a solution with $\alpha \geq 1$. Hence, this optimization problem can be interesting simply to find feasible solutions.

1.2 An Equilibrium-Based Heuristic

The main component of the algorithm is called the best response procedure. Following a game theory analogy, each task is seen as an agent which tries to optimize its own offset t_i, while the other offsets (t_j^*) are fixed. Moreover, the agent only takes into account the constraints in which it is involved, i.e. only the constraints associated with its neighborhood $\mathcal{G}(i)$. We obtain the following program:

$$(BR_i) \quad \max \quad \alpha \tag{6}$$

$$s.t. \quad (t_i - t_j^*) \bmod g_{i,j} \geq l_{j,i}\alpha \qquad \forall j \in \mathcal{G}(i) \tag{7}$$

$$(t_j^* - t_i) \bmod g_{i,j} \geq l_{i,j}\alpha \qquad \forall j \in \mathcal{G}(i) \tag{8}$$

$$\alpha \geq 0 \tag{9}$$

$$t_i \in \mathbb{Z} \tag{10}$$

Definition 1. *A task i is stable if the current offset t_i is optimal for (BR_i). An equilibrium is a solution (t_i) such that all the tasks are stable.*

The heuristic, described in Algorithm 1, counts the number of tasks known to be stable. Starting with an initial solution (for example randomly generated), we choose cyclically a task i and try to optimize its schedule, i.e. we solve (BR_i). If no improvement was found, then one more task is stable, otherwise we update and reinitialize the counter of stable tasks. We continue until N tasks are stable. We refer to [1] for the proof of correction and termination (the latter requires the integrality of the offsets).

Algorithm 1. The heuristic

```
 1: procedure IMPROVESOLUTION((t_j)_{j∈I})
 2:     N_stab ← 0                                          ▷ The number of stabilized tasks
 3:     i ← 0                                               ▷ The task currently optimized
 4:     while N_stab < N do                                 ▷ We run until all the tasks are stable
 5:         (new_t_i, α_i) ← BestResponse(i, (t_j)_{j∈I})   ▷ We optimize the task i
 6:         if new_t_i = t_i then                           ▷ We do not have a strict improvement
 7:             N_stab ← N_stab + 1                         ▷ One more task is stable
 8:             α ← min(α_i, α)
 9:         else                                            ▷ We have a strict improvement
10:             N_stab ← 1                                  ▷ We restart counting the stabilized tasks
11:             α ← α_i; t_i ← new_t_i
12:         end if
13:         i ← (i + 1) mod N                               ▷ We consider the next task
14:     end while
15:     return (α, (t_j)_{j∈I})
16: end procedure
```

---

2 The Best Response Procedure

Since the offsets are integer, and since the problem is periodic, we can always impose t_i to belong to $\{0, \cdots, T_i - 1\}$. Therefore we can trivially solve the best response program (BR_i) by computing the α-value for each $t_i \in \{0, \cdots, T_i - 1\}$, using the following expression, and selecting the best offset:

$$\alpha = \min_{j \in \mathcal{G}(i)} \min \left(\frac{(t_i - t_j^*) \bmod g_{i,j}}{l_{j,i}} \, , \, \frac{(t_j^* - t_i) \bmod g_{i,j}}{l_{i,j}} \right) \qquad (11)$$

This procedure runs in $O(T_i N)$, hence any method should at least be faster. In [1], the authors propose a method consisting in precomputing a set of intersection points to reduce the number of evaluations. In the following, we present a new line-search method which greatly improves (BR_i) solving.

2.1   Structure of the Solution Set of (BR_i)

Each task i is linked with all the tasks $j \in \mathcal{G}(i)$ through non-overlapping constraints. For a given task $j \in \mathcal{G}(i)$, and a fixed offset t_j, the set of solutions (t_i, α) satisfying constraints (7–9) has a shape represented on Fig. 2(a). The set of solutions (t_i, α) for the problem (BR_i), given some fixed offsets $(t_j)_{j \in \mathcal{G}(i)}$, is therefore the intersection of these sets for each $j \in \mathcal{G}(i)$ (see Fig. 2(b)).

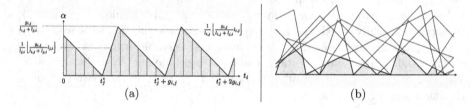

Fig. 2. Possible values for (t_i, α) when constrained by (a) a single task j, and (b) all the tasks $j \in \mathcal{G}(i)$

Hence, this solution set is composed of several adjacent polyhedra. We can give an upper bound on the number n_{poly} of such polyhedra. A polyhedron starts and ends at zero points (i.e. an offset where the curve vanishes). In the case of integer offsets, there is obviously at most T_i zero points in the interval $[0, T_i - 1]$ and therefore, at most T_i polyhedra.

2.2   Principle of the Best-Response Procedure

Graphically, solving the program (BR_i) amounts to finding the solution maximizing the curve described by Fig. 2(b). By periodicity, t_i can be supposed to belong to $\{0, \cdots, T_i - 1\}$. More generally we can start at any initial offset τ, for example the initial value of t_i, and run on the right until we reach the offset $\tau + T_i - 1$, hence we obtain a new solution $t_i \in \{\tau, \cdots, \tau + T_i - 1\}$. If needed, we can then consider $t_i \bmod T_i$ which is an equivalent solution in $\{0, \cdots, T_i - 1\}$.

Given an initial reference offset t_i^{ref}, we can compute the local polyhedron which contains it (see Fig. 3). Using standard LP techniques, we can find the

Fig. 3. Selection of the polyhedron containing a reference offset t_i^{ref}

Fig. 4. Principle of the best response procedure

local fractional optimum, as well as the local integral optimum (see Sect. 2.3). If the latter is better than the current best solution, we update. We now want to reach the next polyhedron. At the local fractional optimum, there are two active lines, an increasing one and a decreasing one. Since the procedure runs from left to right, we follow the decreasing line until we reach the x-axis (i.e. the offset o'_k on Fig. 4). We can use this point as the new reference offset. We continue until a whole period has been traversed. This method is illustrated on Fig. 4.

2.3 Solving the Local Best Response Problem

We now explain more precisely how to optimize on the local polyhedra. Near the reference offset t_i^{ref}, the constraint $(t_i - t_j^*) \bmod g_{i,j} \geq l_{j,i}\alpha$ is locally linear and increasing, of the form (13). In the same way, the constraint $(t_j^* - t_i) \bmod g_{i,j} \geq l_{i,j}\alpha$ is locally linear and decreasing, of the form (14). Thus, the local program has the following form:

$$(Loc{-}BR_i) \qquad \max \quad \alpha \tag{12}$$
$$\text{s.t.} \quad t_i - l_{j,i}\alpha \geq o_j \qquad \forall j \in \mathcal{G}(i) \tag{13}$$
$$t_i + l_{i,j}\alpha \leq o'_j \qquad \forall j \in \mathcal{G}(i) \tag{14}$$
$$t_i \in \mathbb{Z} \tag{15}$$

If the reference point t_i^{ref} is not a zero-point, there is only one possible choice for o_j and o'_j. Otherwise, there is technically two polyhedra, on the left and on the right. As we move to the right, the right one is preferred. This amounts to defining o_j and o'_j by (see [6]):

$$o_j = t_i^{\text{ref}} - (t_i^{\text{ref}} - t_j^*) \bmod g_{i,j} \qquad \text{and} \qquad o'_j = o_j + g_{i,j} \tag{16}$$

In order to solve $(Loc{-}BR_i)$, we can first search for a fractional solution. Since the problem is a particular two dimensional program, Megiddo algorithm allows to find an optimal solution in $O(N)$ [5]. We can then round this fractional solution to the closest smaller and larger integers, compute the α-value associated with these two offsets (using expression (11)), and select the best one. This gives immediately a method to compute an integer solution in $O(N)$. Therefore, the global best response algorithm runs in $O(n_{poly}N)$ with n_{poly} bounded by T_i.

In [6], we give a special implementation of the dual simplex algorithm, which runs in $O(N^2)$, but which outperforms Megiddo algorithm in practice (at least for the sizes of instances we considered). In fact, the integrality assumption allows subsequently to improve the dual simplex algorithm and obtain a complexity in $O(NW)$ where W is the width of the polyhedron. This is enough to give an acceptable complexity in $O(T_iN)$ for the global best response problem.

2.4   The Multiprocessor Best-Response

In order to implement the multiprocessor best response, we solve the best response on each processor. The task is reassign to the processor which gives the best result, and the offset is changed accordingly.

3   Experimental Results

We test the method on non-harmonic instances, with $N = 20$ tasks and $P = 4$ processors, generated using the procedure described in [3]: the periods were choosen in the set $\{2^x 3^y 50 \mid x \in [0,4], y \in [0,3]\}$ and the processing times were generated following an exponential distribution and averaging at about 20 % of the period. Columns 2 and 3 allow to compare our results with a MILP formulation presented in [6] and restricted with a timeout of 200s. Results about the original heuristic [1,2] are presented on columns 4-7. Here, **time**$_{single}$ is the time needed for a single run and **starts**$_{2s}$ = 2/**time**$_{single}$ measures the average number of starts performed by the original heuristic in 2s. Finally, columns 8-11 contain the results for our version of the heuristic. In [1,2], the stopping criterion

Table 1. Results of the MILP, the heuristic of [1], and the new version of the heuristic

id	MILP (200s)		Original heuristic [2] (Bayesian test)				New heuristic (2s)			
	α_{MILP}	time$_{sol}$	$\alpha_{heuristic}$	time$_{single}$	time$_{stop}$	starts$_{2s}$	$\alpha_{heuristic}$	starts$_{sol}$	starts$_{2s}$	time$_{sol}$
0	2.5	159	2.3	1.43	101.33	1.4	2.5	35	3624	0.01932
1	2	18	2.01091	3.27	5064.67	0.61	2.01091	28	4477	0.01251
2	1.6	6	1.40455	1.52	869.45	1.32	1.6	2	2462	0.00162
3	1.6	4	1.6	4.34	8704.45	0.46	1.64324	45	2910	0.03093
4	2	5	1.92	3.48	1115.51	0.57	2	1	2489	0.00080
5*	3	7	1.43413	1.63	1498.21	1.23	3	1	2107	0.00095
6	2.5	54	2.3	1.44	101.25	1.39	2.5	35	3805	0.01840
7	2	19	2	0.23	302.27	8.7	2	3	4431	0.00135
8	2.12222	8	1.75794	1.03	871.8	1.94	2.12222	3	2513	0.00239
9*	2	11	2	2.42	3541.79	0.83	2	3	3575	0.00168
10	1.12	6	0.87	0.72	368.44	2.78	1.12	4	3466	0.00231
11	2.81098	20	0.847368	3.78	478.63	0.53	2.81098	1	3421	0.00058
12	1.5	7	1.5	0.27	313.74	7.4	1.5	4	3645	0.00219
13	1.56833	49	1.5	1.77	3293.33	1.13	1.56833	1	3863	0.00052
14	2	8	2	1.85	3873	1.08	2	2	2331	0.00172

was a bayesian test, but column 4 shows that it often stopped with a solution far from the best solution found by the MILP. Hence for the new heuristic, the stopping criterion is more simply a timeout of $2s$. The value $start_{2s}$ is the number of times the heuristic was started during this period. This number is much greater than the equivalent number for the original heuristic, therefore our heuristic is much faster (about 3100 times on these instances). Moreover, $time_{sol}$ represents approximately the time needed to find the best solution. This is to compare with the column $time_{sol}$ of the MILP formulation, which shows that our version of the heuristic is very competitive compared to the MILP (Table 1).

References

1. Al Sheikh, A.: Resource allocation in hard real-time avionic systems - scheduling and routing problems. Ph.D. thesis, LAAS, Toulouse, France (2011)
2. Al Sheikh, A., Brun, O., Hladik, P.E., Prabhu, B.: Strictly periodic scheduling in IMA-based architectures. Real Time Syst. 48(4), 359–386 (2012)
3. Eisenbrand, F., Kesavan, K., Mattikalli, R.S., Niemeier, M., Nordsieck, A.W., Skutella, M., Verschae, J., Wiese, A.: Solving an avionics real-time scheduling problem by advanced IP-methods. In: de Berg, M., Meyer, U. (eds.) ESA 2010, Part I. LNCS, vol. 6346, pp. 11–22. Springer, Heidelberg (2010)
4. Korst, J.: Periodic multiprocessors scheduling. Ph.D. thesis, Eindhoven University of Technology, Eindhoven, The Netherlands (1992)
5. Megiddo, N.: Linear-time algorithms for linear programming in R^3 and related problems. SIAM J. Comput. 12(4), 759–776 (1983)
6. Pira, C., Artigues, C.: An efficient best response heuristic for a non-preemptive strictly periodic scheduling problem. Technical report LAAS-CNRS, Toulouse, France, October 2012. http://hal.archives-ouvertes.fr/hal-00761345

Finding an Evolutionary Solution to the Game of Mastermind with Good Scaling Behavior

Juan Julian Merelo[1], Antonio M. Mora[1], Carlos Cotta[2],
and Antonio J. Fernández-Leiva[2(✉)]

[1] Department Computer Architecture and Technology + CITIC,
University of Granada, Granada, Spain
{jmerelo,amorag}@geneura.ugr.es
[2] Department of Computer Sciences and Languages,
University of Málaga, Málaga, Spain
{ccottap,afdez}@lcc.uma.es

Abstract. There are two main research issues in the game of Mastermind: one of them is finding solutions that are able to minimize the number of turns needed to find the solution, and another is finding methods that scale well when the size of the search space is increased. In this paper we will present a method that uses evolutionary algorithms to find fast solutions to the game of Mastermind that scale better with problem size than previously described methods; this is obtained by just fixing one parameter.

Keywords: Mastermind · Oracle games · Puzzles · Evolutionary algorithms · Parameter optimization

1 Introduction and State of the Art

Mastermind [1–3] is a puzzle in which one player A hides a combination of κ symbols and length ℓ, while the other player B tries to find it out by playing combinations coded in the same alphabet and length. The answers from player A to every combination include the number of symbols in the combination that are in the correct position and the number of colors that have been guessed correctly. Player B then plays a new combination, until the hidden one is found. The objective of the game is to play repeatedly minimizing the number of turns needed to find the solution.

Most solutions so far [4,5] use the concept of *eligible*, *possible* or *consistent* combinations: those that, according to responses by player A, could still be the hidden combination or, in other words, those that match the played combinations as indicated by the answer. Exhaustive methods [2,6] would eliminate all non-consistent solutions and play a consistent one, while non-exhaustive methods would sample the set of consistent solutions and play one of them. Those solutions are guaranteed to reduce the search space at least by one, but obviously different combinations have a different reduction capability. This *capability*

G. Nicosia and P. Pardalos (Eds.): LION 7, LNCS 7997, pp. 288–293, 2013.
DOI: 10.1007/978-3-642-44973-4_31, © Springer-Verlag Berlin Heidelberg 2013

is reflected by a score. However, scores are heuristic and there is no rigorous way of scoring combinations. To compute these scores, every combination is compared in turn with the rest of the combinations in the set; the number of combinations that get every response (there is a limited amount of possible responses) is noted. Eventually this results in a series of *partitions* in which the set of consistent combinations is divided by its *distance* (in terms of common positions and colors) to every other. This results in a set of combinations with the best score; one of the combinations of this set is chosen deterministically (using lexicographical order, for instance) or randomly. In this paper we use *most parts*, proposed in [7] which takes into account only the number of non-zero partitions.

Currently, the state of the art was established by Berghman et al. in [4]. They obtained a system that is able to find the solution in an average number of moves that is, for all sizes tested, better than previously published. The number of evaluations was not published, but time was. In both cases, their solutions were quite good. However, there were many parameters that had to be set for each size, starting with the first guess and the size of the consistent set, as well as population size and other evolutionary algorithm parameters. In this paper we will try to adapt our previously published Evo method by reducing the number of parameters without compromising too much on algorithm performance, based on the fact that even as you can find a good solution using only a sample of the consistent set size as proved in [4,5], different set sizes do have an influence on the outcome. When you reduce the size to the minimum it is bound to have an influence on the result, in terms of turns needed to win and number of evaluations needed to do it. The effect of the reduction of this sample size will decrease the probability of finding, and thus playing, the hidden combination, and also the probability of finding the combination that maximally reduces the search space size when played. However, in this paper we will prove that good solutions can be found by using a small and, what is more, a common set size across all Mastermind problem sizes.

In the next section we will present the experiments carried out and its results for sizes from $\ell = 4, \kappa = 8$ to $\ell = 7, \kappa = 10$.

2   An Evolutionary Method for Playing MasterMind

This paper uses the method called, simply, *Evo* [8–11]. This method, which has been released as open source code at CPAN (http://search.cpan.org/dist/Algorithm-MasterMind/), is an evolutionary algorithm that has been optimized for speed and to obtain the minimal number of evaluations possible. An evolutionary algorithm [12] is a Nature-inspired search and optimization method that, modeling natural evolution and its molecular base, uses a (codified) population of solutions to find the optimal one. Candidate solutions are scored according to its closeness to the optimal solution (called *fitness*) and the whole population evolved by discarding solutions with the lowest fitness and making those with the highest fitness reproduce via combination (crossover) and random change (mutation).

Evo, which is explained extensively in [11] searches consistent combinations until a prefixed amount of them has been found. It uses *Most Parts* score to assess consistent combinations, and the *distance to consistency* for non-consistent ones, so that the fitness directs search towards finding consistent solutions with better score. The algorithm continues until a pre-fixed number of consistent solutions have been found or until this number does not vary for a number of generations (set to three throughout all experiments).

Evo incorporates a series of methods to decrease the amount of evaluations needed to find the solution, including *endgames* which makes the evolutionary algorithm revert to exhaustive search in the case the search space has been well characterized (for instance, when we know that the solution is a permutation of one of the combinations played or when we have discarded some colors, reverting to a problem of smaller size).

The solutions are quite promising, but the main problem is that the number of evaluations needed to find the solution increases rapidly with problem size (fortunately, not as fast as the problem size itself or this solution would not be convenient) and a new parameter is introduced: the optimal size of the set of consistent combinations, that is, the number of combinations that the algorithm tries to find out before it plays one.

What we do in this paper is testing an *one size fits all* approach by making the size of the consistent set unchanged for any problem size. This reduces the algorithm parameter set by one, but since this parameter set has, a priori, a big influence on result and there is no method to set it other than experimentation, it reduces greatly the amount of experiments needed to obtain a solution.

3   Experiments and Results

The experiments presented in this paper extend those published previously, mainly by [11].

In this paper we will set this size to a common for all sizes and minimal value: 10, that is why we will denominate the method tested Evo10. This value has been chosen to be small enough to be convenient, but not so small that the scoring methods are rendered meaningless. This will reduce the parameters needed by one, leaving only the population size to be set, once, of course, the rest of the evolutionary algorithm parameters have been fixed by experimentation; these parameters are set to crossover rate equal to 80 %, and mutation and permutation rate equal to 10 %; replacement rate is equal to 75 % and tournament size equal to 7.

For every problem size, a fixed set of 5000 combinations were generated randomly. There is at most a single repetition in the smallest size, and no repetition in the rest. The sets can be downloaded from http://goo.gl/6yu16; these sets are the same that have been used in previous papers. A single game is played for every combination.

The results for this fixed parameter setting are shown in Tables 1(a), (b), (c) and (d).

Table 1. Comparison among this approach (*Evo10*) and previous results published by the authors (Evo++) in [11] and Berghman et al. [4].

(a) *Mean number of guesses and the standard error of the mean for $\ell = 4, 5$, the quantities in parentheses indicate population and consistent set size (in the case of the previous results).*

	$\ell = 4$	$\ell = 5$	
	$\kappa = 8$	$\kappa = 8$	$\kappa = 9$
Berghman et al.		5.618	
Evo++	(400,30) 5.15 ± 0.87	(600,40) 5.62 ± 0.84	(800,80) 5.94 ± 0.87
Evo10	(200) 5.209 ± 0.91	(600) 5.652 ± 0.818	(800) 6.013 ± 0.875

(b) *Mean number of guesses and the standard error of the mean for $\ell = 6, 7$, the quantities in parentheses indicate population and consistent set size (in the case of the previous results).*

	$\ell = 6$		$\ell = 7$
	$\kappa = 9$	$\kappa = 10$	$\kappa = 10$
Berghman et al.	6.475		
Evo++	(1000,100) 6.479 ± 0.89		
Evo10	(800) 6.504 ± 0.871	(1000) 6.877 ± 0.013	(1500) 7.425 ± 0.013

(c) *Mean number of evaluations and its standard deviation $\ell = 4, 5$.*

	$\ell = 4$	$\ell = 5$	
	$\kappa = 8$	$\kappa = 8$	$\kappa = 9$
Evo++	6412 ± 3014	14911 ± 6120	25323 ± 9972
Evo10	2551 ± 1367	7981 ± 3511	8953 ± 3982

(d) *Mean number of evaluations and its standard deviation $\ell = 6, 7$.*

	$\ell = 6$		$\ell = 7$
	$\kappa = 9$	$\kappa = 10$	$\kappa = 10$
Evo++	46483 ± 17031		
Evo10	17562 ± 135367	21804 ± 67227	40205 ± 65485

The first of these tables, which represent the average number of moves needed to find the solution, show results that are quite similar. The average for Evo10 is consistently higher (more turns are needed to find the solution) but in half the cases the difference is not statistically significant using Wilcoxon paired test. There is a significant difference for the two smaller sizes ($\ell = 4, \kappa = 8$ and $\ell = 5, \kappa = 8$), but not for the larger sizes $\ell = 5, \kappa = 9$ and $\ell = 6, 7$. This is probably due to the fact that, with increasing search space size, the difference among 10 and other sample size, even if they are in different orders of magnitude, become negligible; the difference between 10 and 1 % of the actual sample size is significant, but the difference 0.001 and 0.0001 % is not.

However, the difference in the number of evaluations (shown in Tables 1(c) and (d)), that is, the total population evaluated to find the solution is quite significant, going from a bit less than half to a third of the total evaluations for the larger size. This means that the time needed scales roughly in the same way, but it is even more interesting to note that it scales better for a fixed size than for the best consistent set size. Besides, in all cases the algorithm does not examine the full set of combinations, while previously the number of combinations evaluated, 6412, was almost 50 % bigger than the search space size for that problem. The same argument can be applied to the comparison with Berghman's results (when they are available); Evo++ was able to find solutions which were quite similar to them, but Evo10 obtains an average number of turns that is slightly worse; since we don't have the complete set of results, and besides they have been made on a different set of combinations, it is not possible to compare, but at any rate it would be reasonable to think that this result is significant.

4   Discussion, Conclusions and Future Work

This paper has shown that using a small and fixed consistent set size when playing mastermind using evolutionary algorithms does not imply a deterioration of results, while cutting in half the number of evaluations needed to find them. This makes the configuration of the algorithm shown quite suitable for real-time games such as mobile apps or web games; the actual time varies from less than one second for the smallest configuration to a few seconds for the whole game in the biggest configuration shown; the time being roughly proportional to the number of evaluations, this is at least an improvement of that order; that is, the time is reduced by 2/3 for the $\kappa = 6, \ell = 9$ problem, attaining an average of 4.7 sec, almost one fourth of the time it can take when we try to achieve the minimum number of turns, 18.6 sec. This number is closer to the one published by [4] for this size, 1.284 s, although without running it under the same conditions we cannot be sure. It is in the same ballpark, anyways. The time needed to find the solution has a strong component in the number of evaluations, but it also depends on the consistent set size, that is why the relation between the time needed (1/4) is smaller than the relation between number of evaluations (roughly 1/3, see Table 1(d)). This allows also to extend the range of feasible sizes, and yields a robust configuration that can be used throughout any Mastermind problem.

As future lines of work, we will try to reduce even more this size and try to check whether it offers good results for bigger sizes such as $\ell = 7, \kappa = 11$ or even $\ell = 8, \kappa = 12$. Several consistent set sizes will be systematically evaluated, looking mainly for a reduction in the number of evaluations, and time, needed. Eventually, what we are looking is for a method that is able to resolve problems with moderate size, but this will need to be tackled from different points of view: implementation, middle-level algorithms used, even the programming language we will be using. We might even have to abandon the paradigm of playing always consistent solutions to settle, sometimes, for non-consistent solutions for the sake of speed.

It is also clear than, when increasing the search space size, the size of the consistent set will become negligible with respect to the actual size of the consistent set. This could work both ways: first, by making the results independent of sample size (for this small size, at least) or by making the strategy of extracting a sample of a particular size indistinguishable from finding a single consistent combination and playing it. As we improve the computation speed, it would be interesting to take measurements to prove these hypotheses.

Acknowledgements. This work is supported by projects TIN2011-28627-C04-02 and TIN2011-28627-C04-01 and -02 (ANYSELF), awarded by the Spanish Ministry of Science and Innovation and P08-TIC-03903 and P10-TIC-6083 (DNEMESIS) awarded by the Andalusian Regional Government.

References

1. Meirovitz, M.: Board game (December 30 1980) US Patent 4,241,923
2. Knuth, D.E.: The computer as master mind. J. Recreational Math. **9**(1), 1–6 (1976–1977)
3. Montgomery, G.: Mastermind: improving the search. AI Expert **7**(4), 40–47 (1992)
4. Berghman, L., Goossens, D., Leus, R.: Efficient solutions for mastermind using genetic algorithms. Compu. Oper. Res. **36**(6), 1880–1885 (2009)
5. Runarsson, T.P., Merelo-Guervós, J.J.: Adapting heuristic mastermind strategies to evolutionary algorithms. In: González, J.R., Pelta, D.A., Cruz, C., Terrazas, G., Krasnogor, N. (eds.) NICSO 2010. SCI, vol. 284, pp. 255–267. Springer, Heidelberg (2010). ArXiV: http://arxiv.org/abs/0912.2415v1
6. Merelo-Guervós, J.J., Mora, A.M., Cotta, C., Runarsson, T.P.: An experimental study of exhaustive solutions for the mastermind puzzle. CoRR abs/1207.1315 (2012)
7. Kooi, B.: Yet another mastermind strategy. ICGA J. **28**(1), 13–20 (2005)
8. Cotta, C., Merelo Guervós, J.J., Mora Garcia, A.M., Runarsson, T.P.: Entropy-driven evolutionary approaches to the mastermind problem. In: Schaefer, R., Cotta, C., Kołodziej, J., Rudolph, G. (eds.) PPSN XI. LNCS, vol. 6239, pp. 421–431. Springer, Heidelberg (2010)
9. Merelo, J., Mora, A., Runarsson, T., Cotta, C.: Assessing efficiency of different evolutionary strategies playing mastermind. In: 2010 IEEE Symposium on Computational Intelligence and Games (CIG), pp. 38–45, August 2010
10. Merelo, J.J., Cotta, C., Mora, A.: Improving and scaling evolutionary approaches to the mastermind problem. In: Di Chio, C., et al. (eds.) EvoApplications 2011, Part I. LNCS, vol. 6624, pp. 103–112. Springer, Heidelberg (2011)
11. Merelo-Guervós, J.J., Mora, A.M., Cotta, C.: Optimizing worst-case scenario in evolutionary solutions to the MasterMind puzzle. In: IEEE Congress on Evolutionary Computation, pp. 2669–2676. IEEE (2011)
12. Eiben, A.E., Smit, J.E.: Introduction to Evolutionary Computing. Springer, Heidelberg (2003)

A Fast Local Search Approach
for Multiobjective Problems

Laurent Moalic, Alexandre Caminada$^{(\boxtimes)}$, and Sid Lamrous

Belfort-Montbéliard University of Technology - UTBM,
F-90010 Belfort, France
{laurent.moalic,sid.lamrous,alexandre.caminada}@utbm.fr

Abstract. In this article, we present a new local method for multiobjective problems. It is an extension of local search algorithms for the single objective case, with specific mechanisms used to build the Pareto set. The performance of the local search algorithm is illustrated by experimental results based on a real problem with three objectives. The problem is issued from electric car-sharing service with a car manufacturer partner. Compared to the Multiobjective Pareto Local Search (PLS) well known in the scientific literature [1], the proposed model aims to improve: the solutions quality and the time computing.

Keywords: Local search algorithm · Multiobjective optimization · Transportation services · Car-sharing

1 Introduction

Many real world problems require to optimize several objectives simultaneously, they are called multiobjective optimization problems (MOP). When it does not exist a unique solution optimizing all objectives in an optimal way, we need to find other decisional mechanisms. The Pareto dominance is one of these; for MOP, the Pareto set is composed of all best compromises between the different objectives. The Pareto set is achieved if there are no other dominant solutions in the search space. The Pareto front is defined as the image of the Pareto set in the objective space [2]. In the past few years, a lot of works were based on multiobjective evolutionary algorithms (MOEA) such NSGA-II [3], SPEA [4] and SPEA2 [5], sometime coupled with local search in memetic approaches [6,7].

To solve single objective combinatorial optimization problems, local search algorithms provide often efficient metaheuristics. They can also be adapted to multiobjective combinatorial problems like in Pareto Local Search algorithm (PLS) [1] with a complete exploration of the neighborhood, or with strategy based on the neighborhood structure [8]. A recent work has been done to unify local search algorithms applied to MOP, known as Dominance-based Multiobjective Local Search (DMLS) [9]. Finally, some algorithms add a Tabu criteria in the local search [10,11].

G. Nicosia and P. Pardalos (Eds.): LION 7, LNCS 7997, pp. 294–298, 2013.
DOI: 10.1007/978-3-642-44973-4_32, © Springer-Verlag Berlin Heidelberg 2013

The local search approach we propose, named FLS-MO, is based on the
Pareto optimality. A neighbor is acceptable if and only if it is not dominated
by the solutions found so far. This criteria was used in other approaches as in
[1] but the originality of our method is to be very intensive while maintaining a
good diversity. With this new tradeoff between intensification and diversification
we get good results in comparison with PLS.

2   Fast Local Search for Multiobjective Problems

The new algorithm is based on a not dominated local search. The initial solution is build randomly, marked as not explored and added to the solutions set.
While it exists a not explored solution in the solutions set, the algorithm chooses
randomly such a solution and use it recursively until being in a local optimum.
At each step the first random neighbor that provides a new non-dominated
solution is accepted. The algorithm stops when all non-dominated solutions are
marked as explored. The result is an approximation of the Pareto set. The approach combines two qualities: a good intensification based on exploration of any
non-dominated solution of the set and a good diversification because all non-
dominated solutions are accepted in the set.

Algorithm 1. Fast Local Search for Multiobjective Problems

1: $S \leftarrow$ init() {init the solution set S with a random individual}
2: $s \leftarrow$ select(S) {select randomly a not explored solution from S}
3: **while** $s \neq \emptyset$ **do**
4:     **repeat**
5:         $s' \leftarrow$ selectNeighbor(s) {select randomly a neighbor of s not dominated by S}
6:         **if** $s' \neq \emptyset$ **then**
7:             $s \leftarrow s'$
8:             addNotDominated(s) {add s in S and remove all dominated solutions}
9:         **end if**
10:     **until** $s' = \emptyset$
11:     mark s as explored
12:     $s \leftarrow$ select(S) {select randomly a not explored solution from S}
13: **end while**

---

3   Study Case: Charging Stations Location for Electric Car-Sharing service

Car-sharing services was first experimented in 1940 [12]. To deploy the service,
we need to locate charging stations where the people take and return the cars. In
our case, it is not necessary to return the vehicle in its starting station. Solving
approaches based on exact methods already exist such as [13] but they consider
simplified problem. We have applied FLS-MO algorithm to approximate the

Pareto set of this problem. The aim is to locate n stations in a given area to maximise several daily requests of population flows.

The area is discretized into a grid and all the flows are set in a 3D matrix $F = (f_{i,j,t})$ where $f_{i,j,t}$ represents the number of displacements from the cell i to the cell j at time period t. We have 3 objectives to locate the charging stations:

f1 : flow maximization i.e. the locations must allow us to maximize the flows between themselves

$$f_1 = \max_{s \in \Omega} \left[\sum_{st_i \in s} \sum_{st_j \in s \setminus \{st_i\}} f(st_i, st_j) \right] \tag{1}$$

f2 : balance maximization i.e. the location must allow us to maximize the balance between inflows and outflows of a station

$$f_2 = \max_{s \in \Omega} \left[\sum_{st_i \in s} \frac{f_r(st_i)}{f_T(st_i)} \right] \tag{2}$$

f3 : minimization of flow standard deviation i.e. the location must allow us to get an uniform flow along the day

$$f_3 = \min_{s \in \Omega} \left[\sum_{st_i \in s} \sqrt{\frac{1}{|T|} \sum_t (f(st_i, t) - \bar{f}(st_i))^2} \right] \tag{3}$$

With,

Ω : set of feasible solutions
s : solution element of Ω corresponding to a network of n charging stations
st_i : charging station i from the solution s
T : set of time periods of the day
t : one time period (for instance 15 minutes)
$f(st_i, st_j)$: number of people moving from st_i to st_j on all time periods
$f(st_i, st_j, t)$: number of people moving from st_i to st_j on time period t
$f(st_i, t)$: number of people moving from/to st_i on time period t
$\bar{f}(st_i)$: average number of people moving from/to st_i on all time periods
$f_r(st_i) = \sum_t \min \left[\sum_{st_j \in s \setminus \{st_i\}} f(st_i, st_j, t), \sum_{st_j \in s \setminus \{st_i\}} f(st_j, st_i, t) \right]$ is the
balanced part of the in/out flow throughout the day
$f_T(st_i) = \sum_t \max \left[\sum_{st_j \in s \setminus \{st_i\}} f(st_i, st_j, t), \sum_{st_j \in s \setminus \{st_i\}} f(st_j, st_i, t) \right]$ is the
total flow going through st_i station

4 Performance Analysis

In multiobjective optimization the comparison of different algorithms is quite difficult. Indeed for two approximations of the Pareto front one can be better for a criteria but worst for another one. Choosing a comparative indicator would be a good way to distinguish these sets. Here we have considered the additive ϵ-indicator [14]. The unary additive ϵ-indicator gives the minimum factor by which a set A has to be translated to dominate the reference set R. As we do not know the optimal reference set of the problem we composed an approximated R with the best solutions obtained with PLS and FLS-MO on many runs.

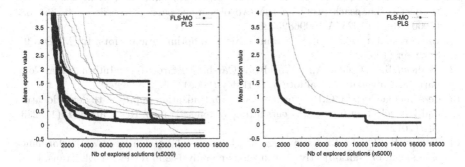

Fig. 1. ϵ-indicator evolution for PLS and FLS-MO algorithms

Figure 1 shows the comparison on 6 runs between PLS and FLS-MO. It reflects the evolution in time of ϵ-indicator. The left side shows 6 runs for each method and the right side shows their mean value on 6 runs. The results given by FLS-MO seems to be very promising. Figure 1 shows that FLS-MO converges twice faster than PLS and provides a better average evaluation.

References

1. Paquete, L., Chiarandini, M., Stützle, T.: Pareto local optimum sets in the biobjective traveling salesman problem: an experimental study. In: Gandibleux, X., Sevaux, M., Sörensen, K., T'kindt, V., Fandel, G., Trockel, W. (eds.) Metaheuristics for Multiobjective Optimisation. Lecture Notes in Economics and Mathematical Systems, vol. 535, pp. 177–199. Springer, Heidelberg (2004)
2. Coello, C., Lamont, G.: Applications of Multi-Objective Evolutionary Algorithms, vol. 1. World Scientific, Singapore (2004)
3. Deb, K., Pratap, A., Agarwal, S., Meyarivan, T.: A fast and elitist multiobjective genetic algorithm: Nsga-2. IEEE Trans. Evol. Comput. **6**, 182–197 (2002)
4. Zitzler, E., Thiele, L.: Multiobjective evolutionary algorithms: a comparative case study and the strength pareto approach. IEEE Trans. Evol. Comput. **3**, 257–271 (1999)
5. Zitzler, E., Laumanns, M., Thiele, L.: Spea2: Improving the strength pareto evolutionary algorithm. TIK-Report 103 (2001)
6. Knowles, J., Corne, D.: M-paes: a memetic algorithm for multiobjective optimization. In: Proceedings of the 2000 Congress on Evolutionary Computation, vol. 1, pp. 325–332 (2000)
7. Jaszkiewicz, A.: Genetic local search for multi-objective combinatorial optimization. Eur. J. Oper. Res. **137**(1), 50–71 (2002)

8. Wu, Z., Chow, T.S.: A local multiobjective optimization algorithm using neighborhood field. Struct. Multi. Optim. **46**, 853–870 (2012)
9. Liefooghe, A., Humeau, J., Mesmoudi, S., Jourdan, L., Talbi, E.-G.: On dominance-based multiobjective local search: design, implementation and experimental analysis on scheduling and traveling salesman problems. J. Heuristics **18**, 317–352 (2012). doi:10.1007/s10732-011-9181-3
10. Gandibleux, X., Freville, A.: Tabu search based procedure for solving the 0-1 multiobjective knapsack problem: the two objectives case. J. Heuristics **6**, 361–383 (2000). doi:10.1023/A:1009682532542
11. Hansen, M.P.: Tabu search for multiobjective optimization: Mots. In: MCDM'97, Springer (1997)
12. Shaheen, S.A., Cohen, A.P.: Worldwide Carsharing Growth: An International Comparison. University of California, Berkeley (2008)
13. de Almeida Correia, G.H., Antunes, A.P.: Optimization approach to depot location and trip selection in one-way carsharing systems. Transp. Res. Part E **48**(1), 233–247 (2012)
14. Zitzler, E., Thiele, L., Laumanns, M., Fonseca, C., da Fonseca, V.: Performance assessment of multiobjective optimizers: an analysis and review. IEEE Trans. Evol. Comput. **7**, 117–132 (2003)

Generating Customized Landscapes in Permutation-Based Combinatorial Optimization Problems

Leticia Hernando, Alexander Mendiburu, and Jose A. Lozano[✉]

Intelligent Systems Group,
Department of Computer Science and Artificial Intelligence,
University of the Basque Country UPV/EHU, San Sebastián, Spain
{leticia.hernando,alexander.mendiburu,ja.lozano}@ehu.es

Abstract. Designing customized optimization problem instances is a key issue in optimization. They can be used to tune and evaluate new algorithms, to compare several optimization algorithms, or to evaluate techniques that estimate the number of local optima of an instance. Given this relevance, several methods have been proposed to design customized optimization problems in the field of evolutionary computation for continuous as well as binary domains. However, these proposals have not been extended to permutation spaces. In this paper we provide a method to generate customized landscapes in permutation-based combinatorial optimization problems. Based on a probabilistic model for permutations, called the Mallows model, we generate instances with specific characteristics regarding the number of local optima or the sizes of the attraction basins.

Keywords: Combinatorial optimization problems · Landscape generator · Mallows model · Permutation space · Local optima

1 Introduction

Generating instances of combinatorial optimization problems (COPs) is an essential factor when comparing and analyzing different metaheuristic algorithms, and when evaluating algorithms that estimate the number of local optima of an instance. The design of a tunable generator of instances is of high relevance as it allows to control the properties of the instances by changing the values of the parameters.

Given the significance of this topic, several proposals have been presented in the literature. For example, a generator for binary spaces is proposed in [1], or a more recent work [2] shows a software framework that generates multimodal test functions for optimization in continuous domains. Particularly, the study developed in [3] has high relevance with our paper. The authors proposed a continuous space generator based on a mixture of Gaussians, which is tunable by

G. Nicosia and P. Pardalos (Eds.): LION 7, LNCS 7997, pp. 299–303, 2013.
DOI: 10.1007/978-3-642-44973-4_33, © Springer-Verlag Berlin Heidelberg 2013

a small number of user parameters. Based on that work we propose a generator of permutation-based COPs instances based on a mixture of a probabilistic model for permutations called the Mallows model.

The rest of the paper is organized as follows. The Mallows model is explained in Sect. 2. In Sect. 3 we present our generator of instances of COPs based on permutations. Finally, future work is given in Sect. 4.

2   Mallows Model

The Mallows model [4] is an exponential probability model for permutations based on a distance. This distribution is defined by two parameters: the central permutation σ_0, and the spread parameter θ. If Ω is the set of all permutations of size n, for each $\sigma \in \Omega$ the Mallows distribution is defined as:

$$p(\sigma) = \frac{1}{Z(\theta)} e^{-\theta d(\sigma_0, \sigma)}$$

where $Z(\theta) = \sum_{\sigma' \in \Omega} e^{-\theta d(\sigma_0, \sigma')}$ is a normalization term and $d(\sigma_0, \sigma)$ is the distance between the central permutation σ_0 and σ. The most commonly used distance is the Kendall tau. Given two permutations σ_1 and σ_2, it counts the minimum number of adjacent swaps needed to convert σ_1 into σ_2. Under this metric the normalization term $Z(\theta)$ has closed form and does not depend on σ_0:

$$Z(\theta) = \prod_{j=1}^{n-1} \frac{1 - e^{-(n-j+1)\theta}}{1 - e^{-\theta}}.$$

Notice that if $\theta > 0$, then σ_0 is the permutation with the highest probability. The rest of permutations $\sigma' \in \Omega - \{\sigma_0\}$ have probability inversely exponentially proportional to θ and their distance to σ_0. So, the Mallows distribution can be considered analogous to the Gaussian distribution on the space of permutations.

3   Instance Generator

In this section we show a generator of instances of COPs where the solutions are in the space of permutations. Our generator defines an optimization function based on a mixture of Mallows models.

The generator proposed in this paper uses $3m$ parameters: m central permutations $\{\sigma_1, ..., \sigma_m\}$, m spread parameters $\{\theta_1, ..., \theta_m\}$ and m weights $\{w_1, ..., w_m\}$. We generate m Mallows models $p_i(\sigma|\sigma_i, \theta_i)$, one for each σ_i and θ_i, $\forall i \in \{1, ..., m\}$. The objective function value for each permutation $\sigma \in \Omega$ is defined as follows:

$$f(\sigma) = \max_{1 \leq i \leq m} \{w_i p_i(\sigma|\sigma_i, \theta_i)\}.$$

Landscapes with different properties, and hence different levels of complexity, are obtained by properly tuning these parameters.

Some of these interesting properties are analyzed here. The first relevant factor we consider is that all central permutations σ_i's were local optima. Clearly, in order to be local optima, $\{\sigma_1, ..., \sigma_m\}$ have to fulfill that $d(\sigma_i, \sigma_j) \geq 2, \forall i \neq j$. A second constraint is that the objective function value of σ_i has to be reached in the ith Mallows model, i.e.:

$$f(\sigma_i) = \max_{1 \leq k \leq m} \{w_k p_k(\sigma_i|\sigma_k, \theta_k)\} = w_i p_i(\sigma_i|\sigma_i, \theta_i) = w_i \frac{e^{-\theta_i d(\sigma_i, \sigma_i)}}{Z(\theta_i)} = \frac{w_i}{Z(\theta_i)} \quad (1)$$

Moreover, in order to be σ_i a local optimum the following constraint has to be fulfilled:

$$f(\sigma_i) > f(\sigma), \quad \forall \sigma \ s.t. \ d(\sigma_i, \sigma) = 1. \tag{2}$$

To satisfy (2), and taking into account the constraint (1), we need to comply with:

$$\forall j = 1, ..., m, \quad \frac{w_i}{Z(\theta_i)} > w_j p_j(\sigma), \quad \forall \sigma \ s.t. \ d(\sigma_i, \sigma) = 1.$$

However, taking into account that if $\sigma \in \Omega$ is s.t. $d(\sigma_i, \sigma) = 1$, then $d(\sigma_j, \sigma) = d(\sigma_j, \sigma_i) - 1$ or $d(\sigma_j, \sigma) = d(\sigma_j, \sigma_i) + 1$, Eq. (2) can be stated as:

$$\frac{w_i}{Z_i(\theta_i)} > \frac{w_j}{Z_j(\theta_j)} e^{-\theta_j(d(\sigma_i, \sigma_j) - 1)}, \quad \forall j \in \{1, 2, ..., m\}, i \neq j. \tag{3}$$

Notice that once the parameters θ_i's have been fixed, the previous inequalities are linear in w_i's. So the values of w_i's could be obtained as the solution of just a linear programming problem. However, we have not defined any objective function to be optimized in our linear programing problem. This function can be chosen taking into account the different desired characteristics for the instance.

For example, one could think about a landscape with similar sizes of attraction basins. In this case, and without loss of generality, we consider that σ_1 is the global optimum and that σ_m is the local optimum with the lowest objective function value. Our objective function tries to minimize the difference between the objective function values of these two permutations (and implicitly minimize the difference of the objective function values of all the local optima). In addition we have to include new constraints to comply with these properties in the objective function values. This landscape can be generated as follows:

1. Choose uniformly at random m permutations in Ω: $\sigma_1, \sigma_2, ..., \sigma_m$, such that $d(\sigma_i, \sigma_j) \geq 2, \quad \forall i, j \in \{1, ..., m\}, \quad i \neq j$.
2. Choose uniformly at random in the interval $[a, b]$ (with $b > a > 0$) m spread parameters : $\theta_1, \theta_2, ..., \theta_m$.
3. Solve the linear programming problem in the weights w_i's:

$$\min \left\{ \frac{w_1}{Z(\theta_1)} - \frac{w_m}{Z(\theta_m)} \right\}$$

$$\frac{w_i}{Z(\theta_i)} > \frac{w_{i+1}}{Z(\theta_{i+1})} \qquad\qquad (\forall i \in \{1, 2, ..., m-1\})$$

$$w_i/Z(\theta_i) > w_j \frac{e^{-\theta_j(d(\sigma_i,\sigma_j)-1)}}{Z(\theta_j)} \quad (\forall i, j \in \{1, 2, ..., m\}, \ ?i > j)$$

4. Assign to each $\sigma \in \Omega$ the objective function value:

$$f(\sigma) = \max_i \{w_i \frac{e^{-\theta_i d(\sigma_i,\sigma)}}{Z(\theta_i)}\}$$

4   Conclusions and Future Work

In this paper we introduce a framework to generate instances of COPs with permutation search spaces that is based on [3]. We create the landscapes based on a Mallows mixture. The aim is to obtain different kinds of instances depending on the central permutations $\sigma_1, ...\sigma_m$, the values of the spread parameters $\theta_1, ..., \theta_m$ and the values of the weights $w_1, ...w_m$.

Once the values of θ_i's are fixed, and the σ_i's are chosen, some linear constraints in w_i's have to be fulfilled in order to be all σ_i's local optima. These constraints can be accompanied by a function to be optimized, and therefore w_i's can be obtained as solutions of a linear programming problem. This optimization function is a key element when creating the instances under desired characteristics. One function is explained in Sect. 4, but obviously one could think of many other functions. For example, if we want to create an instance with a big size of attraction basin of the global optimum σ_i, our intuition leads us to think that we have to maximize the difference between the objective function value of σ_i and the other local optima, where σ_i is the local optimum that is further to the rest of local optima. However, if we want a global optimum σ_i with a small size of attraction basin, we could think about minimizing the difference between the objective function value of σ_i and the value of its neighbors, where σ_i has to be the local optimum that is nearer on average to the other local optima.

A remarkable point is that in the example we have taken the local optima uniformly at random. However, they can be chosen taking into account different criteria, such as the distance between them. For example, we can choose all the local optima as close as possible, or choose them maintaining the same distance, while the global optimum is far from them.

A more tunable model, and therefore more interesting when trying to create instances with different levels of complexity, can be obtained using the Generalized Mallows model [5]. This model uses a decomposition of the Kendall-tau distance and different spread parameters are assigned to each of the index $i \in \{1, 2, ..., n\}$, where n is the size of the permutations. So the parameters of the model ascend to $2m + n * m$. Apart from that, the Mallows model can be used with other distances such as the Hamming distance, the Cayley distance, etc.

We believe that by controlling the parameters we would we able to create instances with similar characteristics to those existing for famous COPs, such as the Traveling Salesman Problem, the Flowshop Scheduling Problem, the Linear Ordering Problem, etc. Moreover, we think that the model could be flexible enough to represent the complexity of real-world problems.

Acknowledgements. This work has been partially supported by the Saiotek, Etortek and Research Groups 2007-2012 (IT- 242-07) programs (Basque Government), TIN2010-14931 (Spanish Ministry of Science and Innovation) and COMBIOMED network in computational biomedicine (Carlos III Health Institute). Leticia Hernando holds a grant from the Basque Government.

References

1. De Jong, K.A., Potter, M.A., Spears, W.M.: Using problem generators to explore the effects of epistasis. Seventh International Conference on Genetic Algorithms, pp. 338–345. Morgan Kaufmann, San Francisco (1997)
2. Rönkkönen, J., Li, X., Kyrki, V., Lampinen, J.: A framework for generating tunable test functions for multimodal optimization. Soft Comput., 1–18 (2010)
3. Gallagher, M., Yuan, B.: A general-purpose tunable landscape generator. IEEE Trans. Evol. Comput. **10**(5), 590–603 (2006)
4. Mallows, C.L.: Non-null ranking models. Biometrika **44**(1–2), 114–130 (1957)
5. Fligner, M.A., Verducci, J.S.: Distance based ranking models. J. Roy. Stat. Soc. **48**(3), 359–369 (1986)

Multiobjective Evolution of Mixed Nash Equilibria

David Iclănzan[✉], Noémi Gaskó, Réka Nagy, and D. Dumitrescu

Babes-Bolyai University, Cluj-Napoca, Romania
david.iclanzan@gmail.com

Abstract. In a mixed strategy equilibrium players randomize between their actions according to a very specific probability distribution, even though with regard to the game payoff, they are indifferent between their actions. Currently, there is no compelling model explaining why and how agents may randomize their decisions is such a way, in real world scenarios.

We experiment with a model for two player games, where the goal of the players is to find robust strategies for which the uncertainty in the outcome of the opponent is reduced as much as possible. We show that in an evolutionary setting, the proposed model converges to mixed strategy profiles, if these exist. The results suggest that only local knowledge of the game is sufficient to attain the adaptive convergence.

1 Introduction

In Game Theory [6] a mixed strategy is a probability distribution over the actions available for the players. This allows a player to randomly choose an action instead of choosing a single pure strategy. If only one action has a probability of one to be selected, the player is said to use a pure strategy.

The most popular solution concept in game theory is Nash equilibrium [5]. A game state is a Nash equilibrium, if no player has the incentive to unilaterally deviate from his/her strategy. Every finite game has a Nash equilibrium. However, not every game has a Nash equilibrium in pure strategies. Therefore, the concept of mixed strategies is a fundamental component in Game Theory, as it can provide Nash equilibria in games where no equilibrium in pure strategies exists.

The empirical relevance of mixed strategies have been often criticized for being "intuitively problematic" [1]. In [9] mixed strategies are viewed from the perspective of evolutionary biology. According to other interpretations players randomize because they think their strategies may be observed in advance by other players [8]. Albeit there are theoretical arguments trying to rationalize this concept [4], it is not clear why and how players randomize their decisions.

Beside the behavioral observation that people seldom make their choices following a lottery, the most puzzling question arises from the "indifference" property of a mixed strategy equilibrium. In mixed equilibrium, given the strategies

G. Nicosia and P. Pardalos (Eds.): LION 7, LNCS 7997, pp. 304–314, 2013.
DOI: 10.1007/978-3-642-44973-4_34, © Springer-Verlag Berlin Heidelberg 2013

chosen by the other players, each player is indifferent among all the actions that he/she may select with positive probability, as they do not affect the resulting payoff. Therefore, there is no direct benefit to select precisely the strategy that induces the opponents to be indifferent, as required for the existence of the equilibrium. Then, in the absence of communication between players, how can a mixed equilibrium arise in a real-world scenario, especially in cases of incomplete information?

Computational computing the Nash equilibrium is a complex problem [2,7]. We experiment with a novel model, that can lead to the emergence of mixed equilibrium. Here, agents aim to develop strategies for which the payoff outcome of the opponent can be predicted.

2   Game Theoretic Prerequisites

Mathematically, a finite strategic non-cooperative game is a system

$$G = (N, S_i, u_i; i = 1, ..., n),$$

where:

- N represents a set of players, and n is the number of players
- for each player $i \in N S_i$ is the set of actions available to him/her;

$$S = S_1 \times S_2 \times ... \times S_N$$

is the set of all possible situations of the game and $s \in S$ is a strategy (or strategy profile) of the game
- for each player $i \in N$, $u_i : S \to R$ represents the payoff function.

Let us denote by (s_i, s^*_{-i}) the strategy profile obtained from s^* by replacing the strategy of player i with s_i :

$$(s_i, s^*_{-i}) = (s^*_1, ..., s_i, ..., s^*_n).$$

Nash equilibrium captures a state in which individual players maximize their own payoffs. A strategy profile is a Nash equilibrium if no player has the incentive to unilaterally deviate. Once all players are playing Nash equilibrium, it is in interest of every player to stick to his/her strategy.

Definition 1. *A strategy profile $s^* \in S$ is a Nash equilibrium if the inequality*

$$u_i(s_i, s^*_{-i}) \leq u_i(s^*),$$

holds $\forall i = 1, ..., n, \forall s_i \in S_i, s_i \neq s^*_i$.

2.1   Mixed Strategies

Games allowing mixed strategies are extensions of standard non-cooperative games where players, instead of choosing a single strategy, they are allowed to choose a probability distribution over their set of actions. Such a probability distribution is called a *mixed strategy.*

A *mixed strategy* of a player i is given by:

$$\sigma_i : S_i \to \Re_+,$$

where

$$\sum_{s_k \in S_i} \sigma_i(s_k) = 1.$$

The strategies available to player i is the set of all probability distributions over S_i. A player may assign a probability 1 to a single action, in this case the player chooses a pure strategy.

The payoff for player i in a game allowing mixed strategies is given by:

$$u_i(\sigma_i) = \sum_{s_i \in S_i} \sigma_i(s_i)u_i(s_{-i}, s_i).$$

The mixed strategy Nash equilibrium (or simply mixed Nash equilibrium) is an extension of the Nash equilibrium. A mixed strategy profile σ^* is a mixed Nash equilibrium is there is no player i that would prefer the lottery over the payoffs generated by the strategy profile $(\sigma_i, \sigma^*_{-i})$.

Definition 2. σ^* *is a mixed Nash equilibrium if,*

$$u_i(\sigma^*) \geq u_i(\sigma_i, \sigma^*_{-i}), \forall i \in N, \forall \sigma_i \in \Sigma_i.$$

Evey finite game has a Nash equilibrium in mixed strategies [5].

Example 1. The game of *Matching Pennies* is a simple two player game. Both players have a penny and that they simultaneously turn to either heads (H) or tails (T). If the two pennies match (both are heads or both are tails), Player 1 wins a penny from Player 2, otherwise Player 2 wins a penny from Player 1. The payoffs for the game are depicted in Table 1.

The game does not have a Nash equilibrium, however there is a Nash equilibrium in mixed strategies, namely when both players choose Heads with a probability of $\frac{1}{2}$ and Tails with a probability of $\frac{1}{2}$. This way both players end up with an expected payoff of 0 and neither can do better by deviating from this strategy.

In most cases a player can benefit from knowing the next move of the opponent, so each player wants to keep his/her opponent guessing. An important feature of mixed Nash equilibrium is, that, given the actions of the opponents, the players are indifferent among the actions chosen with positive possibility. For example in the Matching Pennies game, given that Player 2 chooses Heads

Table 1. The payoff matrix of the mathcing pennies game.

Player2

		H	T
Player1	H	(1,-1)	(-1,1)
	T	(-1,1)	(1,-1)

or Tails with same probability, Player 1 is indifferent among its actions. Thus the goal of each player is to randomize in such a way to keep the opponent indifferent.

3   Proposed Model

Rational agents often build internal models that anticipate the *actions* of the other players and adapt their strategies accordingly. Here, we experiment with a model for two player games where players try to anticipate directly the *game payoff* of the other player. The agents adapt their strategies in order to reduce the uncertainty of this prediction. This is in contrast with the classical scenario, where players foremost objective is to maximize their game utility.

Let (w, p) be a mixed strategy profile for a two player game, where w defines the probability distribution over the actions available for the first player, with p having a similar role for the second player. Let $u_1(w, p)$ and $u_2(w, p)$ denote the game payoff for player one and for player two respectively.

Then, the proposed model is formalized as follows:

$$\begin{cases} o_1 = \underset{w}{\operatorname{argmin}}(\frac{1}{m} \sum_{i=1}^{m}(u_2(w, p) - u_2(w, \delta_i(p)))^2) \\ o_2 = \underset{p}{\operatorname{argmin}}(\frac{1}{m} \sum_{i=1}^{m}(u_1(w, p) - u_1(\delta_i(w), p))^2) \end{cases} \tag{1}$$

where δ_i provides a perturbation to the input probability distribution, and m is the number of perturbations.

If a mixed strategy equilibrium exits, it will optimally satisfy this multiobjective model, as a direct consequence of the "indifference" property, with both objective values equaling 0. Furthermore, the model provides incentive to not deviate from this strategy profile. If an agent deviates from the equilibrium, the other player is not completely indifferent, the squared average difference of different pays might deviate from the zero minimum.

Several important questions arise regarding the proposed model:

- Can the model be used to locate a mixed strategy Nash equilibrium (if it exists), using an adaptive search?
- What should the magnitudes of perturbations provided by δ be? Perturbation magnitudes have a direct consequence on the amount of information about the game, assumed to be available for the players. Arbitrary perturbation equates to a complete information game, where one can internally evaluate

every possible strategy profile, while small perturbations assume only a local knowledge about the game outcomes.
- How many perturbations of the actual strategy profile are required at each step for reliable results (how big should parameter m be)?

We empirically investigate these issues in the next Section.

4   Detection Method and Results

The proposed model is optimized using the NSGA II [3] – a fast multiobjective evolutive algorithm based on the concept of nondomination.

According to this concept, a solution A is dominated by another solution B, if B is better than A in relation to at least one of the objectives, and is better than or equal to A regarding all the other objectives. Two solutions A and B are nondominated in relation to each other if A does not dominate B and neither is B dominating A. The Pareto optimum is the set of nondominated solutions for which any candidate solution that is better than those from the Pareto optimum with respect to one of the objectives, is guaranteed to be worse with respect to at least another objective.

NSGA II segregates the population into layers, according to their domination degree and inside each layer, a diversity enhancing sharing function is employed to assign fitnesses. The elitist selection takes into account both the rank and the diversity maintaining crowding distance factor.

In our setup, an initial population of 60 individuals are generated randomly, where each individual encodes a strategy profile. At each step the nondominated individuals from the actual population can be considered as the approximation of certain equilibria. As genetic operators, crossover and mutation for real values are used, with probability 0.8 and 0.01. For the test problems, selection, recombination, and mutation is repeated in the bound of 500 generations. A run is considered successful, if a strategy profile is located that is very close to the target equilibrium state i.e. the euclidean distance between the two points is less or equal then a preset threshold $\epsilon = 0.0001$.

In the following we describe the two player games used for testing.

4.1   Game 1

We study a game where each player has two actions. The payoff matrix is described in Table 2.

Table 2. The payoff gatrix for Game 1. The game has two pure Nash equilibria ($p = 1, w = 0$) and ($p = 0, w = 1$) and one mixed equilibrium at ($p = \frac{6}{7}, w = \frac{6}{7}$).

		Player 2	
	Strategy	p	1-p
Player 1	w	(9,9)	(6,10)
	1-w	(10,6)	(0,0)

Table 3. The payoff matrix for Game 2. The game has one pure Nash equilibrium and one mixed equilibrium at $(p = 0, w = \frac{1}{2})$.

Player 2

	Strategy	p	1-p
Player 1	w	(0,2)	(4,0)
	1-w	(1,0)	(3,0)

Table 4. Payoff matrix for Game 3. The game has no pure Nash equilibrium and one mixed equilibrium at $(p = \frac{4}{7}, w = \frac{2}{7})$.

Player 2

	Strategy	p	1-p
Player 1	w	(2,-2)	(-1,1)
	1-w	(-3,3)	(1,-1)

The game has two Nash equilibria in pure strategies and one mixed equilibrium at $(\frac{6}{7}, \frac{6}{7})$.

4.2   Game 2

We consider an other two player game, where payoffs and actions are presented in Table 3. The game has one pure Nash equilibrium $(p = 1, w = 0)$ with the corresponding payoff of $(1, 0)$) and one mixed equilibrium at $(0, \frac{1}{2})$.

4.3   Game 3

We consider a bimatrix zero-sum game, having its payoff matrix described in Table 4. The game has no pure Nash equilibrium but it has one mixed equilibrium at $(\frac{4}{7}, \frac{2}{7})$.

4.4   Game 4

Finally, we consider the Rock-Paper-Scissors Game. Both players chose simultaneously a sign: Rock (R), Paper (P) or Scissors (S). The winner gets a dollar from the loser according to the following rule: Paper beats (wraps) Rock, Rock beats (blunts) Scissor and Scissor beats (cuts) Paper. The payoff matrix of the game is depicted in Table 5.

The Rock-Paper-Scissors Game has no Nash equilibrium in pure strategies. There is however a single Nash equilibrium in mixed strategies, namely when both players play all three strategies with equal probability.

4.5   Numerical Results

In a first experiment, for each game, we set parameter m to a constant value of 5 and study the effect of perturbation magnitude on the convergence of the model.

Table 5. The Rock-Paper-Scissors Game. The game has no pure Nash equilibrium and one symmetric mixed equilibrium in $(\frac{1}{3}, \frac{1}{3}, \frac{1}{3})$.

		Player 2		
	Strategy	R	P	S
Player 1	R	(0,0)	(-1,1)	(1,-1)
	P	(1,-1)	(0,0)	(-1,1)
	S	(-1,1)	(1,-1)	(0,0)

In the analyzed approach, a strategy profile of a player is perturbed by adding a Gaussian noise with a standard deviation of σ, where σ takes the following values: [0.00000001, 0.0000001, 0.000001, 0.00001, 0.0001, 0.0005, 0.001, 0.005, 0.01, 0.05, 0.1, 0.15, 0.2]. For each σ 500 independent runs are performed and for the successful runs the average convergence generation and standard deviation is computed.

In the case of Game 1, from the total of 6500 runs only in 9 cases (0.0014 %) the mixed equilibrium is not located with sufficient exactness. For the second game, the method located the mixed equilibrium in all runs while for Game 3 we have 5 cases (0.000769 %) of unsuccessful runs. Therefore, we conclude that

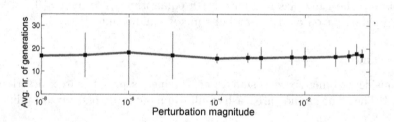

Fig. 1. Semilogarithmic plot of the average convergence speed for various perturbation magnitudes for Game 1.

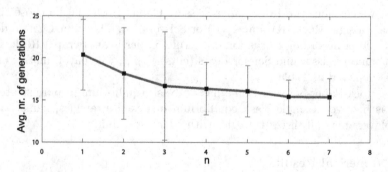

Fig. 2. Average convergence speed when using various number of perturbed states for Game 1.

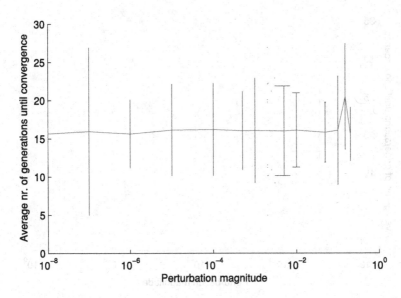

Fig. 3. Semilogarithmic plot of the average convergence speed for various perturbation magnitudes (Game 2).

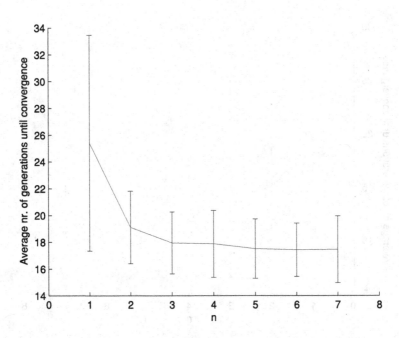

Fig. 4. Average convergence speed when using various number of perturbed states (1-7) (Game 2).

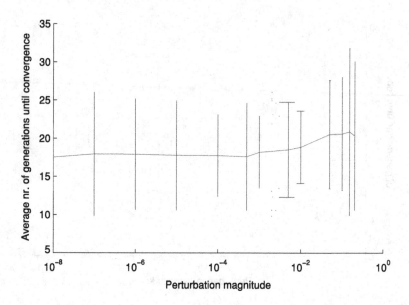

Fig. 5. Semilogarithmic plot of the average convergence speed for various perturbation magnitudes (Game 3).

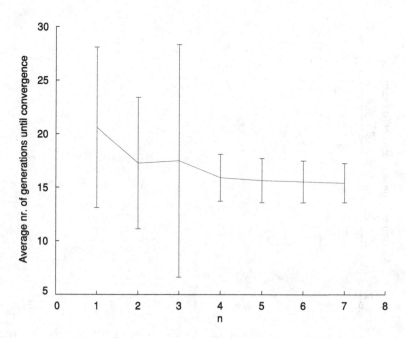

Fig. 6. Average convergence speed when using various number of perturbed states (1-7) (Game 3).

the method displays a robust behavior for finding mixed equilibria, even for very small perturbations.

The obtained averages for the three games are displayed in Figs. 1, 3, respectively Fig. 5 for Game 3.

Surprisingly, as one can see in these figures, the search is mostly insensitive to the amount of perturbation, with a slight better behavior, faster convergence with smaller perturbations.

In a next step, we lock $\sigma = 0.00000001$ and experiment with various number of perturbations (parameter m) used at each evaluation, ranging from 1 to 7. Again, an average of 500 independent run for each game and for each case is computed. The results of this experiment are displayed in Figs. 2, 4 and 6.

In the case of Game 4 the strategy of a player is perturbed by a Gaussian noise with a standard deviation of $\sigma = 0.15$. The number of perturbations is equal to three. 500 different runs are performed, and with a percent of 93.8 % the algorithm finds the mixed Nash equilibrium.

Results suggest that the model is moderately sensitive to the parameter m. The lower this number is, the higher is the required average number of generations until convergence. However, this difference is not very large. For example, for Game 1, the difference between using only one perturbed point and using seven points is on average 4.9627 generations, representing an 32.19 % increase.

5   Conclusions

We propose a model that adaptively converge to mixed strategy Nash equilibria, when optimized via an evolutionary multiobjective search method. The model can work with only a local knowledge about the game, centered around the actual strategy profile, and at each step requires only one evaluation of a slightly perturbed strategy profile.

The results suggest that a player can adaptively develop the strategy that makes the opponent indifferent with regard to his own actions. Interestingly, to obtain this result, it is enough to consider and measure one additional alternative. The players need to have only a local knowledge about the game, where it can internally evaluate the outcome of strategy profiles that are very close to the current profile and its known outcome.

Numerical experiments describe several two player games, with different pure and mixed Nash equilibria. Results indicate the potential of the proposed evolutionary search method.

Future work will extend the model to more than two players.

Acknowledgments. The first author acknowledges the financial support of the Sectoral Operational Program for Human Resources Development 2007-2013, co-financed by the European Social Fund, within the project POSDRU 89/1.5/S/60189 with the title "Postdoctoral Programs for Sustainable Development in a Knowledge Based Society". The second author wishes to thank for the financial support of the national project code TE 252 financed by the Romanian Ministry of Education and Research CNCSIS-UEFISCSU and "Collegium Talentum".

References

1. Aumann, R.J.: What is game theory trying to accomplish? In: Arrow, K., Honkapo-hja, S. (eds.) Frontiers of Economics. Blackwell, Oxford (1985)
2. Daskalakis, C., Goldberg, P.W., Papadimitriou, C.H.: The complexity of computing a nash equilibrium. In: Proceedings of the Thirty-Eighth Annual ACM Symposium on Theory of Computing, STOC '06, pp. 71–78. ACM, New York (2006)
3. Deb, K., Agrawal, S., Pratap, A., Meyarivan, T.: A fast elitist non-dominated sorting genetic algorithm for multi-objective optimisation: NSGA-II. In: Schoenauer, M., Deb, K., Rudolph, G., Yao, X., Lutton, E., Merelo, J.J., Schwefel, H.-P. (eds.) PPSN 2000. LNCS, vol. 1917, pp. 849–858. Springer, Heidelberg (2000)
4. Harsanyi, J.C.: Games with randomly disturbed payoffs: a new rationale for mixed-strategy equilibrium points. Int. J. Game Theor. **2**(1), 1–23 (1973)
5. Nash, J.: Non-cooperative games. Ann. Math. **54**(2), 286–295 (1951)
6. Osborne, M.J.: An introduction to game theory. Oxford University Press, New York (2004)
7. Papadimitriou, C.H.: On the complexity of the parity argument and other inefficient proofs of existence. J. Comput. Syst. Sci. **48**(3), 498–532 (1994)
8. Reny, P.J., Robson, A.J.: Reinterpreting mixed strategy equilibria: a unification of the classical and bayesian views. Games Econ. Behav. **48**(2), 355–384 (2004)
9. Smith, J.M.: Evolution and the Theory of Games. Cambridge University Press, Cambridge (1982)

Hybridizing Constraint Programming and Monte-Carlo Tree Search: Application to the Job Shop Problem

Manuel Loth[1,2]([✉]), Michéle Sebag[2], Youssef Hamadi[1],
Marc Schoenauer[2], and Christian Schulte[3]

[1] Microsoft Research, Cambridge, UK
[2] TAO, CNRS–INRIA–LRI, Université Paris-Sud, Orsay, France
manuel.loth@inria.fr
[3] School of ICT, KTH Royal Institute of Technology, Stockholm, Sweden

Abstract. Constraint Programming (CP) solvers classically explore the solution space using tree search-based heuristics. Monte-Carlo Tree-Search (MCTS), a tree-search based method aimed at sequential decision making under uncertainty, simultaneously estimates the reward associated to the sub-trees, and gradually biases the exploration toward the most promising regions. This paper examines the tight combination of MCTS and CP on the job shop problem (JSP). The contribution is twofold. Firstly, a reward function compliant with the CP setting is proposed. Secondly, a biased MCTS node-selection rule based on this reward is proposed, that is suitable in a multiple-restarts context. Its integration within the Gecode constraint solver is shown to compete with JSP-specific CP approaches on difficult JSP instances.

1 Introduction

This paper focuses on hybridizing Constraint Programming (CP) and Monte-Carlo Tree Search (MCTS) methods. The proof of concept of the approach is given on the job-shop problem (JSP), where JSPs are modelled as CP problem instances, and MCTS is hybridized with the Gecode constraint solver environment [3]. This paper first briefly presents the JSP modeling in constraint programming and the MCTS framework, referring the reader to respectively [2] and [4,5] for a comprehensive presentation. The proposed hybrid approach, referred to as *Bandit-Search for Constraint-Programming (BaSCoP)*, is thereafter described. The first experimental results on the difficult *Taillard 11-20* 20×15 problem instances are presented in Sect. 5. The paper concludes with a discussion w.r.t related work [10], and some perspectives for further research.

2 CP-Based Resolution of JSP

The job-shop problem, one of the classical scheduling problems, is concerned with allocating jobs to machines while minimizing the overall makespan. The

G. Nicosia and P. Pardalos (Eds.): LION 7, LNCS 7997, pp. 315–320, 2013.
DOI: 10.1007/978-3-642-44973-4_35, © Springer-Verlag Berlin Heidelberg 2013

CP modelling of JSP proceeds by considering a sequence of problems: after a given solution with makespan m has been found, the problem is modified by adding the constraint $makespan < m$.

While specific JSP-driven heuristics have been used in CP approaches, e.g. [11], our goal in this paper is to investigate the coupling of a generic CP framework, specifically Gecode [3], with MCTS. The JSP modeling in Gecode, inspired from the reportedly best CP approach to JSP [2], is as follows:

- the restart policy follows a Luby sequence [6]: the search is restarted after the specified number of failures is reached or when a new solution is found;
- the variable ordering is based on the *weighted-degree* heuristics — found as Max Accumulated Failure Count (AfcMax) in Gecode [3];
- the search heuristics is a Depth-First-Search, starting from the last found solution, i.e. the left value associated to a variable is the value assigned to this variable in the previous best solution.

3   Monte-Carlo Tree Search

MCTS [5] is concerned with optimal sequential decision under uncertainty, and is known in particular for its breakthrough in the game of Go [4]. MCTS proceeds by gradually and asymmetrically growing a search tree, carefully balancing the exploitation of the most promising sub-trees and the exploration of the rest of the tree. MCTS iterates N tree-walks, a.k.a simulations, where each tree-walk involves four phases:

- In the so-called bandit phase, the *selection rule* of MCTS selects the child node of the current node —starting from the root— depending on the empirical reward associated to each child node, and the number of times it has been visited. Denoting respectively $\hat{\mu}_i$ and n_i the empirical reward and the number of visits of the i-th child node, the most usual selection rule, inspired from the Multi-Armed Bandit setting [1], is:

$$\text{Select } i^* = \arg\max\left\{\hat{\mu}_i + C\sqrt{\frac{\log \sum_i n_i}{n_i}}\right\} \tag{1}$$

- When MCTS reaches a leaf node, this node is attached a child node —the tree thus involves N nodes after N tree-walks— and MCTS enters the roll-out phase. This expansion may also occur only every k tree-walk, where k can be referred to as an *expand rate*.
- In the roll-out phase, nodes (a.k.a. actions) are selected using a default (usually randomized) policy, until arriving at a terminal state.
- In this terminal state, the overall reward associated to this tree-walk is computed and used to update the reward $\hat{\mu}_i$ of every node in the tree-walk.

MCTS is frequently combined with the so-called RAVE (Rapid Action Value Estimate) heuristic [9], which stores the average reward associated to each action (averaging all rewards received along tree-walks involving this action). In particular, the RAVE information is used to select the new nodes added to the tree.

4   *BaSCoP*

The considered CP setting significantly differs from the MCTS one. Basically, CP relies on the multiple restart strategy, which implies that it deals with many, mostly narrow, trees. In contrast, MCTS proceeds by searching in a single, gradually growing and eventually very large tree. The CP and MCTS approaches were thus hybridized by attaching average rewards (Sect. 4.1) to each value of a variable (Sect. 4.2). Secondly, *BaSCoP* relies on redefining the selection rule in the bandit- and in the roll-out phases (Sect. 4.3, 4.4).

4.1   Reward

Although an optimization problem is addressed, the value to be optimized — the makespan — is of no direct use in the definition of the reward associated to each tree-walk. Indeed, all but a few of these tree-walks are terminated by a failure, that is an early detection of the infeasibility of an improvement over the last-found schedule. No significant information seems to be usable from the makespan's domain at a failure point. Hence, the depth of a failure is used as a base for the reward, as an indication of its closeness to success, with the following argument: the more variables are assigned the correct value, the deeper the failure.

Since the assignments of a given variable can occur at different depths within a tree and through the successive trees, it seemed reasonable, and was empirically validated, to consider the *relative failure depth* rather than the absolute one: after a tree-walk failing at depth d_f, for each variable v that was assigned, letting d_v be its assignment depth and x the assigned value, a reward $(d_f - d_v)$ is added to the statistics of (v, x).

4.2   RAVE

In the line of [4], the most straightforward option would have been to associate to each node in the tree (that is, a (variable, value) assignment conditioned by the former assignment nodes) the average objective associated to this partial assignment. This option is however irrelevant in the considered setting: the multiple restarts make it ineffective to associate an average reward to an assignment *conditioned by other variable assignments*, since there are not enough tree-walks to compute reliable statistics before they are discarded by the change of context (the new tree). Hence, a radical form of RAVE was used, where statistics are computed for each (variable,value) assignment, independently of the context (previous variables assignments).

4.3   Depth-First-Search Roll-Out

The roll-out phase also presents a key difference with the usual MCTS setting. The roll-out policy launched after reaching a leaf of the MCTS tree usually implements a stochastic procedure, e.g. Monte-Carlo sampling (possibly guided using domain knowledge [4]); the roll-out part of the tree-walk is discarded (not stored in memory) after the reward has been computed.

In the considered CP setting, it is however desirable to make the search complete, i.e. exhaustive given enough time. For this reason, the roll-out policy is set to a depth-first search. As mentioned, in each node the left branch corresponds to setting the variable to its value in the last solution. DFS thus implements a search in the neighborhood of this last solution.

Contrary to random roll-outs, DFS requires node storage; yet only the last path of one DFS need be stored. DFS thus provides a simple and light-storage roll-out policy from which to gather reward statistics. As desired, the use of DFS as roll-out policy within MCTS enforces a complete search, provided that the restart sequence includes a "sufficiently long" epoch.

Overall, the coupling of MCTS with DFS in *BaSCoP* is similar in spirit to the *Interleaved Depth-First Search* [8]; the difference is that *BaSCoP* adaptively explores different regions of the search tree (within a restart).

4.4   Selection Rules

As the left branch associated to each variable corresponds to the value assigned to this variable in the previous best solution, the selection rules determine the neighborhood of the previous solution which is explored in the current tree.

Several rules have been considered:

- **Balanced**: selects alternatively the left and the right node;
- $\epsilon -$ **left**: selects the left node with probability $1 - \epsilon$, and can be seen as a stochastic emulation of Limited Discrepancy Search;
- **UCB**: selects a node according to Eq. (1) with no bias towards the left branch;
- **UCB-Left**: same as UCB, where different constants C_{right} and C_{left} are used to enforce the bias toward the left branch.

Figure 1 illustrates the domains and shapes designed by the selection rule and roll-out policies, by an example using a *Balanced* selection rule and DFS roll-outs.

5   Experimental Results

Figure 2 depicts the overall results in terms of mean relative error w.r.t. the best (non CP-based) solution found in the literature, on the *Taillard 11-20* problem suite (20 × 15), averaged on 11 independent runs, versus the number of tree-walks. The computational cost is ca. 30 mn on a PC with Intel dual-core CPU 2.66 GHz. Compared to DFS, a simple diversification improves only on the early stages, while a left-biased one yields a significant improvement, of the same order as a failure-depth one, and improvements seem to add up when combining both biases.

Overall, *BaSCoP* is shown to match the CP-based state of the art [2]: the use of MCTS was found to compensate for the lack of JSP-specific variable ordering.

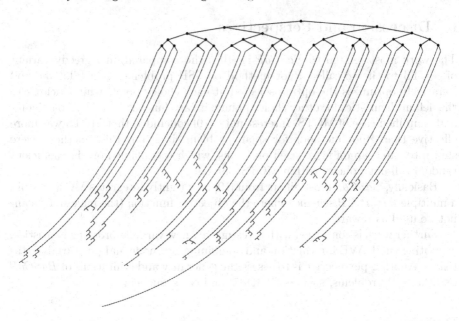

Fig. 1. Balanced + DFS search tree. Concurrent DFS are run under each leaf of the
−growing− MCTS tree (dotted nodes).

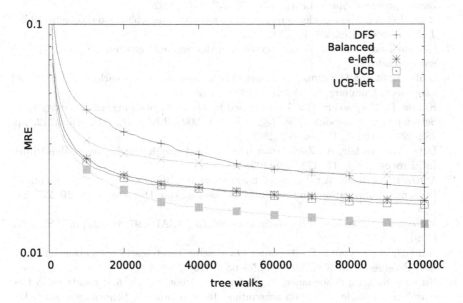

Fig. 2. Mean relative error, for 11 runs on the 10 20×20 Taillard instances.

6   Discussion and Perspectives

The work most related to *BaSCoP* is [10], who compared an ϵ-greedy variant of MCTS to the so-called Pilot method on JSP problems. The Pilot method iteratively optimizes the option selected at each choice point, while sticking to the default heuristics for other choice points: it can thus be viewed as a particular and simplified case of MCTS. Interestingly, [10] concluded that MCTS was more effective than Pilot methods for small problem sizes; but Pilot methods were shown to catch up on large-sized problems, which was blamed on the inefficient random roll-out policies within MCTS.

Basically, *BaSCoP* most differs from [10] as it tightly integrates MCTS within a multiple-restart CP scheme, where the objective function (the makespan) cannot be used as reward.

Further work is concerned with investigating new variable-ordering heuristics, exploiting the RAVE information and combining per-node and per-variable statistics. Another perspective is to assess the generality and limitations of *BaSCoP* on other CP problems, such as BIBD [7] and car sequencing.

References

1. Auer, P., Cesa-Bianchi, N., Fischer, P.: Finite-time analysis of the multiarmed bandit problem. Mach. Learn. **47**(2–3), 235–256 (2002)
2. Beck, J.C.: Solution-guided multi-point constructive search for job shop scheduling. J. Artif. Intell. Res. **29**, 49–77 (2007)
3. Gecode Team: Gecode: Generic constraint development environment, www.gecode.org
4. Gelly, S., et al.: The grand challenge of computer go: monte carlo tree search and extensions. Commun. ACM **55**(3), 106–113 (2012)
5. Kocsis, L., Szepesvári, C.: Bandit based monte-carlo planning. In: Fürnkranz, J., Scheffer, T., Spiliopoulou, M. (eds.) ECML 2006. LNCS (LNAI), vol. 4212, pp. 282–293. Springer, Heidelberg (2006)
6. Luby, M., Sinclair, A., Zuckerman, D.: Optimal speedup of las vegas algorithms. Inf. Process. Lett. **47**, 173–180 (1993)
7. Mathon, R., Rosa, A.: Tables of parameters for BIBD's with $r \leq 41$ including existence, enumeration, and resolvability results. Ann. Discrete Math. **26**, 275–308 (1985)
8. Meseguer P.: Interleaved depth-first search. In: IJCAI 1997, vol. 2, pp. 1382–1387 (1997)
9. Rimmel, A., Teytaud, F., Teytaud, O.: Biasing monte-carlo simulations through RAVE values. In: ICCG 2010, pp. 59–68 (2010)
10. Runarsson, T.P., Schoenauer, M., Sebag, M.: Pilot, rollout and monte carlo tree search methods for job shop scheduling. In: Hamadi, Y., Schoenauer, M. (eds.) LION 2012. LNCS, vol. 7219, pp. 160–174. Springer, Heidelberg (2012)
11. Watson, J.-P., Beck, J.C.: A hybrid constraint programming/local search approach to the job-shop scheduling problem. In: Trick, M.A. (ed.) CPAIOR 2008. LNCS, vol. 5015, pp. 263–277. Springer, Heidelberg (2008)

From Grammars to Parameters: Automatic Iterated Greedy Design for the Permutation Flow-Shop Problem with Weighted Tardiness

Franco Mascia[✉], Manuel López-Ibáñez, Jérémie Dubois-Lacoste, and Thomas Stützle

IRIDIA, CoDE, Université Libre de Bruxelles, Brussels, Belgium
{fmascia,manuel.lopez-ibanez,jeremie.dubois-lacoste,stuetzle}@ulb.ac.be

Abstract. Recent advances in automatic algorithm configuration have made it possible to configure very flexible algorithmic frameworks in order to fine-tune them for particular problems. This is often done by the use of automatic methods to set the values of algorithm parameters. A rather different approach uses grammatical evolution, where the possible algorithms are implicitly defined by a context-free grammar. Possible algorithms may then be instantiated by repeated applications of the rules in the grammar. Through grammatical evolution, such an approach has shown to be able to generate heuristic algorithms. In this paper we show that the process of instantiating such a grammar can be described in terms of parameters. The number of parameters increases with the maximum number of applications of the grammar rules. Therefore, this approach is only practical if the number of rules and depth of the derivation tree are bounded and relatively small. This is often the case in the heuristic-generating grammars proposed in the literature, and, in such cases, we show that the parametric representation may lead to superior performance with respect to the representation used in grammatical evolution. In particular, we first propose a grammar that generates iterated greedy (IG) algorithms for the permutation flow-shop problem with weighted tardiness minimization. Next, we show how this grammar can be represented in terms of parameters. Finally, we compare the quality of the IG algorithms generated by an automatic configuration tool using the parametric representation versus using the codon-based representation of grammatical evolution. In our scenario, the parametric approach leads to significantly better IG algorithms.

Keywords: Automatic algorithm configuration · Grammatical evolution · Iterated greedy · Permutation flow-shop problem

1 Introduction

Designing an effective stochastic local search (SLS) algorithm for a hard optimisation problem is a time-consuming, creative process that relies on the experience

G. Nicosia and P. Pardalos (Eds.): LION 7, LNCS 7997, pp. 321–334, 2013.
DOI: 10.1007/978-3-642-44973-4_36, © Springer-Verlag Berlin Heidelberg 2013

and intuition of the algorithm designer. Recent advances in automatic algorithm configuration have shown that this process can be partially automated, thus reducing human effort. This allows algorithm designers to explore a larger number of algorithm designs than it was previously feasible and, by relying less on human intuition, to explore many design choices that would have never been implemented and tested because they were regarded as supposedly poor design alternatives.

Nowadays, methods for automatic algorithm configuration are able to handle large parameter spaces composed of both categorical and numerical parameters with complex interactions. This capability has enabled researchers to automatically configure flexible frameworks of state-of-the-art SAT solvers [7], and to automatically design multi-objective algorithms [9]. The development of these flexible frameworks follows a *top-down* approach, in which a framework for generating SLS algorithms is built starting from algorithmic components already known to perform well for the problem at hand.

Instead, we consider here a *bottom-up* approach, where an algorithm is assembled from simple components without a priori fixing how they could be combined. There are two recent works that follow such a bottom-up approach. Vázquez-Rodríguez and Ochoa [13] automatically generate by using genetic programming an initial order for the NEH algorithm, a well-known constructive heuristic for the PFSP. More recently, Burke et al. [2] automatically generate iterated greedy (IG) algorithms for the one-dimensional bin packing problem. Both works instantiate algorithms bottom-up from a context-free grammar.

In this paper, we propose to use a parametric representation, using categorical, numerical and conditional parameters, to instantiate algorithms from a grammar. In particular, we show how grammars can be represented in terms of a parametric space, using categorical, numerical and conditional parameters. Such parametric representation exploits the abilities of automatic configuration tools, which are mentioned above. Moreover, the proposed parametric representation avoids known disadvantages of GE, such as low fine-tuning behaviour due to the low locality of the operators used by GE [10]. We apply our proposed approach to the automatic generation of IG algorithms for the permutation flowshop problem (PFSP). The PFSP models many variants of a common kind of production environment in industries. Because of its relevance in practice, the PFSP has attracted a large amount of research since its basic version was formally described decades ago [6]. Moreover, with the exception of few special cases, most variants of the PFSP are $\mathcal{NP}$-hard [5], and, hence, tackling real-world instances often requires the use of heuristic algorithms. For these reasons, the PFSP is an important benchmark problem for the design and comparison of heuristics. When tackling new PFSP variants, the automatic generation of heuristics can save a significant effort.

Finally, we also compare our proposed parametric representation with the codon-based representation used in GE. Our experiments show that, for the particular grammar considered in this paper, the parametric representation produces better heuristics than the GE representation.

This paper is structured as follows. In Sect. 2 we introduce the PFSP problem we tackle. Section 3 presents the methodology we use and describes the mapping from a grammar to a parametric representation. Next, we present our experimental results in Sect. 4 and we conclude in Sect. 5.

2   Permutation Flowshop Scheduling

The flowshop scheduling problem (FSP) is one of the most widely studied scheduling problems, as it models a very common kind of production environment in industries. The goal in the FSP is to find a schedule to process a set of n jobs $(J_1, \ldots, J_n)$ on m machines $(M_1, \ldots, M_m)$. The specificity of flowshop environments is that all jobs must be processed on the machines in the same order, i.e., all jobs have to be processed on machine M_1, then machine M_2, and so on until machine M_m. A common restriction in the FSP is to forbid job passing between machines, i.e., to restrict to solutions that are permutations of jobs. The resulting problem is called permutation flowshop scheduling problem (PFSP). In the PFSP, all processing times p_{ij} for a job J_i on a machine M_j are fixed, known in advance, and non-negative. In what follows, C_{ij} denotes the completion time of a job J_i on machine M_j and C_i denotes the completion time of a job J_i on the last machine, M_m.

In many practical situations, for instance when products are due to arrive at customers at a specific time, jobs have an associated *due date*, denoted here by d_i for a job J_i. Moreover, some jobs may be more important than others, which can be expressed by a weight associated to them representing their priority. Thus, the so-called *tardiness* of a job J_i is defined as $T_i = \max\{C_i - d_i, 0\}$ and the *total weighted tardiness* is given by $\sum_{i=1}^{n} w_i \cdot T_i$, where w_i is the priority assigned to job J_i.

We consider the problem of minimizing the total weighted tardiness (WT). This problem, which we call *PFSP-WT*, is $\mathcal{NP}$-hard in the strong sense even for a single machine [3]. Let π_i denote the job in the ith position of a permutation π. Formally, the *PFSP-WT* consists of finding a given job permutation π as to

$$
\begin{aligned}
\text{minimize} \quad & F(\pi) = \sum_{i=1}^{n} w_i \cdot T_i \\
\text{subject to} \quad & C_{\pi_0 j} = 0 \quad j \in \{1, \ldots, m\}, \\
& C_{\pi_i 0} = 0 \quad i \in \{1, \ldots, n\}, \\
& C_{\pi_i j} = \max\{C_{\pi_{i-1} j}, C_{\pi_i j-1}\} + p_{ij}, \\
& \qquad i \in \{1, \ldots, n\} \quad j \in \{1, \ldots, m\}. \\
& T_i = \max\{C_i - d_i, 0\} \quad i \in \{1, \ldots, n\}.
\end{aligned}
\tag{1}
$$

We tackle the *PFSP-WT* by means of iterated greedy (IG), which has been shown to perform well in several PFSP variants [11]. Our goal is to *automatically* generate IG algorithms in a bottom-up manner from a grammar description. The next section explains our approach in more detail.

3   Methods

The methodology used in this work is the following. Given a problem to be tackled, we first define a set of algorithmic components for the problem at hand, avoiding as much as possible assumptions on which component will contribute the most to the effectiveness of the algorithm or which is the best way of combining the components in the final algorithm. Once the building blocks are defined, we use tools for automatic algorithm configuration to explore the large design space of all possible combinations and select the best algorithm for the problem at hand.

The size and complexity of the building blocks is set at a level that is below a full-fledged heuristic, but still allows us to easily combine them in a modular way to generate a very large number of different algorithms. This is in contrast with the more standard way of designing SLS algorithms for a given problem, in which the algorithm designer defines the full-fledged heuristics and leaves out some parameters to tune specific choices within the already defined structure.

In this paper, the building blocks and the way in which they can be combined will be described by means of context-free grammars. Grammars comprise a set of rules that describe how to construct sentences in a language given a set of symbols. The grammar discussed in this paper generates algorithm descriptions in pseudo-code, the actual grammar used in the experiments is equivalent to the one presented in this paper but generates directly C++ code.

3.1   The Grammar for PFSP

The PFSP has been the object of many studies over the past decades, and it is still attracting a significant amount of research nowadays. Recent studies [4, 11] have shown that many high-performing algorithms to tackle the PFSP (whatever is the objective to optimize) are based on the *iterated greedy* (IG) principle. IG consists of the iterative partial "destruction" of the current solution, and its "reconstruction" into a full solution afterwards. The term "greedy" comes from the fact that the reconstruction of the solution is often done using a greedy heuristic. In the case of the PFSP, the destruction phase removes a number of jobs from the schedule. The reconstruction phase inserts these jobs back in the solution, to obtain again a complete solution.

In this work, we define a grammar for generating IGs for the PFSP in which we allow several underlying heuristic choices. State-of-the-art IG algorithms for the PFSP from the literature always apply a local search step to each full solution after the reconstruction phase. However, such local search step would hide performance differences when using different choices for the other components of IG. Our goal is not to generate a state-of-the-art IG for the PFSP, but rather to study different methods for the automatic generation of algorithms, and, thus, in this paper we do not apply any local search step.

Figure 1 shows the grammar for generating the main step of the IG algorithm in Backus–Naur Form (BNF). In BNF, production rules are in the form `<non-terminal> ::= `*expression*. Each rule describes how the non-terminal

```
       <program> ::= procedure ig_step()
                         <select_jobs>
                         <select_more>
                         remove_selected()
                         sort_removed_jobs(<order_criteria> <tie_breaking>)
                         insert_jobs(insert_criteria)

   <select_more> ::= <select_jobs> <select_more>
                   | ""

   <select_jobs> ::= select_jobs(<heuristic>, <num>, <low_range>, <high_range>)

     <heuristic> ::= priority | position | sumProcessingTimes | dueDate
                   | tardiness | waitingTime | idleTime

          <num> ::= [0..100]

    <low_range> ::= [0..99]

   <high_range> ::= [0..100]

 <tie_breaking> ::= , <order_criteria> <tie_breaking>
                   | ""

<order_criteria> ::= order(<comparator>, <heuristic>)

   <comparator> ::= "<" | ">"

<insert_criteria> ::= weightedTardiness
                    | weightedTardiness, sumCompletionTimes
                    | weightedTardiness, sumCompletionTimes, weightedEarlyness
                    | weightedTardiness, weightedEarlyness
                    | weightedTardiness, weightedEarlyness, sumCompletionTimes
```

Fig. 1. Grammar that describes the rules for generating IG algorithms for the PFSP.

symbol on the left-hand side can be replaced by the expression on the right-hand side. Expressions are strings of terminal and/or non-terminal symbols. If there are alternative strings of symbols for the replacement of the non terminal on the left-hand side, the alternative strings are separated with the symbol "|".

In Fig. 1, the non-terminal symbol <program> defines the main step of the algorithm. First one or more jobs are marked for removal from the current solution, then the selected jobs are removed and sorted, and finally the solution is reconstructed inserting the jobs back in the current solution. Implementing an IG algorithm for the PFSP requires to make some design choices. In particular, (i) which jobs and how many are selected for removal, (ii) in which order the jobs are reinserted; and (iii) which criteria should be optimized when deciding the insertion point. All the possibilities that we consider in this paper are described by the grammar in Fig. 1. Next, we explain these components in detail.

Heuristics for the Selection of Jobs. The selection of jobs for removal (rule <select_jobs>) consists in the application of one or more selection rules. In particular this is done with the function select_jobs (<heuristic>, <num>, <low_range>, <high_range>) that selects <num> jobs from the current solution according to the rule specified in <heuristic>. Each rule computes a numerical value for each job J_i, which may be one of the following properties:

- Priority: the weight w_i that defines its priority;
- DueDate: its due date d_i;
- SumProcessingTimes: the sum of its processing times, $\sum_{j=1}^{m} p_{ij}$;
- Position: its position in the current solution;
- Tardiness: its tardiness in the current solution;
- WaitingTime: its waiting time between machines computed as $\sum_{j=2}^{m} C_{\pi_i j} - C_{\pi_i j-1} - p_{\pi_i j}$;
- IdleTime: the time during which machines are idle because the job is still being processed on a previous machine, that is, $\sum_{j=1}^{m} C_{\pi_i j} - C_{\pi_{i-1} j} - p_{\pi_i j}$, for $i \neq \pi_1$.

After the heuristic values are computed, they are normalized in the following way: the minimum for each heuristic value among all jobs is normalized to 0, the maximum one to 100, and values in-between are normalized linearly to the range $[0, 100]$. Only jobs whose normalized heuristic value is between a certain range $[low, high]$ are considered for selection. The range is computed from the values given by the grammar as $high = $ <high_range> and $low = $ <low_range> $\cdot high/100$. Finally, from the jobs considered for selection, at most <num> percent (computed as <num> $\cdot n/100$) of the jobs are actually selected, where n is the total number of jobs. An example of selection rule would be select_jobs(DueDate,20,10,50), which means that, from those jobs that have a normalized due date in the range $[10, 50]$, at most $0.2 \cdot n$ jobs are selected.

Rules for Ordering the Jobs. The function that sorts jobs for re-insertion (sort_removed_jobs) is composed by one or more order criteria (<order_criteria>), where each additional order criterion is used for breaking ties. Each order criterion sorts the removed jobs by a particular heuristic value, in either ascending or descending order, according to <comparator>. The result is a permutation of the removed jobs according to the order criteria.

Rules for Inserting the Jobs. In this paper we consider the minimization of the weighted tardiness of the solution, thus, it is natural to optimize primarily this objective when choosing the position for re-inserting each job. However, it often happens that the weighted tardiness is the same for any insertion position of a job (in particular, when the solution is partial: all jobs can easily respect due dates and therefore the weighted tardiness is 0).

Thus, we consider the possibility of breaking ties according to additional criteria, namely, the minimization of the *sum of completion times* and the maximization of the *weighted earliness*, computed as $\sum_{i=1}^{n} w_i \cdot (d_i - C_i)$. Both are correlated with the minimization of the weighted tardiness and allow us to differentiate between partial schedules with zero weighted tardiness because none of the jobs is tardy. In total, we consider five alternatives for the insertion criteria (<insert_criteria>), corresponding to breaking ties with any combination of either, none or both sum of completion times and weighted earliness.

Table 1. Parametric representation of the grammar in Fig. 1

Parameter	Domain	Condition
$select\_jobs_1$	{Priority, Position, SumProcessingTimes, DueDate, Tardiness, WaitingTime, IdleTime}	
num_1	[0,100]	
low_range_1	[0,99]	
$high\_range_1$	[0,100]	
$select\_jobs_2$	{Priority, Position, SumProcessingTimes, DueDate, Tardiness, WaitingTime, IdleTime, ""}	
num_2	[0,100]	if $select\_jobs_2 \neq$ ""
low_range_2	[0,99]	if $select\_jobs_2 \neq$ ""
$high\_range_2$	[0,100]	if $select\_jobs_2 \neq$ ""
...	...	
$select\_jobs_i$	{Priority, Position, SumProcessingTimes, DueDate, Tardiness, WaitingTime, IdleTime, ""}	if $select\_jobs_{i-1} \neq$ ""
num_i	[0,100]	if $select\_jobs_i \neq$ ""
low_range_i	[0,99]	if $select\_jobs_i \neq$ ""
$high\_range_i$	[0,100]	if $select\_jobs_i \neq$ ""
$order\_criteria_1$	{Priority, Position, SumProcessingTimes, DueDate, Tardiness, WaitingTime, IdleTime}	
$comparator_1$	{"<", ">"}	
$order\_criteria_2$	{Priority, Position, SumProcessingTimes, DueDate, Tardiness, WaitingTime, IdleTime, ""}	
$comparator_2$	{"<", ">"}	if $order\_criteria_2 \neq$ ""
...		
$order\_criteria_j$	{Priority, Position, SumProcessingTimes, DueDate, Tardiness, WaitingTime, IdleTime, ""}	if $order\_criteria_{j-1} \neq$ ""
$comparator_j$	{"<", ">"}	if $order\_criteria_j \neq$ ""
$insert\_criteria$	{"WaitingTime", "WaitingTime, SumCompletionTimes", "WaitingTime, SumCompletionTimes, WeightedEarlyness", "WeightedEarlyness, WeightedEarlyness", "WaitingTime, WeightedEarlyness, SumCompletionTimes"}	

3.2 From Grammars to Parameters

To tune the algorithms with a tool for automatic algorithm configuration, we need to define the process of instantiating a grammar as a choice between alternative parameter settings. Table 1 is a possible parametric representation of the grammar given in Fig. 1. We now explain in detail how the parametric representation was obtained.

First, rules that do not contain alternatives do not require a parameter. Second, numeric terminals, such as <num>, <low_range> and <high_range> in Fig. 1 can be naturally represented by numerical parameters with a defined range. Third, rules with alternative choices are represented as categorical parameters. This is especially natural in the case of rules that consist only of alternative terminals, such as <insert_criteria>.

The only difficulty appears if the same rule can be applied more than once, for example, rules <select_jobs> and <tie_breaking>. In such a case, each application requires its own parameter. Some of these rules might be applied an infinite number of times, and, thus, they might seem to require an infinite number of parameters. However, when generating algorithms from grammars, such rules are never applied more than a small number of times. We use this consideration and explicitly limit the number of parameters that describe such rules; thus, in this way we also limit the length of the generated algorithm.

Converting rules that can be derived an unbounded number of times is the non trivial case, and we will explain it here with an example. Assume we want to map the following rule to a set of categorical parameters:

```
<select_jobs> ::= <a_job> | <a_job> <select_jobs>
      <a_job> ::= criterion1 | criterion2
```

What is expressed by the rule is that a valid program contains a list of at least one criterion. Suppose we want to limit the number of rule-applications to five, then the rule could be converted into five categorical parameters with possible values `criterion1` or `criterion2`. This mapping leads to exactly five criteria. To have *at most* five, the parameters should consider also the empty string among the possible values. The corresponding grammar would be the following:

```
<select_jobs> ::= <a_job> <a_job> <a_job> <a_job> <a_job>
      <a_job> ::= criterion1 | criterion2 | ""
```

In order to have *at least* one job, the first parameter should not have the empty string among the possible values. This would more directly map to the following equivalent grammar:

```
<select_jobs>  ::= <a_job> <further_jobs>
<further_jobs> ::= "" | <a_job> <further_jobs>
      <a_job>  ::= criterion1 | criterion2
```

Table 1 shows the mapping of Fig. 1 to parameters. Both rules `<select_jobs>` and `<order_criteria>` can be applied up to i and j times respectively. Moreover, each parameter used in those rules has to be duplicated for each possible application of the rules.

3.3   From Grammars to Sequences of Integers

How to search for the best algorithm in the design space defined by the grammar and how to represent the sequence of derivation rules that represent an algorithm is the goal of different methods in grammar based genetic programming (GBGP) [10]. Among the GBGP techniques proposed in the literature, we consider recent works in grammatical evolution (GE) [2].

In GE, the instantiation of a grammar is done by starting with the `<program>` non-terminal symbol, and successively applying the derivation rules in the grammar, until there are no non-terminal symbols left. Every time that a non-terminal symbol can be replaced following more than one production rule, a choice has to be made. The sequence of specific choices made during the derivation, which leads to a specific program, is encoded in a sequence of integers.

This linearisation of the derivation tree, leads to a high decoupling between the sequence of integers and the programs being generated. For example, when a derivation is complete and there are still numbers left in the sequence, these numbers are discarded. Conversely, if the derivation is not complete and there are no numbers left in the sequence, the sequence is read again from the beginning. This operation is called wrapping and is repeated for a limited number of times. If after a given number of wrappings the derivation is not complete, the sequence of strings is considered to lead to an invalid program. Moreover, since the integers are usually in a range which is bigger than the possible choices for the derivation of a non terminal, a modulo operation is applied at each choice.

In GE, the sequences of integers are used as chromosomes in a genetic algorithm that is used to derive the best algorithm for a given problem. The high decoupling between the programs and their representation, has it drawbacks when used within a genetic algorithm. The decoupling translates to non locality in the mutation and crossover operators [10]. Wrapping operations are clearly responsible of this decoupling, but even without wrapping, the way in which an algorithm is derived from a grammar and the sequence of integer values leads to non locality in the operation. In fact, since the integer values are used to transform the left-most non terminal symbol, a choice in one of the early transformations can impact on the structure of the program being generated and on the meaning of all subsequent integers in the sequence. Therefore since a mutation on one integer in the sequence (a codon) impacts on the meaning of all the following codons, one-bit mutations in different positions of the individual genotype have impacts of different magnitude on the phenotype of the individual. For the same reason the offspring of two highly fit parents is not necessarily composed of highly fit individuals. On the contrary a one-point cross-over of the best individuals in the population could lead to individuals whose genotype can not be translated to any algorithm, because of the upper-bound on the wrapping operations. But, regardless of the specific issues when used in a genetic algorithm, we are interested to see if this representation presents similar drawbacks also when used with a tool for automatic algorithm configuration. In fact, this linearisation of the grammar, can easily be used within a tool for algorithmic configuration by mapping all codons to categorical parameters. The choice here between integer and categorical parameters is due to the high non linear response between the values of the codons and the algorithm they are decoded into.

Both the parameters and the sequence of codons limit the length of the algorithms that can be generated. In fact, a grammar can represent an arbitrarily long algorithm, but in practice the length is limited by the number of parameters in one case, and in the other case by the number of possible wrapping operations.

4   Experimental Results

4.1   Experiments

The automatic configuration procedure used in this work is irace [8], a publicly available implementation of Iterated F-Race [1]. Iterated F-Race starts by sampling a number of parameter configurations uniformly at random. Then, at each iteration, it selects a set of elite configurations using a racing procedure and the non-parametric Friedman test. This racing procedure runs the configurations iteratively on a sequence of (training) problem instances, and discards configurations as soon as there is enough statistical evidence that a configuration is worse than the others. After the race, the elite configurations are used to bias a local sampling model. The next iteration starts by sampling new configurations from this model, and racing is applied to these configurations together with the previous elite configurations. This procedure is repeated until a given budget of runs is exhausted. The fact that irace handles categorical, numerical and surrogate parameters with complex constraints makes it ideal to instantiate algorithms from grammars in the manner proposed in this paper.

Benchmark Sets. We generated two benchmark sets of PFSP instances: 100 instances of 50 jobs and 20 machines (50x20), and 100 other instances of 100 jobs and 20 machines (100x20). These two sizes are nowadays the most common ones in the literature to evaluate heuristic algorithms on various PFSP variants. The processing times of the jobs on each machine are drawn from a discrete uniform distribution $\mathcal{U}\{1, \ldots, 99\}$ [12]. The weights of the jobs are generated at random from $\mathcal{U}\{1, \ldots, 10\}$, and each due date d_i is generated in a range proportional to the sum of processing times of the job J_i as: $d_i = \lfloor r \cdot \sum_{j=1}^{m} p_{ij} \rfloor$, where r is a random number sampled from the continuous uniform distribution $\mathcal{U}(1, 4)$.

Experimental Setup. We compare the quality of the heuristics generated by irace when using either the grammar representation used by GE (irace-ge) or the parametric representation given in Table 1 (irace-param). In irace-ge an algorithm is derived from the grammar by means of 30 codons, which are mapped to 30 integer parameters that can assume values in the range [0, 100]. For the parametric representation given in Table 1, we need to specify the number of times the select_jobs and order_criteria rules are applied (i and j, respectively). Large values give more flexibility to the automatic configuration tool to find the best heuristics, however, they also enlarge the space of potential heuristics. We study three possibilities: irace-param5, which uses $i = 5$, $j = 3$; irace-param3, which uses $i = 3$, $j = 3$, and irace-param1, which uses $i = 1$, $j = 1$. The first variant is larger than what we expect to be necessary, and its purpose is to test if irace can find shorter heuristics than the maximum bounds. The purpose of the last variant is to verify that more than one application per rule is necessary to generate good results.

Each run of irace has a maximum budget of 2 500 runs of IG, and each run of IG is stopped after $0.001 \cdot n \cdot m$ s.

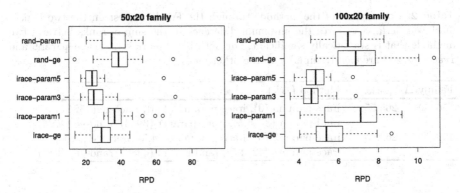

Fig. 2. Mean relative percentage deviation (RPD) obtained by the heuristics generated by each tuning method. Results are given separately for the heuristics trained and tested on 50x20 instances and on 100x20 instances.

Using the same computational budget, we also consider two additional methods that generate heuristics randomly, to use as a baseline comparison. These methods generate 250 IG heuristics randomly, run them on 10 randomly selected training instances and select the heuristic that obtains the lowest mean value. Method `rand-ge` uses the grammar representation, while method `rand-param` uses the parametric representation.

Each method (`irace-ge`, `irace-param5`, `irace-param3`, `irace-param1`, `rand-ge`, `rand-param`) is repeated 30 times with different random seeds for each benchmark set, that is, in total, each method generates 60 IG heuristics. The training set used by all methods are the first 90 instances of each size. A grammar equivalent to the one in Fig. 1 is used to generate directly C++ code, which is in turn compiled with GCC 4.4.6 with optimization level -O3. Experiments were run on a single core of an AMD Opteron 6272 CPU (2.1 GHz, 16 MB L2/L3 cache size) running under Cluster Rocks Linux version 6/CentOS 6.3, 64 bits.

4.2   Results

For assessing the quality of the generated heuristics, we run them on 10 test instances (that are distinct from the ones used for the training), repeating each run 10 times with different random seeds. Next, we compute the relative percentage deviation (RPD) from the best solution ever found by any run for each instance. The RPD is averaged over the 10 runs and over the 10 test instances.

Figure 2 compares the quality of the heuristics generated by the four methods described above on each test set for 50x20 and 100x20 benchmark sets. For each method, we plot the distribution of the mean RPD of each heuristic generated by it. In both benchmark sets, `irace-param5` and `irace-param3` obtain the best heuristics. The heuristics generated by `irace-param1` are typically worse than those generated by `irace-ge`.

Pairwise comparisons using the Wilcoxon signed rank test indicate that all pair-wise differences are statistically significant with the only exception being the

Table 2. Comparison of the methods through the Friedman test on the two benchmark sets. $\Delta R_{\alpha=0.95}$ gives the minimum difference in the sum of ranks between two methods that is statistically significant. For both benchmark sets, `irace-param5` and `irace-param3` are clearly superior to the other methods.

Family	$\Delta R_{\alpha=0.95}$	Method (ΔR)
50x20	265.55	irace-param5 (0), irace-param3 (453), irace-ge (1574.5), rand-param (3706.5), irace-param1 (3976), rand-ge (4705)
100x20	262.85	irace-param3 (0), irace-param5 (474.5), irace-ge (2106), irace-param1 (4214.5), rand-param (4380), rand-ge (4917)

pair `rand-param` and `irace-param1`. Moreover, we compare the different methods using the Friedman test in Table 2. Both `irace-param5` and `irace-param3` are ranked much better than `irace-ge`, thus confirming the superiority of the parameterised representation in our case studies.

When analysing the heuristics generated by each method, we observe that both `irace-ge` and `rand-ge` generate on average around three `<select_jobs>` rules and no more than 2.23 `<order_criteria>` rules. On the other hand, `irace-param5` and `rand-param` generate on average more than 4.5 `<select_jobs>` rules and more than 2.5 `<order_criteria>` rules. This suggests to us that the GE-based representation has trouble generating programs with as many rules as the parametric representation. The results obtained by `irace-param1` and `irace-param3` suggest also that at least three rules are necessary to obtain good results. In terms of the particular heuristics generated, we observe that most heuristics contain a rule that selects jobs according to idle time. Perhaps more surprising is that the most common order criteria for sorting removed jobs is by position. On the other hand, there is no clear winner among the `insert_criteria` methods, which suggests that breaking ties in some particular order is not advantageous. A complete analysis of the heuristics is not the purpose of this paper, but our results indicate that the heuristics generated by irace are quite different from what a human expert would consider when designing a similar algorithm.

5   Conclusions

The main conclusion from our work is that existing automatic configuration tools may be used to generate algorithms from grammars.

Defining algorithmic components to be combined in an SLS algorithm presents several advantages over the design of a full fledged heuristic, where some design choices are left to be tuned automatically. Most importantly, less intuition, and therefore less bias of the designer goes in the definition of the separate blocks with respect to the classical top-down approach. But there are also more practical advantages of following a bottom-up strategy. In fact, every instantiation of the grammar is a minimal SLS algorithm designed and implemented to have a very specific behaviour. Less programming abstractions are needed, and a simpler code may be optimised more easily by the compilers. Even the parameters become constant values in the source code, and there is no need to pass them to various parts of the algorithm. On the contrary, when following a top-down approach, the designer tackles the hard engineering task of designing a full-fledged framework where all possible combinations of design choices have to be defined beforehand. This leads to a reduced number of possible combinations with respect to a modular bottom-up approach, and also to the added complexity of intricate conditional expressions required to instantiate only the parts of the framework needed to express a specific algorithm.

We have shown that it is possible to represent the instantiation of the grammar by means of a parametric space. The number of parameters required is proportional to the number of times a production rule can be applied, and, hence, our approach is more appropriate for grammars where this number is bounded and not excessively large. It is an open research question for which kind of grammars the number of parameters required to represent applications of production rules becomes prohibitively expensive and other representations are more appropriate. Nonetheless, the grammar used in this work is similar in this respect to others that can be found in the literature, and, hence, we believe that grammars, where production rules are to be applied only a rather limited number of times are common in the development of heuristic algorithms.

From our experimental results, the heuristics generated by irace when using the parametric representation achieve better results than those generated when using the GE representation. This indicates that the parametric representation can help to avoid disadvantages of grammatical evolution such as a low fine-tuning behaviour due to a low locality of the used operators. Furthermore, our approach is not limited to irace and it can be applied using other automatic configuration tools, as long as they are able to handle categorical and conditional parameters.

In future work, we plan to compare our approach with a pure GE algorithm, which is the algorithm used in previous similar works. Moreover, our intention is to test the proposed method on different grammars and benchmark problems to investigate its benefits and limitations.

Acknowledgments. This work was supported by the META-X project, an *Action de Recherche Concertée* funded by the Scientific Research Directorate of the French Community of Belgium. Franco Mascia, Manuel López-Ibáñez and Thomas Stützle acknowledge support from the Belgian F.R.S.-FNRS. Jérémie Dubois-Lacoste acknowledges support from the MIBISOC network, an Initial Training Network funded by the European Commission, grant PITN–GA–2009–238819. The authors also acknowledge support from the FRFC project *"Méthodes de recherche hybrids pour la résolution de problèmes complexes"*. This research and its results have also received funding from the COMEX project within the Interuniversity Attraction Poles Programme of the Belgian Science Policy Office.

References

1. Balaprakash, P., Birattari, M., Stützle, T.: Improvement strategies for the F-Race algorithm: sampling design and iterative refinement. In: Bartz-Beielstein, T., Blesa Aguilera, M.J., Blum, C., Naujoks, B., Roli, A., Rudolph, G., Sampels, M. (eds.) HCI 2007. LNCS, vol. 4771, pp. 108–122. Springer, Heidelberg (2007)
2. Burke, E.K., Hyde, M.R., Kendall, G.: Grammatical evolution of local search heuristics. IEEE Trans. Evol. Comput. **16**(7), 406–417 (2012)
3. Du, J., Leung, J.Y.T.: Minimizing total tardiness on one machine is NP-hard. Math. Oper. Res. **15**(3), 483–495 (1990)
4. Dubois-Lacoste, J., López-Ibáñez, M., Stützle, T.: A hybrid TP+PLS algorithm for bi-objective flow-shop scheduling problems. Comput. Oper. Res. **38**(8), 1219–1236 (2011)
5. Garey, M.R., Johnson, D.S., Sethi, R.: The complexity of flowshop and jobshop scheduling. Math. Oper. Res. **1**, 117–129 (1976)
6. Johnson, D.S.: Optimal two- and three-stage production scheduling with setup times included. Naval Res. Logistics Quart. **1**, 61–68 (1954)
7. KhudaBukhsh, A.R., Xu, L., Hoos, H.H., Leyton-Brown, K.: SATenstein: automatically building local search SAT solvers from components. In: Boutilier, C. (ed.) Proceedings of the Twenty-First International Joint Conference on Artificial Intelligence (IJCAI-09), pp. 517–524. AAAI Press/International Joint Conferences on Artificial Intelligence, Menlo Park (2009)
8. López-Ibáñez, M., Dubois-Lacoste, J., Stützle, T., Birattari, M.: The irace package, iterated race for automatic algorithm configuration. Technical report TR/IRIDIA/2011-004, IRIDIA, Université Libre de Bruxelles, Belgium (2011)
9. López-Ibáñez, M., Stützle, T.: The automatic design of multi-objective ant colony optimization algorithms. IEEE Trans. Evol. Comput. **16**(6), 861–875 (2012)
10. Mckay, R.I., Hoai, N.X., Whigham, P.A., Shan, Y., O'Neill, M.: Grammar-based genetic programming: a survey. Genet. Program. Evolvable Mach. **11**(3–4), 365–396 (2010)
11. Ruiz, R., Stützle, T.: A simple and effective iterated greedy algorithm for the permutation flowshop scheduling problem. Eur. J. Oper. Res. **177**(3), 2033–2049 (2007)
12. Taillard, É.D.: Benchmarks for basic scheduling problems. Eur. J. Oper. Res. **64**(2), 278–285 (1993)
13. Vázquez-Rodríguez, J.A., Ochoa, G.: On the automatic discovery of variants of the NEH procedure for flow shop scheduling using genetic programming. J. Oper. Res. Soc. **62**(2), 381–396 (2010)

Architecture for Monitoring Learning Processes Using Video Games

N. Padilla-Zea[✉], J. R. Lopez-Arcos, F. L. Gutiérrez-Vela,
P. Paderewski, and N. Medina-Medina

GEDES Research Group, University of Granada, Granada, Spain
{npadilla,jrlarco,fgutierr,patricia,nmedina}@ugr.es

Abstract. PLAGER-VG is an architecture used to design, execute, analyze and adapt educational processes supported by video games, especially those that include collaborative activities and which use collaborative learning techniques. In this paper, we have focused on the monitoring and adaptive processes in order to customize activities both within the game and the learning process to improve the results obtained from using these collaborative video games. To perform these processes we propose a mechanism based on the use of a set of specialized agents included in this architecture to collect relevant information and to process it in order to obtain the necessary adaptation actions.

Keywords: Video games · Architecture · Learning processes

1 Introduction

The incorporation of Game Based Learning (GBL) into learning processes has already become a reality, both in schools and for research purposes. In particular, playing these games in groups has been shown to be a desirable way of accomplishing this, as players are used to playing commercial video games in groups. By combining these two aspects, we have focused our research on the use of group activities within Educational Video Games (EVG) in order to promote collaborative skills in students and to promote the many advantages that both elements involve.

Although several aspects of designing and using EVG have been developed in our research, in this paper we have fixed the focus on the personalization of the learning/playing process. Assuming that an educational video game includes recreational activities that hide some educational content, the monitoring and adapting of the game are closely related to the monitoring and adapting of the learning process. Starting from this assumption, we think that it is necessary to monitor relevant activities in the game in order to analyze and adapt it according to the features of the player or the group who is playing.

In this context of using collaborative learning and EVG as tools for teaching, we think our previously proposed architecture PLAGER-VG [1] (PLAtform for managinG Educational multiplayeR Video Games) needs to be modified. PLAGER-VG helps teachers and designers to obtain more efficient video games and is able to monitor and to adapt the learning processes underpinned by them.

G. Nicosia and P. Pardalos (Eds.): LION 7, LNCS 7997, pp. 335–340, 2013.
DOI: 10.1007/978-3-642-44973-4_37, © Springer-Verlag Berlin Heidelberg 2013

The problem of monitoring and analyzing learning processes is of particular interest in Computer Supporter Collaborative Learning (CSCL) environments where learning is usually based on interaction and communication among students.

Many CSCL systems offer functions to study the way in which collaboration takes place. For this purpose, they usually record the actions carried out with the interactions with the system [2], the communication between collaborators [3], and/or the changes carried out in the shared workspaces [4]. However, we have found that more information is needed to contextualize the performed actions and that automatic mechanisms are needed to collect this information and to adapt the games.

Thus, we propose a modification of the architecture PLAGER-VG using agents as active entities in order to collect information generated by the collaborative activities performed by the players during the game. In addition, these agents help teachers to improve the learning process by proposing adaptation actions. An adaptation action allows some aspects in the game to be changed, for example, modifying (decreasing) the difficulty level if the player or the group is unable to overcome a challenge in the stated time.

In this paper, we use the terms "game" and "video game" as synonyms.

2 Architecture PLAGER-VG

The architecture PLAGER-VG [1] is composed of five interrelated and interconnected subsystems: Personalization Sub-system, Design Sub-system, Groups Sub-system, Monitoring Sub-system and Game Sub-system. Both the educational and recreational contents are designed using functionalities included in the Design Sub-system. Components designed as a result of this process are stored in a central repository, as can be seen in Fig. 1.

The *Personalization Sub-system* accesses the designed elements and customizes them according to the needs of each of the users. Such changes, which customize both the learning and game processes, are also reflected in the central repository.

The *Personalization Sub-system* communicates with the *Game Sub-system* which, given a set of educational specifications for a student or group, generates a personal *Game Instance*, which is run in accordance with the educational restrictions.

During the game, a set of *"interesting events"* occurs, from both an educational and recreational viewpoint, which are collected for processing by the *Monitoring Sub-system*. As a result, the *Monitoring Sub-system* generates a set of recommendations, which are reported to both the *Personalization Sub-system* and the central repository.

The *Personalization Sub-system* adapts the game and learning features to each of the users and is therefore responsible for implementing these recommendations.

Finally, the *Groups Sub-system* manages both the design and creation of groups and stores this information in the central repository. Information about groups allows both the *Personalization Sub-system*, *Monitoring Sub-system* and *Game Sub-system* to manage collaborative activities.

Although the PLAGER-VG architecture is composed of five interrelated and interconnected subsystems, in this paper we are going to focus our attention on the *Monitoring* and the *Personalization Sub-system*.

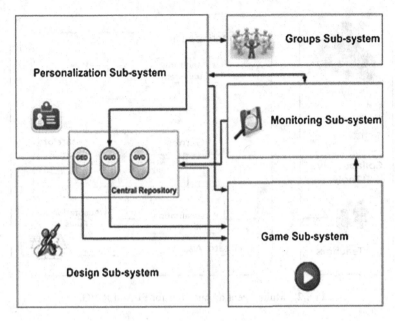

Fig. 1. PLAGER-VG architecture

3   Using Agents to Retrieve Relevant Information

In order to analyze learning processes and to improve them accordingly, we need to select the information to be studied carefully, but also to decide how to process and to use it.

Based on our previous works [5, 6], we have decided to use agents in order to obtain the functions to monitor and adapt the game.

In those works we presented an architecture for dynamic and evolving cooperative software agents. We defined a model that allowed communication between agents and preserved system activity to take place while it was running. A central blackboard was used to communicate and coordinate agents and to store the information needed by those agents. This blackboard was controlled by special agents with specific functions to store and retrieve information when needed and to perform evolution actions over a software system.

Following this idea, and to facilitate the analysis of the learning processes, we propose to include two types of specialized agents in the architecture (PLAGER-VG), which will be responsible for monitoring (*Monitoring Agents*) and for providing information to teachers about the activities to be performed in the game and by the groups to adapt and improve the learning process (*Facilitator Agent*). Figure 2 shows the integration of these agents with other elements in the architecture.

Since EVG aim to provide implicit learning, we need to establish a relationship between what a student is doing in the game and the learning implicit in such an activity. To do so, we need to determine what activities are relevant in the game and which information is relevant for each of them.

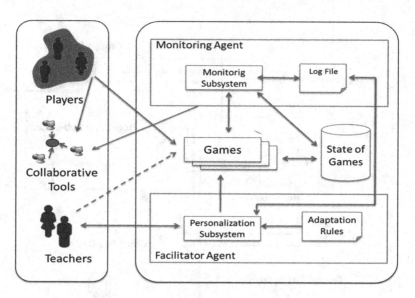

Fig. 2. Multi-Agent architecture for PLAGER-VG.

Thus, we use *interesting events*. These events can be individual, if information is only related to one player; or for a group, if the activity is being done by several players and includes information about interaction, in addition to that related to the learning process. The information that the agent has to collect regarding each event is: event identification, game identification, game mode, video game task, educational task, educational goal, task starts, success/failure, added score, player, group and task end.

Information collected by agents while the game is running is stored in a global structure, called *State of Game*. This structure contains details about every interesting event and allows us to analyze sequences of the game in order to study patterns of behavior or repetitive actions.

In our platform, players have a set of tools (Synchronous Communication Tools, Asynchronous Communication Tools, Information about the group members, Scheduler, Voting system, Map of the game, Common Warehouse) that enable group communication, coordination and improve awareness.

Using these types of tools, players can maintain contact with their group partners to do the group activities but also to obtain advice or information while performing individual activities, if they need help to perform them. Therefore, all the previously stated elements could appear in both individual and group activities.

One of the most important problems is how to identify interesting events. The Monitoring Agent uses a set of events based on the 3 C's model [7], and classifies them according to a message classification that we have proposed be adapted to interaction analysis in EVG. This classification falls outside the scope of this paper, but can be found in [8].

In general, making an automatic classification of messages produced during the game is difficult. A way of reducing this difficulty is to define the set of tools to be used and how this classification can be made by using them. In our system, defining specialized agents with specific information associated to them makes the process easier, because the frame of information that every agent collects includes what kind of event it is.

Facilitator Agents use some analysis mechanisms based on Social Network Analysis [9] (SNA), focusing the analysis on the educational process that students are engaged in, specifically, on the collaborative process. From the results of this analysis, we can make adaptations to the learning process to improve the process itself and, therefore, the skills of students and learning outcomes.

There are different types of adaptation actions. Depending on how they are performed, they can be automatic, performed by the teacher (directed) or semi-automatic, if the teacher's agreement is required. Depending on the duration, they can be temporary or permanent. Finally, depending on the applicability scope, adaptations can affect only the current game, or the next game or until another change is made.

4 Conclusions and Further Works

In this paper we have presented our proposal of using agents (1) to analyze interaction between players of EVG in order to assess their learning processes and (2) to control the adaptation of the game. These agents have been included as an extension of the architecture PLAGER-VG.

We have defined the concept of the interesting event and we have presented the information to be collected in order to classify and analyze the learning process during the game based on the widely known model of the 3 C's.

To monitor the interaction during the game, we have proposed the inclusion of additional information (context) referring to the conditions under which events occur. This information is collected by a set of special agents of a Monitoring Agent type.

Once the previously mentioned information has been analyzed, another special agent called Facilitator Agent decides whether some adaptation actions have to be performed or whether the teacher has to be informed. In the latter case, the teacher could accept or reject the proposed changes.

Our immediate future work is to refine and to implement the modifications of the PLAGER-VG prototype with the defined agents. We want to integrate the prototype with a modular design which allows the design, execution and analysis of educational video games with group activities.

We also intend to improve the adaptation mechanism. We are going to complete the list of adaptation actions and to include the corresponding pre and post conditions to them. Pre-conditions and Post-conditions will guarantee the integrity of the game when these new adaptation actions are carried out. This process is dynamically performed while the game is running.

Acknowledgments. This study has been financed by the Ministry of Science and Innovation, Spain, as part of the VIDECO Project (TIN2011-26928) and Vice-Rector's Office for Scientific Policy and Research of University of Granada (Spain).

References

1. Padilla Zea, N.: Metodología para el diseño de videojuegos educativos sobre una arquitectura para el análisis del aprendizaje colaborativo. Ph.D. thesis, University of Granada (2011)
2. Gutwin, C., Stark, G., Greenberg, S.: Support for workspace awareness in educational groupware. In: Proceedings of CSCL'95. The First International Conference on Computer Support for Collaborative Learning, pp. 147–156 (1995)
3. Baker, M., Lund, K.: Promoting reflective interactions in a CSCL environment. J. Comput. Assist. Learn. 3(13), 175–193 (1997)
4. Collazos, C., Guerrero, L.A., Pino, J., Ochoa, S.F.: Evaluating collaborative learning processes. In: Haake, J., Pino, J. (eds.) CRIWG 2002. LNCS, vol. 2440, pp. 203–221. Springer, Heidelberg (2002)
5. Paderewski-Rodríguez, P., Rodríguez-Fortiz, M.J., Parets-Llorca, J.: An architecture for dynamic and evolving cooperative software agents. Comput. Stand. Interfaces 25(3), 261–269 (2003)
6. Paderewski-Rodríguez, P., Torres-Carbonell, J., Rodríguez-Fortiz, M.J., Medina-Medina, N., Molina-Ortiz, F.: A software system evolutionary and adaptive framework: application to agent-based systems. J Syst Archit. Elsevier 50, 407–416 (2004)
7. Ellis, C.A., Gibbs, S.J., Rein, G.L.: Groupware: some issues and experiences. Commun. ACM 34(1), 39–58 (1991)
8. Padilla, N., González, J.L., Gutiérrez, F.L.: Collaborative learning by means of video games: an entertainment system in the learning processes. In: Proceedings of 9th IEEE International Conference on Advanced Learning Technologies (ICALT), pp. 215–217 (2009)
9. Hanneman, R.A., Riddle, M.: Introduction to social network methods. Free online textbook. http://www.faculty.ucr.edu/~hanneman/nettext/ (2005). Accessed 2010

Quality Measures of Parameter Tuning for Aggregated Multi-Objective Temporal Planning

M.R. Khouadjia[1], M. Schoenauer[1], V. Vidal[2], J. Dréo[3], and P. Savéant[3(✉)]

[1] TAO Project, INRIA Saclay & LRI Paris-Sud University, Orsay, France
{mostepha-redouane.khouadjia, marc.schoenauerg}@inria.fr
[2] ONERA-DCSD, Toulouse, France
Vincent.Vidal@onera.fr
[3] THALES Research & Technology, Palaiseau, France
{johann.dreo, pierre.saveantg}@thalesgroup.com

Abstract. Parameter tuning is recognized today as a crucial ingredient when tackling an optimization problem. Several meta-optimization methods have been proposed to find the best parameter set for a given optimization algorithm and (set of) problem instances. When the objective of the optimization is some scalar quality of the solution given by the target algorithm, this quality is also used as the basis for the quality of parameter sets. But in the case of multi-objective optimization by aggregation, the set of solutions is given by several single-objective runs with different weights on the objectives, and it turns out that the hypervolume of the final population of each single-objective run might be a better indicator of the global performance of the aggregation method than the best fitness in its population. This paper discusses this issue on a case study in multi-objective temporal planning using the evolutionary planner DaEYAHSP and the meta-optimizer PARAMILS. The results clearly show how PARAMILS makes a difference between both approaches, and demonstrate that indeed, in this context, using the hypervolume indicator as PARAMILS target is the best choice. Other issues pertaining to parameter tuning in the proposed context are also discussed.

1 Introduction

Parameter tuning is now well recognized as a mandatory step when attempting to solve a given set of instance of some optimization problem. All optimization algorithms behave very differently on a given problem, depending on their parameter values, and setting the algorithm parameters to the correct value can make the difference between failure and success. This is equally true for deterministic complete algorithms [1] and for stochastic approximate algorithms [2,3]. Current approaches range from methods issued from racing-like methods [4,5] to

This work is being partially funded by the French National Research Agency under the research contract DESCARWIN (ANR-09-COSI-002).

G. Nicosia and P. Pardalos (Eds.): LION 7, LNCS 7997, pp. 341–356, 2013.
DOI: 10.1007/978-3-642-44973-4_38, © Springer-Verlag Berlin Heidelberg 2013

meta-optimization, using Gaussian Processes [6], Evolutionary Algorithms [7] or Iterated Local Search [8]. All these methods repeatedly call the target algorithm and record their performance on the given problem instances.

Quality criteria for parameter sets usually involve the solution quality of the target algorithm and the time complexity of the algorithm, and, in the case of a set of problem instances, statistics of these quantities over the whole set. The present work is concerned with the case of instance-based parameter tuning (i.e. a single instance is considered), and the only goal is the quality of the final solution, for a fixed computational budget. In this context, the objective of the meta-optimizer is generally also directly based on the quality of the solution.

However, things are different in the context of multi-objective optimization, when using an aggregation method, i.e. optimizing several linear combinations of the objectives, gathering all results into one single set, and returning the non-dominated solutions within this set as an approximation of the Pareto front. Indeed, the objective of each single-objective run is the weighted sum of the problem objectives, and using this weighted sum as the objective for parameter tuning seems to be the most straightforward approach. However, the objective of the whole algorithm is to approximate the Pareto front of the multi-objective problem. And the hypervolume indicator [9] has been proved to capture into a single real value the quality of a set as an approximation of the Pareto front. Hence an alternative strategy could be to tune each single-objective run so as to optimize the hypervolume of its final population, as a by-product of optimizing the weighted sum of the problem objectives. This paper presents a case study of the comparison of both parameter-tuning approaches described above for the aggregated multi-objective approach, in the domain of AI planning [10]. This domain is rapidly introduced in Sect. 2. In particular, MULTIZENO, a tunable multi-objective temporal planning benchmark inspired by the well-known zeno IPC logistic domain benchmark, is described in detail. Section 3 introduces Divide-and-Evolve (DAE_{YAHSP}), a single-objective evolutionary AI planning algorithm that has obtained state-of-the-art results on different planning benchmark problems [11], and won the deterministic temporal satisficing track at IPC 2011 competition.[1] Section 4 details the experimental conditions of the forthcoming experiments, introduces the parameters to be optimized, the aggregation method, the meta-optimizer PARAMILS, the parameter tuning method that has been chosen here [8], and precisely defines the two quality measures to be used by PARAMILS in the experiments: either the best fitness or the global hypervolume of its final population. Section 5 details the experimental results obtained by DAE_{YAHSP} for solving MULTIZENO instances using these two quality measures. The values of the parameters resulting from the PARAMILS runs are discussed, and the quality of the approximations of the Pareto front given by both approaches are compared, and the differences analyzed.

---

[1] See http://www.plg.inf.uc3m.es/ipc2011-deterministic

2   AI Planning

An AI Planning problem (see e.g. [10]) is defined by a set of predicates, a set of actions, an initial state and a goal state. A state is a set of non-exclusive instantiated predicates, or (Boolean) atoms. An action is defined by a set of *pre-conditions* and a set of *effects*: the action can be executed only if all pre-conditions are true in the current state, and after an action has been executed, the effects of the action modify the state: the system enters a new state. A plan is a sequence of actions, and a *feasible plan* is a plan such that executing each action in turn from the initial state puts the systems into the goal state. The goal of (single objective) AI Planning is to find a feasible plan that minimizes some quantity related to the actions: number of actions for STRIPS problems, sum of action costs in case actions have different costs, or makespan in the case of temporal planning, when actions have a duration and can eventually be executed in parallel. All these problems are P-SPACE.

A simple planning problem in the domain of logistics (inspired by the well-known ZENO problem of IPC series) is given in Fig. 1: the problem involves cities, passengers, and planes. Passengers can be transported from one city to another, following the links on the figure. One plane can only carry one passenger at a time from one city to another, and the flight duration (number on the link) is the same whether or not the plane carries a passenger (this defines the *domain* of the problem). In the simplest non-trivial *instance* of such domain, there are 3 passengers and 2 planes. In the initial state, all passengers and planes are in `city` `0`, and in the goal state, all passengers must be in `city` `4`. The not-so-obvious optimal solution has a total makespan of 8 and is left as a teaser for the reader.

AI Planning is a very active field of research, as witnessed by the success of the ICAPS series of yearly conferences (`http://icaps-conferences.org`), and its biannual competition IPC, where the best planners in the world compete on a set of problems. This competition has lead the researchers to design a common language to describe planning problems, PDDL (Planning Domain Definition Language). Two main categories of planners can be distinguished: *exact planners* are guaranteed to find the optimal solution ... if given enough time; *satisficing planners* give the best possible solution, but with no optimality guarantee.

2.1   Multi-Objective AI Planning

Most existing work in AI Planning involves one single objective, even though real-world problems are generally multi-objective (e.g., optimizing the makespan while minimizing the cost, two contradictory objectives). An obvious approach to Multi-Objective AI planning is to aggregate the different objectives into a single objective, generally a fixed linear combination (weighted sum) of all objectives. The single objective is to be minimized, and the weights have to be positive (resp. negative) for the objectives to be minimized (resp. maximized) in the original problem. The solution of one aggregated problem is Pareto optimal if all weights are non-zero, or the solution is unique [12]. It is also well-known that

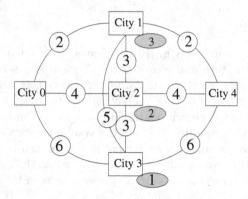

Fig. 1. A schematic view of MULTIZENO, a simple benchmark transportation domain: Flight durations of available routes are attached to the corresponding edges, costs are attached to landing in the central cities (in grey circles).

whatever the weights, the optimal solution of an aggregated problem is always on the convex parts of the Pareto front. However, some adaptive techniques of the aggregation approach have been proposed, that partially address this drawback [13] and are able to identify the whole Pareto front by maintaining an archive of non-dominated solutions ever encountered during the search.

Despite the fact that pure multi-objective approaches like e.g., dominance-based approaches, are able to generate a diverse set of Pareto optimal solutions, which is a serious advantage, aggregation approaches are worth investigating, as they can be implemented seamlessly from almost any single-objective algorithm, and rapidly provide at least part of the Pareto front at a low man-power cost.

This explains why all works in multi-objective AI Planning used objective aggregation, to the best of our knowledge.[2] Early works used some twist in PDDL 2.0 [16–18]. PDDL 3.0, on the other hand, explicitly offered hooks for several objectives [19], and a new track of IPC was dedicated to aggregated multiple objectives: the "net-benefit" track took place in 2006 [20] and 2008 [21], ... but was canceled in 2011 because of a too small number of entries.

2.2   Tunable Benchmarks for Multi-Objective Temporal Planning

For the sake of understandability, it is important to be able to experiment with instances of tunable complexity for which the exact Pareto fronts are easy to determine, and this is the reason for the design of the MULTIZENO benchmark family. The reader will have by now solved the little puzzle illustrated in Fig. 1, and found the solution with makespan 8, whose rationale is that no plane ever stays idle. In order to turn this problem into a not-too-unrealistic logistics multi-objective problem, some costs are added to all 3 central cities (1 to 3). This leads

---

[2] With the exception of an early proof-of-concept for DAEX [14] and its recently accepted follow-up [15].

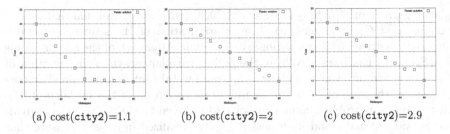

(a) cost(city2)=1.1   (b) cost(city2)=2   (c) cost(city2)=2.9

Fig. 2. The exact Pareto Fronts for the MULTIZENO6 problem for different values of cost(city2) (all other values as in Fig. 1).

to the MULTIZENO$_{Cost}$ problems, where the second objective is additive: each plane has to pay the corresponding tax every time it lands in that city.[3]

In the simplest instance, MULTIZENO3, involving 3 passengers only, there are 3 obvious points that belong to the Pareto Front, using the small trick described above, and going respectively through city1, city 2 or city 3. The values of the makespans are respectively 8, 16 and 24, and the values of the costs are, for each solution, 4 times the value of the single landing tax. However, different cities can be used for the different passengers, leading to a Pareto Front made of 5 points, adding points (12,10) and (20,6) to the obvious points (8,12), (16,8), and (24,4).

There are several ways to make this first simple instance more or less complex, by adding passengers, planes and central cities, and by tuning the different values of the makespans and costs. In the present work, only additional bunches of 3 passengers have been considered, in order to be able to easily derive some obvious Pareto-optimal solutions as above, using several times the little trick to avoid leaving any plane idle. This lead to the MULTIZENO6, and MULTIZENO9 instances, with respectively 6 and 9 passengers. The Pareto front of MULTIZENO6 on domain described by Fig. 1 can be seen on Fig. 2b. The other direction for complexification that has been investigated in the present work is based on the modification of the cost value for city 2, leading to different shapes of the Pareto front, as can be seen on Fig. 2a, c. Further work will investigate other directions of complexification of this very rich benchmark test suite.

3   Divide-and-Evolve

Let $\mathcal{P}_D(I, G)$ denote the planning problem defined on domain D (the predicates, the objects, and the actions), with initial state I and goal state G. In STRIPS representation model [22], a state is a list of Boolean atoms defined using the predicates of the domain, instantiated with the domain objects.

In order to solve $\mathcal{P}_D(I, G)$, the basic idea of DAE$_X$ is to find a sequence of states $S_1, \ldots, S_n$, and to use some embedded planner X to solve the series of

---

[3] In the MULTIZENO$_{Risk}$ problem, not detailed here, the second objective is the risk: its maximal value ever encountered is to be minimized.

planning problems $\mathcal{P}_D(S_k, S_{k+1})$, for $k \in [0, n]$ (with the convention that $S_0 = I$ and $S_{n+1} = G$). The generation and optimization of the sequence of states $(S_i)_{i \in [1,n]}$ is driven by an evolutionary algorithm. After each of the sub-problems $\mathcal{P}_D(S_k, S_{k+1})$ has been solved by the embedded planner, the concatenation of the corresponding plans (possibly compressed to take into account possible parallelism in the case of temporal planning) is a solution of the initial problem. In case one sub-problem cannot be solved by the embedded solver, the individual is said *unfeasible* and its fitness is highly penalized in order to ensure that feasible individuals always have a better fitness than unfeasible ones, and are selected only when there are not enough feasible individual. A thorough description of DAE$_X$ can be found in [11]. The rest of this section will briefly recall the evolutionary parts of DAE$_X$.

3.1   Representation, Initialization, and Variation Operators

Representation: An individual in DAE$_X$ is a variable-length list of states of the given domain. However, the size of the space of lists of complete states rapidly becomes untractable when the number of objects increases. Moreover, goals of planning problems need only to be defined as partial states, involving a subset of the objects, and the aim is to find a state such that all atoms of the goal state are true. An individual in DAE$_X$ is thus a variable-length list of partial states, and a partial state is a variable-length list of atoms (instantiated predicates).

Initialization: Previous work with DAE$_X$ on different domains of planning problems from the IPC benchmark series have demonstrated the need for a very careful choice of the atoms that are used to build the partial states [23]. The method that is used today to build the partial states is based on a heuristic estimation, for each atom, of the earliest time from which it can become true [24], and an individual in DAE$_X$ is represented by a variable-length time-consistent sequence of partial states, and each partial state is a variable-length list of atoms that are not pairwise mutually exclusive (aka *mutex*), according to the partial mutex relation computed by the embedded planner.

Crossover and Mutation Operators are applied with respective user-defined probabilities `Proba-cross` and `Proba-mut`. They are defined on the DAE$_X$ representation in a straightforward manner - though constrained by the heuristic chronology and the partial mutex relation between atoms. **One-point crossover** is adapted to variable-length representation: both crossover points are independently chosen, uniformly in both parents. Only one offspring is kept, the one that respects the approximate chronological constraint on the successive states.

Four mutation operators are included, and operate either at the individual level, by adding (`addGoal`) or removing (`delGoal`) an intermediate state, or at the state level by adding (`addAtom`) or removing (`delAtom`) some atoms in a uniformly chosen state. The choice among these mutations is made according to user-defined relative weights, named `w-MutationName` - see Table 1.

3.2 Hybridization and Multi-Objectivization

DAE_X uses an external embedded planner to solve in turn the sequence of sub-problems defined by the ordered list of partial states. Any existing planner can in theory be used. However, there is no need for an optimality guarantee when solving the intermediate problems in order for DAE_X to obtain good quality results [11]. Hence, and because a very large number of calls to this embedded planner are necessary for a single fitness evaluation, a sub-optimal but fast planner was found to be the best choice: YAHSP [25] is a lookahead strategy planning system for sub-optimal planning which uses the actions in the relaxed plan to compute reachable states in order to speed up the search process. Because the rationale for DAE_X is that all sub-problems should hopefully be easier than the initial global problem, and for computational performance reason, the search capabilities of the embedded planner YAHSP are limited by setting a maximal number of nodes that it is allowed to expand to solve any of the sub-problems (see again [11] for more details).

However, even though YAHSP, like all known planners to-date, is a single-objective planner, it is nevertheless possible since PDDL 3.0 to add in a PDDL domain file other quantities (aka *Soft Constraints* or *Preferences* [19]) that are simply computed throughout the execution of the final plan, without interfering with the search. Two strategies are then possible for YAHSP in the two-objective context of MULTIZENO: it can optimize either the makespan or the cost, and simply compute the other quantity (cost or makespan) along the solution plan. The corresponding strategie will be referred to as $YAHSP_{makespan}$ and $YAHSP_{cost}$.

In the multi-objective versions of DAE_{YAHSP} the choice between both strategies is governed by user-defined weights, named respectively W-makespan and W-cost (see Table 1). For each individual, the actual strategy is randomly chosen according to those weights, and applied to all subproblems of the individual.

4 Experimental Conditions

The Aggregation Method for multi-objective optimization runs in turn a series of single-objective problems. The fitness of each of these problems is defined using a single positive parameter α. In the following, F_α will denote $\alpha * \text{makespan} + (1 - \alpha) * \text{cost}$, and DAE_{YAHSP} run optimizing F_α will be called the α-run. Because the range of the makespan values is approximately twice as large as that of the cost, the following values of α have been used instead of regularly spaced values: 0, 0.05, 0.1, 0.3, 0.5, 0.55, 0.7, 1.0. One "run" of the aggregation method thus amounts to running the corresponding eight α-runs, and returns as the approximation of the Pareto front the set of non-dominated solutions among the merge of the eight final populations.

ParamILS [8] is used to tune the parameters of DAE_{YAHSP}. PARAMILS uses the simple Iterated Local Search heuristic [26] to optimize parameter configurations, and can be applied to any parameterized algorithm whose parameters can be discretized. PARAMILS repeats local search loops from different random starting

Table 1. Set of DaE parameters and their discretizations for PARAMILS, leading to approx. $1.5 \cdot 10^9$ possible configurations.

Parameters	Range	Description
W-makespan	0,1,2,3,4,5	Weighting for optimizing makespan during the search
W-cost		Weighting for optimizing cost during the search
Pop-size	30,50,100,200,300	Population size
Proba-cross	0.0,0.1,0.2,0.5,0.8,1.0	Probability (at population level) to apply crossover
Proba-mut		Probability (at population level) to apply one mutation
w-addAtom	0,1,3,5,7,10	Relative weight of the addAtom mutation
w-addGoal		Relative weight of the addGoal mutation
w-delAtom		Relative weight of the delAtom mutation
w-delGoal		Relative weight of the delGoal mutation
Proba-change	0.0,0.1,0.2,0.5,0.8,1.0	Probability to change an atom in addAtom mutation
Proba-delatom		Average probability to delete an atom in delAtom mutation
Radius	1,3,5,7,10	Number of neighbour goals to consider in addGoal mutation

points, and during each local search loops, modifies one parameter at a time, runs the target algorithm with the new configuration and computes the quality measure it aims at optimizing, accepting the new configuration if it improves the quality measure over the current one.

The most prominent parameters of DAE_{YAHSP} that have been subject to optimization can be seen in Table 1.

Quality Measures for ParamILS: The goal of the experiments presented here is to compare the influence of two quality measures of PARAMILS for the aggregated DAE_{YAHSP} on MULTIZENO instances. In $Aggreg_{Fitness}$, the quality measure used by PARAMILS to tune the α-run of DAE_{YAHSP} is F_α, the fitness also used by the target α-run. In $Aggreg_{Hyper}$, PARAMILS uses, for each of the α-run, the same quality measure, i.e., the unary hypervolume [27] of the final population of the α-run w.r.t. the exact Pareto front of the problem at hand (or its best available approximation when it is not available). The lower the better (a value of 0 indicates that the exact Pareto front has been reached).

Implementation: Algorithms have been implemented within the PARADISEO-MOEO framework.[4] All experiments were performed on the MULTIZENO3, MULTIZENO6, and MULTIZENO9 instances. The first objective is the makespan, and the second objective is the cost. The values of the different flight durations (makespans) and costs are those given on Fig. 1 except otherwise stated.

Performance Assessment and Stopping Criterion: For all experiments, 11 independent runs were performed. Note that all the performance assessment procedures, including the hypervolume calculations, have been achieved using the PISA performance assessment tool suite.[5] The main quality measure used here to compare Pareto Fronts is, as above, the unary hypervolume I_{H-} [27] of the set of non-dominated points output by the algorithms with respect to the complete true Pareto front. For aggregated runs, the union of all final populations of the α-runs for the different values of α is considered the output of the complete 'run'.

---

[4] http://paradiseo.gforge.inria.fr/

[5] http://www.tik.ee.ethz.ch/pisa/

However, and because the true front is known exactly, and is made of a few scattered points (at most 17 for MULTIZENO9 in this paper), it is also possible to visually monitor, for each point of the front, the ratio of actual runs (out of 11) that discovered it at any given time. This allows some other point of view on the comparison between algorithms, even when none has found the whole Pareto front. Such *hitting plots* will be used in the following, together with more classical plots of hypervolume vs computational effort. In any case, when comparing different approaches, statistical significance tests are made on the hypervolumes, using Wilcoxon signed rank test with 95 % confidence level.

Finally, because different fitness evaluations involve different number calls to YAHSP – and because YAHSP runs can have different computational costs too, depending on the difficulty of the sub-problem being solved – the computational efforts will be measured in terms of CPU and not number of function evaluations – and that goes for the stopping criterion: The absolute limits in terms of computational efforts were set to 300, 600, and 1800 seconds respectively for MULTIZENO3, MULTIZENO6, and MULTIZENO9. The stopping criterion for PARAMILS was likewise set to a fixed wall-clock time: 48 h (resp. 72 h) for MULTIZENO3 and 6 (resp. MULTIZENO9), corresponding to 576, 288, and 144 parameter configuration evaluations per value of α for MULTIZENO3, 6 and 9 respectively.

5   Experimental Results

5.1   ParamILS Results

Table 2 presents the optimal values for DAE_{YAHSP} parameters of Table 1 found by PARAMILS in both experiments, for all values of α - as well as for the multi-objective version of DAE_{YAHSP} presented in [15] (last column, entitled $IBEA_H$).

The most striking and clear conclusion regards the weights for the choice of YAHSP strategy (see Sect. 3.2) W-makespan and W-cost. Indeed, for the

Table 2. PARAMILS results: Best parameters for DAE_{YAHSP} on MULTIZENO6

	Hypervolume								Fitness								IBEA$_H$
α	0.0	0.05	0.1	0.3	0.5	0.55	0.7	1.0	0.0	0.05	0.1	0.3	0.5	0.55	0.7	1.0	
W-makespan	3	3	3	2	2	2	0	0	0	0	0	0	5	5	1	4	1
W-cost	0	0	0	4	3	3	3	4	2	4	4	2	1	1	0	1	1
Pop-size	100	100	200	200	100	100	200	300	200	300	300	100	100	100	100	100	30
Proba-cross	0.5	1.0	1.0	1.0	1.0	1.0	1.0	0.8	0.8	0.1	0.2	0.2	0.5	0.8	0.2	0.1	0.2
Proba-mut	0.8	0.2	0.2	0.2	0.2	0.2	0.5	1.0	0.5	1.0	0.5	1.0	1.0	1.0	0.8	1.0	0.2
w-addatom	1	1	5	5	5	5	5	5	3	10	3	10	3	5	5	3	7
w-addgoal	5	1	5	7	7	7	0	0	3	7	10	10	10	10	10	10	10
w-delatom	3	3	1	5	10	1	7	0	3	5	10	0	10	10	3	1	5
w-delgoal	5	5	5	7	10	3	7	10	1	7	1	1	0	1	10	1	5
Proba-change	0.5	0.5	0.5	1.0	0.8	0.5	0.5	0.8	0.8	0.8	0.8	0.5	0.0	1.0	0.8	0.5	1.0
Proba-delatom	0.1	0.0	0.0	0.1	0.5	0.5	0.5	0.0	0.1	0.8	0.0	0.8	1.0	0.8	0.5	0.5	1.0
Radius	3	3	10	1	7	7	1	5	3	3	1	3	5	5	10	3	5

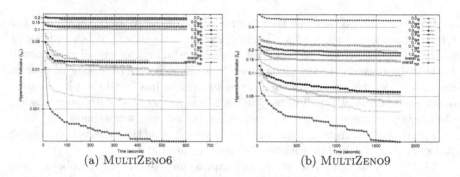

(a) MULTIZENO6   (b) MULTIZENO9

Fig. 3. Evolution of the Hypervolume for both approaches, for all α-runs and overall, on MULTIZENO instances. Warning: Hypervolume is in log scale, and the X-axis is not the value 0, but $6.7 \; 10^{-5}$ for MULTIZENO6 and 0.0125 for MULTIZENO9.

Aggreg$_{Hyper}$ approach, PARAMILS found out that YAHSP should optimize only the makespan (W-cost = 0) for small values of α, and only the cost for large values of α while the exact opposite is true for the Aggreg$_{Fitness}$ approach. Remember that small (resp. large) values of α correspond to an aggregated fitness having all its weight on the cost (resp. the makespan). Hence, during the 0- or 0.5-runs, the fitness of the corresponding α-run is pulling toward minimizing the cost: but for the Aggreg$_{Hyper}$ approach, the best choice for YAHSP strategy, as identified by PARAMILS, is to minimize the makespan (i.e., setting W-cost to 0): as a result, the population has a better chance to remain diverse, and hence to optimize the hypervolume, i.e., PARAMILS quality measure. In the same situation (small α), on the opposite, for Aggreg$_{Fitness}$, PARAMILS has identified that the best strategy for YAHSP is to also favor the minimization of the cost, setting W-makespan to zero. The symmetrical reasoning can be applied to the case of large values of α. For the multi-objective version of DAE$_{YAHSP}$ (IBEA column in Table 2), the best strategy that PARAMILS came up with is a perfect balance between both strategies, setting both weights to 1.

The values returned by PARAMILS for the other parameters are more difficult to interpret. It seems that large values of **Proba-mut** are preferable for Aggreg$_{Hyper}$ for α set to 0 or 1, i.e. when the DAE$_{YAHSP}$ explores the extreme sides of the objective space – more mutation is needed to depart from the boundary of the objective space and cover more of its volume. Another tendancy is that PARAMILS repeatedly found higher values of **Proba-cross** and lower values of **Proba-mut** for Aggreg$_{Hyper}$ than for Aggreg$_{Fitness}$. Together with large population sizes (compared to the one for IBEA for instance), the 1-point crossover of DAE$_{YAHSP}$ remains exploratory for a long time, and leads to viable individuals that can remain in the population even though they don't optimize the α-fitness, thus contributing to the hypervolume. On the opposite, large mutation rate is preferable for Aggreg$_{Fitness}$ as it increases the chances to hit a better fitness, and otherwise generates likely non-viable individuals that will be quickly eliminated by selection, making DAE$_{YAHSP}$ closer from a local search. The values found for IBEA, on the other hand, are rather small – but the small population size also has to be considered here: because it aims at exploring the whole objective space in one go, the most efficient strategy for IBEA is to make more but smaller steps, in all possible directions.

5.2   Comparative Results

Figure 3 represents the evolution during the course of the runs of the hypervolumes (averaged over the 11 independent runs) of some of the (single-objective) α-runs, for both methods together (labelled α_{hyp} or α_{fit}), as well as the evolution of the overall hypervolume, i.e., the hypervolume covered by the union of all populations of the different α-runs as a function of CPU time. Only the results on MULTIZENO6 and MULTIZENO9 are presented here, but rather similar behaviors can be observed for the two approaches on these two instances, and similar results were obtained on MULTIZENO3, though less significantly different.

First of all, Aggreg_{Hyper} appears as a clear winer against $\text{Aggreg}_{Fitness}$, as confirmed by the Wilcoxon test with 95 % confidence: On both instances, the two lowest lines are the results of the overall hypervolume for, from bottom to top, Aggreg_{Hyper} and $\text{Aggreg}_{Fitness}$, that reach respectively values of 6.7 10^{-5} and 0.015 on MULTIZENO6 and 0.0127 and 0.03155 on MULTIZENO9. And for each value of α, a similar difference can be seen. Another remark is that the central values of α (0.5, 0.7 and 0.3, in this order) outperform the extreme values (1 and 0, in this order): this is not really surprising, considering that these runs, that optimize a single objective (makespan or cost), can only spread in one direction, while more 'central' values allow the run to cover more volume around their best solutions. Finally, in all cases, the 0-runs perform significantly worse than the corresponding 1-runs, but this is probably only due to the absence of normalization between both objectives.

Another comparative point of view on the convergence of both aggregation approaches is given by the hitting plots of Fig. 4. These plots represent, for each point of the true Pareto front, the ratio along evolution of the runs (remember that one 'run' represent the sum of the eight α-runs, see Sect. 4) that reached that point, for all three instances MULTIZENO{3,6,9}. On MULTIZENO3 (results not shown here for space reasons), only one point, (20, 6), is not found by 100 % of the runs. But it is found by 10/11 runs by Aggreg_{Hyper} and only by 6/11 runs by $\text{Aggreg}_{Fitness}$. On MULTIZENO6, the situation is even clearer in favor of Aggreg_{Hyper}: Most points are found very rapidly by Aggreg_{Hyper}, and only point (56, 12) is not found by 100 % of the runs (it is missed by 2 runs); on the other hand, only 4 points are found by all α-runs of $\text{Aggreg}_{Fitness}$, the extreme makespan (60, 10), and the 3 extreme costs (20, 30), (24, 28), and (28, 26). The other points are discovered by different runs . . . but overall, not a single run discovers all 11 points. Finally, the situation is even worse in the MULTIZENO9 case: only 6 points (out of 17) are ever discovered by $\text{Aggreg}_{Fitness}$, while Aggreg_{Hyper} somehow manages to hit 12 different points. Hence again, no method does identify the full Pareto front.

But take a look at Fig. 5, that displays the union of the 11 Pareto front returned by the aggregated runs, for both Aggreg_{Hyper} and $\text{Aggreg}_{Fitness}$. No big

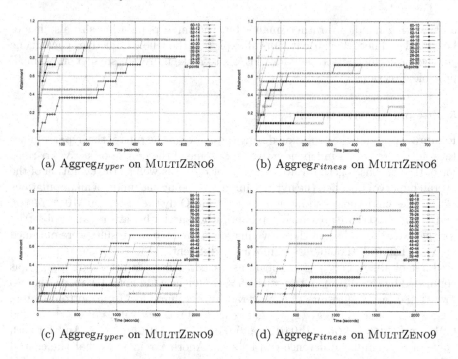

(a) Aggreg$_{Hyper}$ on MULTIZENO6   (b) Aggreg$_{Fitness}$ on MULTIZENO6

(c) Aggreg$_{Hyper}$ on MULTIZENO9   (d) Aggreg$_{Fitness}$ on MULTIZENO9

Fig. 4. Hitting plots on the 3 MULTIZENO instances.

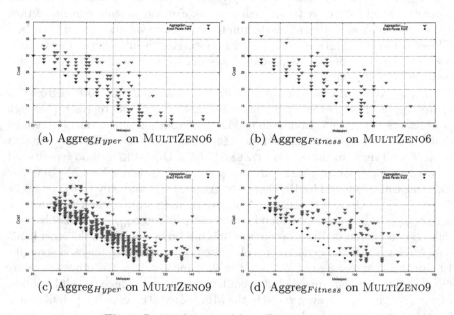

(a) Aggreg$_{Hyper}$ on MULTIZENO6   (b) Aggreg$_{Fitness}$ on MULTIZENO6

(c) Aggreg$_{Hyper}$ on MULTIZENO9   (d) Aggreg$_{Fitness}$ on MULTIZENO9

Fig. 5. Pareto fronts on MULTIZENO instances.

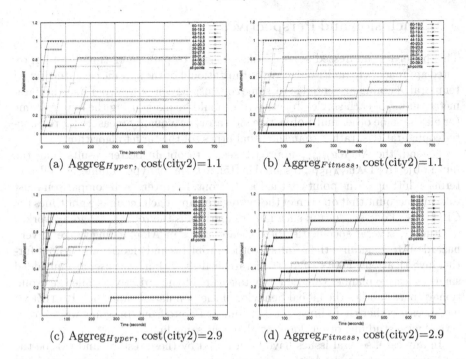

(a) Aggreg$_{Hyper}$, cost(city2)=1.1

(b) Aggreg$_{Fitness}$, cost(city2)=1.1

(c) Aggreg$_{Hyper}$, cost(city2)=2.9

(d) Aggreg$_{Fitness}$, cost(city2)=2.9

Fig. 6. Hitting plots for different Pareto fronts for MULTIZENO6. See Sect. 2.2 and compare with Fig. 4(a), (b).

difference is observed on MULTIZENO6, except maybe a higher diversity away from the Pareto front for Aggreg$_{Hyper}$. On the other hand, the difference is clear on MULTIZENO9, where Aggreg$_{Fitness}$ completely misses the center of the Pareto-delimited region of the objective space.

Preliminary runs have been made with the two other instances presented in Sect. 2.2, where the costs of city2 have changed, respectively to 1.1 and 2.9, giving the Pareto fronts that are displayed in Fig. 2. However, no specific parameter tuning was done for these instances, and all parameters have been carried on from the PARAMILS runs on the corresponding MULTIZENO instance where the cost of city2 is 2. First, it is clear that the overall performance of both aggregation methods is rather poor, as none ever finds the complete Pareto front in the 1.1 case, and only one run out of 11 finds it in the 2.9 case. Here again, only the extreme points are reliably found by both methods. Second, the advantage of Aggreg$_{Hyper}$ over Aggreg$_{Fitness}$ is not clear any more: some points are even found more often by the latter. Finally, and surprisingly, the ankle point in the case 1.1 (Fig. 2a) is not found as easily as it might have seemed; and the point on the concave part of the case 2.9 (point (56,22.8), see Fig. 2c) is nevertheless found by respectively 9 and 4 runs, whereas aggregation approaches should have difficulties to discover such points.

6   Conclusion and Perspectives

This paper has addressed several issues related to parameter tuning for aggregated approaches to multi-objective optimization. For the specific case study in AI temporal planning presented here, some conclusions can be drawn. First, the parameter tuning of each single-objective run should be made using the hypervolume (or maybe some other multi-objective indicator) as a quality measure for parameter configurations, rather than the usual fitness of the target algorithm.

Second, the $Aggreg_{Hyper}$ approach seems to obtain better results than the multi-objective DAE_{YAHSP} presented in [15], in terms of hypervolume, as well as in terms of hitting the points of the Pareto front. However, such comparison must take into account that one run of the aggregated approach requires eight times the CPU time of one single run: such fair comparison is the topic of on-going work.

Finally, several specificities of the case study in AI planning make it very hazardous to generalize the results to other problems and algorithms without further investigations: DAE_{YAHSP} is a hierarchical algorithm, that uses an embedded single objective planner that can only take care of one objective, while the evolutionary part handles the global behavior of the population; and the MULTIZENO instances used here have linear, or quasi-linear Pareto front; on-going work is concerned with studying other domains along the same lines.

In any case, several issues have been raised by these results, and will be the subject of further work. At the moment, only instance-based parameter tuning was performed – and the preliminary results on the other instances with different Pareto front shapes (see Fig. 6) suggest that the best parameter setting is highly instance-dependent (as demonstrated in a similar AI planning context in [28]). But do the conclusions drawn above still apply in the case of class-driven parameter tuning? Another issue that was not discussed here is that of the delicate choice of the values for α. Their proper choice is highly dependent on the scales of the different objectives. Probably some adaptive technique, as proposed by [13], would be a better choice.

References

1. Hutter, F., Hoos, H.H., Leyton-Brown, K.: Automated configuration of mixed integer programming solvers. In: Lodi, A., Milano, M., Toth, P. (eds.) CPAIOR 2010. LNCS, vol. 6140, pp. 186–202. Springer, Heidelberg (2010)
2. Eiben, A., Michalewicz, Z., Schoenauer, M., Smith, J.: Parameter control in evolutionary algorithms. In: Lobo, F.G., Lima, C.F., Michalewicz, Z. (eds.) Parameter Setting in Evolutionary Algorithms. SCI, vol. 54, pp. 19–46. Springer, Heidelberg (2007)
3. Yuan, Z., de Oca, M.A.M., Birattari, M., Stützle, T.: Modern continuous optimization algorithms for tuning real and integer algorithm parameters. In: Dorigo, M., et al. (eds.) ANTS 2010. LNCS, vol. 6234, pp. 203–214. Springer, Heidelberg (2010)
4. Birattari, M., Yuan, Z., Balaprakash, P., Stützle, T.: Automated algorithm tuning using F-Races: recent developments. In: Caserta, M., et al. (eds.) Proceedings of MIC'09. University of Hamburg (2009)

5. Dubois-Lacoste, J., López-Ibáñez, M., Stützle, T.: Automatic configuration of state-of-the-art multi-objective optimizers using the TP+PLS framework. In: Krasnogor, N., Lanzi, P.-L. (eds.) Proceedings of 13th ACM-GECCO, pp. 2019–2026 (2011)
6. Bartz-Beielstein, T., Lasarczyk, C., Preuss, M.: Sequential parameter optimization. In: McKay, B., et al. (eds.) Proceedings of CEC'05, pp. 773–780. IEEE (2005)
7. Nannen, V., Eiben, A.: Relevance estimation and value calibration of evolutionary algorithm parameters. In: Veloso, M., et al. (eds.) Proceedings of IJCAI'07, pp. 975–980 (2007)
8. Hutter, F., Hoos, H., Leyton-Brown, K., Stützle, T.: ParamILS: an automatic algorithm configuration framework. J. Artif. Intel. Res. 36(1), 267–306 (2009)
9. Zitzler, E., Thiele, L., Laumanns, M., Fonseca, C., da Fonseca, V.: Performance assessment of multiobjective optimizers: an analysis and review. IEEE Trans. Evol. Comput. 7(2), 117–132 (2003)
10. Ghallab, M., Nau, D., Traverso, P.: Automated Planning, Theory and Practice. Morgan Kaufmann, San Francisco (2004)
11. Bibaï, J., Savéant, P., Schoenauer, M., Vidal, V.: An evolutionary metaheuristic based on state decomposition for domain-independent satisficing planning. In: Brafman, R., et al. (eds.) Proceedings of 20th ICAPS, pp. 18–25. AAAI Press (2010)
12. Miettinen, K.: Nonlinear Multiobjective Optimization, vol. 12. Springer, Heidelberg (1999)
13. Jin, Y., Okabe, T., Sendhoff, B.: Adapting weighted aggregation for multiobjective evolution strategies. In: Zitzler, E., Deb, K., Thiele, L., Coello Coello, C.A., Corne, D.W. (eds.) EMO 2001. LNCS, vol. 1993, pp. 96–110. Springer, Heidelberg (2001)
14. Schoenauer, M., Savéant, P., Vidal, V.: Divide-and-Evolve: a new memetic scheme for domain-independent temporal planning. In: Gottlieb, J., Raidl, G. (eds.) EvoCOP 2006. LNCS, vol. 3906, pp. 247–260. Springer, Heidelberg (2006)
15. Khouadjia, M.R., Schoenauer, M., Vidal, V., Dréo, J., Savéant, P.: Multi-objective AI planning: evaluating DAEYAHSP on a tunable benchmark. In: Purshouse, R.C., Fleming, P.J., Fonseca, C.M., (eds.) Proceedings of EMO'2013 (2013, to appear)
16. Do, M., Kambhampati, S.: SAPA: a multi-objective metric temporal planner. J. Artif. Intell. Res. (JAIR) 20, 155–194 (2003)
17. Refanidis, I., Vlahavas, I.: Multiobjective heuristic state-space planning. Artif. Intell. 145(1), 1–32 (2003)
18. Gerevini, A., Saetti, A., Serina, I.: An approach to efficient planning with numerical fluents and multi-criteria plan quality. Artif. Intell. 172(8–9), 899–944 (2008)
19. Gerevini, A., Long, D.: Preferences and soft constraints in PDDL3. In: ICAPS Workshop on Planning with Preferences and Soft, Constraints, pp. 46–53 (2006)
20. Chen, Y., Wah, B., Hsu, C.: Temporal planning using subgoal partitioning and resolution in SGPlan. J. Artif. Intell. Res. 26(1), 323–369 (2006)
21. Edelkamp, S., Kissmann, P.: Optimal symbolic planning with action costs and preferences. In: Proceedings of 21st IJCAI, pp. 1690–1695 (2009)
22. Fikes, R., Nilsson, N.: STRIPS: a new approach to the application of theorem proving to problem solving. Artif. Intell. 1, 27–120 (1971)
23. Bibai, J., Savéant, P., Schoenauer, M., Vidal, V.: On the benefit of sub-optimality within the Divide-and-Evolve scheme. In: Cowling, P., Merz, P. (eds.) EvoCOP 2010. LNCS, vol. 6022, pp. 23–34. Springer, Heidelberg (2010)
24. Haslum, P., Geffner, H.: Admissible heuristics for optimal planning. In: Proceedings of AIPS 2000, pp. 70–82 (2000)
25. Vidal, V.: A lookahead strategy for heuristic search planning. In: Proceedings of the 14th ICAPS, pp. 150–159. AAAI Press (2004)

26. Lourenço, H., Martin, O., Stützle, T.: Iterated local search. In: Glover, F., Kochen-berger, G.A. (eds.) Handbook of Metaheuristics, pp. 320–353. Kluwer Academic, New York (2003)
27. Zitzler, E., Künzli, S.: Indicator-based selection in multiobjective search. In: Yao, X., et al. (eds.) PPSN VIII. LNCS, vol. 3242, pp. 832–842. Springer, Heidelberg (2004)
28. Bibaï, J., Savéant, P., Schoenauer, M., Vidal, V.: On the generality of parameter tuning in evolutionary planning. In: Proceedings of 12th GECCO, pp. 241–248. ACM (2010)

Evolutionary FSM-Based Agents
for Playing Super Mario Game

R.M. Hidalgo-Bermúdez[1,2], M.S. Rodríguez-Domingo[1,2]([✉]), A.M. Mora[1,2],
P. García-Sánchez[1,2], Juan Julian Merelo[1,2], and Antonio J. Fernández-Leiva[1,2]

[1] Depto. Arquitectura y Tecnología de Computadores, University of Granada,
Granada, Spain
[2] Depto. Lenguajes y Ciencias de la Computación,
University of Málaga, Málaga, Spain
rosa.hb84@gmail.com, zandra@correo.ugr.es,
{amorag,pgarcia,jmerelo}@geneura.ugr.es, afdez@lcc.uma.es

Abstract. Most of game development along the years has been focused
on the technical part (graphics and sound), leaving the artificial intelligence aside. However computational intelligence is becoming more significant, leading to much research on how to provide non-playing characters
with adapted and unpredictable behaviour so as to afford users a better gaming experience. This work applies strategies based on Genetic
Algorithms mixed with behavioural models, to obtain an agent (or bot)
capable of completing autonomously different scenarios on a simulator
of Super Mario Bros. game. Specifically, the agent follows the rules of
the *Gameplay* track of Mario AI Championship. Different approaches
have been analysed, combining Genetic Algorithms with Finite State
Machines, yielding agents which can complete levels of different difficulties playing much better than an expert human player.

1 Introduction

Mario Bros. games series were created by Shigeru Miyamoto[1], and appeared in
early 80s. The most famous so far is the platform game Super Mario Bros. and
its sequels (for instance the blockbuster Super Mario World).

All of them follow a well-known plot: the plumber Mario must rescue the
princess of Mushroom Kingdom, Peach, who has been kidnapped by the king
of the koopas, Bowser. The main goal is to go across lateral platforming levels,
trying to avoid different types of enemies and obstacles and using some useful
(but limited) items, such as mushrooms or fire flowers.

Due to their success, amusement and attractiveness, Mario series have
become a successful researching environment in the field of Computational Intelligence (CI) [1,5,6]. The most used framework is Mario AI, a modified version
of the game known as Infinite Mario Bros.,[2] an open-code application where the

---

[1] Designer and producer of Nintendo Ltd., and winner of the 2012 Príncipe de Asturias
Prize in Humanities and Communication
[2] http://www.mojang.com/notch/mario/

G. Nicosia and P. Pardalos (Eds.): LION 7, LNCS 7997, pp. 357–363, 2013.
DOI: 10.1007/978-3-642-44973-4_39, © Springer-Verlag Berlin Heidelberg 2013

researchers can implement, for instance interactive and autonomous behavioural routines, using the set of functions and variables it offers. Moreover, in order to motivate the scientific community to perform these studies, a competition is proposed three times a year, inside several famous conferences, it is called the *Mario AI Championship*,[3] and is composed by some tracks: Learning, Level generation, Turing test, and Gameplay. The latter is devoted to create the best autonomous agent (also known as bot) as possible for automatically playing and pass sequential levels with a growing difficulty.

This work presents different approaches of autonomous agents aimed to this track, which consider a behavioural model created by means of a finite state machine (FSM) [2]. This model, based on expert knowledge, has been latter improved by applying offline (not during game) optimisation using Genetic Algorithms (GAs) [3], in two different schemes.

These approaches have been widely tested and analysed, getting an optimal set of parameters for the EA and thus, very competent agents in a number of difficulty levels.

2   Mario AI: Competition and Environment

The proposed agents follow the rules of the Mario AI Championship, considering the GamePlay track (complete as many levels as possible). The game consists in moving the character, Mario, through bi-dimensional levels. He can move left and right, down (crouch), run (letting the button pushed), jump and shoot fireballs (when in "fire" mode).

The main goal is complete the level, whereas secondary goals could be killing enemies and collecting coins or other items. These items may be hidden and may cause Mario to change his state (for instance a fire flower placed 'inside' a block). The difficulty of the game lies in the presence of cliffs/gaps and enemies. Mario loses power (i.e., its status goes down one level) when touched by an enemy and dies if he falls off a cliff.

The Mario AI simulator provides information about Mario's surrounding areas. According to the rules of the competition, two matrices give this information, both of them are 19×19 cells size, centred in Mario. One contains the positions of surrounding enemies, and the other provides information about the objects in the area (scenery objects and items).

Every tick (40 ms), Mario's next *action* must be indicated. This action consist in a combination of the five possible movements that Mario can do (left, right, down, fire/speed, jump). This information is encoded into a boolean array, containing a true value when a movement must be done.

The action to perform depends, of course, in the scenery characteristics around Mario, but it is also important to know where the enemies are and their type. Thus, the agent could know if it is best to jump, shoot or avoid them. We have defined four main enemies groups according to what the agent needs

---

[3] http://www.marioai.org/

to do to neutralize them: one for enemies who die by a fireball/jump/Koopa shell, other for those who only die by a fireball, others which only die jumping on them, and finally others which just die by a Koopa shell.

3 Evolutionary FSM-Based Agent

The proposed agent, *evoFSM-Mario*, is based in a FSM which models a logical behaviour, and which has been designed following an expert player knowledge. We decided to combine this technique with EAs, since they have proved being an excellent adapting and optimisation method, very useful for improving pre-defined behavioural rules, as the FSMs model.

We have defined a table of possible states for the agent, including all the possible (valid) actions the agent can perform in a specific instant, i.e. feasible combinations of moving left, moving right, crouch (going down), jump and fire/run. Table 1 shows the codification of the states in boolean values.

Table 1. Codification of the feasible states of the FSM which will model the Mario agent's AI. 1 is *true/active*, 0 is *false/non − active*.

	St 0	St 1	St 2	St 3	St 4	St 5	St 6	St 7	St 8	St 9	St 10	St 11	St 12	St 13
Right	1	1	1	1	0	0	0	0	0	0	0	0	0	0
Left	0	0	0	0	1	1	1	1	0	0	0	0	0	0
Fire/Run	0	0	1	1	0	0	1	1	0	0	0	0	1	1
Jump	0	1	0	1	0	1	0	1	0	0	1	1	0	1
Down	0	0	0	0	0	0	0	0	0	1	0	1	0	0

Depending on the input string, the state changes (or remains), so the transition is decided. Inputs are the possible situations of the agent in the environment: for example, find an enemy or being near a cliff/gap. Each input is represented as a boolean string, having a *true (1)* value in a position if a specific situation or event has happened.

These possible states along with the possible inputs and transitions will be evolved by means of the GA, considering that the output state for a new entry is randomly set, but according to a probability which models the preference of the states, in order to improve the convergence of the algorithm. Due to the huge search space a parameter indicating the percentage of new individuals that will be added in each generation is included, in order to control the diversity rate.

Every individual in the population is represented by a set of tables, one per state, and every table contains an output state for every possible input. The *fitness function* is calculated for each individual by setting the FSM represented in a chromosome as the AI of one agent. It is then placed in a level, and then, it plays for obtaining a fitness value. Two different schemes have been implemented: *mono-seed*, where all the individuals are tested in the same level (with the same difficulty), which grows in length with the generations; and a *multi-seed* approach, where every individual is tested in 30 levels (in the same difficulty) generated randomly (using different seeds). In both cases every agent plays until it pass the level, dies or gets stacked.

The aim of the latter scheme is: first, avoid the usual noise [4] present in this type of problems (videogames), i.e. it tries to get a fair valuation for an individual, since the same configuration could represent an agent which is very good sometimes and quite bad some others, due to the stochasticity present in every play (just in the agent's behaviour); and second, get individuals prepared to a wide set of situations in every level and difficulty, since 30 levels should present a high amount of different scenarios and configurations.

Thus, there is a *generic fitness* which has as restriction completely finish the level to be set to positive. On the contrary, individuals that have not finished the level start from the lowest fitness possible and their negativity is reduced according the behaviour during the level run. This generic fitness is a weighted aggregation based in the values: *marioWinner* (1 if the agent finish the level), *marioSize* (0 small, 1 big, and 2 fire), *numKilledEnemies*, *numTotalEnemies*, *numCellsPassed*, *remainingTime*, *timeSpent*, *coinsGathered*, *totalCoins*, *numCollisions* (number of times the agent has bumped with an enemy), *numGatheredPowerUps*, *causeOfDeath* (value representing how the agent has died).

This fitness is considered as the result of the evaluation for the individuals in the mono-seed approach, meanwhile multi-seed considers a *hierarchical fitness*, where the population is ordered according the next criteria: First, taking into account the percentage of levels where the individuals have been stacked or fallen from a cliff. Then, they are ordered considering the average percentage of levels completed. Finally, the individuals are ordered by the average generic fitness.

The *selection mechanism* considers the best individual and a percentage of the best ones, selected by tournament according to their fitness. The percentage of individuals to consider as parents follows to schemes: in mono-seed it is low at the beginning and will be increased when the number of generations grows; in multi-seed it is constant.

Uniform *Crossover* is performed considering the best individual of the present generation as one of the parents, and one of the individuals with positive fitness as the other parent. They generate a number of descendents which depends on the percentage of population to complete with the crossover.

The *Mutation operator* selects a percentage of individuals to be mutated, and a random set of genes to be changed in every individual, then the output state is randomly changed for an input in the table.

There is a 1 − *elitism replacement* to form the new population (the best individual survives). The rest of the population is composed by the offspring generated in the previous generation (a percentage of the global population) and a set of random individuals, in order to increase the diversity.

4   Experiments and Results

In order to test the approaches, several experiments have been conducted. The aim was to find good enough agents for completing any level in any difficulty.

Table 2. Optimal parameter values for mono- and multi-seed approaches.

	Mono-seed	Multi-seed
Population size	1000 (difficulty 0)	2000 (difficulty 4)
Number of generations	30 (difficulty 0)	500 (difficulty 4)
Crossover percentage	95 %	95 %
Mutation percentage	2 % (individuals)	2 % (individuals)
Mutation rate	1 % (genes)	1 % (genes)
Percentage of random individuals	5 % (decreased with the generations)	5 % (constant)
Fitness function	generic (aggregation)	hierarchical

Previously it was performed a hard fine-tuning stage (through systematic experimentation), where several different values for the parameters were tested, searching for the best configuration. The best values are presented in Table 2.

The values in the table (mono-seed) show that a small number of generations is required to get a competent agent in difficulty level 0, along with a population size quite high since every individual is just test once in this approach, so it is needed to ensure the evolution, even considering that none of the individuals may not complete a level. In that cases, some other generations are run to improve them and give another step in the evolution process.

It is important to remark that a single play of an agent could spend around 40–50 s on average (depending on the level length and difficulty), because it must be played in real-time, not simulated. So a single evaluation for one individual in the multi-seed approach could take around 25 min.

Moreover, when the experiments were conducted, it could be noticed than the most difficult levels strongly limited the algorithmic behaviour of the approaches, making it hard to converge and even ending the execution abruptly. In addition to the high computational cost (one run may take several days), there was a problem with the structure that stores the FSM of the individuals, since it is huge (in memory terms) and grows exponentially during the run (new inputs and outputs are added to the tables in crossovers), so in levels higher than 4, the program frequently crashes. Thus, this structure implementation was redesigned to an optimal one, letting to evolve agents in all the possible levels, in the mono-seed approach, but with a shorter number of generations than recommended.

In multi-seed case the memory problem still remained, so the number of states where reduced to 12, by deleting those considered as non-useful (in Table 1): State 8 (no action is done) and State 11 (Mario jumps and crouch, since these actions are not possible simultaneously in this implementation of Infinite Mario). With this change, it was possible optimising competent agents from difficulty levels 0 to 4, which are, in turn, enough for the GamePlay competition.

The fitness evolution was studied, showing a grow (improvement) with the generations, as expected in a GA. Moreover there are always enough individuals with positive fitness to assure the offspring generation (at least in the easiest levels). Some examples of the obtained evolved FSM-based agents can be seen in action (in a video) from the next urls:

- *Difficulty level 1* (*completed*): http://www.youtube.com/watch?v=6Pj6dZCEO7O
- *Difficulty level 2* (*completed*): http://www.youtube.com/watch?v=gtfuY-LOWDA
- *Difficulty level 3* (*completed*): http://www.youtube.com/watch?v=qQVQ43sWwYY
- *Difficulty level 12* (*stacked*): http://www.youtube.com/watch?v=zNGfBApX7sk

The last one was evolved for some generations (not all the desired) in that level of difficulty, due to the commented problems, so it cannot complete this hard level in the simulator. However, as it can be seen, it is quite good in the first part of the play. Thus, if we could finish the complete evolution process in this difficulty level we think the agent could complete any possible level.

5   Conclusions

In this work, two different approaches for evolving, by means of Genetic Algorithms (GAs), agents which play Super Mario Bros. game have been proposed and analysed. They have been implemented using Finite State Machine (FSM) models, and considering different schemes: mono-seed and a multi-seed evaluation approaches, along with two different fitness functions. Both algorithms have been tested inside a simulator named Mario AI, implemented for the Mario AI Competition, focusing on the GamePlay Track.

Several experiments have been conducted to test the algorithms and a deep analysis has been performed in each case, in order to set the best configuration parameters for the GA. Some problems have arisen such as the high memory requirements, which have done it hard to complete the optimisation process in several cases. However, very competent agents have been obtained for the difficulty levels 0 to 4 in both approaches, which are, in turn, enough for the Game Play competition requirements.

In the comparison between the approaches, mono-seed can yield excellent agents for the level where they were 'trained' (evolved), having a quite bad behaviour in a different level. Multi-seed takes much more computational time and has higher resource requirements, but the agents it yields are very good playing in any level of the considered difficulty (in the evolution). All these agents play much better than an expert human player and can complete the levels in a time impossible to get for the human.

Acknowledgements. This work has been supported in part by the P08-TIC-03903 and P10-TIC-6083 projects awarded by the Andalusian Regional Government, the FPU Grant 2009-2942 and the TIN2011-28627-C04-01 and TIN2011-28627-C04-02 projects, awarded by the Spanish Ministry of Science and Innovation.

References

1. Bojarski, S., Bates-Congdon, C.: REALM: A rule-based evolutionary computation agent that learns to play mario. In: Proceedings of the IEEE CIG 2011, pp. 83–90. IEEE Press (2011)

2. Booth, T.L.: Sequential Machines and Automata Theory, 1st edn. Wiley, New York (1967)
3. Goldberg, D.E., Korb, B., Deb, K.: Messy genetic algorithms: motivation, analysis, and first results. Complex Syst. **3**(5), 493–530 (1989)
4. Mora, A.M., Fernández-Ares, A., Merelo-Guervós, J.-J., García-Sánchez, P.: Dealing with noisy fitness in the design of a RTS game bot. In: Di Chio, C., et al. (eds.) EvoApplications 2012. LNCS, vol. 7248, pp. 234–244. Springer, Heidelberg (2012)
5. Pedersen, C., Togelius, J., Yannakakis, G.: Modeling player experience in super mario bros. In: Proceedings 2009 IEEE Symposium on Computational Intelligence and Games (CIG'09), pp. 132–139. IEEE Press (2009)
6. Togelius, J., Karakovskiy, S., Koutnik, J., Schmidhuber, J.: Super mario evolution. In: Proceedings 2009 IEEE Symposium on Computational Intelligence and Games (CIG'09), pp. 156–161. IEEE Press (2009)

Identifying Key Algorithm Parameters and Instance Features Using Forward Selection

Frank Hutter, Holger H. Hoos, and Kevin Leyton-Brown[✉]

University of British Columbia, 2366 Main Mall, Vancouver BC, V6T 1Z4, Canada
{hutter,hoos,kevinlb}@cs.ubc.ca

Abstract. Most state-of-the-art algorithms for large-scale optimization problems expose free parameters, giving rise to combinatorial spaces of possible configurations. Typically, these spaces are hard for humans to understand. In this work, we study a model-based approach for identifying a small set of both algorithm parameters and instance features that suffices for predicting empirical algorithm performance well. Our empirical analyses on a wide variety of hard combinatorial problem benchmarks (spanning SAT, MIP, and TSP) show that—for parameter configurations sampled uniformly at random—very good performance predictions can typically be obtained based on just two key parameters, and that similarly, few instance features and algorithm parameters suffice to predict the most salient algorithm performance characteristics in the combined configuration/feature space. We also use these models to identify settings of these key parameters that are predicted to achieve the best overall performance, both on average across instances and in an instance-specific way. This serves as a further way of evaluating model quality and also provides a tool for further understanding the parameter space. We provide software for carrying out this analysis on arbitrary problem domains and hope that it will help algorithm developers gain insights into the key parameters of their algorithms, the key features of their instances, and their interactions.

1 Introduction

State-of-the-art algorithms for hard combinatorial optimization problems tend to expose a set of parameters to users to allow customization for peak performance in different application domains. As these parameters can be instantiated independently, they give rise to combinatorial spaces of possible parameter configurations that are hard for humans to handle, both in terms of finding good configurations and in terms of understanding the impact of each parameter. As an example, consider the most widely used mixed integer programming (MIP) software, IBM ILOG CPLEX, and the manual effort involved in exploring its 76 optimization parameters [1].

By now, substantial progress has been made in addressing the first sense in which large parameter spaces are hard for users to deal with. Specifically, it has been convincingly demonstrated that methods for *automated algorithm*

G. Nicosia and P. Pardalos (Eds.): LION 7, LNCS 7997, pp. 364–381, 2013.
DOI: 10.1007/978-3-642-44973-4_40, © Springer-Verlag Berlin Heidelberg 2013

configuration [2–7] are able to find configurations that substantially improve the state of the art for various hard combinatorial problems (e.g., SAT-based formal verification [8], mixed integer programming [1], timetabling [9], and AI planning [10]). However, much less work has been done towards the goal of explaining to algorithm designers which parameters are important and what values for these important parameters lead to good performance. Notable exceptions in the literature include experimental design based on linear models [11,12], an entropy-based measure [2], and visualization methods for interactive parameter exploration, such as contour plots [13]. However, to the best of our knowledge, none of these methods has so far been applied to study the configuration spaces of state-of-the-art highly parametric solvers; their applicability is unclear, due to the high dimensionality of these spaces and the prominence of discrete parameters (which, e.g., linear models cannot handle gracefully).

In the following, we show how a generic, model-independent method can be used to:

- identify key parameters of highly parametric algorithms for solving SAT, MIP, and TSP;
- identify key instance features of the underlying problem instances;
- demonstrate interaction effects between the two; and
- identify values of these parameters that are predicted to yield good performance, both unconditionally and conditioned on instance features.

Specifically, we gather performance data by randomly sampling both parameter settings and problem instances for a given algorithm. We then perform *forward selection*, iteratively fitting regression models with access to increasing numbers of parameters and features, in order to identify parameters and instance features that suffice to achieve predictive performance comparable to that of a model fit on the full set of parameters and instance features. Our experiments show that these sets of sufficient parameters and/or instance features are typically very small—often containing only two elements—even when the candidate sets of parameters and features are very large. To understand what values these key parameters should take, we find performance-optimizing settings given our models, both unconditionally and conditioning on our small sets of instance features. We demonstrate that parameter configurations that set as few as two key parameters based on the model (and all other parameters at random) often substantially outperform entirely random configurations (sometimes by up to orders of magnitude), serving as further validation for the importance of these parameters. Our qualitative results still hold for models fit on training datasets containing as few as 1 000 data points, facilitating the use of our approach in practice. We conclude that our approach can be used out-of-the-box by algorithm designers wanting to understand key parameters, instance features, and their interactions. To facilitate this, our software (and data) is available at http://www.cs.ubc.ca/labs/beta/Projects/EPMs.

2   Methods

Ultimately, our forward selection methods aim to identify a set of the k_{max} most important algorithm parameters and m_{max} most important instance features (where k_{max} and m_{max} are user-defined), as well as the best values for these parameters (both on average across instances and on a per-instance basis). Our approach for solving this problem relies on predictive models, learned from given algorithm performance data for various problem instances and parameter configurations. We identify important parameters and features by analyzing which inputs suffice to achieve high predictive accuracy in the model, and identify good parameter values by optimizing performance based on model predictions.

2.1   Empirical Performance Models

Empirical Performance Models (EPMs) are statistical models that describe the performance of an algorithm as a function of its inputs. In the context of this paper, these inputs comprise both features of the problem instance to be solved and the algorithm's free parameters. We describe a problem instance by a vector of m features $z = [z_1, \ldots, z_m]^\intercal$, drawn from a given *feature space* $\mathcal{F}$. These features must be computable by an automated, domain-specific procedure that efficiently extracts features for any given problem instance (typically, in low-order polynomial time w.r.t. the size of the given problem instance). We describe the *configuration space* of a parameterized algorithm with k parameters $\theta_1, \ldots, \theta_k$ and respective domains $\Theta_1, \ldots, \Theta_k$ by a subset of the cross-product of parameter domains: $\Theta \subseteq \Theta_1 \times \cdots \times \Theta_k$. The elements of Θ are complete instantiations of the algorithm's k parameters, and we refer to them as *configurations*. Taken together, the configuration and the feature space define the *input space* $\mathcal{I} := \Theta \times \mathcal{F}$.

EPMs for predicting the "empirical hardness" of instances have their origin over a decade ago [14–17] and have been the preferred core reasoning tool of early state-of-the-art methods for the algorithm selection problem (which aim to select the best algorithm for a given problem, dependent on its features [18–20]), in particular of early iterations of the SATzilla algorithm selector for SAT [21]. Since then, these predictive models have been extended to model the dependency of performance on (often categorical) algorithm parameters, to make probabilistic predictions, and to work effectively with large amounts of training data [11,12,22,23].

In very recent work, we comprehensively studied EPMs based on a variety of modeling techniques that have been used for performance prediction over the years, including ridge regression [17], neural networks [24], Gaussian processes [22], regression trees [25], and random forests [23]. Overall, we found random forests and approximate Gaussian processes to perform best. Random forests (and also regression trees) were particularly strong for very heterogeneous benchmark sets, since their tree-based mechanism automatically groups similar inputs together and does not allow widely different inputs to interfere with the predictions for a given group. Another benefit of the tree-based methods is apparent from the fact that hundreds of training data points could be shown to suffice to yield competitive performance predictions in joint input spaces induced by as many as 76 algorithm parameters and 138 instance features [23]. This strong performance suggests that the functions being modeled must be relatively simple, for example, by depending at most very weakly on most inputs. In this paper, we ask whether this is the case, and to the extent that this is so, aim to identify the key inputs.

2.2 Forward Selection

There are many possible approaches for identifying important input dimensions of a model. For example, one can measure the model coefficients w in ridge regression (large coefficients mean that small changes in a feature value have a large effect on predictions, see, e.g., [26]) or the length scales λ in Gaussian process regression (small length scales mean that small changes in a feature value have a large effect on predictions, see, e.g., [27]). In random forests, to measure the importance of input dimension i, Breiman suggested perturbing the values in the ith column of the out-of-bag (or validation) data and measuring the resulting loss in predictive accuracy [28].

All of these methods run into trouble when input dimensions are highly correlated. While this does not occur with randomly sampled parameter configurations, it does occur with instance features, which cannot be freely sampled. Our goal is to build models that yield good predictions but yet depend on as few input dimensions as possible; to achieve this goal, it is not sufficient to merely find important parameters, but we need to find a set of important parameters that are as uncorrelated as possible.

Forward selection is a generic, model-independent tool that can be used to solve this problem [17,29].[1] Specifically, this method identifies sets of model inputs that are jointly *sufficient* to achieve good predictive accuracy; our variant of it is defined in Algorithm 1. After initializing the complete input set I and the subset of important inputs S in lines 1–2, the outer **for**-loop incrementally adds one input at a time to S. The **forall**-loop over inputs i not yet contained in S (and not violating the constraint of adding at most k_{max} parameters and m_{max} features) uses validation data to compute err(i), the root mean squared error (RMSE) for a model containing i and the inputs already in S. It then adds the input resulting in lowest RMSE to S. Because inputs are added one at a time, highly correlated inputs will only be added if they provide large *marginal* value to the model.

Note that we simply call procedure *learn* with a subset of input dimensions, regardless of whether they are numerical or categorical (for models that require a so-called "1-in-K encoding" to handle categorical parameters, this means we introduce/drop all K binary columns representing a K-ary categorical input at once). Also note that, while here, we use prediction RMSE on the validation set

---

[1] A further advantage of forward selection is that it can be used in combination with arbitrary modeling techniques. Although here, we focus on using our best-performing model, random forests, we also provide summary results for other model types.

Algorithm 1: Algorithm 1: Forward Selection

In line 10, *learn* refers to an arbitrary regression method that fits a function f to given training data. Note that input dimensions $1, \ldots, k$ are parameters, $k+1, \ldots, k+m$ are features.

Input : Training data $\mathcal{D}_{train} = \langle (\mathbf{x}_1, y_1), \ldots, (\mathbf{x}_n, y_n) \rangle$; validation data
$\mathcal{D}_{valid} = \langle (\mathbf{x}_{n+1}, y_{n+1}), \ldots, (\mathbf{x}_{n+n'}, y_{n+n'}) \rangle$; number of parameters, k; number of features, m; desired number $K \leq d = k + m$ of key inputs; bound on number of key parameters, $k_{max} \geq 0$; bound on number of key features, $m_{max} \geq 0$, such that $k_{max} + m_{max} \geq K$

Output: Subset of K feature indices $S \subseteq \{1, \ldots, d\}$

1 $I \leftarrow \{1, \ldots, d\}$;

2 $S \leftarrow \emptyset$;

3 **for** $j = 1, \ldots, K$ **do**

4      $I_{\text{allowed}} \leftarrow I \setminus S$;

5   **if** $|S \cap \{1, \ldots, k\}| \geq k_{max}$ **then** $I_{\text{allowed}} \leftarrow I_{\text{allowed}} \setminus \{1, \ldots, k\}$;

6   **if** $|S \cap \{k+1, \ldots, k+m\}| \geq m_{max}$ **then**
    $I_{\text{allowed}} \leftarrow I_{\text{allowed}} \setminus \{k+1, \ldots, k+m\}$;

7   **forall** $i \in I_{allowed}$ **do**

8          $S \leftarrow S \cup \{i\}$;

9   **forall** $(\mathbf{x}_j, y_j) \in \mathcal{D}_{train}$ **do** $\mathbf{x}_j^S \leftarrow \mathbf{x}_j$ restricted to input dimensions in S;

10         $f \leftarrow learn(\langle (\mathbf{x}_1^S, y_1), \ldots, (\mathbf{x}_n^S, y_n) \rangle)$;

11         $\text{err}(i) \leftarrow \sqrt{\sum_{(\mathbf{x}_j, y_j) \in \mathcal{D}_{valid}} (f(\mathbf{x}_j) - y_j)^2}$;

12         $S \leftarrow S \setminus \{i\}$;

13     $\hat{\imath} \leftarrow$ random element of $\arg \max_i \text{err}(i)$;

14     $S \leftarrow S \cup \{\hat{\imath}\}$;

15 **return** S;

---

to assess the value of adding input i, forward selection can also be used with any other objective function.[2]

Having selected a set S of inputs via forward selection, we quantify their relative importance following the same process used by Leyton-Brown et al. to determine the importance of instance features [17], which is originally due to [31]: we simply drop one input from S at a time and measure the increase in predictive RMSE. After computing this increase for each feature, we normalize by dividing by the maximal RMSE increase and multiplying by 100.

We note that forward selection can be computationally costly due to its need for repeated model learning: for example, to select 5 out of 200 inputs via forward selection requires the construction and validation of $200 + 199 + 198 + 197 + 196 = 990$ models. In our experiments, this process required up to a day of CPU time.

---

[2] In fact, it also applies to classification algorithms and has, e.g., been used to derive classifiers for predicting the solubility of SAT instances based on 1–2 features [30].

2.3 Selecting Values for Important Parameters

Given a model f that takes k parameters and m instance features as input and predicts a performance value, we identify the best values for the k parameters by optimizing predictive performance according to the model. Specifically, we predict the performance of the partial parameter configuration $\mathbf{x}$ (instantiating k parameter values) on a problem instance with m selected instance features $\mathbf{z}$ as $f([\mathbf{x}^\mathsf{T}, \mathbf{z}^\mathsf{T}]^\mathsf{T})$. Likewise, we predict its average performance across n instances with selected instance features $\mathbf{z}_1, \ldots, \mathbf{z}_n$ as $\sum_{j=1}^{n} \frac{1}{n} \cdot f([\mathbf{x}^\mathsf{T}, \mathbf{z}_j^\mathsf{T}]^\mathsf{T})$.

3 Algorithm Performance Data

In this section, we discuss the algorithm performance data we used in order to evaluate our approach. We employ data from three different combinatorial problems: propositional satisfiability (SAT), mixed integer programming (MIP), and the traveling salesman problem (TSP). All our code and data is available online: instances and their features (and feature computation code & binaries), parameter specification files and wrappers for the algorithms, as well as the actual runtime data upon which our analysis is based.

3.1 Algorithms and Their Configuration Spaces

We employ peformance data from three algorithms: CPLEX for MIP, SPEAR for SAT, and LK-H for TSP. The parameter configuration spaces of these algorithms are summarized in Table 1.

IBM ILOG CPLEX [32] is the most-widely used commercial optimization tool for solving MIPs; it is used by over 1 300 corporations (including a third of the Global 500) and researchers at more than 1 000 universities. We used the same configuration space with 76 parameters as in previous work [1], excluding all CPLEX settings that change the problem formulation (e.g., the optimality gap below which a solution is considered optimal). Overall, we consider 12 preprocessing parameters (mostly categorical); 17 MIP strategy parameters (mostly

Table 1. Algorithms and their parameter configuration spaces studied in our experiments.

Algorithm	Parameter type	# parameters of this type	# values considered	Total # configurations
	Boolean	6	2	
CPLEX	Categorical	45	3–7	1.90×10^{47}
	Integer	18	5–7	
	Continuous	7	5–8	
	Categorical	10	2–20	
SPEAR	Integer	4	5–8	8.34×10^{17}
	Continuous	12	3–6	
	Boolean	5	2	
LK-H	Categorical	8	3–10	6.91×10^{14}
	Integer	10	3–9	

categorical); 11 categorical parameters deciding how aggressively to use which types of cuts; 9 real-valued MIP "limit" parameters; 10 simplex parameters (half of them categorical); 6 barrier optimization parameters (mostly categorical); and 11 further parameters. In total, and based on our discretization of continuous parameters, these parameters gave rise to 1.90×10^{47} unique configurations.

SPEAR [33] is a state-of-the-art SAT solver for industrial instances. With appropriate parameter settings, it was shown to be the best available solver for certain types of SAT-encoded hardware and software verification instances [8] (the same IBM and SWV instances we use here). It also won the quantifier-free bit-vector arithmetic category of the 2007 Satisfiability Modulo Theories Competition. We used exactly the same 26-dimensional parameter configuration space as in previous work [8]. SPEAR's categorical parameters mainly control heuristics for variable and value selection, clause sorting, resolution ordering, and also enable or disable optimizations, such as the pure literal rule. Its numerical parameters mainly deal with activity, decay, and elimination of variables and clauses, as well as with the randomized restart interval and percentage of random choices. In total, and based on our discretization of continuous parameters, SPEAR has 8.34×10^{17} different configurations.

LK-H [34] is a state-of-the-art local search solver for TSP based on an efficient implementation of the Lin-Kernighan heuristic. We used the LK-H code from Styles et al. [35], who first reported algorithm configuration experiments with LK-H; in their work, they extended the official LK-H version 2.02 to allow several parameters to scale with instance size and to make use of a simple dynamic restart mechanism to prevent stagnation. The modified version has a total of 23 parameters governing all aspects of the search process, with an emphasis on parameterizing moves. In total, and based on our discretization of continuous parameters, LK-H has 6.91×10^{14} different configurations.

3.2 Benchmark Instances and Their Features

We used the same benchmark distributions and features as in previous work [23] and only describe them on a high level here. For MIP, we used two instance distributions from computational sustainability (RCW and CORLAT), one from winner determination in combinatorial auctions (REG), two unions of these (CR := CORLAT ∪ RCW and CRR := CORLAT ∪ REG ∪ RCW), and a large and diverse set of publicly available MIP instances (BIGMIX). We used 121 features to characterize MIP instances, including features describing problem size, the variable-constraint graph, the constraint matrix, the objective function values, an LP programming relaxation, various probing features extracted from short CPLEX runs and timing features measuring the computational expense required for various groups of features.

For SAT, we used three sets of SAT-encoded formal verification benchmarks: SWV and IBM are sets of software and hardware verification instances, and SWV-IBM is their union. We used 138 features to characterize SAT instances, including features describing problem size, three graph representations, syntactic features, probing features based on systematic solvers (capturing unit propagation and clause learning) and local search solvers, an LP relaxation, survey propagation, and timing features.

For TSP, we used TSPLIB, a diverse set of prominent TSP instances, and computed 64 features, including features based on problem size, cost matrix, minimum spanning trees, branch & cut probing, local search probing, ruggedness, and node distribution, as well as timing features.

3.3 Data Acquisition

We gathered a large amount of runtime data for these solvers by executing them with various configurations and instances. Specifically, for each combination of solver and instance distribution (CPLEX run on MIP, SPEAR on SAT, and LK-H on TSP instances), we measured the runtime of each of $M = 1\,000$ randomly-sampled parameter configurations on each of the P problem instances available for the distribution, with P ranging from 63 to 2\,000. The resulting runtime observations can be thought of as a $M \times P$ matrix. Since gathering this runtime matrix meant performing $M \cdot P$ (i.e., between 63\,000 and 2\,000\,000) runs per dataset, we limited each single algorithm run to a cutoff time of 300 CPU seconds on one node of the Westgrid cluster Glacier (each of whose nodes is equipped with two 3.06 GHz Intel Xeon 32-bit processors and 2–4 GB RAM). While collecting this data required substantial computational resources (between 1.3 CPU years and 18 CPU years per dataset), we note that this much data was only required for the thorough empirical analysis of our methods; in practice, our methods are often surprisingly accurate based on small amounts of training data. For all our experiments, we partitioned both instances and parameter configurations into training, validation, and test sets; the training sets (and likewise, the validation and test sets) were formed as subsamples of training instances and parameter configurations. We used 10\,000 training subsamples throughout our experiments but demonstrate in Sect. 4.3 that qualitatively similar results can also be achieved based on subsamples of 1\,000 data points.

We note that sampling parameter configurations uniformly at random is not the only possible way of collecting training data. Uniform sampling has the advantage of producing unbiased training data, which in turn gives rise to models that can be expected to perform well on average across the entire configuration space. However, because algorithm designers typically care more about regions of the configuration space that yield good performance, in future work, we also aim to study models based on data generated through a biased sequential sampling approach (as is implemented, e.g., in model-based algorithm configuration methods, such as SMAC [6]).

4   Experiments

We carried out various computational experiments to identify the quality of models based on small subsets of features and parameters identified using forward selection, to quantify which inputs are most important, and to determine good values for the selected parameters. All our experiments made use of the algorithm performance data described in Sect. 3, and consequently, our claims hold on average across the entire configuration space. Whether they also apply to biased samples from the configuration space (in particular, regions of very strong algorithm performance) is a question for future work.

4.1 Predictive Performance for Small Subsets of Inputs

First, we demonstrate that forward selection identifies sets of inputs yielding low predictive root mean squared error (RMSE), for predictions in the feature space, the parameter space, and their joint space. Figure 1 shows the root mean squared error of models fit with parameter/feature subsets of increasing size. Note in particular the horizontal line, giving the RMSE of a model based on *all* inputs, and that the RMSE of subset models already converges to this performance with few inputs. In the feature space, this has been observed before [17, 29] and is intuitive, since the features are typically very correlated, allowing a subset of them to represent the rest. However, the same cannot be said for the parameter

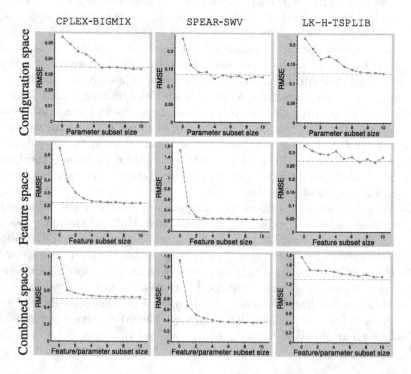

Fig. 1. Predictive quality of random forest models as a function of the number of allowed parameters/features selected by forward selection for 3 example datasets. The inputless prediction (subset size zero) is the mean of all data points. The dashed horizontal line in each plot indicates the final performance of the model using the full set of parameters/features.

space: in our experimental design, parameter values have been sampled uniformly at random and are thus independent (i.e., uncorrelated) by design. Thus, this finding indicates that some parameters influence performance much more than others, to the point where knowledge of a few parameter values suffices to predict performance just as well as knowledge of all parameters.

Figure 2 focuses on what we consider to be the most interesting case, namely performance prediction in the joint space of instance features and parameter configurations. The figure qualitatively indicates the performance that can be achieved based on subsets of inputs of various sizes. We note that in some cases, in particular in the SPEAR scenarios, predictions of models using all inputs closely resemble the true performance, and that the predictions of models based on a few inputs tend to capture the salient characteristics of the full models. Since the instances we study vary widely in hardness, instance features tend to be more predictive than algorithm parameters, and are thus favoured by forward selection. This sometimes leads to models that *only* rely on instance features, yielding predictions that are constant across parameter configurations; for example, see the predictions with up to 10 inputs for dataset CPLEX-CORLAT (the second row in Fig. 2). While these models yield low RMSE, they are uninformative about parameter settings; this observation caused us to modify forward selection as discussed in Sect. 2.2 to limit the number of features/parameters selected.

4.2   Relative Importance of Parameters and Features

As already apparent from Fig. 1, knowing the values of a few parameters is sufficient to predict marginal performance across instances similarly well as when knowing all parameter values. Figure 3 shows *which* parameters were found to be important in different runs of our procedure. Note that the set of selected key parameters was remarkably robust across runs.

The most extreme case is SPEAR-SWV, for which SPEAR's variable selection heuristic (sp-var-dec-heur) was found to be the most important parameter every single time by a wide margin, followed by its phase selection heuristic (sp-phase-dec-heur). The importance of the variable selection heuristic for SAT solvers is well known, but it is surprising that the importance of this choice dominates so clearly. Phase selection is also widely known to be important for the performance of modern CDCL SAT solvers like SPEAR. As can be seen from Fig. 1 (top middle), predictive models for SPEAR-SWV based on 2 parameters essentially performed as well as those based on all parameters, as is also reflected in the very low importance ratings for all but these two parameters.

In the case of both CPLEX-BIGMIX and LK-H-TSPLIB, up to 5 parameters show up as important, which is not surprising, considering that predictive performance of subset models with 5 inputs converged to that of models with all inputs (see Fig. 1, top left and right). In the case of CPLEX, the key parameters included two controlling CPLEX's cutting strategy (mip_limits_cutsfactor and mip_limits_cutpasses, limiting the number of cuts to add, and the number of cutting plane passes,

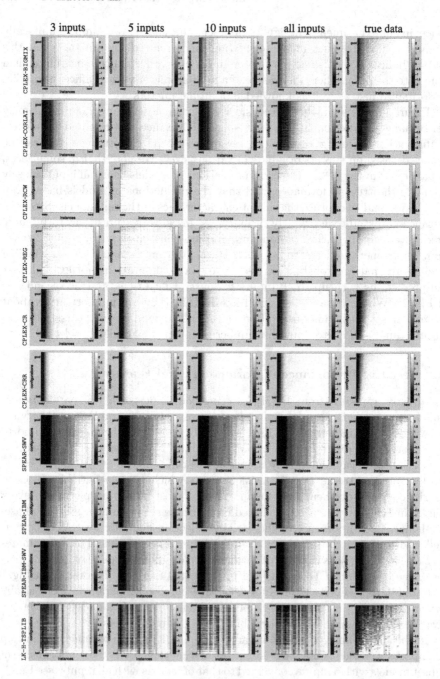

Fig. 2. Performance predictions by random forest models based on subsets of features and parameters. To generate these heatmaps, we ordered configurations by their average performance across instances, and instances by their average hardness across configurations; the same ordering (based on the true heatmap) was used for all heatmaps. All data shown is test data.

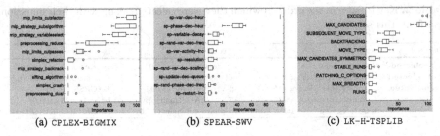

(a) CPLEX-BIGMIX   (b) SPEAR-SWV   (c) LK-H-TSPLIB

Fig. 3. Parameter importance for 3 example datasets. We show boxplots over 10 repeated runs with different random training/validation/test splits.

respectively), two MIP strategy parameters (mip_strategy_subalgorithm and mip_strategy_variableselect, determining the continuous optimizer used to solve subproblems in a MIP, and variable selection, respectively), and one parameter determining which kind of reductions to perform during preprocessing (preprocessing_reduce). In the case of LK-H, all top five parameters are related to moves, parameterizing candidate edges (EXCESS and MAX_CANDIDATES, limiting the maximum alpha-value allowed for any candidate edge, and the maximum number of candidate edges, respectively), and move types (MOVE_TYPE, BACK-TRACKING, SUBSEQUENT_MOVE_TYPE, specifying whether to use sequential k-opt moves, whether to use backtracking moves, and which type to use for moves following the first one in a sequence of moves).

To demonstrate the model independence of our approach, we repeated the same analysis based on other empirical performance models (linear regression, neural networks, Gaussian processes, and regression trees). Although overall, these models yielded weaker predictions, the results were qualitatively similar: for SPEAR, all models reliably identified the same two parameters as most important, and for the other datasets, there was an overlap of at least three of the top five ranked parameters. Since random forests yielded the best predictive performance, we focus on them in the remainder of this paper.

As an aside, we note that the fact that a few parameters dominate importance is in line with similar findings in the machine learning literature on the importance of hyperparameters, which has informed the analysis of a simple hyperparameter optimization algorithm [36] and the design of a Bayesian optimization variant for optimizing functions with high extrinsic but low intrinsic imensionality [37]. In future work, we plan to exploit this insight to design better automated algorithm configuration procedures.

Next, we demonstrate how we can study the joint importance of instance features and algorithm parameters. Since foward selection by itself chose mostly instance features, for this analysis we constrained it to select 3 features and 2 parameters. Table 2 lists the features and parameters identified for our 3 example datasets, in the order forward selection picked them. Since most instance features are strongly correlated with each other, it is important to measure and understand our importance metric in the context of the specific subset of inputs it is computed for. For example, consider the set of important features for dataset

Table 2. Key inputs, in the order in which they were selected, along with their omission cost from this set.

Dataset	CPLEX-BIGMIX	SPEAR-SWV	LK-H-TSPLIB
1st selected	cplex_prob_time (10.1)	Pre_featuretime (35.9)	tour_const_heu_avg (0.0)
2nd selected	obj_coef_per_constr2_std (7.7)	nclausesOrig (100.0)	cluster_distance_std (0.8)
3rd selected	vcg_constr_weight0_avg (30.2)	sp-var-dec-heur (32.6)	EXCESS (10.0)
4th selected	mip_limits_cutsfactor (8.3)	VCG_CLAUSE_entropy (34.5)	bc_nols_q25 (100.0)
5th selected	mip_strategy_subalgorithm (100.0)	sp-phase-dec-heur (27.6)	BACKTRACKING (0.0)

Table 3. Key parameters and their best fixed values as judged by an empirical performance model based on 3 features and 2 parameters.

Dataset	1st selected param	2nd selected param
CPLEX-BIGMIX	mip_limits_cutsfactor = 8	mip_strategy_subalgorithm = 2
CPLEX-CORLAT	mip_strategy_subalgorithm = 2	preprocessing_reduce = 3
CPLEX-REG	mip_strategy_subalgorithm = 2	mip_strategy_variableselect = 4
CPLEX-RCW	preprocessing_reduce = 3	mip_strategy_lbheur = no
CPLEX-CR	mip_strategy_subalgorithm = 0	preprocessing_reduce = 1
CPLEX-CRR	preprocessing_coeffreduce = 2	mip_strategy_subalgorithm = 2
SPEAR-IBM	sp-var-dec-heur = 2	sp-resolution = 0
SPEAR-SWV	sp-var-dec-heur = 2	sp-phase-dec-heur = 0
SPEAR-SWV-IBM	sp-var-dec-heur = 2	sp-use-pure-literal-rule = 0
LK-H-TSPLIB	EXCESS = −1	BACKTRACKING = NO

CPLEX-BIGMIX (Table 2, left). While the single most important feature in this case was cplex_prob_time (a timing feature measuring how long CPLEX probing takes), in the context of the other four features, its importance was relatively small; on the other hand, the input selected 5th, mip_strategy_subalgorithm (CPLEX's MIP strategy parameter from above) was the most important input in the context of the other 4. We also note that all algorithm parameters that were selected as important in this context of instance features (mip_limits_cutsfactor and mip_strategy_subalgorithm for CPLEX; sp-var-dec-heur and sp-phase-dec-heur for SPEAR; and EXCESS and BACKTRACKING for LK-H) were already selected and labeled important when considering only parameters. This finding increases our confidence in the robustness of this analysis.

4.3   Selecting Values for Key Parameters

Next, we used our subset models to identify which values the key parameters identified by forward selection should be set to. For each dataset, we used the same subset models of 3 features and 2 parameters as above; Table 3 lists the best predicted values for these 2 parameters. The main purpose of this experiment was to demonstrate that this analysis can be done automatically, and we thus

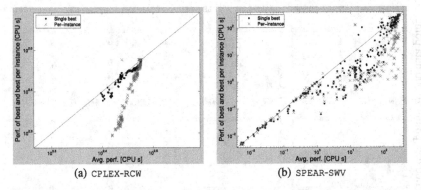

(a) CPLEX-RCW   (b) SPEAR-SWV

Fig. 4. Performance of random configurations *vs* configurations setting almost all parameters at random, but setting 2 key parameters based on an empirical performance model with 3 features and 2 parameters.

only summarize the results at a high level; we see them as a starting point that can inform domain experts about empirical properties of their algorithm in a particular application context and trigger further in-depth studies. At a high level, we note that CPLEX's parameter mip_strategy_subalgorithm (determining the continuous optimizer used to solve subproblems in a MIP) was important for most instance sets, the most prominent values being 2 (use CPLEX's dual simplex optimizer) and 0 (use CPLEX's auto-choice, which also defaults to the dual simplex optimizer). Another important choice was to set preprocessing_reduce to 3 (use both primal and dual reductions) or 1 (use only primal reductions), depending on the instance set. For SPEAR, the parameter determining the variable selection heuristic (sp-var-dec-heur) was the most important one in all 3 cases, with an optimal value of 2 (select variables based on their activity level, breaking ties by selecting the more frequent variable). For good average performance of LK-H on TSPLIB, the most important choices were to set EXCESS to −1 (use an instance-dependent setting of the reciprocal problem dimension), and to not use backtracking moves.

We also measured the performance of parameter configurations that actually set these parameters to the values predicted to be best by the model, both on average across instances and in an instance-specific way. This serves as a further way of evaluating model quality and also facilitates deeper understanding of the parameter space. Specifically, we consider parameter configurations that instantiate the selected parameters according to the model and assign all other parameter to randomly sampled values; we compare the performance of these configurations to that of configurations that instantiate *all* values at random. Figure 4 visualizes the result of this comparison for two datasets, showing that the model indeed selected values that lead to high performance: by just controlling two parameters, improvements of orders of magnitude could be achieved for some instances. Of course, this only compares to random configurations; in contrast to our work on algorithm configuration, here, our goal was to gain a better understanding of an

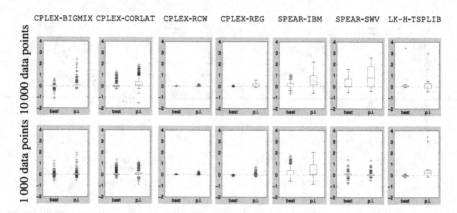

Fig. 5. $\log_{10}$ speedups over random configurations by setting almost all parameters at random, except 2 key parameters, values for which (fixed best, and best per instance) are selected by an empirical performance model with 3 features and 2 parameters. The boxplots show the distribution of $\log_{10}$ speedups across all problem instances; note that, e.g., a $\log_{10}$ speedup of 0, -1, and 1 mean identical performance, a 10-fold slowdown, and a 10-fold speedup, respectively. The dashed green lines indicate where two configurations performed the same, points above the line indicate speedups. Top: based on models trained on 10 000 data points; bottom: based on models trained on 1 000 data points.

algorithms' parameter space rather than to improve over its manually engineered default parameter settings.[3] However, we nevertheless believe that the speedups achieved by setting only the identified parameters to good values demonstrate the importance of these parameters. While Fig. 4 only covers 2 datasets, Fig. 5 (top) summarizes results for a wide range of datasets. Figure 5 (bottom) demonstrates that predictive performance does not degrade much when using sparser training data (here: 1 000 instead of 10 000 training data points); this is important for facilitating the use of our approach in practice.

5   Conclusions

In this work, we have demonstrated how forward selection can be used to analyze algorithm performance data gathered using randomly sampled parameter configurations on a large set of problem instances. This analysis identified small sets of key algorithm parameters and instance features, based on which the performance of these algorithms could be predicted with surprisingly high accuracy. Using this fully automated analysis technique, we found that for high-performance solvers for some of the most widely studied NP-hard combinatorial problems, namely SAT, MIP and TSP, only very few key parameters (often just two of dozens) largely determine algorithm performance. Automatically constructed performance models, in our case based on random forests, were of sufficient

---

[3] In fact, in many cases, the best setting of the key parameters were their default values.

quality to reliably identify good values for these key parameters, both on average across instances and dependent on key instance features. We believe that our rather simple importance analysis approach can be of great value to algorithm designers seeking to identify key algorithm parameters, instance features, and their interaction.

We also note that the finding that the performance of these highly parametric algorithms mostly depends on a few key parameters has broad implications on the design of algorithms for NP-hard problems, such as the ones considered here, and of future algorithm configuration procedures.

In future work, we aim to reduce the computational cost of identifying key parameters; to automatically identify the relative performance obtained with their possible values; and to study which parameters are important in high-performing regions of an algorithm's configuration space.

References

1. Hutter, F., Hoos, H.H., Leyton-Brown, K.: Automated configuration of mixed integer programming solvers. In: Proceedings of CPAIOR-10, pp. 186–202 (2010)
2. Nannen, V., Eiben, A.E.: Relevance estimation and value calibration of evolutionary algorithm parameters. In: Proceedings of IJCAI-07, pp. 975–980 (2007)
3. Ansotegui, C., Sellmann, M., Tierney, K.: A gender-based genetic algorithm for the automatic configuration of solvers. In: Proceedings of CP-09, pp. 142–157 (2009)
4. Birattari, M., Yuan, Z., Balaprakash, P., Stützle, T.: F-race and iterated F-race: an overview. In: Bartz-Beielstein, T., Chiarandini, M., Paquete, L., Preuss, M. (eds.) Empirical Methods for the Analysis of Optimization Algorithms. Springer, Heidelberg (2010)
5. Hutter, F., Hoos, H.H., Leyton-Brown, K., Stützle, T.: ParamILS: an automatic algorithm configuration framework. JAIR **36**, 267–306 (2009)
6. Hutter, F., Hoos, H.H., Leyton-Brown, K.: Sequential model-based optimization for general algorithm configuration. In: Coello, C.A.C. (ed.) LION 5. LNCS, vol. 6683, pp. 507–523. Springer, Heidelberg (2011)
7. Hutter, F., Hoos, H.H., Leyton-Brown, K.: Parallel algorithm configuration. In: Hamadi, Y., Schoenauer, M. (eds.) LION 2012. LNCS, vol. 7219, pp. 55–70. Springer, Heidelberg (2012)
8. Hutter, F., Babić, D., Hoos, H.H., Hu, A.J.: Boosting verification by automatic tuning of decision procedures. In: Proceedings of FMCAD-07, pp. 27–34 (2007)
9. Chiarandini, M., Fawcett, C., Hoos, H.: A modular multiphase heuristic solver for post enrolment course timetabling. In: Proceedings of PATAT-08 (2008)
10. Vallati, M., Fawcett, C., Gerevini, A.E., Hoos, H.H., Saetti, A.: Generating fast domain-optimized planners by automatically configuring a generic parameterised planner. In: Proceedings of ICAPS-PAL11 (2011)
11. Ridge, E., Kudenko, D.: Sequential experiment designs for screening and tuning parameters of stochastic heuristics. In: Proceedings of PPSN-06, pp. 27–34 (2006)
12. Chiarandini, M., Goegebeur, Y.: Mixed models for the analysis of optimization algorithms. In: Bartz-Beielstein, T., Chiarandini, M., Paquete, L., Preuss, M. (eds.) Experimental Methods for the Analysis of Optimization Algorithms, pp. 225–264. Springer, Berlin (2010)

13. Bartz-Beielstein, T.: Experimental Research in Evolutionary Computation: The New Experimentalism. Natural Computing Series. Springer, Berlin (2006)
14. Finkler, U., Mehlhorn, K.: Runtime prediction of real programs on real machines. In: Proceedings of SODA-97, pp. 380–389 (1997)
15. Fink, E.: How to solve it automatically: selection among problem-solving methods. In: Proceedings of AIPS-98, pp. 128–136. AAAI Press (1998)
16. Howe, A.E., Dahlman, E., Hansen, C., Scheetz, M., Mayrhauser, A.: Exploiting competitive planner performance. In: Biundo, S., Fox, M. (eds.) ECP 1999. LNCS, vol. 1809, pp. 62–72. Springer, Heidelberg (2000)
17. Leyton-Brown, K., Nudelman, E., Shoham, Y.: Learning the Empirical Hardness of Optimization Problems. In: Hentenryck, P. (ed.) CP 2002. LNCS, vol. 2470, pp. 556–572. Springer, Heidelberg (2002)
18. Rice, J.R.: The algorithm selection problem. Adv. Comput. 15, 65–118 (1976)
19. Smith-Miles, K.: Cross-disciplinary perspectives on meta-learning for algorithm selection. ACM Comput. Surv. 41(1), 6:1–6:25 (2009)
20. Smith-Miles, K., Lopes, L.: Measuring instance difficulty for combinatorial optimization problems. Comput. Oper. Res. 39(5), 875–889 (2012)
21. Xu, L., Hutter, F., Hoos, H.H., Leyton-Brown, K.: SATzilla: portfolio-based algorithm selection for SAT. JAIR 32, 565–606 (2008)
22. Hutter, F., Hamadi, Y., Hoos, H.H., Leyton-Brown, K.: Performance Prediction and Automated Tuning of Randomized and Parametric Algorithms. In: Benhamou, F. (ed.) CP 2006. LNCS, vol. 4204, pp. 213–228. Springer, Heidelberg (2006)
23. Hutter, F., Xu, L., Hoos, H.H., Leyton-Brown, K.: Algorithm runtime prediction: the state of the art. CoRR, abs/1211.0906 (2012)
24. Smith-Miles, K., van Hemert, J.: Discovering the suitability of optimisation algorithms by learning from evolved instances. AMAI 61, 87–104 (2011)
25. Bartz-Beielstein, T., Markon, S.: Tuning search algorithms for real-world applications: a regression tree based approach. In: Proceedings of CEC-04, pp. 1111–1118 (2004)
26. Bishop, C.M.: Pattern recognition and machine learning. Springer, New York (2006)
27. Rasmussen, C.E., Williams, C.K.I.: Gaussian Processes for Machine Learning. MIT Press, Cambridge (2006)
28. Breiman, L.: Random forests. Mach. Learn. 45(1), 5–32 (2001)
29. Leyton-Brown, K., Nudelman, E., Shoham, Y.: Empirical hardness models: methodology and a case study on combinatorial auctions. J. ACM 56(4), 1–52 (2009)
30. Xu, L., Hoos, H.H., Leyton-Brown, K.: Predicting satisfiability at the phase transition. In: Proceedings of AAAI-12 (2012)
31. Friedman, J.: Multivariate adaptive regression splines. Ann. Stat. 19(1), 1–141 (1991)
32. IBM Corp.: IBM ILOG CPLEX Optimizer. http://www-01.ibm.com/software/integration/optimization/cplex-optimizer/ (2012). Accessed 27 Oct 2012
33. Babić, D., Hutter, F.: Spear theorem prover. Solver description SAT competition (2007)
34. Helsgaun, K.: General k-opt submoves for the Lin-Kernighan TSP heuristic. Math. Program. Comput. 1(2–3), 119–163 (2009)
35. Styles, J., Hoos, H.H., Müller, M.: Automatically configuring algorithms for scaling performance. In: Hamadi, Y., Schoenauer, M. (eds.) LION 6. LNCS, vol. 7219, pp. 205–219. Springer, Heidelberg (2012)

36. Bergstra, J., Bengio, Y.: Random search for hyper-parameter optimization. JMLR **13**, 281–305 (2012)
37. Wang, Z., Zoghi, M., Hutter, F., Matheson, D., de Freitas, N.: Bayesian optimization in a billion dimensions via random embeddings. ArXiv e-prints, January (2013). arXiv:1301.1942

Using Racing to Automatically Configure Algorithms for Scaling Performance

James Styles[(✉)] and Holger H. Hoos

University of British Columbia, 2366 Main Mall, Vancouver, BCV6T 1Z4, Canada
{jastyles,hoos}@cs.ubc.ca

Abstract. Automated algorithm configuration has been proven to be an effective approach for achieving improved performance of solvers for many computationally hard problems. Following our previous work, we consider the challenging situation where the kind of problem instances for which we desire optimised performance are too difficult to be used during the configuration process. In this work, we propose a novel combination of racing techniques with existing algorithm configurators to meet this challenge. We demonstrate that the resulting algorithm configuration protocol achieves better results than previous approaches and in many cases closely matches the bound on performance obtained using an oracle selector. An extended version of this paper can be found at www.cs.ubc. ca/labs/beta/Projects/Config4Scaling.

1 Introduction

High performance algorithms for computationally hard problems often have numerous parameters which control their behaviour and performance. Finding good values for these parameters, some exposed to end users and others hidden as hard-coded design choices, can be a challenging problem for algorithm designers. Recent work on automatically configuring algorithms has proven to be very effective. These automatic algorithm configurators rely on the use of significant computational resource to explore the design space of an algorithm.

In previous work [7], we examined a limitation of the basic protocol for using automatic algorithm configurators in scenarios where the intended use case of an algorithm is too expensive to be feasibly used during configuration. We proposed a new protocol for using algorithm configurators, referred to as train-easy select-intermediate (TE-SI), which uses so-called easy instances during the configuration step of the protocol and so-called intermediate instances during the selection step. Through a large empirical study we were able to show that TE-SI reliably out performed the basic protocol.

In this work, we show how even better configurations can be found using two novel configuration protocols that combine the idea of using intermediate instances for validation with the concept of racing. One of these protocols uses a new variant of F-Race [1] and the other is based on a novel racing procedure dubbed *ordered permutation race*. We show that both racing-based protocols reliably outperform our previous protocol [7] and are able to produce configurations up to 25 % better within the same time budget or configurations of the same quality in up to 45 % less total time and up to 90 % less time for validation.

G. Nicosia and P. Pardalos (Eds.): LION 7, LNCS 7997, pp. 382–388, 2013.
DOI: 10.1007/978-3-642-44973-4_41, © Springer-Verlag Berlin Heidelberg 2013

To assess the effectiveness of our new protocols, we performed a empirical study across five configuration scenarios, described in Sect. 3. All scenarios use the freely available algorithm configurators ParamILS [5] and SMAC [4].

2   Validation Using Racing

Racing, as applied to algorithm configuration, evaluates a set of candidate configurations of a given target algorithm on a set of problem instances, one of which is presented in each stage of the race, and eliminates configurations from consideration once there is sufficient evidence that they are performing significantly worse than the current leader of the race. The race ends when either a single configuration remains, when all problem instances have been used, or when an overall time budget has been exhausted. There are three important aspects to racing strategies: (1) how the set of candidate configurations is constructed, (2) what metric is used to evaluate configurations, and (3) what method is used to determine if a configuration can be eliminated from further consideration.

The first and most prominent racing procedure for algorithm configuration is F-Race [1], which uses the non-parametric, rank-based Friedman test to determine when to eliminate candidate configurations. A major limitation of this basic version of F-Race stems from the fact that in the initial steps, all given configurations have to be evaluated. This property of basic F-Race severely limits the size of the configuration spaces to which the procedure can be applied effectively. Basic F-Race and its variants select the instance used to evaluate configurations for each round of the race at random from the given training set.

Slow Racers Make for Slow Races. In each round of a race, every candidate configuration must be evaluated. If the majority of candidate configurations have poor performance, then much time is spent performing costly evaluations of bad configurations before anything can be eliminated. This is problematic, because good configurations are often quite rare, so that the majority of configurations in the initial candidate set are likely to exhibit poor performance. Therefore, we perform racing on a set of candidate configurations obtained from multiple runs of a powerful configurator rather than for the configuration task itself; this way, we start racing from a set of configurations that tend perform to well which significantly speeds up the racing process.

It Doesn't Take a Marathon to Separate the Good from the Bad. The first few stages of racing are the most expensive. Yet, during this initial phase, there is not yet enough information to eliminate any of the configurations, so the entire initial candidate set is being considered. We know how the default configuration of an algorithm performs on each validation instance, which gives us an idea for the difficulty of the instance for all other configurations of the target algorithm. By using instances in ascending order of difficulty, we reserve the most difficult (i.e., costly) evaluations for later stages of the race, when there are the fewest configurations left to be evaluated.

Judge the Racers by What Matters in the End. The configuration scenarios examined in this work involve minimising a given target algorithm's runtime. While rank-based methods may indirectly lead to a reduction in runtime they are more appropriate for scenarios where the magnitude of performance differences does not matter. We therefore propose the use of a permutation test instead of the rank-based Friedman test, focused on runtime, for eliminating configurations.

In detail, our testing procedure works as follows. Given n configurations $c_1, \ldots c_n$, and m problem instances $i_1, \ldots, i_m$ considered at stage m of the race, we use $p_{k,j}$ to denote the performance of configuration c_k on instance i_j, and p_k to denote the aggregate performance of configuration c_k over $i_1, \ldots, i_m$. In this work, we use penalised average run time, PAR10, to measure aggregate performance, and our goal is to find a configuration with minimal PAR10. Let c_1 be the current leader of the race, i.e., the configuration with the best aggregate performance among $c_1, \ldots, c_n$, We now perform pairwise permutation tests between the leader, c_1, and all other configurations c_k. Each of these tests assesses whether c_1 performs significantly better than c_k; if so, c_k is eliminated from the race. To perform this one-sided pairwise permutation test between c_1 and c_k, we generate 100,000 resamples of the given performance data for these two configurations. Each resample is generated from the original performance data by swapping the performance values $p_{1,j}$ and $p_{k,j}$ with probability 0.5 and leaving them unchanged otherwise; this is done independently for each instance $j = 1, \ldots, m$. We then consider the distribution of the aggregate performance ratios p_1'/p_k' over these resamples and determine the q-quantile of this distribution that equals the p_1/p_k ratio for the original performance data. Finally, if, and only if, $q > \alpha_2$, where α_2 is the significance of the one-sided pairwise test, we conclude that c_1 performs significantly better than c_k. Different from F-race, where the multi-way Friedman test is used to gate a series of pairwise post-tests, we only perform pairwise tests and therefore need to perform multiple testing correction. While more sophisticated corrections could be applied, we decided to use the simple, but conservative Bonferroni correction and set $\alpha_2 := \frac{\alpha}{n-1}$ for an overall significance level α.

We refer to the racing procedure that considers problem instances in order of increasing difficulty for the default configuration of the given target algorithm and in each stage eliminates configurations using the previously described series of pairwise permutation tests as *ordered permutation race (op-race)*, and the variant of basic F-race that uses the same ordering as *ordered F-race (of-race)*.

The TE-FRI and TE-PRI Configuration Protocols. We now return to the application of racing in the context of a configuration protocol that starts from a set of configurations obtained from multiple independent runs of a configurator. In this context, we start op-race and of-race from the easiest intermediate difficulty instance and continue racing with increasingly difficult instances until either a single configurations remains, the time budget for validation has been exhausted, or all available intermediate instances have been used.

This yields two new protocols for using algorithm configurators: (1) train-easy validate-intermediate with of-race (TE-FRI) and (2) train-easy validate-intermediate with op-race (TE-PRI). We have observed that both protocols are quite robust with respect to the significance level α (see extended version) and generally use $\alpha = 0.01$ for TE-FRI and $\alpha = 0.1$ for TE-PRI.

3   Experimental Setup and Protocol

The result of a single, randomized, configuation experiment (i.e., a set of configurator runs and the corresponding global validation step) may be misleading when trying to assess the quality of a configuation procedure. We therefore performed a large number of configurator runs, up to 300, for each scenario, and fully evaluated the configuration found by each run on the training, validation and testing sets. For a specific protocol and a target number n of configurator runs, we generated 100,000 bootstrap samples by selecting, with replacement, the configurations obtained from the n runs. For each such sample R, we chose a configuration with the selection criteria of the protocol under investigation and used the performance of that configuration on the testing set as the result of R.

For all experiments, we measured the performance of configurations on a given instance, using penalised average runtime required for reaching the optimal solution and a penalty factor of 10 times the scenario-specific cutoff for every run that failed to reach the optimal solution. For all scenarios, we configured the target algorithm for minimised PAR-10 using a set of easy training instances defined as being solvable by the default configuration within the per-instance cutoff used during training. We then defined the set of intermediate instances as being in the 12.5th to 20th percentile difficulty of the testing set. The easy, intermediate and testing instance sets are disjoint for each scenario. Each scenario can then be defined by: the target algorithm, the instance set, the configurator time budgets and the per-instance cutoffs enforced during training and testing.

TSP Solving Using LKH. The first scenario we considered involves configuring Keld Helsgaun's implementation of the Lin-Kerninghan algorithm (LKH), the state-of-the art incomplete solver for the traveling salesperson problem (TSP) [3], to solve structured instances similar to those found in the well known TSPLIB benchmark collection [6,7]. Each run of ParamILS and SMAC was given a time budget of 24 h. A 120 second per-instance cutoff was enforced during configuration and a 2 hour per-instance cutoff was enforced during testing.

MIP Solving Using CPLEX. The final three scenarios we considered involve configuring CPLEX, one of the best-performing and most widely used industrial solvers for mixed integer programming (MIP), for solving instances based on real data modeling either wildlife corridors for grizzly bears in the Northern Rockies [2] (CORLAT instances) or the spread of endangered red-cockaded woodpeckers based on decisions to protect certain parcels of land (RCW instances).

The first CPLEX scenario considered configuring CPLEX 12.1 for CORLAT instances. Each run of ParamILS was given a time budget of 20 h. A 120 second per-instance cutoff was enforced during configuation and a 2 hour per-instance cutoff was enforced during testing. The second CPLEX scenario considered configuring CPLEX 12.3 for CORLAT instances. Each run of ParamILS and SMAC was given a time budget of 3456 s. A 15 second per-instance cutoff was enforced during configuation and a 346 second cutoff was enforced during testing. The third CPLEX scenario considered configuring CPLEX 12.3 for RCW instances. Each run of ParamILS and SMAC was given a time budget of 48 h. A 180 second per-instance cutoff was enforced during configuration and a 10 hour cutoff was enforced during testing.

Execution Environment. All our computational experiments were performed on the 384 node DDR partition of the Westgrid Orcinus cluster; Orcinus runs 64-bit Red Hat Enterprise Linux Server 5.3, and each DDR node has two quad-core Intel Xeon E5450 64-bit processors running at 3.0 GHz with 16 GB of RAM.

4 Results

Using the methods described in Sect. 3 we evaluated each of the four protocols on all five configuration scenarios. The results are shown in Table 1, where we report bootstrapped median quality (in terms of speedup over the default configurations, where run time was measured using PAR10 scores) of the configurations found within various time budgets as well as bootstrap [10 %, 90 %] percentile confidence intervals (i.e., 80 % of simulated applications of the respective protocol fall within these ranges; note that these confidence intervals are *not* for median speedups, but for the actual speedups over simulated experiments).

As can be seen from these results, TE-PRI is the most effective configuration protocol, followed by TE-FRI and TE-SI. These three protocols tend to produce very similar [10 %, 90 %] confidence intervals, but the two racing approaches achieve better median speedups, especially for larger time budgets.

To further investigate the performance differences between the protocols, we compared them against a hypothetical protocol with an oracle selection mechanism. This mechanism uses the same configurator runs as the other protocols, but always selects the configuration from this set that has the best *testing* performance, without incurring any additional computational burden. This provides a upper bound of the performance that could be achieved by *any* method for selecting from a set of configurations obtained for a given training set, configurator and time budget. These results, shown in Table 1, demonstrate that for some scenarios (e.g., CPLEX 12.1 for CORLAT) the various procedures, particularly TE-PRI, provide nearly the same performance as the oracle, while for others (e.g., CPLEX 12.3 for RCW), there is a sizable gap.

Table 1. Speedups obtained by our configuration protocols, using ParamILS, on configuration scenarios with different overall time budgets. An increase in overall configuration budget corresponds to an increase in the number of configuration runs performed, rather than to an increase in the time budget for individual runs of the configurator. This means larger time budgets can be achieved by increasing either wall-clock time or the number of concurrent parallel configurator runs. The highest median speedups, excluding the oracle selector, for each configuration scenario and time budget are boldfaced.

	Median [10 %, 90 %] Speedup (PAR10)			
Time Budget (CPU Days)	TE-SI	TE-FRI	TE-PRI	Oracle Selector
Configuring LKH for TSPLIB, using ParamILS				
20	1.33 [0.96, 2.29]	**1.34** [1.00, 2.11]	**1.34** [0.95, 2.11]	1.71 [1.33, 3.11]
50	1.52 [1.06, 3.10]	1.60 [1.25, 3.10]	**1.85** [1.25, 3.10]	2.11 [1.46, 3.19]
100	2.10 [1.24, 3.19]	2.11 [1.46, 3.19]	**2.29** [1.38, 3.19]	2.29 [1.85, 3.19]
Configuring LKH for TSPLIB, using SMAC				
20	0.99 [0.71, 1.23]	1.00 [0.73, 1.23]	**1.08** [0.89, 1.23]	1.12 [0.89, 1.25]
50	**1.08** [0.89, 1.23]	**1.08** [0.92, 1.23]	**1.08** [0.89, 1.23]	1.23 [1.08, 1.25]
100	1.08 [0.89, 1.23]	**1.23** [1.00, 1.23]	**1.23** [0.89, 1.25]	1.25 [1.23, 1.25]
Configuring CPLEX 12.3 for RCW, using ParamILS				
40	1.11 [0.97, 1.39]	**1.12** [0.96, 1.39]	1.08 [0.98, 1.42]	1.23 [1.08, 1.42]
100	1.12 [1.03, 1.42]	**1.16** [1.06, 1.42]	**1.16** [0.98, 1.42]	1.39 [1.16, 1.42]
200	1.13 [1.11, 1.42]	1.37 [1.06, 1.42]	**1.42** [0.98, 1.42]	1.42 [1.37, 1.42]
Configuring CPLEX 12.3 for RCW, using SMAC				
40	**0.79** [0.54, 1.01]	**0.79** [0.54, 1.24]	**0.79** [0.54, 1.01]	0.95 [0.77, 1.24]
100	0.79 [0.77, 1.24]	**0.84** [0.54, 1.24]	0.82 [0.77, 1.24]	1.01 [0.84, 1.24]
200	0.79 [0.77, 1.24]	0.84 [0.54, 1.24]	**1.24** [0.77, 1.24]	1.24 [0.98, 1.24]
Configuring CPLEX 12.1 for CORLAT, using ParamILS				
40	54.5 [42.2, 61.1]	53.8 [42.9, 61.1]	**55.8** [48.3, 61.1]	60.0 [48.8, 68.3]
100	60.1 [49.0, 68.3]	60.6 [53.4, 68.3]	**61.1** [50.3, 68.3]	61.3 [60.0, 68.3]
200	61.5 [53.8, 68.3]	**68.3** [60.1, 68.3]	**68.3** [60.6, 68.3]	68.3 [60.6, 68.3]
Configuring CPLEX 12.3 for CORLAT, using ParamILS				
1.0	2.00 [1.02, 2.64]	1.93 [1.19, 2.64]	**2.24** [1.00, 3.04]	2.36 [1.94, 3.04]
2.5	**2.36** [1.95, 3.04]	**2.36** [1.95, 3.04]	**2.36** [1.93, 3.04]	2.64 [2.24, 3.04]
5.0	2.64 [2.24, 3.04]	**3.02** [1.95, 3.04]	**3.02** [2.24, 3.04]	3.04 [2.64, 3.04]
Configuring CPLEX 12.3 for CORLAT, using SMAC				
1.0	2.41 [1.46, 3.66]	2.41 [1.39, 3.66]	**2.89** [1.54, 3.66]	2.89 [2.16, 3.84]
2.5	**3.26** [1.94, 3.84]	**3.26** [2.19, 3.84]	**3.26** [2.41, 3.66]	3.66 [2.93, 3.84]
5.0	**3.66** [2.89, 3.84]	**3.66** [3.26, 3.84]	**3.66** [2.41, 3.66]	3.84 [3.66, 3.84]

5 Conclusion

In this work, we have addressed the problem of using automated algorithm configuration in situations where instances in the intended use case of an algorithm are too difficult to be used directly during the configuration process. Building on the idea of selecting from a set of configurations optimised on easy training instances by validating on instances of intermediate difficulty recently, we have introduced two novel protocols for using automatic configurators by leveraging racing techniques to improve the efficiency of validation. The first of these protocols, TE-FRI, uses a variant of F-Race [1], and the second, TE-PRI, uses a novel racing method based on permutation tests. Through a large empirical study we have shown that these protocols are very effective and reliably outperform the TE-SI protocol we previously introduced across every scenario we have tested. This is the case for SMAC [4] and ParamILS [5], two fundamental different configuration procedures (SMAC is based on predictive performance models while ParamILS performs model-free stochastic local search), which suggests that our new racing protocols are effective independently of the configurator used.

References

1. Birattari, M., Stützle, T., Paquete, L., Varrentrapp, K.: A racing algorithm for configuring metaheuristics. In: GECCO '02: Proceedings of the Genetic and Evolutionary Computation Conference, pp. 11–18 (2002)
2. Gomes, C.P., van Hoeve, W.-J., Sabharwal, A.: Connections in networks: A hybrid approach. In: Perron, L., Trick, M. (eds.) CPAIOR 2008. LNCS, vol. 5015, pp. 303–307. Springer, Heidelberg (2008)
3. Helsgaun, K.: An effective implementation of the Lin-Kernighan traveling salesman heuristic. EJOR **126**, 106–130 (2000)
4. Hutter, F., Hoos, H.H., Leyton-Brown, K.: Sequential model-based optimization for general algorithm configuration. In: Coello Coello, C.A. (ed.) LION 2011. LNCS, vol. 6683, pp. 507–523. Springer, Heidelberg (2011)
5. Hutter, F., Hoos, H.H., Leyton-Brown, K., Stützle, T.: ParamILS: An automatic algorithm configuration framework. J. Artif. Intell. Res. **36**, 267–306 (2009)
6. Reinelt, G.: TSPLIB. http://www.iwr.uni-heidelberg.de/groups/comopt/software/TSPLIB95. Version visited in October 2011
7. Styles, J., Hoos, H.H., Müller, M.: Automatically configuring algorithms for scaling performance. In: Hamadi, Y., Schoenauer, M. (eds.) LION 2012. LNCS, vol. 7219, pp. 205–219. Springer, Heidelberg (2012)

Algorithm Selection
for the Graph Coloring Problem

Nysret Musliu[✉] and Martin Schwengerer[✉]

Institut für Informationssysteme 184/2, Technische Universität Wien,
Favoritenstraße 9-11, 1040, Vienna, Austria
musliu@dbai.tuwien.ac.at, mschweng@kr.tuwien.ac.at

Abstract. We present an automated algorithm selection method based
on machine learning for the graph coloring problem (GCP). For this
purpose, we identify 78 features for this problem and evaluate the performance of six state-of-the-art (meta) heuristics for the GCP. We use the
obtained data to train several classification algorithms that are applied
to predict on a new instance the algorithm with the highest expected performance. To achieve better performance for the machine learning algorithms, we investigate the impact of parameters, and evaluate different
data discretization and feature selection methods. Finally, we evaluate
our approach, which exploits the existing GCP techniques and the automated algorithm selection, and compare it with existing heuristic algorithms. Experimental results show that the GCP solver based on machine
learning outperforms previous methods on benchmark instances.

Keywords: Algorithm selection · Graph coloring · Machine learning

1 Introduction

Many heuristic algorithms have been developed to solve combinatorial optimization problems. Usually, such techniques show different behavior when solving
particular instances. According to the *no free lunch* theorems [45], no algorithm
can dominate all other techniques on each problem. In practice, this raises new
issues, as selecting the best (or most appropriate) solver for a particular instance
may be challenging. Often, the "winner-take-all" strategy is applied and the
algorithm with the best average performance is chosen to solve all instances.
However, this methodology has its drawbacks, because the distribution of tested
instances effects the average performance, and usually in practice only a special
class of instances are solved.

One possible approach to obtain better solutions on average is to select for
each particular instance the algorithm with the highest expected performance.
This task is known as algorithm selection (AS) and one emerging and very
promising approach that is used for AS is based on *machine learning* methods.
These techniques are able to learn a model based on previous observations an
then predict on a new and unseen instance the best algorithm.

G. Nicosia and P. Pardalos (Eds.): LION 7, LNCS 7997, pp. 389–403, 2013.
DOI: 10.1007/978-3-642-44973-4_42, © Springer-Verlag Berlin Heidelberg 2013

In this paper, we address AS using *classification* algorithms for the well-known Graph Coloring Problem (GCP) *machine learning* techniques. The GCP is a classical NP-hard problem in computer science. The task for this problem is to assign a color to each node of a given graph such that (a) no adjacent nodes received the same color and (b) the number of colors used is minimized. Various heuristic algorithms to solve GCP have been developed in the literature. However, recent studies [7,26] show that the performance of different heuristics highly depend on attributes of the graph like for example the density or the size. Therefore, the aim of this paper is to apply automated algorithm selection for this problem. We evaluate experimentally different heuristics and classification algorithms and show that our solver that includes algorithm selection is able to achieve much better performance than the underlying heuristics.

The rest of this paper is organized as follows: Sect. 2 gives a short introduction into the GCP, AS and the related work. In Sect. 3, we present features of a GCP instance and describe our AS approach for the GCP. The experimental results are given in Sect. 4 while Sect. 5 concludes our work and describes the future work.

2   Background and Related Work

2.1   The Graph Coloring Problem

Given a graph $G = (V, E)$, a *coloring* of G is an assignment of a color $c \leq k$ to each vertex $v \in V$ such that no vertices sharing an edge $e \in E$ receive the same color. The Graph Coloring Problem (GCP) deals with finding a coloring for G whereby it can occur as decision problem (also known as *k-coloring problem*), where the number of colors k is fixed, or as optimization problem (the *chromatic number problem*), where k has to be minimized. Instances of the k-coloring problem are, unlike other *NP-complete* problems (e.g. the *Hamilton path problem*), "hard on average" [43], meaning that also random instances tend to be difficult to solve. Moreover, approximating the chromatic number itself is very hard [14], although many different approaches for this task exist (see [36] for more details). Graph coloring itself has many applications like scheduling [25,48], references: register allocation [5], circuit testing [17] etc.

There exist many exact methods for solving the GCP (see [29] for more details). However, all these approaches are only usable in general on small graphs up to 100 vertices [7]. Consequently, many solvers apply heuristic algorithms. Early approaches in this context are greedy constructive heuristics (e.g. DSATUR [3] or RLF [25]) while recent algorithms use more sophisticated techniques. Especially, local search methods like tabu search [1,21] provide good results. Moreover, also several population-based and hybrid algorithms have been proposed [16,28,46]. For a survey on different heuristics, we refer to [26,29,35].

2.2   Algorithm Selection

The *Algorithm Selection Problem* postulated by Rice [37] deals with this question: Given different algorithms to solve a problem, which one should be selected for a particular instance? For this purpose, Rice identified four important components, namely

- the set of candidate algorithms A,
- the instances of a problem, the problem space P,
- measurable attributes of an instance, denoted as feature space F, and
- the performance space Y.

For solving this problem, it is necessary to use relevant features $f(x) \in F$ of an instance $x \in P$ that model the performance of an algorithm $a \in A$ with respect to a performance criteria Y. Designing the algorithm portfolio A is also usually not so hard, because in most cases the available algorithms are limited and a good selection procedure will not use suboptimal solvers anyway. More challenging is the choice of appropriate features F and to find a good selection procedure (denoted as S).

Unfortunately, there exist no automatic way to find good features [34], as this requires usually deep domain knowledge and analytical skills. Nevertheless, some approaches seem to be useful across different problems and sometimes, even features of related problems can be reused. Regarding concrete features for the GCP, in [40] various properties of a graph that may be useful are introduced. We also adapted some other features that can be found in [47] and additionally introduced some new features.

Regarding the selection procedure, there exist different methods that for example use analytical aspects or complexity parameters [12]. One successfully and widely used solution is the application of machine learning techniques. Usually, either *classification* or *regression* techniques are used. Classification techniques classify the new instances into one category, which is the recommended algorithm. In contrast to this, regression techniques model the behavior of each algorithm by using an regression function and predict the result (e.g. runtime, solution quality) on a new instance. Based on this prediction the algorithm with the best performance is selected. Both paradigms have been successfully applied for algorithm selection. Applications of regression include [4,31,47]. Classification techniques have been used among others in [18,19,23,30,39,41]. However, none of these approaches is specially designed for the GCP and although some also consider graph properties, only one paper, [42], deals explicitly with graph coloring. This work, which was published just recently, utilizes 16 different features and a decision tree to chose between two algorithms for graph coloring: DSATUR and tabu search. Although this approach shows some similarities concerning features and algorithms, there are several differences to our work. We consider additional features (including the greedy algorithm DSATUR) and evaluate the performance of several classifiers. In addition, we focus on using multiple and more sophisticated heuristics and investigate the effect of data preparation. In this paper we present new attributes of a graph that can be calculated in polynomial time and are suitable to predict the most appropriate heuristic for GCP.

3   Algorithm Selection for the GCP

First step in algorithm selection is to identify characteristic features that can be calculated in reasonable time. Furthermore, we collect performance information about each algorithm on a representative set of benchmark instances and determine for each graph the most suited algorithm. Then, we use machine learning to train classification algorithms that act as selection procedure. To predict the best algorithm on a new instance, the proposed system extracts the features of that instance and then determines the corresponding class, which corresponds to the most appropriate algorithm.

3.1   Instance Features

We identify 78 features that are grouped in eight categories: graph size, node degree, maximal clique, clustering coefficient, local search probing features, greedy coloring, tree decomposition, and lower- and upper bound. Figure 1 gives a more detailed view on the different attributes. The first two groups, *graph size* and *node degree* contain classical features that are also used in other systems (e.g. [47]). For the *maximal clique* features, we calculate for each node a *maximal*

Graph Size Features:
1: no. of nodes: n
2: no. of edges: m
3,4: ratio: $\frac{n}{m}, \frac{m}{n}$
5: density: $\frac{2 \cdot m}{n \cdot (n-1)}$

Node Degree:
6-13: **nodes degree statistics:** min, max, mean, median, $Q_{0.25}$, $Q_{0.75}$, variation coefficient, entropy

Maximal Clique:
14-20: **normalized by** n: min, max, median, $Q_{0.25}$, $Q_{0.75}$, variation coefficient, entropy
21:    **computation time**
22:    **maximum cardinality**

Clustering Coefficient
23: **global clustering coefficient** [27]
24-31: **local clustering coefficient:** min, max, mean, median, $Q_{0.25}$, $Q_{0.75}$, variation coefficient, entropy
32-39: **weighted local clustering coefficient:** min, max, mean, median, $Q_{0.25}$, $Q_{0.75}$, variation coefficient, entropy
40: **computation time**

Local Search Probing Features:
41, 42: **avg. impr.:** per iteration, per run
43: **avg no. iterations to LO**[a] per a run
44, 45: **no. conflict nodes:** at LO, at end
46, 47: **no. conflict edges:** at LO, at end
48: **no. LO found**
49: **computation time**

Greedy Coloring:
50,51: **no. colors needed:** k_{DSAT}, k_{RLF}
52, 53: **computation time:** t_{DSAT}, t_{RLF}
54, 55: **ratio:** $\frac{k_{DSAT}}{k_{RLF}}$, $\frac{k_{RLF}}{k_{RLF}}$
56: **best coloring:** $\min(k_{DSAT}, k_{RLF})$
57-72: **independent-set size:** min, max, mean, median, $Q_{0.25}$, $Q_{0.75}$, variation coefficient, entropy

Tree Decomposition:
73: **width of decomposition**
74: **computation time**

Lower- and Upper Bound:
75, 76: **distance:** $\frac{(B_l - B_u)}{B_l}$, $\frac{(B_u - B_l)}{B_u}$
77, 78: **ratio:** $\frac{B_l}{B_u}$, $\frac{B_u}{B_l}$

[a] local optima

Fig. 1. Basic features for an instance of the GCP.

clique by using a simple greedy algorithm and take statistical information about the size of these cliques as relevant attributes. Regarding the local clustering coefficient [44] of a node, we use, besides the classical value, a modified version denoted as *weighted clustering coefficient*, where the coefficient of the node is multiplied with its degree. The *local search probing features* are extracted from 10 executions of a simple 1-opt best-improvement local search on the *k-coloring* problem. The *greedy coloring* attributes are based on the application of DSATUR and RLF. For these features, we take, besides the number of used colors, also the sizes of the independent sets into account and calculate statistical information like the average size or the variation coefficient. Furthermore, we consider attributes of a *tree decomposition* obtained by a *minimum-degree* heuristic. Such features have been used successfully by [32] for AS in Answer Set Programming. The last category builds on a lower bound of k, denoted as B_l, which is the cardinality of the greatest maximal clique found, and an upper bound B_u, which is the minimum number of colors needed by the two greedy algorithms. Apart from features we described above, we also take the computation times of some feature classes as additional parameters. Note that we also experimented with attributes based on the *betweenness centrality* [15] and the *eccentricity* [20] of the nodes. Unfortunately, the algorithms we implemented to calculate these features required during our tests much time, for which reasons we did not use them in our approach.

It is widely accepted that the performance of learning algorithms depend on the choice of features, and that using irrelevant features may lead to suboptimal results. Therefore, we apply a *feature subset selection* using a *forward selection* with limited backtracking and a *genetic search* to reduce the set of basic features. Both techniques are applied with the CfsSubsetEval criteria as evaluation function. Only features that are selected by one of these methods are used further. Additionally, for each pair of features $x_j, x_k, k > j$ we create two new features that represent the product $x_j \cdot x_k$ and the quotient x_j / x_k, respectively. This idea is based on a similar technique used in [47], where also the product of two features is included as an additional attribute. Finally, we apply feature selection on these expanded attributes to eliminate unnecessary attributes. In the end, we obtain 90 features, including 8 basic features and 82 composed attributes.

3.2 Algorithm Portfolio

To demonstrate our approach, we use six state-of-the-art (meta)heuristics for the GCP, namely: HEA [16], ILS [8], MAFS [46], MMT [28] (only the component containing the genetic algorithm), FOO-PARTIALCOL [1] (further abbreviated to FPC), and TABUCOL [21] (further denoted as TABU).

For each of these algorithms, we use parameter settings proposed in the original publications and that are suggested by their developers. The main reason for selecting the TABU solver is the fact that this technique is one of the most-studied heuristics and is often used as local search in various population-based algorithms for the GCP. In addition, according to a comparison by Chiarandini [6], TABU is besides HEA and ILS the most effective algorithm for random graphs. HEA is chosen because it shows good performance on *flat* graphs and it is used as basis for many other evolutionary heuristics that are applied for GCP. We selected FPC and MMT because we also wanted to use algorithms working with *partial* colorings and these two candidates are the correspondent versions of TABU and HEA. The last competitor, MAFS, is included because it shows good performance on large graphs.

3.3   Benchmark Instances

As training instances, we take three different publicly available sets: The first set, further denoted as dimacs, consists of 174 graphs from the *Graph Coloring and its Generalizations*-series (COLOR02/03/04)[1] which builds up on the well-established Dimacs Challenge [22]. This set includes instances from the *coloring* and the *clique* part of the Dimacs Challenge. The second and third set of instances, denoted as chi500 and chi1000, are used by a comparative study [9] of several heuristics for the GCP and contain 520 instances with 500 nodes and 740 instances with 1000 nodes respectively.[2] These instances are created using Culberson's [10] random instance generator by controlling various parameters like the *edge density* ($p = \{0.1, 0.5, 0.9\}$) or the *edge distribution* (resulting in three groups of graphs: *uniform graphs* (G), *geometric graphs* (U) and *weight biased graphs* (W)).

For the final evaluation of our algorithm selection approach with the underlying algorithms, we use a test set comprising complete new instances of different size, density and type, generated with Culberson's instance generator. We constructed *uniform* (G), *geometric* (U) and *weight biased* (W) graphs of different sizes $n = \{500, 750, 1000, 1250\}$ and density values $p = \{0.1, 0.5, 0.9\}$. For each parameter setting we created 5 graphs, leading to a total of 180 instances.

In order to ensure practicable results and prevent excessive computational effort, we use a maximal time limit per color $t_{max} = \min(3600, \sqrt{|E|} \cdot x)$ where $|E|$ is the number of edges and x is 15, 5 and 3 for the sets dimacs, chi500 and chi1000, respectively. For the test set which contains graphs of different size, we stick to the values used for chi1000 ($x = 3$). These values for x are obtained experimentally. In this context, we want to note that the average time needed for the best solution on the *hard* instances is only 21.58 % of the allowed value t_{max} and 90 % of the best solutions are found within 62.66 % of t_{max}.

Regarding the feature computation, we do not use any time limitations except for the local search probing, although this might be reasonable for practical implementations. However, for our test data the median calculation time is 2 s, the 95th percentile is 18 s and the 99th percentile is 53 s.

In total, we collected 1434 graphs of variable size and density as training data. We removed instances where an optimal solution has been found by one of

---

[1] Available at http://mat.gsia.cmu.edu/COLOR04/, last visited on 22.10.2012
[2] Available at www.imada.sdu.dk/~marco/gcp-study/, last visited on 28.10.2012

the two greedy algorithms or where the heuristics did not find better colorings than obtained by the greedy algorithms. We further excluded all instances where at least four heuristics (more than 50 %) yield the best solution in less than five seconds. These seem to be *easy* instances which can be solved efficiently by most heuristics. Therefore, they are less interesting for algorithm selection. In the end, our training data consist of 859 *hard* instances.

Note that during our experiments, we discovered instances where several heuristics obtain best result. For machine learning, this is rather uncomfortable, as the training data should contain only one recommended algorithm per instance. One solution for this issue is using multi-labeled classification [23]. However, we follow a different strategy where we prioritize the algorithms according to their average rank on all instances. Thus, in case of a tie, we prefer the algorithm with the lower rank. Concerning the performance evaluation of the classifiers, we have to take into account that there might be several "best" algorithms. For that reason, we introduce a new performance measurement, called *success rate* (*sr*), that is defined as follows: Given for each instance $i \in I$ a set of algorithms B^i that obtains best result on i. Then, the *sr* of a classifier c on a set of instances I is $sr = \frac{|\{i \in I : c(i) \in B^i\}|}{|I|}$ where $c(i)$ is the predicted algorithm for the instance i. Furthermore, the success rate of a solver is the ratio between the number of instances for which the solver achieves the best solution and the total number of instances.

3.4 Classification Algorithms

For the selection procedure itself, we test six popular classification algorithms: Bayesian Networks (BN), C4.5 Decision Trees (DT), k-Nearest Neighbor (kNN), Multilayer Perceptrons (MLP), Random Forests (RF), and Support-Vector Machines (SVM). For all these techniques, we use the implementation included in the *Weka* software collection [2], version *3.6.6*. Furthermore, we manually identify important parameters of these learning algorithms and experimented with different settings. We refer the reader to [38] for more details regarding different parameter settings that we used for classification algorithms.

3.5 Data Discretization

Apart from selection of relevant features, a different, but also important issue is whether to use the original numeric attributes or to apply a *discretization* step to transform the values into nominal attributes. Besides the fact that some classification algorithms can not deal with numeric features, research has clearly shown that some classifiers achieve significant better results when applied with discretized variables [11]. In this work, we experimented with two different supervised discretization techniques. The first one is the classical minimum-descriptive length (MDL) method [13], while the second method is a derivation of MDL using a different criteria [24] (further denoted as Kononenko's criteria (KON)).

4   Experimental Results

All our experiments have been performed on a Transtec CALLEO 652 Server containing 4 nodes, each with 2 AMD Opteron Magny-Cours 6176 SE CPUs $(2 \cdot 12 = 24$ cores with 2.3 GHz) and 128 GB memory.

Concerning the heuristic for the GCP, we execute each algorithm $n = 10$ times ($n = 20$ for the dimacs instances) using different random seeds. The result of each algorithm is the lowest number of colors that has been found in more than 50 % of the trials. Furthermore, we take the median time needed within the n executions as required computation time. In cases of a timeout, we take t_{max} as computation time. Detailed results of the experiments can be found at http://www.kr.tuwien.ac.at/staff/mschweng/gcp/.

4.1   Parameter Configuration and Discretization

Regarding the effect of data discretization, we compare the success rate of the best parameter configuration for each of the three methods on several data sets (e.g. using different feature subsets). The experimental results clearly show that most classifiers achieve a higher accuracy on data with nominal attributes.

Table 1 gives an overview regarding the impact of discretization. The column *avg* shows the improvement regarding the average success rate, whereas the column *best* represents the gap between the best value obtained using numerical values and the best value achieved with the discretized data sets. Both discretization variants improve the best reached success rate. The classical MDL method improves *sr* on average by 5.14 %, while Kononenko's criteria by 4.35 %. However, for some classifiers, the benefits of discretized values are up to +9.41 % with MDL and even +11.09 % using Kononenko's criteria (KON). The only classifier which does not benefit from a discretization is MLP. Its training time increases dramatically (up to several hours). Even more, when using Kononenko's criteria (KON), the average success rate decreases by 20.33 %. Nevertheless, as Kononenko's criteria (KON) provides for the most classifiers slightly better results than MDL, we decided to use Kononenko's criteria for all further experiments.

As mentioned before, we experimented with different parameter configurations for each classifier. Based on these tests, we selected for the remaining tests

Table 1. Improvements of *sr* (in percent) when using discretized data in relation to results achieved with non-discretized data on the training set using cross validation.

Method	BN		C4.5		kNN	
	Avg	Best	Avg	Best	Avg	Best
MDL	+2.40	+2.30	+6.34	+7.15	+9.41	+7.00
KON	+4.93	+4.85	+5.78	+6.23	+11.09	+8.92
	MLP		RF		SVM	
MDL	+4.16	+5.42	+2.25	+2.25	+4.57	+1.75
KON	−20.33	+4.37	+3.94	+4.38	+5.33	+4.20

the most successful configuration. In detail, the maximum number of parent nodes that we used for BN is 3. For the DT the minimum number of objects per leave was set to 3. Regarding the kNN, the size of the neighborhood is set to 5 and for the RF, we set the number of trees to 15. For the MPL and SVM, and other remaining parameters, we used the default settings from the *Weka* system.

4.2 Results on the Training Data

To show the performance on the training data set, we tested each classifier 20 times using a 10-fold cross validation. The results of these experiments are given in Fig. 2, which shows the average number of correct predictions for each classifier and instance set. The figure also gives a comparison with the existing solvers for the GCP regarding the number of instances on which the best solution is achieved. The diagram shows that 5 of 6 tested classifiers achieve good results. Only the MLP gives very weak results. This method requires more than 24 h for one run of cross-validation and its results are even below those of the existing heuristics. Nevertheless, other approaches show very good performance by obtaining on up to 625.9 (72.86 %) instances the best solution. Compared with MMT, which is the best heuristic for the GCP, an improvement on 259 instances (30.15 %) is reached. Even more, this performance increase can be observed on all three instance sets.

For a more detailed statistical analysis, we applied a *corrected resampled T-test* [33] on the results of the 10-fold cross-validation (except the MLP). These experiments, applied with a level of significance of $\alpha = 0.05$, reveal that BN, kNN and RF are significant better than DT while all other pairwise comparisons do not show significant differences.

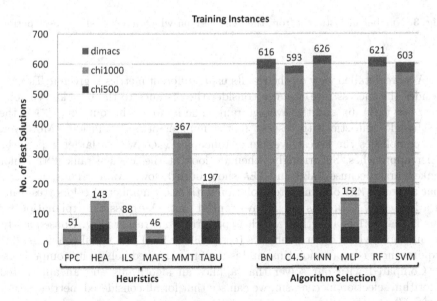

Fig. 2. Prediction of the best algorithm by different classifiers on the training data and their comparison with the existing (meta)heuristics.

4.3   Evaluation on the Test Set

In the next step, we trained the classifiers with the complete training set and evaluate the performance of them on the test set. The corresponding results are shown in Fig. 3, which shows the number of instances on which the solvers show the best performance. From this figure, we can see that all learning strategies except MLP accomplish a higher number of best solutions than any existing solver for the GCP. The most successful classifiers are RF, BN and kNN which predict on up to 71.71 % of the 152 graphs the most appropriate algorithm.

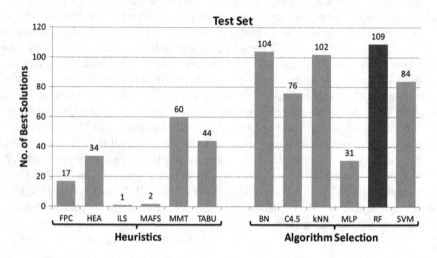

Fig. 3. Number of instances from the test set on which a solver shows best performance.

A more detailed view on the results using different metrics is given in Table 2. Besides the success rate, we also consider the distance to the best known solution, $err(\widehat{\chi}, G)$ [6], and the average rank. The figures point out that MMT is the best single heuristic with respect to the number of best solutions. Moreover, it accomplishes the lowest average distance $err(\widehat{\chi}, G)$ with a larger gap to the other approaches. Surprisingly, when we look at the average rank, MMT is not ranked first because TABU and HEA show both a lower value. Thus, it seems that although MMT obtains often solution with a low number of colors (resulting in a low $err(\widehat{\chi}, G)$), it is not always ranked first. One possible explanation for this is that MMT is a method which is powerful, but rather slow. Consequently, on instances where other heuristics (e.g. TABU or HEA) find equal colorings, MMT requires more computation time and is therefore, ranked behind its competitors.

Compared with our solver that applies all algorithms and an automated algorithm selection mechanism, we can see that for all considered metrics except $err(\widehat{\chi}, G)$ at least one system shows a stronger performance than the best single heuristic. The best selection mechanism provides clearly RF, which is on all

Table 2. Performance metrics of the algorithm selection and the underlying heuristics on the *test set*.

Solver	No. best solution	sr (%)	$err(\widehat{\chi}, G)$ (%)	Rank avg	σ
Heuristics (H)					
FPC	17	11.18	25.42	3.29	1.42
HEA	34	22.37	14.91	2.66	1.38
ILS	1	0.66	21.73	3.82	1.36
MAFS	2	1.32	30.17	4.62	1.52
MMT	60	39.47	**3.78**	2.76	1.84
TABU	44	28.95	19.23	2.58	1.29
Algorithm Selection (AS)					
BN	104	68.42	5.16	1.59	1.08
C4.5	76	50.00	5.86	2.21	1.50
kNN	102	67.11	3.82	1.52	0.91
MLP	31	20.39	24.90	3.14	1.66
RF	**109**	**71.71**	5.44	**1.41**	**0.78**
SVM	84	55.26	8.32	1.97	1.38
Best (H)	60	39.47	**3.78**	2.58	1.29
Best (AS)	**109**	**71.71**	3.82	**1.41**	**0.78**

criteria except $err(\widehat{\chi}, G)$ better than the other classifiers. In detail, this system achieves a success rate of 71.71 % (+32.24 % compared with MMT) and an average rank of 1.41 (−1.17 compared with TABU). Only on the metric $err(\widehat{\chi}, G)$, MMT shows with 3.78 % a lower value than RF, which colorings of the choosen algorithms have an average distance of 5.44 % to the best known solution. The best classifier in this context is kNN, which achieves with 3.82 a slightly higher value than MMT. The worst performance among the classifiers shows clearly MLP, which results concerning the number of instances where it finds the best solution are even below those of MMT or TABU. These data confirm that this machine learning technique in combination with Kononenko's criteria (KON) is not suited for the GCP. This does not imply that MLP is in general inappropriate for AS. The results using data sets with continuous attributes show that this classifier can achieve competitive results compared to the other tested classifiers. However, when using nominal features, its training time usually increases dramatically while the accuracy decreases.

For a more detailed analysis, we group the graphs according to their density and graph class and compare the performance of RF, which is the best classifier, with the existing heuristics. Figure 4 presents the number of graphs on which the different methods show the best performance. The figures show that our solver based on algorithm selection is on 5 of the 9 subsets better or equal compare to the best heuristic. On the group U-0.9 the algorithm selection fails by predicting on only 6 of 10 graphs the correct algorithm. The reason for this bad results might be that for almost all graphs with high density, either TABU or MMT is the best heuristic whereby the decision criteria is based on the *edge distribution*. In

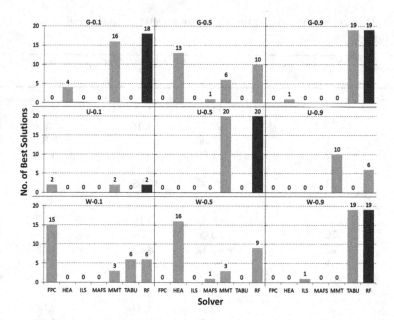

Fig. 4. Number of instances of the test set on which a solver shows the best performance, grouped by the graph type (edge distribution) and the density. The dark bar denotes that our approach is at least as successful as the best single solver.

detail, TABU is to prefer on *uniform* and *weight biased* graphs while MMT is better on *geometric* graphs. Unfortunately this information is not part of the feature space and it seems that the classifier is not able to distinguish this based on other features, which leads to mispredictions. The suboptimal prediction rate on W-0.1 is hard to explain, as FPC is also in the related subset W-0.1 of the training data the best algorithm. Thus, is seems that the classifier is just not able to learn this pattern correctly. On the groups G-0.5 and W-0.5 our approach is also not able to achieve competitive results compare to the best single solver. This is surprising as the best heuristic on these instances is HEA, which shows also on the corresponding training data good results. One possible explanation is that, as for instances with high density, the classifier is unable to detect the type of edge distribution. Nevertheless, we can see that in many cases, the classifier is able to predict the most appropriate algorithm, which leads to a better average performance compare to any single heuristic.

5   Conclusion

In this paper, we presented a novel approach based on machine learning to automate algorithm selection for the GCP. Given a set of algorithms and a set of specific features of a particular instance, such a system selects the algorithm which is predicted to show the best performance on that instance. Our proposed approach applies a classification algorithm as selection procedure that assigns a new instance to one of the available algorithms based on a previously learned model. For this purpose, we identified 78 attributes for the GCP that can be calculated in reasonable time and that have impact on solving of the GCP.

To demonstrate our approach, we evaluated the performance of six state-of-the-art (meta)heuristics on three publicly available sets of instances and showed that no algorithm is dominant on all instances. We further applied machine learning to build an automated selection procedure based on the obtained data. For that purpose, we experimented with six well known classification algorithms that are used to predict on a new instance the most appropriate algorithm. Our experiments clearly showed that a solver that applies machine learning yield a significant better performance compared with any single heuristic. We further demonstrated that using data discretization increases the accuracy of most classifiers.

Regarding future work, we plan to investigate a *regression*-based approach using runtime and solution quality predictions. This technique, which is successfully used for other systems, is an alternative to our classification-based approach. Also worth considering is a hybridization of our method with *automated parameter selection* and the combination of heuristic and exact techniques for the GCP in a system that applies automated algorithm selection.

Acknowledgments. The work was supported by the Austrian Science Fund (FWF): P24814-N23. Additionally, the research herein is partially conducted within the competence network Softnet Austria II (www.soft-net.at, COMET K-Projekt) and funded by the Austrian Federal Ministry of Economy, Family and Youth (bmwfj), the province of Styria, the Steirische Wirtschaftsförderungsgesellschaft mbH. (SFG), and the city of Vienna in terms of the center for innovation and technology (ZIT).

References

1. Blöchliger, I., Zufferey, N.: A graph coloring heuristic using partial solutions and a reactive tabu scheme. Comput. Oper. Res. **35**(3), 960–975 (2008)
2. Bouckaert, R.R., Frank, E., Hall, M., Kirkby, R., Reutemann, P., Seewald, A., Scuse, D.: Weka manual (3.6.6), October 2011
3. Brélaz, D.: New methods to color the vertices of a graph. Commun. ACM **22**, 251–256 (1979)
4. Brown, K.L., Nudelman, E., Shoham, Y.: Empirical hardness models: methodology and a case study on combinatorial auctions. J. ACM **56**(4), 1–52 (2009)
5. Chaitin, G.: Register allocation and spilling via graph coloring. SIGPLAN Not. **39**(4), 66–74 (2004)
6. Chiarandini, M.: Stochastic local search methods for highly constrained combinatorial optimisation problems. Ph.D. thesis, TU Darmstadt, August 2005
7. Chiarandini, M., Dumitrescu, I., Stützle, T.: Stochastic local search algorithms for the graph colouring problem. In: Gonzalez, T.F. (ed.) Handbook of Approximation Algorithms and Metaheuristics. Chapman & Hall/CRC, Boca Raton (2007)
8. Chiarandini, M., Stützle, T.: An application of iterated local search to graph coloring. In: Johnson, D.S., Mehrotra, A., Trick, M.A. (eds.) Proceedings of the Computational Symposium on Graph Coloring and its Generalizations (2002)

9. Chiarandini, M., Stützle, T.: An analysis of heuristics for vertex colouring. In: Festa, P. (ed.) SEA 2010. LNCS, vol. 6049, pp. 326–337. Springer, Heidelberg (2010)

10. Culberson, J.C., Luo, F.: Exploring the k-colorable landscape with iterated greedy. In: Dimacs Series in Discrete Mathematics and Theoretical Computer Science, pp. 245–284. American Mathematical Society, Providence (1995)

11. Dougherty, J., Kohavi, R., Sahami, M.: Supervised and unsupervised discretization of continuous features. In: Machine Learning: Proceedings of the Twelfth International Conference, pp. 194–202. Morgan Kaufmann, San Francisco (1995)

12. Ewald, R.: Experimentation methodology. In: Ewald, R. (ed.) In: Automatic Algorithm Selection for Complex Simulation Problems, pp. 203–246. Vieweg+Teubner Verlag, Wiesbaden (2012)

13. Fayyad, U.M., Irani, K.B.: Multi-interval discretization of continuous-valued attributes for classification learning. In: Bajcsy, R. (ed.) IJCAI. Morgan Kaufmann, San Mateo (1993)

14. Feige, U., Kilian, J.: Zero knowledge and the chromatic number. J. Comput. Syst. Sci. **57**(2), 187–199 (1998)

15. Freeman, L.C.: A set of measures of centrality based on betweenness. Sociometry **40**(1), 35–41 (1977)

16. Galinier, P., Hao, J.-K.: Hybrid evolutionary algorithms for graph coloring. J. Comb. Optim. **3**, 379–397 (1999)

17. Garey, M.R., Johnson, D.S., Hing, S.C.: An application of graph coloring to printed circuit testing. IEEE Trans. Circ. Syst. (1976)

18. Guerri, A., Milano, M.: Learning techniques for automatic algorithm portfolio selection. In: de Mántaras, R.L., Saitta, L. (eds.) In: Conference on Artificial Intelligence, ECAI'2004, pp. 475–479. IOS Press, Amsterdam (2004)

19. Guo, H., Hsu, W.H.: A machine learning approach to algorithm selection for NP-hard optimization problems: a case study on the MPE problem. Ann. Oper. Res. **156**, 61–82 (2007)

20. Hage, P., Harary, F.: Eccentricity and centrality in networks. Soc. Netw. **17**(1), 57–63 (1995)

21. Hertz, A., de Werra, D.: Using tabu search techniques for graph coloring. Computing **39**(4), 345–351 (1987)

22. Johnson, D.J., Trick, M.A. (eds) Cliques, Coloring, and Satisfiability: Second DIMACS Implementation Challenge, 11–13 October 1993. American Mathematical Society (1996)

23. Kanda, J., de Carvalho, A.C.P.L.F., Hruschka, E.R., Soares, C.: Selection of algorithms to solve traveling salesman problems using meta-learning. Int. J. Hybrid Intell. Syst. **8**(3), 117–128 (2011)

24. Kononenko, I.: On biases in estimating multi-valued attributes. In: IJCAI. Morgan Kaufmann, San Francisco (1995)

25. Leighton, F.T.: A graph coloring algorithm for large scheduling problems. J. Res. Natl Bur. Stand. **84**(6), 489–506 (1979)

26. Lewis, R., Thompson, J., Mumford, C.L., Gillard, J.W.: A wide-ranging computational comparison of high-performance graph colouring algorithms. Comput. Oper. Res. **39**(9), 1933–1950 (2012)

27. Luce, D.R., Perry, A.D.: A method of matrix analysis of group structure. Psychometrika **14**, 95–116 (1949)

28. Malaguti, E., Monaci, M., Toth, P.: A metaheuristic approach for the vertex coloring problem. INFORMS J. Comput. **20**(2), 302–316 (2008)

29. Malaguti, E., Toth, P.: A survey on vertex coloring problems. Int. Trans. Oper. Res. **17**, 1–34 (2010)

30. Malitsky, Y., Sabharwal, A., Samulowitz, H., Sellmann, M.: Non-model-based algorithm portfolios for SAT. In: Sakallah, K.A., Simon, L. (eds.) SAT 2011. LNCS, vol. 6695, pp. 369–370. Springer, Heidelberg (2011)

31. Messelis, T., De Causmaecker, P.: An algorithm selection approach for nurse rostering. In: Proceedings of BNAIC 2011, Nevelland, pp. 160–166, November (2011)

32. Morak, M., Musliu, N., Pichler, R., Rümmele, S., Woltran, S.: Evaluating tree-decomposition based algorithms for answer set programming. In: Hamadi, Y., Schoenauer, M. (eds.) LION 2012. LNCS, vol. 7219, pp. 130–144. Springer, Heidelberg (2012)

33. Nadeau, C., Bengio, Y.: Inference for the generalization error. Mach. Learn. **52**(3), 239–281 (2003)

34. Nudelman, E.: Empirical approach to the complexity of hard problems. Ph.D. thesis, Stanford University, Stanford, CA, USA (2006)

35. Pardalos, P., Mavridou, T., Xue, J.: The Graph Coloring Problem: A Bibliographic Survey, pp. 331–395. Kluwer Academic Publishers, Boston (1998)

36. Paschos, V.T.: Polynomial approximation and graph-coloring. Computing **70**(1), 41–86 (2003)

37. Rice, J.R.: The algorithm selection problem. Adv. Comput. **15**, 65–118 (1976)

38. Schwengerer, M.: Algorithm selection for the graph coloring problem. Vienna University of Technology, Master's thesis, October 2012

39. Smith-Miles, K.: Towards insightful algorithm selection for optimisation using meta-learning concepts. In: IEEE International Joint Conference on Neural Networks. IEEE, New York (2008)

40. Smith-Miles, K., Lopes, L.: Measuring instance difficulty for combinatorial optimization problems. Comput. OR **39**(5), 875–889 (2012)

41. Smith-Miles, K., van Hemert, J., Lim, X.Y.: Understanding TSP difficulty by learning from evolved instances. In: Blum, C., Battiti, R. (eds.) LION 2010. LNCS, vol. 6073, pp. 266–280. Springer, Heidelberg (2010)

42. Smith-Miles, K., Wreford, B., Lopes, L., Insani, N.: Predicting metaheuristic performance on graph coloring problems using data mining. In: El Talbi, G. (ed.) Hybrid Metaheuristics. SCI, pp. 3–76. Springer, Heidelberg (2013)

43. Venkatesan, R., Levin, L.: Random instances of a graph coloring problem are hard. Proceedings of the Twentieth Annual ACM Symposium on Theory of Computing, STOC '88, pp. 217–222. ACM, New York (1988)

44. Watts, D.J., Strogatz, S.M.: Collective dynamics of 'small-world' networks. Nature **393**(6684), 440–442 (1998)

45. Wolpert, D.H., Macready, W.G.: No free lunch theorems for optimization. IEEE Trans. Evol. Comput. **1**(1), 67–82 (1997)

46. Xie, X.F., Liu, J.: Graph coloring by multiagent fusion search. J. Comb. Optim. **18**(2), 99–123 (2009)

47. Xu, L., Hutter, F., Hoos, H.H., Leyton-Brown, K.: SATzilla: portfolio-based algorithm selection for sat. J. Artif. IntelL. Res. **32**, 565–606 (2008)

48. Zufferey, N., Giaccari, P.: Graph colouring approaches for a satellite range scheduling problem. J. Schedul. **11**(4), 263–277 (2008)

Batched Mode Hyper-heuristics

Shahriar Asta, Ender Özcan, and Andrew J. Parkes[(⊠)]

School of Computer Science, University of Nottingham, Nottingham NG8 1BB, UK
{sba,exo,ajp}@cs.nott.ac.uk
http://www.cs.nott.ac.uk/~{sba,exo,ajp}

Abstract. A primary role for *hyper-heuristics* is to control search processes based on moves generated by neighbourhood operators. Studies have shown that such hyper-heuristics can be effectively used, without modification, for solving unseen problem instances not only from a particular domain, but also on different problem domains. They hence provide a general-purpose software component to help reduce the implementation time needed for effective search methods. However, hyper-heuristic studies have generally used time-contract algorithms (i.e. a fixed execution time) and also solved each problem instance independently. We consider the potential gains and challenges of a hyper-heuristic being able to treat a set of instances as a batch; to be completed within an overall joint execution time. In batched mode, the hyper-heuristic can freely divide the computational effort between the individual instances, and also exploit what it learns on one instance to help solve other instances.

Keywords: Combinatorial optimisation · Metaheuristics · Hyper-heuristics

1 Introduction

A goal of hyper-heuristic [1] research is to raise the level of generality of search methods by providing high level strategies, and associated directly-usable software components, that are useful across different problem domains rather than for a single one. (Note that this general goal is not unique to hyper-heuristics but also occurs in other forms, e.g. in memetic computation [2].) There are two main types of hyper-heuristics depending on whether they do *generation* or *selection* of heuristics [3]. In this paper, we focus on selection hyper-heuristics, and in particular, those that combine heuristic selection and move acceptance processes under a single point search (i.e. not population-based) framework [4]. A candidate solution is improved iteratively by selecting and applying a heuristic (neighbourhood operator) from a set of low level heuristics and then using some acceptance criteria to decide if it should replace the incumbent. We also use the HyFlex (Hyper-heuristics Flexible framework)[1] [5] software tool associated with the CHeSC2011[2] hyper-heuristic competition.

---

[1] http://www.hyflex.org/

[2] http://www.asap.cs.nott.ac.uk/chesc2011/

G. Nicosia and P. Pardalos (Eds.): LION 7, LNCS 7997, pp. 404–409, 2013.
DOI: 10.1007/978-3-642-44973-4_43, © Springer-Verlag Berlin Heidelberg 2013

Usually hyper-heuristics are used to individually and independently solve single instances. However, in some cases, it might well be that a batch of instances need to be solved; where the "batching" simply means that a whole set of instances are to be solved within an overall time limit, but there is no a priori restriction on how much time should be spent on each instance, or even that they need to be treated entirely separately (unlike within the CHeSC2011 competition). A real-world application of this is when many different instances, or maybe many variants of a few instances but with different choices for available resources, need to be solved "as well as possible overnight" so that a decision can be made next day as to which one(s) to use. In this case, it is reasonable to consider that hyper-heuristics should be extended so as to treat the instances collectively as a batch. This batching has two immediate potential advantages: (i) "Effort balancing". Better balancing of computational effort across the instances. If some are much easier than others then it seems reasonable that they should be allocated less computational time, and more time allocated to those that will benefit most. (ii) "Inter-instance learning". If some of the instances are from the same domain (as would be the case in most practical applications) then it makes sense that the hyper-heuristic should be able to learn from the earlier instances in order to perform better on the later instances. This gives an intermediate between online and offline learning.

In this study, we do not consider the interesting challenge of the inter-instance learning. Instead, we provide evidence that there is a significant potential benefits of the better balancing of computational effort between instances. Note that, although we do not here provide a mechanism that would be able to directly exploit the potential, the aim is to show that it would be worthwhile for hyper-heuristics research to develop such effort balancing schemes.

After a brief discussion of HyFlex and the CHeSC competition, we study some statistics of the performance of a particular hyper-heuristic on the competition instances; showing that there is a wide variation in their properties, and that this can lead to about half the computational effort effectively being wasted.

2   Background

HyFlex is an interface supporting development and research of hyper-heuristics and other metaheuristics in which the domain level is separated from the hyper-heuristic level. In order to discriminate between the interface from its implementation, we will refer to its first version as HyFlex v1.0 which was developed at the University of Nottingham by a team of researchers of the ASAP research group during 2009-2011. HyFlex v1.0 was used for the "Cross-domain Heuristic Search Challenge" (CHeSC) in 2011. CHeSC2011 used the following problem domains: Boolean Satisfiability (SAT), One Dimensional Bin Packing (BP), Permutation Flow Shop (FS), Personnel Scheduling (PS), Vehicle Routing Problem (VRP) and the Traveling Salesman Problem (TSP). For each domain ten different instances were provided. A hyper-heuristic was given 10 minutes of execution time for each instance on a specified machine. The winner of CHeSC,

AdapHH[3] was a learning hyper-heuristic which uses a learning adaptive heuristic selection method in combination with an adaptive iteration limited list-based threshold move accepting method [6]. All the problem domain implementations compatible with HyFlex v1.0 have been serving as benchmarks to test the generality of hyper-heuristics.

3   Performance Properties of the Instances

A standard property of a search is the Performance Profile "PP" or the curve of quality versus time. This is used heavily within the area of anytime reasoning [7] but is also relevant to the case of balancing of computational effort between optimisation tasks. To study the PP on the CHeSC2011 instances, we used *AdapHH* and Robinhood hyper-heuristic (RHH) [8] which is not as successful as *AdapHH* on the CHeSC benchmark. Some examples of the PP on the SAT domain using *AdapHH* are given in Fig. 1 and from these we immediately see that those are cases for which improvements in quality cease well before the standard time deadline of 10 (nominal) minutes.[4] This suggests that it could be worth transferring time from such instances to others in which the search does not stagnate. In order to quantify this we performed experiments with the overall deadline extended to 30 (nominal) minutes per instance. For each run, we determined "t(LastImp)" the time at which the last improvement occurred in each run. The results of these are analysed in two ways. Firstly, in Table 1 we compare the fraction of time that is spent before the last improvement against the overall time using *AdapHH*. We see that, on average, around half the run time is actually "wasted" in the sense that it is after the last improvement.

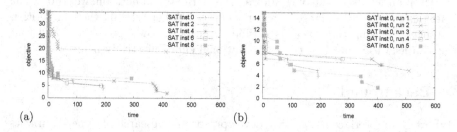

Fig. 1. Performance Profiles for runs on instances from the SAT domain using *AdapHH*. The Best-So-Far (BSF) quality is plotted against running time; and the lines stop when no further improvements are made within the 600 nsec limit. (a) A single run on each of 5 separate instances. (b) 5 separate runs on a single instance.

---

[3] http://code.google.com/p/generic-intelligent-hyper-heuristic/downloads/list

[4] In experiments, the "10 minute" is a "nominal" (or normalised) standardised time as determined by a benchmarking program available via the CHeSC website. To aid future comparisons, we always report results using nominal seconds (nsecs).

Table 1. For each domain, and then the aggregate of all domains, we give the average "non-stagnant fraction of the runtime" that was taken to reach the last improvement. (Based on runs of 1800 s per instance).

Domain	SAT	BP	PS	FS	VRP	TSP	ALL
Non-stagnant fraction of runtime	0.24	0.78	0.62	0.57	0.52	0.81	0.59

In Fig. 2(a) we give the results of ranking the instances in a domain by their value of t(LastImp) and then plotting t(LastImp) against this rank (achieved by *AdapHH*). In the SAT domain we see that most instances stagnate fairly early. In other domains there are a wide range of these stagnation times. In Fig. 2(b) we use the same data, but give the ranking over the domains aggregated together. We see that the instances show widely different behaviours. In particular, around 10 instances stagnate well before the usual 600 nsecs deadline; in contrast, many other instances would potentially benefit from a longer runtime.

The dispersion of stagnation times suggests that each instance should be given a new time limit. There are many potential ways to do this, however, here we use two simple schemes. In the first scheme, some instances are selected to be 'down' by reducing their time to a new limit arbitrarily chosen as 500 nsecs, and others are 'up' with their time limit increased to 700 nsecs. Equal numbers of up and down instances are taken so that the average time will still be 600 nsecs. The up and down instances are (heuristically) selected based on their performance during trial runs with down being those that do not need the need the extra time, and the up the ones that are most likely to benefit. Specifically, 8 and 6 instances were selected for *AdapHH* and RHH respectively. We then tested the effects of the new time limits on the selected instances using separate test runs that compare the qualities achieved at 500 vs. 600 nsecs for the down instances, and 600 vs. 700 nsecs for the up instances. For efficiency, we use the "Retrospective Parameter Variation" technique in the style of [9] (originally used for the loosely-related topic of restarts) so as to re-use data from long runs to quickly simulate shorter runs of different lengths. A run on a down (up) instance is called a *loser* (*winner*) if it improved between 500 and 600 (600 and 700) nsecs,

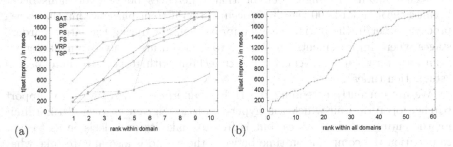

Fig. 2. Times till last improvement on the instances after ranking using *AdapHH*. (a) For each domain separately, (b) for all the domains together.

as the reduced (increased) time would have caused a loss (an improvement) in quality. The net gain is just the numbers of winners minus losers per run; we found net gains of around 7 and 5 instances for *AdapHH* and RHH, respectively. Although this first scheme uses knowledge of the performances that means it is not directly implementable, it does show that simple transfers of runtimes between instances have a potential for significant improvements.

In the second experimental scheme, a two phase *on-the-fly* approach is followed. In the first phase, each instance is given a standard time, however if it makes no progress within a given of duration 100 nsecs, it is terminated and the remaining time is added to a time budget. In the second phase, the time budget is shared between instances that they were not terminated in the first phase. The point of this simple and common scheme, unlike the first scheme, is that it can be implemented directly without pre-knowledge of the performance profile on an instance. The success rate, denoting the percentage of runs on which extra time allocation leads to improvement in the solution, was 75 and 69% for *AdapHH* and RHH, respectively.

4   Conclusion and Future Work

We proposed a batch mode in which the hyper-heuristic is given many instances and an overall time limit, but is not unrestricted to treating them independently and with the same run-time. We found large variations in run times for the CHeSC2011 benchmark instances, and provided simple changes to time limits that lead to significant improvement. A key question for future work is whether such decisions as to better timing can be made reliably, dynamically and in advance, by using the properties of the performance profiles. However, this paper should be taken as indication that if such predictions can be made, then potentially significant runtime can be saved. We also remark, that it seems that the 6 different CHeSC2011 domains have rather different properties with respect to the standard time limit of 600 nsecs. This wide distribution was presumably good for the competition as it made it more likely that there would be a good differentiation between hyper-heuristics, though it does suggest some caution needs to be taken when interpreting results of comparisons of hyper-heuristics between domains. It might well be that differences between domains occur because the standard 600 nsec limit occurs at a different phase within the search process; either in the initial "easy" improvements or during the later "harder" stages where improvements are harder to find. Future analyses might benefit from longer run-times to classify the time limit with respect to the expected "stagnation times".

We are currently extending the HyFlex interface and software to support batched operation, in which the hyper-heuristic is given an overall time limit for the entire set of instances, but is free to make its own decision as to how to partition the computation time between them, and is also able to take what it has learned on one instance and use it for others. During a run, HyFlex will have the utilities to allow saving a solution into a file so that the hyper-heuristic could choose to continue improving it at a later stage if time allows. Since part of the goal of batched mode is to allow inter-instance learning, and learning might naturally start with the smallest instances first, then we will also add a method returning an indicative measure of the size of a given instance. A further planned improvement, "delayed commitment" arises from the observation that moves are often rejected, and the decision to reject relies only on the objective function which is often obtainable with a fast incremental computation with no immediate need to also update internal data structures. It may be more efficient if "virtual-states" were created, with a full construction of the state and data structures only occurring after a separate call to a commit method, and only if it is decided to accept the move.

Besides the extensions specific to batch mode, we hold the view (shared, we believe, by many others in the community) that the classical strict domain barrier returning only the fitness is too restrictive; that more information needs to be passed through the barrier, but still ensuring that there is no loss of the modularity of splitting the search control from the domain details (for example, a common and natural request seems to be to allow multiple objectives rather than one). However, we expect that such relaxations of the domain barrier can generally be made independently of changes for the batched mode.

References

1. Cowling, P., Kendall, G., Soubeiga, E.: A hyperheuristic approach to scheduling a sales summit. In: Burke, E., Erben, W. (eds.) PATAT 2000. LNCS, vol. 2079, pp. 176–190. Springer, Heidelberg (2001)
2. Chen, X., Ong, Y.S.: A conceptual modeling of meme complexes in stochastic search. IEEE Trans. Syst. Man Cybern. Part C: Appl. Rev. **42**(5), 612–625 (2012)
3. Burke, E.K., Hyde, M., Kendall, G., Ochoa, G., Özcan, E., Qu, R.: Hyper-heuristics: a survey of the state of the art. Technical Report NOTTCS-TR-SUB-0906241418-2747, School of Computer Science, University of Nottingham (2010)
4. Özcan, E., Bilgin, B., Korkmaz, E.E.: A comprehensive analysis of hyper-heuristics. Intell. Data Anal. **12**(1), 3–23 (2008)
5. Ochoa, G., et al.: HyFlex: a benchmark framework for cross-domain heuristic search. In: Hao, J.-K., Middendorf, M. (eds.) EvoCOP 2012. LNCS, vol. 7245, pp. 136–147. Springer, Heidelberg (2012)
6. Mısır, M., Verbeeck, K., De Causmaecker, P., Vanden Berghe, G.: A new hyper-heuristic implementation in HyFlex: a study on generality. In: Fowler, J., Kendall, G., McCollum, B. (eds.) Proceedings of the MISTA'11, pp. 374–393 (2011)
7. Zilberstein, S., Russell, S.J.: Approximate reasoning using anytime algorithms. In: Natarajan, S. (ed.) Imprecise and Approximate Computation. Kluwer Academic Publishers, The Netherlands (1995)
8. Kheiri, A., Özcan, E.: A Hyper-heuristic with a round robin neighbourhood selection. In: Middendorf, M., Blum, C. (eds.) EvoCOP 2013. LNCS, vol. 7832, pp. 1–12. Springer, Heidelberg (2013)
9. Parkes, A.J., Walser, J.P.: Tuning local search for satisfiability testing. In: Proceedings of AAAI 1996, pp. 356–362 (1996)

Tuning Algorithms for Tackling Large Instances: An Experimental Protocol

Franco Mascia[✉], Mauro Birattari, and Thomas Stützle

IRIDIA, CoDE, Université Libre de Bruxelles, Brussels, Belgium
{fmascia,mbiro,stuetzle}@ulb.ac.be

Abstract. Tuning stochastic local search algorithms for tackling large instances is difficult due to the large amount of CPU-time that testing algorithm configurations requires on such large instances. We define an experimental protocol that allows tuning an algorithm on small tuning instances and extrapolating from the obtained configurations a parameter setting that is suited for tackling large instances. The key element of our experimental protocol is that both the algorithm parameters that need to be scaled to large instances and the stopping time that is employed for the tuning instances are treated as free parameters. The scaling law of parameter values, and the computation time limits on the small instances are then derived through the minimization of a loss function. As a proof of concept, we tune an iterated local search algorithm and a robust tabu search algorithm for the quadratic assignment problem.

Keywords: Automatic algorithm configuration · Scaling of parameters · Iterated local search · Robust tabu search · Quadratic assignment problem

1 Introduction

Many applications require the solution of very large problem instances. If such large instances are to be solved effectively, the algorithms need to operate at appropriate settings of their parameters. As one intriguing way of deriving appropriate algorithm parameters, the automatic configuration of algorithms has shown impressive advances [1]. However, tuning algorithms for very large instances directly is difficult, a main reason being the high computation times that even a single algorithm run on very large instances requires. There are two main reasons for these high computation times. First, the computational cost of a single search step scales with instance size; second, larger instances usually require a much larger number of search steps to find good quality solutions. From a theoretical side, the tuning time would scale linearly with the number of configurations tested or linearly with the computation time given to each instance. However, even if a limited number of algorithm configurations are tested during the tuning of the algorithm, the sheer amount of time required to test a single

G. Nicosia and P. Pardalos (Eds.): LION 7, LNCS 7997, pp. 410–422, 2013.
DOI: 10.1007/978-3-642-44973-4_44, © Springer-Verlag Berlin Heidelberg 2013

algorithm configuration on a very large instance makes it a problem of practical relevance also for the tuning.

Here, we define a protocol for tuning a stochastic local search (SLS) algorithm on small instances that allows us to make predictions on the behaviour of the tuned SLS algorithm on very large instances. To do so, we optimise the value of free variables of the experimental setting that allow us to make this kind of extrapolations. To give a concrete example, in this paper we present a case study on an iterated local search (ILS) [2,3] algorithm and a robust tabu search (RoTS) [4] algorithm for the quadratic assignment problem (QAP). In the ILS case, we intend to study the strength of the perturbation while in the case of RoTS, we intend to define the appropriate tabu list setting (or better said, finding an appropriate range for the tabu list length settings). Hence, we need to identify the scaling law for these two variables in the respective algorithms to very large instances.

For this we define an experimental protocol where we actually allow two free variables in our experimental setting: a policy for the cut-off time and a policy for the actual parameter configuration. The rationale of having also the cut-off time as a free variable (in addition to the actual parameter setting) is to find an appropriate cut-off time for training on small instances that allows us to extrapolate the parameter configuration for a target cut-off time on a very large instance. As an illustrative example, consider an ILS algorithm that exposes a single parameter to the tuning, for example, the one that controls the amount of perturbation of the current solution. This parameter acts on the balance between intensification and diversification of the algorithm. A reasonable assumption is that the amount of intensification and diversification determined by the parameter value depends on the instance size, and more specifically, that on very large instances the amount of diversification required is smaller due to the fact that the search space to be explored is already large and the algorithm will have to spend most of the time intensifying. In such cases a small cut-off time when tuning on small instances, can lead to algorithm configurations that imply stronger intensification and that therefore allow for a more realistic prediction for very large instances.

The structure of the paper is as follows. Section 2 presents the formalisation of the experimental setting. Section 3 presents a proof of concept with an ILS algorithm and a RoTS algorithm for the QAP. In Sect. 4 we draw the conclusions.

2   Modelling

In the most general case, we want to define a protocol for tuning an SLS algorithm on a set of small training instances $s \in S$ and make predictions on the behaviour of the tuned algorithm on a very large instance $s^\star$. There are two free variables in our experimental setting. The first one is the maximum cut-off time on the small instances, which we define as a policy $t(s)$ that depends on the instance size s. The second one is the parameter setting that we define as the policy $\hat{m}(s; t(s))$ that depends on the instance size and the cut-off time $t(s)$.

Our aim is to optimise policies $t(s)$ and $\hat{m}(s;t(s))$ to predict a good parameter setting $\hat{m}(s^\star;T^\star)$ when executing the algorithm on a very large instance $s^\star$ with cut-off time $T^\star$, where $T^\star$ is the maximum time-budget available for solving the target instance $s^\star$.

We cast this problem as a parameter estimation for the minimisation of a loss function. More in detail, we select a priori a parametric family for the policies $t(s)$ and $\hat{m}(s;t(s))$. The value defined by the policies for an instance size s will be determined by the sets of parameters π_t and $\pi_{\hat{m}}$. The number and type of parameters in π_t and $\pi_{\hat{m}}$ depend on the parametric family chosen for the respective policies. We further constrain the policies by requiring that the maximum cut-off time is larger than a specified threshold $t(s) > \delta$ and that the policy defines a specific cut-off time for the target instance $t(s^\star) = T^\star$.

Very small instances should have a smaller impact on the optimisation of the policies than have small or small-to-medium training instances. The latter are in fact closer and more similar to the target instance size $s^\star$. Therefore, in the most general case, we also use a weighting policy $\omega(s)$ of a specific parametric family with parameters π_ω. The only constraint on this policy is that $\sum_{s \in S} \omega(s) = 1$.

We define the loss function in Eq. 1 as the difference between $C_{\hat{m}}(s;t(s))$, which is the cost obtained when executing the algorithm with the parameter setting determined by $\hat{m}(s;t(s))$; and $C_B(s;t(s))$, which is the cost function obtained when executing the algorithm with the best possible parameter setting $B(s;t(s))$ given the same maximum run-time $t(s)$, and try to determine:

$$\underset{\pi_\omega,\pi_{\hat{m}},\pi_t}{\arg\min} \sum_{s \in S} \omega(s) \left[C_{\hat{m}}(s;t(s)) - C_B(s;t(s)) \right]. \tag{1}$$

By finding the optimal settings for π_ω, $\pi_{\hat{m}}$, and π_t, we effectively find the best scaling of the examples in S, and the best cut-off time, which allow us to find the policy that best describes how the parameter setting scales with the sizes in S. The same policy can be used to extrapolate a parameter setting for a target instance size $s^\star$ and a target cut-off time $T^\star$.

3   A Proof of Concept

In this paper, we present a proof of concept in which we concretely use the parameter estimation in Eq. 1 to tune an ILS algorithm and a RoTS algorithm for the QAP [5].

The QAP models the problem of finding a minimal cost assignment between a set P of facilities and a set L of locations. Between each pair of facilities there is a flow defined by a flow function $w : P \times P \to \mathbb{R}$, and locations are at distance $d : L \times L \to \mathbb{R}$. To simplify the notation, flow and distance functions can be seen as two real-valued square matrices W and D respectively. The QAP is to find a bijective function $\pi : P \in L$ that assigns each facility to a location and that minimises the cost functional:

$$\sum_{i,j \in P} w_{i,j} d_{\pi(i),\pi(j)}.$$

3.1   Iterated Local Search

In our ILS algorithm for the QAP [3], after generating a random initial solution, a first improvement local search is applied until a local minimum is reached. Then the algorithm undergoes a series of iterations until the maximum cut-off time is reached. At each iteration the current solution is perturbed by a random k-exchange move. After the perturbation, an iterative improvement algorithm is applied until a local optimum is reached. The new solution obtained is accepted if and only if it improves over the current solution. A parameter k specifies the size of the perturbation. It is the only parameter exposed to the tuning, and it can assume values from 2 up to the instance size.

The Experimental Setting. For each size in $S = \{40, 50, \ldots, 100\}$, we generate 10 random instances of Taillard's structured asymmetric family [6]. We then measure the average percentage deviation from the best-known solution for each size $s \in S$, by running the ILS algorithm 10 times on each instance with 100 values of the perturbation parameter and by taking the mean value. The maximum cut-off time for these experiments is set to a number of CPU-seconds that is larger than a threshold $max_t(s)$, which allows at least for 1000 iterations of the ILS algorithm with the perturbation strength set to $k = 0.5s$.

We fix the scaling policy as $\omega(s) = \frac{s^3}{\sum_{s \in S} s^3}$ with no parameters π_ω to be optimised. The parametric family of the parameter policy is the linear function $\hat{m}(s; t(s)) = c + ms$ with the parameters $\pi_{\hat{m}} = (c, m)$. The cut-off time policy $t(s)$ is defined as a function $t(s) = c_0 + c_1 s^\alpha$, with the constraint that $t(s) > \delta \, \forall s > s'$, where s' is the smallest $s \in S$. The constant δ is set to 20 ms as the minimum amount of time that can be prescribed for an experiment. Moreover, since we pre-computed the cost function for a given maximum cut-off time, we set also an upper-bound $t(s) < 3 \, max_t(s)$. Finally, the policy has to pass through the point (s^*, T^*), hence one of the parameters can be determined as function of the others and π_t can be restricted to the two parameters (c_0, α):

$$\underset{c,m,c_0,\alpha}{\arg\min} \sum_{s \in S} \frac{s^3}{\sum_{s \in S} s^3} \left[C_{\hat{m}}(s; t(s)) - C_B(s; t(s)) \right]. \tag{2}$$

To minimize the loss in Eq. 2 we implemented, for this case study, an ad hoc local search procedure that estimates the parameter values within predefined ranges. The local search starts by generating random solutions until a feasible one is obtained. This initial random solution is then minimised until a local optimum is reached. The algorithm is repeated until the loss function is equal to zero or a maximum number of iterations is reached. To minimise the incumbent solution, the algorithm selects an improving neighbour by systematically selecting a parameter and increasing or decreasing its value by a step l. For integer-valued parameters l is always equal to 1. For real-valued parameters the value of l is set as in a variable neighbourhood search (VNS) [10, 11]. Initially, l is set to 1^{-6}, then as soon as the the local search is stuck in a local minimum, its value l is first increased to 1^{-5}, then 1^{-4} and so on until l is equal to 1. As soon as the local search escapes the local minimum, the value of l is reset to 1^{-6}. With this local search, we do not want to define an effective procedure for the parameter estimation, the aim here is mainly to have some improvement algorithm for finding reasonable parameter settings for our policy and to present a proof of concept of the experimental protocol presented in this paper. In the future, we plan to replace our ad hoc local search with more performing local search algorithms for continuous optimization such as CMA- ES [7] or Mtsls1 [8]. The parameters c and m are evaluated in the range $[-0.5, 0.5]$, and the parameter α in the range $[3, 4, \ldots, 7]$. The parameter c_0 is initialised to $max_t(s')$, where s' is the smallest $s \in S$, and evaluated in the range $[0, 3 \, max_t(s')]$.

Results. For each target size $s^\star$ and target cut-off time $T^\star$, we first let the $t(s)$ policy pass through the point $(s^\star, T^\star)$, and then we find the optimal policies by minimising Eq. 2. We optimise and test our policies for the target sizes $s^\star$ 150, 200, 300, 400 and 500. For the minimisation of the loss function, we allow our ad hoc local search algorithm to restart at most 25 times.

To evaluate the policies obtained, we compute two metrics: the loss obtained by the predicted value and a normalised score inspired by [9]. The loss is the difference between the cost obtained when running the algorithm with the parameter prescribed by the policy and the best parameter for the given size and cut-off time. The normalised score, is computed as:

$$\frac{E_{unif}C(s; t(s)) - C_{\hat{m}}(s; t(s))}{E_{unif}C(s; t(s)) - C_B(s; t(s))},$$

where $E_{unif}C(s; t(s))$ is the expectation of an uniform choice of the parameter setting. This score is equal to zero when the cost of the parameter prescribed by the policy is the same as the one expected by an uniform random choice of the parameter. It is equal to one when the cost of the prescribed parameter corresponds to the cost attainable by an oracle that selects the best parameter setting. Negative values of the score indicate that the prescribed parameter is worse than what could be expected by an uniform random choice.

To calculate the two metrics, we pre-compute on the target instance sizes, the values of the cost function for 100 possible parameter configurations. Then, to evaluate the predicted values, we round them to the closest pre-computed ones.

In Fig. 1 we present the results for the largest of the test instances, with $s^\star = 500$ and $T^\star = 6615.44$ CPU-seconds. The plot on top shows the cut-off time policy with exponent $\alpha = 4$ that passes through the target size and target cut-off time. The second plot from the top shows the linear policy for the parameter setting that prescribes a perturbation of size 177, while the optimal perturbation value for the specified cut-off time is 171. The third plot shows the loss. The predicted parameter is rounded to the closest precomputed one, which is 176. The difference in the average deviation from the best known solution amounts to 0.044977. The last plot at the bottom shows the normalised score that for the prediction on target size 500 is equal to 0.841609. In Table 1 we summarise similar results also for 150, 200, 300 and 400.

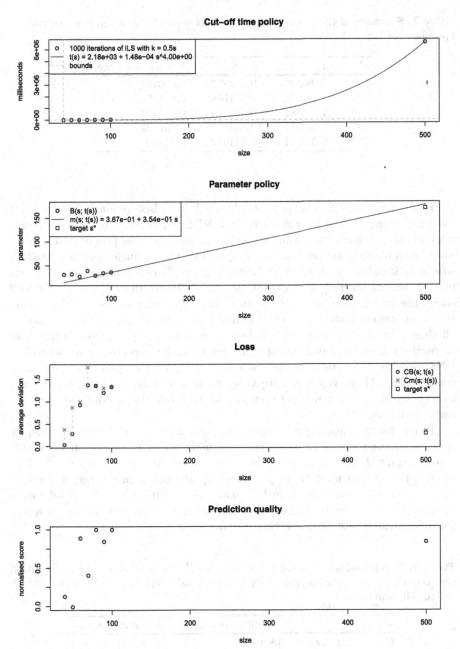

Fig. 1. Cut-off time policy, parameter policy for ILS, loss, and prediction quality on target instance size $s^* = 500$.

Table 1. Summary of the loss and normalised score on the target sizes of the policies optimised for ILS.

$s^\star$	$T^\star$	Loss	Normalised score
150	102.85	0.058957	0.880454
200	257.50	0.084706	0.814831
300	1 047.82	0.039883	0.887496
400	2 596.33	0.039318	0.881627
500	6 615.44	0.044977	0.841609

To further evaluate the policies obtained, we also compare them with a dynamic setting of the parameter as in a VNS algorithm. This comparison is relevant, as a dynamic variation of the perturbation size as propagated in VNS would be a reasonable way of addressing the fact that a single best perturbation size value is unknown for the very large instances. For each target size $s^\star$ in our test set, we run both algorithms 10 times on the 10 instances of size $s^\star$. Figure 2 shows the average deviation of the two algorithms with respect to the results that are obtained with the a posteriori best parameter for the given size and cut-off time. Also in this case, the policies obtained for the parameter setting lead to results which are much better than what can be expected from a baseline VNS algorithm. A stratified rank-based permutation test (akin to a stratified version of the Mann-Whitney U test) rejects at a 0.05 significance level the null hypothesis of no difference between the average deviations obtained with the two algorithms.

To test for the importance of optimising also a policy for the cut-off time, we tested a tuning protocol in which the parameter policy is the only free variable being optimised. In this case we fixed the cut-off time to a number of CPU-seconds that allows for 1000 steps of the ILS algorithm on the target instance size $s^\star$ with a perturbation size $k = 0.5s$. As shown in Table 2, on all target instance there is a clear advantage of leaving the cut-off time policy as a free variable of the experimental setting.

Table 2. Normalised score on the target sizes for ILS in the case in which the cut-off time is optimised as a free variable of the experimental setting, and in the case in which the cut-off time is fixed.

$s^\star$	$T^\star$	Cut-off time policy	Fixed cut-off time
150	102.85	0.880454	0.742731
200	257.50	0.814831	0.769348
300	1 047.82	0.887496	0.661414
400	2 596.33	0.881627	0.606192
500	6 615.44	0.841609	0.701780

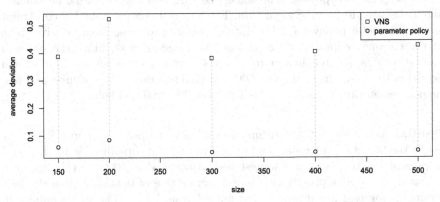

Fig. 2. Comparison between VNS and the parameter policy for ILS on target instances $s^\star$ at time $T^\star$.

3.2   Robust Tabu Search

The RoTS algorithm for the QAP [4] is a rather straightforward tabu search algorithm that is put on top of a best improvement algorithm making use of the usual 2-exchange neighbourhood for the QAP, where the location of two facilities are exchanged at each iteration. A move is tabu, if at least the two facilities involved are assigned to a location they were assigned in the last tl iterations, where tl is the tabu list length. Diversification is ensured by enforcing specific assignments of facilities to locations if such an assignment was not considered for a rather large number of local search moves. In addition, an aspiration criterion is used that overrides the tabu status of a move if it leads to a new best solution. The term *robust* in RoTS stems from the random variation of the tabu list length within a small interval; this mechanism was intended to increase the robustness of the algorithm by making it less dependent on one fixed setting of tl [4]. Hence, instead of having a fixed tabu list length, at each iteration the value for the tabu tenure is selected uniformly random in the range $\max(2, \text{unif}(\mu - 0.1, \mu + 0.1) \cdot s))$. Thus, μ is the expected value of the tabu list length and it is the only parameter exposed to the tuning. In the original paper, a setting of $\mu = 1.0$ was proposed.

The Experimental Setting. For instance sizes $S = \{40, 50, \dots, 100\}$, we generate 10 Taillard's instances with uniformly random flows between facilities and uniformly random distances between locations [6]. For each parameter setting of $\mu \in \{0.0, 0.1, \dots, 2.5\}$, and for each instance size, we compute the mean deviation from the best-known solution. The mean is computed over 10 runs of the RoTS algorithm on each of the 10 instances. The maximum cut-off time for these experiments is set to a number of CPU-seconds that allow for at least $100 \cdot s$ iterations of the RoTS algorithm.

We keep for this problem the same free variables and the same parametric families we used for the ILS algorithm. The only difference is a further constraint on the parameter policy $\hat{m}(s; t(s))$ that is required to prescribe a positive value for the parameter for all $s \in S$ and on the target size $s^\star$. Since the problem is more constrained and harder to minimise, we allow our ad hoc local search algorithm to restart from at most 5000 random solutions. We optimise and test the policies on target instance sizes 150, 200, 300, 400 and 500.

Results. As for the ILS algorithm, we evaluate the policies by measuring the loss on the target instance sizes and by computing the normalised score. In Fig. 3 we present the results on the largest test instance $s^\star = 500$. On this instance, the parameter policy prescribes a parameter setting of 0.040394 while the best parameter for this instance size and cut-off time is 0.1. The loss amounts to 0.051162 and the normalised score is 0.653224.

Table 3 summarises the results on all test instances. On instance $s^\star = 400$ the parameter policy obtains a loss of 0 and a normalised score equal to 1. This is due to the fact that for evaluating the policies, and hence knowing the best parameter setting given the size and cut-off time, we pre-computed the cost of a fixed number of parameter configurations. When computing the cost of the parameter configuration prescribed by the policy, the value of the parameter is rounded to the closest value for which the cost has been pre-computed. For instance, for size $s^\star = 400$ the prescribed parameter value is 0.086961. This value is rounded to 0.1, which corresponds to the best parameter setting.

To further evaluate the policy we evaluate it with a default setting of the parameter that, in the case of RoTS, would be $\mu = 1.0$. Figure 4 shows the average deviation obtained with the parameter setting prescribed by the policy and the default parameter setting. The average deviation is computed with respect to the solution quality obtained when using the best a posteriori parameter setting given the instance size and the cut-off time. On all instances the policy obtained for the parameter setting lead to results which are much better than what we could expect from the default parameter setting. Also in this case, a stratified rank-based permutation test rejects at a 0.05 significance level the null hypothesis of no difference between the average deviations obtained with the two algorithms.

Table 3. Summary of the loss and normalised score on the target sizes of the policies optimised for RoTS.

$s^\star$	$T^\star$	Loss	Normalised score
150	120.09	0.038592	0.827993
200	278.84	0.062164	0.697857
300	954.89	0.021499	0.861288
400	2 517.97	0	1
500	5 473.06	0.051162	0.653224

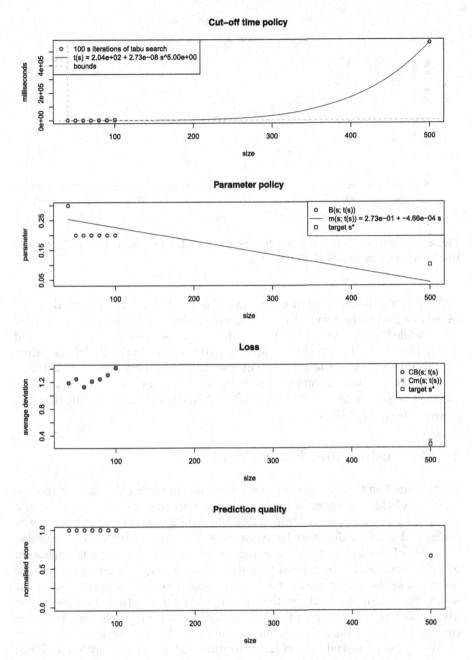

Fig. 3. Cut-off time policy, parameter policy for RoTS, loss, and prediction quality on target instance size $s^* = 500$.

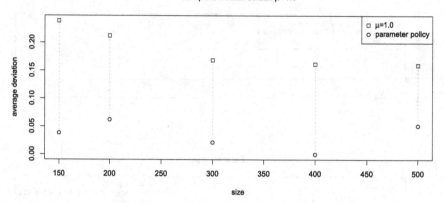

Fig. 4. Comparison between the default value $\mu = 1.0$ and the parameter policy for RoTS on target instances s^* at time T^*.

To test the importance of optimising also a policy for the cut-off time, we tested also for this problem a tuning protocol in which we optimise the parameter policy while keeping the cut-off time fixed. In this case the cut-off time was fixed to a number of CPU-seconds that allow for $100 \cdot s$ steps of the RoTS algorithm. On this problem there was no clear-cut result, in fact on instance sizes 300, 400, and 500, with a fixed cut-off time policy the score remains the same; on instance 150 the score drops from 0.827993 to 0.354483; and on instance size 200 the score increases from 0.697857 to 1.

4   Conclusions and Future Work

We presented an experimental protocol in which optimising the value of the free variables of the experimental setting allows for tuning an SLS algorithm on a set of small instances and extrapolating the results obtained on the parameter configuration for tackling very large instances. We cast the problem of optimising the value of the free variables as a parameter estimation for the minimisation of a loss function. In the general formulation as well as in the proofs of concept presented in this paper, we suggested as possible free variables: (i) a policy for scaling the parameter configuration, (ii) a policy for selecting the cut-off time when tuning the algorithm on the small instances, and (iii) a policy for weighting the small instances during the minimisation of the loss function.

We presented a study on an ILS algorithm and a RoTS algorithm with one parameter for the QAP. On both problems we obtained promising results, with the extrapolated parameter setting being close to best a posteriori parameter setting for the instances being tackled. We also showed that results obtained by our method are much better than default static or dynamic settings of the para-meters. We believe that our approach may be a viable way of tuning algorithms for very large instances if SLS algorithms rely on few key parameters such as the algorithms tested here.

One key element of our contribution is the optimisation of a policy for the cut-off time that prescribes how long a configuration should be tested during the tuning on small instances. We showed experimentally, that at least for the ILS algorithm, optimising a cut-off time policy allows for better extrapolations of the parameter setting on large instances.

As future work, we plan to extend the approach to algorithms with (many) more than one parameter and to extrapolate to much larger instance sizes. In both cases we also need to define an extended protocol for assessing the performance of our method since pre-computing the cost function may become prohibitive. Furthermore, an automatic selection of the parametric models, and a comparisons to other recent approaches for tuning for large instances such as [12] would be interesting.

Acknowledgments. This work was supported by the META-X project, an *Action de Recherche Concertée* funded by the Scientific Research Directorate of the French Community of Belgium. Franco Mascia, Mauro Birattari, and Thomas Stützle acknowledge support from the Belgian F.R.S.-FNRS. The authors also acknowledge support from the FRFC project *"Méthodes de recherche hybrids pour la résolution de problèmes complexes"*. This research and its results have also received funding from the COMEX project within the Interuniversity Attraction Poles Programme of the Belgian Science Policy Office.

References

1. Hoos, H.H.: Programming by optimization. Commun. ACM **55**(2), 70–80 (2012)
2. Lourenço, H.R., Martin, O., Stützle, T.: Iterated local search: framework and applications. In: Gendreau, M., Potvin, J.Y. (eds.) Handbook of Metaheuristics. International Series in Operations Research & Management Science, 2nd edn, pp. 363–397. Springer, New York (2010)
3. Stützle, T.: Iterated local search for the quadratic assignment problem. Eur. J. Oper. Res. **174**(3), 1519–1539 (2006)
4. Taillard, É.D.: Robust taboo search for the quadratic assignment problem. Parallel Comput. **17**(4–5), 443–455 (1991)
5. Koopmans, T.C., Beckmann, M.J.: Assignment problems and the location of economic activities. Econometrica **25**, 53–76 (1957)
6. Taillard, E.D.: Comparison of iterative searches for the quadratic assignment problem. Location Sci. **3**(2), 87–105 (1995)
7. Hansen, N., Ostermeier, A.: Completely derandomized self-adaptation in evolution strategies. Evol. Comput. **9**(2), 159–195 (2001)
8. Tseng, L.Y., Chen, C.: Multiple trajectory search for large scale global optimization. In: Proceedings of IEEE Congress on Evolutionary Computation, Piscataway, NJ, IEEE, pp. 3052–3059 June 2008
9. Birattari, M., Zlochin, M., Dorigo, M.: Towards a theory of practice in metaheuristics design: a machine learning perspective. Theor. Inform. Appl. **40**(2), 353–369 (2006)

10. Mladenovic, N., Hansen, P.: Variable neighbourhood search. Comput. Oper. Res. **24**(11), 71–86 (1997)
11. Hansen, P., Mladenovic, N.: Variable neighborhood search: principles and applications. Eur. J. Oper. Res. **130**(3), 449–467 (2001)
12. Styles, J., Hoos, H.H., Müller, M.: Automatically configuring algorithms for scaling performance. In: Hamadi, Y., Schoenauer, M. (eds.) LION 6. LNCS, vol. 7219, pp. 205–219. Springer, Heidelberg (2012)

Automated Parameter Tuning Framework for Heterogeneous and Large Instances: Case Study in Quadratic Assignment Problem

Lindawati[1], Zhi Yuan[1,2], Hoong Chuin Lau[1], and Feida Zhu[1(✉)]

[1] School of Information Systems, Singapore Management University,
Singapore, Singapore
{lindawati,zhiyuan,hclau,fdzhu}@smu.edu.sg
[2] IRIDIA, CoDE, Université Libre de Bruxelles, Brussels, Belgium

Abstract. This paper is concerned with automated tuning of parameters of algorithms to handle heterogeneous and large instances. We propose an automated parameter tuning framework with the capability to provide instance-specific parameter configurations. We report preliminary results on the Quadratic Assignment Problem (QAP) and show that our framework provides a significant improvement on solutions qualities with much smaller tuning computational time.

Keywords: Automated parameter tuning · Instance-specific parameter configuration · Parameter search space reduction · Large instance parameter tuning

1 Introduction

Good parameter configurations are critically important to ensure algorithms to be efficient and effective. *Automated parameter tuning* (also called *automated algorithm configuration* or *automated parameter optimization*) is concerned with finding good parameter configurations based on training instances. Existing approaches for *automated parameter tuning* mainly differ in their applicability to different types of parameters: tuners that handle both categorical and numerical parameters include ParamILS [16], GGA [1], iterated F-Race [4] and SMAC [15] etc.; and there exist approaches specially designed for tuning numerical parameters, e.g. by Kriging models (SPO [2]) and continuous optimizers [32].

In this paper, we are concerned with two specific challenges of automated parameter tuning:

1. Heterogeneity. This refers to the phenomenon that different problem instances require different parameter configurations on the same target algorithm to solve. Hutter et al. [14] defines "inhomogeneous" instances as those for which algorithm rankings are unstable across instances. They state that inhomogeneous instance sets are problematic to address with both manual and automated methods for offline algorithm configuration. Schneider and Hoos [26]

G. Nicosia and P. Pardalos (Eds.): LION 7, LNCS 7997, pp. 423–437, 2013.
DOI: 10.1007/978-3-642-44973-4_45, © Springer-Verlag Berlin Heidelberg 2013

provides two quantitative measures of homogeneity and observes that homogeneity increases when partitioning instance sets by means of clustering based on observed runtimes. In this paper, we are concerned with the notion of heterogeneity in the sense that instances perform differently when run with different configurations. Table 1 gives an illustration of this phenomenon for benchmark QAP instances on robust tabu search algorithm as presented in [9].

2. Large problem instances. By "large", we mean instances that require a prohibitively long computation time [29]. This notion of largeness is typically tied to the size of the problem instances (i.e. larger problem instances typically take longer time to run), but one needs to be mindful that this quantity varies across problems and target algorithms.

Table 1. Effect of Three Different Parameter Configurations on 4 QAP instances performance. The performance is an average of percentage deviation from optimum or best known solution.

Instances	Configuration 1	Configuration 2	Configuration 3
tai40a	1.4	**1.0**	2.0
tai60a	1.7	**1.6**	2.2
tai40b	9.0	9.0	**0.0**
tai60b	2.1	2.9	**0.3**

There have been approaches that attempt to find instance-specific configurations for a heterogeneous set of instances (see for example, [17, 21, 30]). Unfortunately, finding *instance features* itself is often tedious and domain-specific [8], requiring a discovery of good features for each new problem. Similarly, tuning algorithms for large instances is a frustrating experience, as the tuning algorithm typically requires a large number of evaluations on training instances. This quickly makes automatic tuning suffer computationally prohibitive run time.

In this paper, we attempt to tackle the above challenges by proposing a new automated parameter tuning framework AutoParTune that attempt to bring together components that are helpful for automated parameter tuning. Having a unified software framework allows algorithm designers to readily experiment with different mixes of parameter tuning components in deriving good parameter configurations. Our emphasis in this paper is heterogeneity and large instances, and we briefly describe our approach as follows.

To handle heterogeneous instances, we propose a generic instance clustering technique SufTra that mines generic features from instance search trajectory patterns. For this purpose, we make use of a novel data structure "suffix tree" [7]. A search trajectory is defined as a path of solutions discovered by the target algorithm as it searches through its neighborhood search space [13]. A nice characteristic of our work is that we can obtain these trajectories from the target local search algorithm with minimal additional computation effort. Our approach improves the work of [21] that captures similarity using a single (and relatively short) segment through out the entire sequence, and works only on short and small number of sequences due to its inherit computational bottleneck. In contrast, our approach is capable of retrieving similarity across multiple segments with linear time complexity. Using a Suffix Tree data structure, our approach can efficiently and effectively form better and tighter clusters and hence improve the overall performance of the underlying target algorithm.

To handle large instances, we propose ScaLa that automatically finds computationally less expensive instances as surrogate to large instances. ScaLa detects similarity among different instances with different runtime using performance-based similarity measures [26] and clustering techniques. In this work, we experimentally explore the feasibility of this approach.

We apply our approach on the Quadratic Assignment Problem (QAP), since it is a notoriously hard problem that has been shown to have heterogeneous instances, and the run time required to solve grows rapidly with instance size. The major contributions (and thus the flow) of this paper are summarized as follows:

- We propose a new generic automated parameter tuning framework AutoParTune for handling heterogeneous and large instances.
- We present SufTra, a novel technique for clustering heterogeneous instances.
- We present ScaLa that performs runtime analysis for scaling large instances.

2   AutoParTune

AutoParTune is a realization of the concept proposed in [20]. As shown in Fig. 1(a), the framework consists of four components: (I) parameter search space reduction; (II) instance-specific tuning to handle heterogeneous instances; (III) scaling large instances; and (IV) global tuning. The first components are considered as pre-processing steps which can be executed in any combination and in any order. The detail of the AutoParTune subsystems is shown in Fig. 1(b).

Parameter Search Space Reduction. This component allows us to drastically reduce the size of the parameter space before tuning. We implement the method presented in [6], which is based on Design of Experiment (DoE), that involves full factorial design and Response Surface Methodology (RSM).

Instance-Specific Tuning. For the instance-specific tuning, we construct a generic approach for instance-specific parameter tuning: SufTra. The detail will be presented in Sect. 4.

Scaling Large Instances. For handling large instances, we present a novel technique based on the computational time analysis: ScaLa, and empirically explore its feasibility in Sect. 5.

Global Tuning. There are many global tuning methods in the literature. Here, we embed ParamILS [16] which utilizes Iterated Local Search (ILS) to explore the parameter space in order to find a good parameter configuration based on the given training instances. ParamILS has been very successfully applied to tune a broad range of high-performing algorithms for several hard combinatorial problems with large number of parameters.

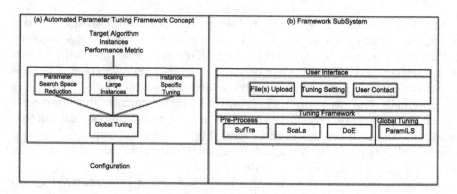

Fig. 1. Automated parameter tuning framework concept and subsystem

3   Target Algorithm and Experimental Setup

To avoid confusion, we refer the algorithm whose performance is being optimized as *target algorithm* and the one that is used to tune it as the *tuner*. As target algorithm, we use a hybrid Simulated Annealing and Tabu Search (SA-TS) algorithm [22]. It uses the Greedy Randomized Adaptive Search Procedure (GRASP) [31] to obtain an initial solution, and then uses a combined Simulated Annealing (SA) [18] and Tabu Search (TS) [5] algorithm to improve the solution. There are four numerical parameters, real-valued or integer-valued, to be tuned as described in Table 2.

In SufTra, we use a set of instances from two generators in [19] for single-objective QAP as in [23]. The first generator generates uniformly random instances where all flows and distances are integers sampled from uniform distributions. The second generator generates flow entries that are non-uniform random values, having the so called real-like structure since it resemble the structure of QAP problems found in practical applications. We generated 500 instances with size from 10 to 150 from each generator and randomly choose 100 as training

Table 2. Parameters for SA-TS on QAP

Parameter	Description	Type	Range	Default
Temp	Initial temperature of SA algorithm	Continuous	[100, 5000]	5000
Alpha	Cooling rate	Continuous	[0.1, 0.9]	0.1
Length	Length of tabu list	Integer	[1, 10]	10
Limit	maximum number of non-improving iterations prior to intensification strategy (re-start the search from best-found solution)	Integer	[1, 10]	10

instances and 400 as testing instances. In ScaLa, we use 120 training instances with size 50, 70, 90, and 150 and 100 testing instances with size 150 from the first generator.

All experiments were performed on a 1.7 GHz Pentium-4 machine running Windows XP for SufTra and on a 3.30 GHz Intel Core i5-3550 running Windows 7 for ScaLa. We measured runtime as the CPU time needed by these machine.

4   SufTra: Clustering Heterogeneous Instances

SufTra is premised on the assumption that an algorithm configuration is correlated with its fitness landscape, i.e. configuration that performs well on a problem instance of certain fitness landscape will also perform well on another instance of a similar topology [24]. Furthermore, since fitness landscape is difficult to compute, it can be approximated by a search trajectory [10,11] which is deemed a probe through the landscape under a given algorithm configuration. SufTra is an extension of the work on search trajectory clustering CluPaTra [21], and overcomes the following major limitations of CluPaTra.

1. **Scalability**
   The pair-wise sequence alignment used in CluPaTra is implemented using standard dynamic programming with a complexity $O(m^2)$, where m is the maximum sequence length. Hence, the total time complexity for all instances is $O(n^2 \times m^2)$, where n is the number of instances. This poses a serious problem for instances with long sequences and when the number of instances is large.
2. **Flexibility**
   The nature of sequence alignment is to align a sequence pair that gives us the highest alignment score. A matched symbol contributes a positive score $(+1)$, while a gap contributes a negative score (-1). The sum of the scores is taken as the maximal similarity score of the two sequences. However, it is possible that sequences share similarity on more than one segment, especially for long sequences. Sequence alignment is not flexible enough to capture multi-segment alignment with an acceptable time complexity.

SufTra works by transforming search trajectory into a string of symbols based on its solution attributes. A suffix tree is constructed on these strings to extract frequent substrings. Substrings may occur in multiple segments along the search trajectory, it allows us to consider **multi-segment** similarities to improve the accuracy of the clusters. Using these frequent substrings as features to cluster the strings, we calculate the similarity and cluster the instances.

4.1   Search Trajectory Representation and Extraction

Search Trajectory. The search trajectory is obtained by running a local search procedure and keep track of all the solutions visited. These solutions are then transformed into a sequence of symbols based on its attributes. Each symbol encodes a combination of two solution attributes: (1) position type, based on the topology of the local neighborhood as given in Table 3 [13]; and (2) percentage deviation from optimum or best known. These two attributes are combined: the first two digits are the deviation from optimum and the last digit is the position type. To handle target algorithms with cycles and (random) restarts, SufTra adds two additional symbols: 'CYCLE' and 'JUMP'; 'CYCLE' is used when the target algorithm returns to a previously-visited position, while 'JUMP' is used when the local search is restarted.

Table 3. Position types of solution

Position type label	Symbol	<	=	>
SLMIN (strict local min)	S	+	-	-
LMIN (local min)	M	+	+	-
IPLat (interior plateau)	I	-	+	-
SLOPE	P	+	-	+
LEDGE	L	+	+	+
LMAX (local max)	X	-	+	+
SLMAX (strict local max)	A	-	-	+

'+' = present, '-' = absent; referring to the presence of neighbors with larger ('<'), equal ('=') and smaller ('>') objective values

Note that in a search trajectory, several consecutive solutions may have similar solution properties before final improvement and reaching local optimum. We therefore compress the search trajectory sequence to a *Hash String* by removing the consecutive repetition symbols and store the number of repetitions in a *Hash Table* to be used later in pair-wise similarity calculation.

Suffix Tree Construction. The search trajectory sequences found in the previous section is used to build a suffix tree. Suffix tree is a data structure that exposes internal structure of a string for a particularly fast implementation of many important string operations. The construction of a suffix tree proves to have a linear time complexity w.r.t. the input string length [7]. A suffix tree T for an m-character string S is a rooted directed tree having exactly m leaves numbered 1 to m. Each internal node, except for the root, has at least two children and each edge is labeled with a substring (including the empty substring) of S. No two edges out of a node has edge-labels beginning with the same character. To represent suffixes of a set $\{S_1, S_2, ...S_n\}$ of strings, we use a *generalized* suffix tree. *Generalized* suffix tree is built by appending a different end of string marker (which is a symbol not in used in any of the string) to each string in the set, then concatenate all the strings together, and build a suffix tree for the concatenated string [7].

We construct the suffix tree for the *hash strings* derived from search trajectories using the Ukkonen's algorithm [7]. We build a single *generalized* suffix tree by concatenating all the *Hash Strings* together to cover all training instances. Length of the concatenate string is proportional to the sum of all the *Hash String* lengths. The Ukkonen's algorithm works by first building an implicit suffix tree containing the first character of the string and then adding successive characters until the tree is complete. Details of Ukkonen's algorithm can be found in [7]. Our Ukkonen's algorithm implementation requires $O(n \times l)$, where n is the number of instances and l is the maximum length of the *Hash String*.

Features Extraction. After constructing the suffix tree, we extract frequent substrings as features. A substring is considered as frequent if it has a length greater than min_{length} and it occurs in at least $min_{support}$ number of strings. min_{length} and $min_{support}$ value is different for each problem. To find a good value of min_{length} and $min_{support}$, we apply one run of a first-improvement local search, starting from the best of five random initial points, using a 1-flip neighborhood.

4.2 Similarity Score Calculation

After extracting the features, we calculate instance's scores for each feature and construct an instance-feature metric using the following rules:

1. if the instance does not contain the feature, the score is 0.
2. Otherwise, the score is calculated by summing numbers of repetitions for each symbol in feature from previously constructed *Hash Table*. A frequent substring may occur few times in one string. We calculate the score for each occurrence and choose the maximum score as a score for instance-feature metric.

Using the metric, we calculate similarity for each pair of instances by applying cosine similarity. Cosine similarity has been widely used to measure similarity between two vectors by measuring cosine angle between them [7]. Cosine similarity result is equal to 1 when the angle is 0, and it is less than 1 when the angle is of any other value. Cosine similarity is formulated as follows.

$$similarity = \frac{\sum_{i=0}^{n}(Inst_1(feature_i) \times Inst_2(feature_i))}{\sqrt{\sum_{i=0}^{n} Inst_1(feature_i)^2} \times \sqrt{\sum_{i=0}^{n} Inst_2(feature_i)^2}} \qquad (1)$$

with $Inst_1(feature_i)$ and $Inst_2(feature_i)$ as score from instance-feature metric for instance 1 and 2, and feature i.

4.3 Clustering

Similar to [21], we cluster the instances by a well-known clustering approach AGNES [12] with L method [25]. AGNES or AGglomerative NESting is a hierarchical clustering approach that works by creating clusters for each individual instance and then merging two closest clusters (i.e., a pair of clusters with the smallest distance) resulting in fewer clusters of larger sizes until all instances belong to one cluster or a termination condition is satisfied (e.g. a prescribed number of clusters is reached). We implement the L method [25] to automatically find the optimal number of clusters.

4.4 Experimental Result

To evaluate SufTra's effectiveness, we first compared the time needed (in seconds) for SufTra and CluPaTra to form the clusters in training phase and to map the testing instances in testing phase. Table 4 (I) shows the result. From the table, we observe that SufTra is 18 times faster then CluPaTra.

Next, we compared the *target algorithm* performance using parameter configuration from SufTra, CluPaTra and ISAC as well as the one-size-fits-all tuner ParamILS. Since ISAC requires problem-specific features, we used 2 features: *flow dominance* and *sparsity* of flow metric which is believed to have significant influence on the performance [28].

For the three instance-specific methods, we used the same one-size-fits-all tuner, ParamILS [16]. Since ParamILS works only with discrete parameters, we first discretized the values of the parameters. We measured the performance as the average of percentage deviation from optimum or best known solution. We set the cutoff runtime of ParamILS to 100 s. For CluPaTra and SufTra, we allowed each tuner to execute the target algorithm for a maximum of two CPU hours for each cluster. To ensure fair comparison, we set the time budget for ISAC and ParamILS to be equal to the total time needed to run SufTra. For reducing evaluation error, we used five different set of training instances and testing instances, and measured the average performance on these instance sets. We also performed *t-test* on the significance of our result where a *p-value* below 0.05 is deemed to be statistically significant.

In Table 4 (II), we show the performance comparison results. From the table, we observe that SufTra performs better on training and testing instances compare to other approaches. But the result for training instances is not statistically significant compared to ISAC.

Table 4. QAP experiment result

	Training	Testing
I. Computational time		
CluPaTra	1,051 s	2,718 s
SufTra	**56 s**	**146 s**
II. Performance result		
ParamILS	1.07	2.12
CluPaTra	0.87	1.54
ISAC	0.83	1.21
SufTra	**0.81**	**1.16**
p-value*	0.061	0.042

*based on statistical test on ISAC and SufTra

5   ScaLa: Scaling Large Instances

As mentioned in the introduction, tuning on instances that require long computation time to solve is a challenging task. These instances are referred to as *large* instances in this work, since typically, instances with larger size require longer computation time to solve. However, the *largeness* referred here may also depend on other instance features and whether the target algorithm is effective in tackling them. These features may include, e.g. problem-independent features such as ruggedness of the fitness landscape, fitness distance correlation, and so on; and problem-specific instance features such as dominance and sparsity of the instance matrix in QAP, etc. More detailed survey on measuring instance difficulty can be found in [27]. Our preliminary experiments here consider only instance size as a measure of instance "largeness", but incorporating other features is straightforward.

To make automatic tuning applicable for large instances, one idea is to tune on smaller instances [3,29]. Styles et al. proposed in [29] to run multiple tuning processes on small instances, validate the independently tuned configurations on medium instances, and use the best validated configuration for solving the large instances. Our current work takes a different direction. The goal is to tune on small instances with short runtime, such that the tuned configuration performs similarly on large instances with long runtime. In order to realize this goal, a number of questions have to be addressed:

1. How to measure similarity among different instances?
2. Does there exist similarity between small instances and large instances at all?
3. Do good configurations on small instances perform well on similar large instances?

Each of the three questions above will be addressed in one of the following subsections.

5.1   Measuring Instance Similarity

To answer question 1, there exist two different approaches to finding similarities among instances. One is based on instance features, e.g. instance size, fitness distance correlation, search trajectory patterns [21] etc. Both instance-specific tuners ISAC [17] and CluPaTra [21] cluster instances based on features, then tune on each cluster, and confirm that tuning on separate clustered instance set leads to better performance than tuning on all instances. Unlike our approach in this work, they did not take into account the computation time as instance feature. Another approach is based on empirical algorithm performance [26]. Scheider et al. introduced in [26] two measures, a *ratio measure* and a *variance measure*, for measuring instance similarity based on relative performance of different algorithms (or same algorithm with different configurations). However, the performance-based similarity measure depends on two folds: computation time and solution quality. Although [26] considered computation time, but did not consider scaling among different instances by, e.g. considering different solution quality threshold. In this work, we adopt the performance-based similarity measures proposed in [26], more specifically, the *variance measure* that is described in more details in the next section, and use them to find similarities among different instances with different runtime.

5.2   Finding Similarities Between Large and Small Instances

To answer question 2, we set up experiments to test the hypothesis whether large instances could be similar to small instances at all given different computation time. Unlike in SufTra (Sect. 4) where we try to separate heterogeneous instances, here the goal is to join instances with different features, given different runtime. We take QAP as our target problem, and an implementation of SA-TS algorithm as our target algorithm (see Sect. 3 for a description and parameter ranges). 30 instances are generated for each of the four instance sizes 50, 70, 90, and 150. The SA-TS has four parameters as described in Table 2. In order to use the performance-based measure, 100 parameter configurations are sampled uniformly within the parameter range. Each parameter configuration runs once on each instance. The solution cost $c_\theta(n, t_n)$ of a configuration $\theta \in \Theta$ on an instance size $n \in N = \{50, 70, 90, 150\}$ with a given runtime t_n is computed by taking the mean solution cost across the 30 instances with size n, and $C_\Theta(n, t_n) = \{c_\theta(n, t_n), \theta \in \Theta\}$. For each instance size n, a set of runtime T_n is determined as follows: let minimum runtime $t_{min} = 0$, maximum runtime t_{max} takes value of the maximum natural stopping time of the algorithm (no restart), and T_n takes values in a logarithmically spaced sequence between t_{min} and t_{max}, excluding t_{min}. Following [26], we perform a standardized z-score normalization for each cost vector $C_\Theta(n, t_n)$, and use the variance measure

$$Q_{var}(\Theta, N', T_{N'}) = \frac{1}{|\Theta|} \sum_{\theta \in \Theta} Var(c_\theta(N', T_{N'})), \text{ for } N' \subseteq N \qquad (2)$$

for measuring similarity (more precisely, dissimilarity) among different pairs of $(n, t_n) \in (N, T_N)$. Based on this measure Q_{var}, instances of different size and computation time can be clustered with the goal of optimizing similarity of the resulting subsets. A classical clustering approach Hierarchical Agglomerative Clustering or AGNES [12] is adopted in our preliminary experiments (an alternative clustering method K-mean also gives very similar clustering result). For illustrative purpose, 5 logarithmic time intervals for each instance size n are used, excluding t_{min}, this makes $|T_n| = 4$. The clustering results are shown in Fig. 2. Interestingly, the most similar two subgroups turn out to be the longest runtime (natural stopping time) of each of the four instance sizes $n \in N$, and the second longest logarithmic runtime level of each $n \in N$. More specifically, the four (n, t_n)-pairs (50, 23.6), (70, 29.8), (90, 39.2), (150, 127.6) form the most similar group, while (50, 6.7), (70, 8.0), (90, 9.8), (150, 23.8) comprise the second most similar group. In the two shorter levels of runtime, the similarities across the four instance sizes are less obvious. Nevertheless, this interesting clustering result confirms our hypothesis raised in question 2: using performance-based similarity measure, given the right runtime, different instances, small or large, can become similar to each other.

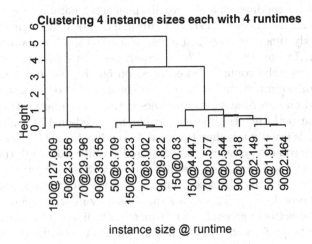

Fig. 2. Clustering four different instance sizes each with four different computation times by hierarchical agglomerative based on variance measure.

5.3 Solving Large Instances by Tuning on Small Instances

How can automatic tuning benefit from this automatically detected instance similarity? One straightforward follow-up idea is to use the best parameter configuration tuned on small instances with short runtime to solve similar large instances with long runtime. However, it remains unjustified that how good these tuned-on-small parameter configurations are, compared with, for example, parameter configuration tuned directly on instances of the same size with the same runtime. In this experiment, two most similar groups of size-runtime pairs (see Fig. 2 of Sect. 5.2) are used: the first group includes (50, 6.709), (70, 8.002), (90, 9.822), and (150, 24.823); the second group includes (50, 23.556), (70, 29.796), (90, 39.156), (150, 127.609). For each of the two groups, two sets of experiments are set up: (1) tuned by `oracle`: as a quick proof-of-concept, we take the best configuration from 100 configurations based on 30 instances on each instance size as found in Sect. 5.2, and test them on another 100 testing instances of size 150 with corresponding runtime; (2) tuned by `ParamILS` [16]: we tune the target algorithm using three independent runs of `ParamILS` for each instance size using the size-runtime pairs mentioned above, each run was assigned maximum 300 calls of target algorithm on new randomly generated training instances, and each tuned configuration is tested on 100 same testing instances as in (1). The second experiment set is to test generalizability of the similarity information detected in Sect. 5.2. The goal is to see how good these best configurations tuned on small instances such as 50, 70, and 90 with shorter runtime, compared with the best configuration tuned on instance size 150, when tested on instance size 150 with the same runtime.

The results are listed in Table 5. The results confirm that, firstly as expected, a large amount of tuning time is saved by tuning on small instances, ranging from 59 to 81 % in our experiments; and secondly, in general, parameter configurations tuned on smaller instances with shorter runtime do not differ significantly from the ones directly tuned on large instances, as long as similarities between them are confirmed. In both groups of both experiment sets, there is no statistical difference between the configurations tuned on 50, 70, 90, and 150, tested by Wilcoxon's rank-sum test. In the first experiment set tuned by oracle, configurations tuned on size 70 and 90 sometimes perform even better than tuned on 150. The mean performance difference from the tuned-on-150 configuration in the first group is usually less than 0.1 %, and even less than 0.01 % in the second group. In the second experiment set tuned by ParamILS, although configuration tuned on size 150 performs best, the difference is not significant: the mean performance difference is usually less than 0.1 % in the first group, and less than 0.05 % in the second group. This shows the similarity information detected from Sect. 5.2 can be actually generalized to tuners with different training instances. As reference, the performance of the default parameter configuration (listed in Table 2) is presented in Table 5, and it is statistically significantly outperformed by almost all the above tuned configurations in both groups, which proves the necessity and success of tuning process. We also include as reference the best configuration tuned on instance size 150 with runtime 23.556 (127.609) s to be tested

Table 5. Results for the performance of the best parameter configurations tuned on sizes 50, 70, 90, 150, and tested on instances of size 150. Two most similar groups of size-runtime pairs (see text or Fig. 2) are used. Two experiment sets are presented, oracle and ParamILS (see text). Each column of %oracle and %ParamILS shows the mean percentage deviation from the reference cost. In each column, $+x$ $(-x)$ means that the tuned configuration performance is x% more (less) than the reference cost. %time.saved shows the percentage of tuning time saved comparing with tuning on instances of size 150. The performance of default parameter configuration is shown in row "def.". The last row 150' used the best parameter configuration tuned on instance size 150 with runtime 127.609 (23.556) s, and tested on instance size 150 with runtime 23.556 (127.609) s, respectively. Results marked with † refers to statistically significantly worse results compared to tuned-on-150 using Wilcoxon's rank-sum test.

23.556 s				127.609 s		
tuned.on	%oracle	%ParamILS	%time.saved	%oracle	%ParamILS	%time.saved
50	−0.48	−0.047	72	−0.048	−0.048	81
70	−0.65	−0.053	66	−0.060	−0.027	76
90	−0.61	−0.093	59	−0.057	−0.040	69
150	−0.58	−0.151	0	−0.060	−0.070	0
def.	+1.17†	+0.150†	-	−0.008†	−0.024	-
150'	+1.16†	+0.195†	-	+0.232†	+0.208†	-

on instance size 150 with different runtime, i.e. 127.609 (23.556) s, respectively (in row 150' of Table 5). Although the tuning and testing instances are of the same size, different runtime makes a great performance difference, resulting in almost one order of magnitude worse than tuning on the small instances with appropriate runtime. The 150' performance is statistically significantly worse than all the above tuned configurations belonging to the same group, and it is even significantly worse than the default configuration in the second group. This contrasts with the fact that the difference among the similar size-runtime pairs (the first four rows of Table 5) is indeed very minor, and it also shows the risk of tuning on algorithm solution quality without assigning the right runtime, which in fact proves the necessity of our automatic similarity detection procedure in ScaLa.

6   Conclusion and Future Work

In this paper, we proposed an automated parameter tuning framework for heterogeneous and large instances and tested it on Quadratic Assignment Problem (QAP). We constructed SufTra for tuning heterogeneous instances and ScaLa for large instances. We verified SufTra's performance and observed a significant improvement compared to a vanilla one-size-fits-all approach (ParamILS) and other generic instance-specific approach CluPaTra. We claim that: (1) SufTra is a suitable approach for instance-specific configuration that significantly improves the performance with minor additional computational time; and (2) SufTra has overcome CluPaTra limitations with a new efficient method for feature extraction and similarity computation using suffix tree. In the development of ScaLa, we used performance-based similarity measure and clustering technique to automatically detect and group similar instances with different sizes by assigning different runtime, such that one can tune on small instances with much less runtime and apply the tuned configuration to solve large instances with long runtime. This greatly reduces computation time required when tuning large instances. Through our preliminary experiments, we empirically showed that small instances and large instances can be similar when given the right runtime, and in such case, the good configurations tuned on small instances can also perform well on large instances.

Up to this stage of our work, the SufTra and ScaLa are not yet integrated. In near future, we plan to integrate those two components on the AutoParTune framework, in particular, we plan to integrate also features extracted from trajectory patterns of SufTra into ScaLa. As future works on SufTra, we will investigate how to generate clusters from population-based-algorithm using generic features pertaining to population dynamics, since currently SufTra can only be applied to target algorithms which are local-search-based due to the search trajectory. On the other hand, ScaLa is still an actively ongoing work. Future works include largely extending the amount of experiments, consider also testing on problems other than QAP, and extend our studies to other state-of-the-art algorithms. The correlation between computation time and instance size may be algorithm-specific, therefore, an automatic approach to detecting it is practically valuable. Our current approach is still a proof-of-concept, since it is computationally expensive for computing the performance-based measure. In future work, we plan to investigate how to reduce the computation expenses by, e.g. taking fewer instances and fewer but good configurations found during the tuning process. In particular, we plan to investigate the possibility of predicting the "right" runtime for an unseen instance such that it is similar to a known group of instances.

Acknowledgments. We thank Saifullah bin Hussin, Thomas Stützle, Mauro Birattari, Matteo Gagliolo for valuable discussion on scaling large instances, and Aldy Gunawan for allowing us to use his DoE codes.

References

1. Ansótegui, C., Sellmann, M., Tierney, K.: A gender-based genetic algorithm for the automatic configuration of algorithms. In: 15th International Conference on Principles and Practice of Constraint Programming, pp. 142–157 (2009)
2. Bartz-Beielstein, T., Lasarczyk, C., Preuss, M.: Sequential parameter optimization. In: Congress on Evolutionary Computation 2005, pp. 773–780. IEEE Press (2005)
3. Birattari, M., Gagliolo, M., Saifullah bin Hussin, Stützle, T., Yuan, Z.: Discussion in IRIDIA coffee room, October 2008
4. Birattari, M., Yuan, Z., Balaprakash, P., Stützle, T.: F-race and iterated f-race: an overview. In: Bartz-Beielstein, T., Chiarandini, M., Paquete, L., Preuss, M. (eds.) Experimental Methods for the Analysis of Optimization Algorithms, pp. 311–336. Springer, Heidelberg (2010)
5. Glover, F.: Tabu search - part I. ORSA J. Comput. **1**, 190–206 (1989)
6. Gunawan, A., Lau, H.C., Lindawati, : Fine-tuning algorithm parameters using the design of experiments approach. In: Coello Coello, C.A. (ed.) LION 5. LNCS, vol. 6683, pp. 278–292. Springer, Heidelberg (2011)
7. Gusfield, D.: Algorithms on Strings, Trees and Sequences. Cambridge University Press, Cambridge (1997)
8. Guyon, I., Gunn, S., Nikravesh, M., Zadeh, L.A. (eds.): Feature Extraction: Foundations and Applications. Springer, Heidelberg (2006)
9. Halim, S., Yap, Y.: Designing and tuning sls through animation and graphics an extended walk-through. In: Stochastic Local Search, Workshop (2007)
10. Halim, S., Yap, Y., Lau, H.C.: Viz: a visual analysis suite for explaining local search behavior. In: 19th ACM Symposium on User Interface Software and Technology, pp. 57–66 (2006)
11. Halim, S., Yap, R.H.C., Lau, H.C.: An integrated white+black box approach for designing and tuning stochastic local search. In: Bessière, C. (ed.) CP 2007. LNCS, vol. 4741, pp. 332–347. Springer, Heidelberg (2007)
12. Han, J., Kamber, M.: Data Mining: Concept and Techniques, 2nd edn. Morgan Kaufman, San Francisco (2006)
13. Hoos, H.H., Stützle, T.: Stochastic Local Search: Foundation and Application. Morgan Kaufman, San Francisco (2004)
14. Hutter, F., Hoos, H., Leyton-Brown, K.: Tradeoffs in the empirical evaluation of competing algorithm designs. Ann. Math. Artif. Intell. (AMAI), Spec. Issue Learn. Intell. Optim. **60**, 65–89 (2011)

15. Hutter, F., Hoos, H.H., Leyton-Brown, K.: Sequential model-based optimization for general algorithm configuration. In: Coello Coello, C.A. (ed.) LION 5. LNCS, vol. 6683, pp. 507–523. Springer, Heidelberg (2011)

16. Hutter, F., Hoos, H.H., Leyton-Brown, K., Stützle, T.: Paramils: an automatic algorithm configuration framework. J. Artif. Intell. Res. **36**, 267–306 (2009)

17. Kadioglu, S., Malitsky, Y., Sellmann, M., Tierney, K.: Isac: instance-specific algorithm configuration. In: 19th European Conference on Artificial Intelligence (2010)

18. Kirkpatrick, S., Gelatt, C.D., Vecchi, M.P.: Optimization by simulated annealing. Science **200**, 671–680 (1983)

19. Knowles, J.D., Corne, D.W.: Instance generators and test suites for the multiobjective quadratic assignment problem. In: Fonseca, C.M., Fleming, P.J., Zitzler, E., Thiele, L., Deb, K. (eds.) EMO 2003. LNCS, vol. 2632, pp. 295–310. Springer, Heidelberg (2003)

20. Lau, H.C., Xiao, F.: Enhancing the speed and accuracy of automated parameter tuning in heuristic design. In: 8th Metaheuristics International Conference (2009)

21. Lindawati, Lau, H.C., Lo, D.: Clustering of search trajectory and its application to parameter tuning. JORS Special Edition: Systems to Build Systems (to appear)

22. Ng, K.M., Gunawan, A., Poh, K.L.: A hybrid algorithm for the quadratic assignment problem. In: International Conference on Scientific Computing, pp. 14–17 (2008)

23. Ochoa, G., Verel, S., Daolio, F., Tomassini, M.: Clustering of local optima in combinatorial fitness landscapes. In: Coello Coello, C.A. (ed.) LION 5. LNCS, vol. 6683, pp. 454–457. Springer, Heidelberg (2011)

24. Reeves, C.R.: Landscapes, operators and heuristic search. Ann. Oper. Res. **86**(1), 473–490 (1999)

25. Salvador, S., Chan, P.: Determining the number of clusters/segments in hierarchical clustering/segmentation algorithms. In: 16th IEEE International Conference on Tools with Artificial Intelligence, pp. 576–584 (2004)

26. Schneider, M., Hoos, H.H.: Quantifying homogeneity of instance sets for algorithm configuration. In: Hamadi, Y., Schoenauer, M. (eds.) LION 6. LNCS, vol. 7219, pp. 190–204. Springer, Heidelberg (2012)

27. Smith-Miles, K., Lopes, L.: Measuring instance difficulty for combinatorial optimization problems. Comput. Oper. Res. **39**(5), 875–889 (2012)

28. Stützle, T., Fernandes, S.: New benchmark instances for the QAP and the experimental analysis of algorithms. In: Gottlieb, J., Raidl, G. (eds.) EvoCOP 2004. LNCS, vol. 3004, pp. 199–209. Springer, Heidelberg (2004)

29. Styles, J., Hoos, H.H., Müller, M.: Automatically configuring algorithms for scaling performance. In: Hamadi, Y., Schoenauer, M. (eds.) LION 6. LNCS, vol. 7219, pp. 205–219. Springer, Heidelberg (2012)

30. Xu, L., Hoos, H.H., Leyton-Brown, K.: Hydra: automatically configuring algorithms for portfolio-based selection. In: Conference of the Association for the Advancement of Artificial Intelligence (AAAI-10) (2010)

31. Yong, L., Pardalos, P.M., Resende, M.G.C.: A greedy randomized adaptive search procedure for the quadratic assignment problem. In: Pardalos, P.M., Wolkowicz, H. (eds.) Quadratic Assignment and Related Problems. DIMACS Series in Discrete Mathematics and Theoretical Computer Science, vol. 16, pp. 237–261. American Mathematical Society, Providence (1994)

32. Yuan, Z., Montes de Oca, M., Birattari, M., Stützle, T.: Continuous optimization algorithms for tuning real and integer parameters of swarm intelligence algorithms. Swarm Intell. **6**(1), 49–75 (2012)

Practically Desirable Solutions Search on Multi-Objective Optimization

Natsuki Kusuno[1], Hernán Aguirre[1], Kiyoshi Tanaka[1], and Masataka Koishi[2(✉)]

[1] Faculty of Engineering, Shinshu University, 4-17-1 Wakasato,
Nagano 380-8553, Japan
{nsk0936@iplab.,ahernan@,ktanaka@}shinshu-u.ac.jp
[2] R&D Center, The Yokohama Rubber Co. Ltd., 2-1 Oiwake,
Hiratsuka, Kanagawa 254-8601, Japan
koishi@hpt.yrc.co.jp

Abstract. This work investigates a method to search practically desirable solutions expanding the objective space with additional fitness functions associated to particular decision variables. The aim is to find solutions around preferred values of the chosen variables while searching for optimal solutions in the original objective space. Solutions to be practically desirable are constrained to be within a certain distance from the instantaneous Pareto optimal set computed in the original objective space. Our experimental results show that the proposed method can effectively find practically desirable solutions.

1 Introduction

Evolutionary multi-objective algorithms [1] optimize simultaneously two or more objective functions that are usually in conflict with each other. The aim of the algorithm is to find the set of Pareto optimal solutions that capture the trade-offs among objective functions. In the presence of several solutions, a decision maker often considers preferences in objective space to choose one or few candidate solutions for implementation. This approach is valid when there is no concern about the buildability of candidate solutions.

In many practical situations, however, the decision maker has to pay special attention to decision space in order to determine the constructability of a potential solution. In manufacturing applications, for example, preferences for particular values of variables could appear due to unexpected operational constraints, such as the availability or lack of materials with particular specifications. Or simple because physical processes that determine a particular value for a decision variable have become easier to perform than those required to determine another value. When these situations arise the decision maker is interested in knowing how far these possible solutions are from optimality. Furthermore, in design optimization and innovation related applications, rather than precise values of decision variables of few candidate solutions, the extraction of useful design knowledge is more relevant. In these cases, analysis of what-if scenarios in decision space, without losing sight of optimality, are important.

G. Nicosia and P. Pardalos (Eds.): LION 7, LNCS 7997, pp. 438–443, 2013.
DOI: 10.1007/978-3-642-44973-4_46, © Springer-Verlag Berlin Heidelberg 2013

From this standpoint, in this work, we investigate an approach that incorporates additional fitness functions associated to particular decision variables, aiming to find solutions around preferred values of the chosen variables while searching for optimal solutions in the original objective space. In addition to expanding the space, we also constraint the distance that solutions could be from the instantaneous Pareto optimal set computed in the original space. We call these solutions as practically desirable solutions. We test the algorithm using DTLZ3 function with two and three objectives in the original space and two additional objectives for the expanded space. Our results show that the proposed method can effectively find practically desirable solutions.

2   Proposed Method

2.1   Concept

We pursue an approach that incorporates additional fitness functions associated to particular decision variables, aiming to find solutions around preferred values of the chosen variables while searching for optimal solutions in the original objective space.

Let us define the original objective space $f^{(m)}$ as the vector of functions

$$f^{(m)}(x) = (f_1(x), f_2(x), \cdots, f_m(x)), \tag{1}$$

where x is a vector of variables and $m \geq 2$ the number of functions. The extended objective space $f^{(M)}$ with $M > m$ objectives is given by

$$f^{(M)}(x) = (f_1(x), f_2(x), \cdots, f_m(x), f_{m+1}(x), \cdots, f_M(x)) \tag{2}$$

where $f_{m+1}(x), \cdots, f_M(x)$ are the additional $M - m$ functions used to evaluate solutions with preferred values in one or more decision variables.

Extending the objective space would work to bias selection to include solutions with particular desired values for some decision variables. However, it is also expected that evolution in an expanded objective space would substantially increase diversity of solutions, which could jeopardize convergence of the algorithm in the original space and the expanded space as well. Thus, in addition to an expanded space, we also constraint the distance that solutions could be from the instantaneous Pareto optimal set computed in the original space, as illustrated in Fig. 1(a). We call these solutions as practically desirable solutions. In the following we describe a method that implements this concept.

2.2   Two Populations, Concurrent Evolution

This method evolves concurrently two populations in different objective spaces. Population A evolves in the extended objective space $f^{(M)}$ using an enhanced ranking of solutions that prefers practically desirable solutions for survival selection and parent selection. On the other hand, Population B evolves in the original objective space $f^{(m)}$. The instantaneous set of Pareto optimal solutions

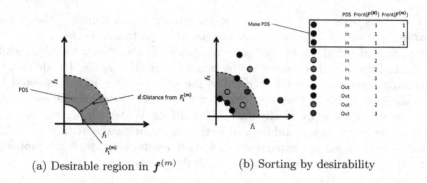

(a) Desirable region in $f^{(m)}$ (b) Sorting by desirability

Fig. 1. (a) Region of practically desirable solutions with preferred values in variable space located at most a distance d from the Pareto optimal set computed in the original objective space $f^{(m)}$. (b) Sorting by desirability respect to the original space $f^{(m)}$ and front number in the extended space $f^{(M)}$

computed in $f^{(m)}$ from the Population B is incorporated into Population A and used as a reference to establish the desirability of solutions in Population A. Ranking for Population A is enhanced by making it dependant on both front number in the extended space $f^{(M)}$ and desirability respect to the original space $f^{(m)}$. This new ranking is used for survival selection and parent selection as well.

In this work Population A evolves using NSGA-II with the enhanced ranking and survival selection, whereas Population B evolves using conventional NSGA-II [2]. In the following we explain survival selection and ranking procedure used to evolve Population A, illustrated in Fig. 2.

Step 1 Get a copy of the set of non-dominated solutions from Population B that evolves in the original space $f^{(m)}$. Let us call this set $F_1^{(m)}$.

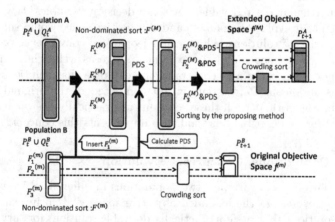

Fig. 2. Two evolving populations, concurrent search method

Step 2 Apply non-dominated sorting to $R_t^A \cup F_1^{(m)}$ in the space $F^{(M)}$, where $R_t^A = P_t^A \cup Q_t^A$ is the combined population of parents P_t^A and offspring Q_t^A evolving in the expanded space $\boldsymbol{f}^{(M)}$. Classify solutions into fronts $F_i^{(M)}$ and rank solutions according to the i-th front they belong to, where $i = 1, 2, \cdots, NF$. Solutions in $F_1^{(m)}$ will be part of $F_1^{(M)}$.

Step 3 Calculate the Euclidean distances from solutions in the fronts $F_i^{(M)}$ to the set $F_1^{(m)}$. The distance from solution $\boldsymbol{x} \in F_i^{(M)}$ to $F_1^{(m)}$ is given by $\delta(\boldsymbol{x}) = \min \parallel \boldsymbol{f}^{(m)}(\boldsymbol{x}) - \boldsymbol{f}^{(m)}(\boldsymbol{y}) \parallel$, $\boldsymbol{y} \in F_1^{(m)}$. If the distance $\delta(\boldsymbol{x})$ is smaller than a threshold distance d then solution $\boldsymbol{x}$ is marked as *desirable*. Otherwise, it is marked as *undesirable*.

Step 4 Sort solutions by front rank and *desirability*. The front number (rank) of desirable solutions remains the same, while the front number of an undesirable solution initially classified in front i is modified to $i + NF$, where NF is the number of fronts initially obtained by non-dominated sorting. That is, undesirable solutions are penalized so that no undesirable solution is assigned better rank than a desirable one, while still differentiating among undesirable ones. Sorting by front number and desirability is illustrated in Fig. 1(b).

Step 5 Form the population P_{t+1}^A for the next generation by copying to it fronts $F_i^{(M)}$ in ascending order, starting with front $F_1^{(M)}$. If all solutions in $F_i^{(M)}$ do not fit in P_{t+1}^A ($|P_{t+1}^A| = |R_t^A|/2$), select the required number according to their crowding distance (less crowded is better). Since undesirable solutions are penalized, as explained above, desirable solutions are given priority for survival and reproduction as well (better rank than undesirable solutions).

3   Simulation Results and Discussion

We study the performance of the algorithms in continuous DTLZ3 functions [3]. In our experiments we set the number of objectives to $m = \{2, 3\}$ varying the number of variables $n = \{5, 10, 15\}$. Thus, the original objective space is given by $\boldsymbol{f}^{(m=2)} = (f_1, f_2)$ and $\boldsymbol{f}^{(m=3)} = (f_1, f_2, f_3)$, respectively. The original objective space is extended by adding two functions to form $\boldsymbol{f}^{(M)}$, where $M = m + 2$. The two additional functions are as follows

$$f_{m+1} = |x_5 - 0.3| \tag{3}$$
$$f_{m+2} = |x_5 - 0.4| \tag{4}$$

Here, the assumed desirable values for variable x_5 are 0.3 and 0.4. Note that in this problem it is known that the optimal value for x_5 is 0.5. We set the threshold distance $d = 10$ to determine the desirability of solutions respect to Pareto optimal solutions in $\boldsymbol{f}^{(m)}$.

To evaluate convergence of solutions we use the Generational Distance (GD), which measures the distance of solutions to the true Pareto front. Smaller values of GD mean better convergence of solutions.

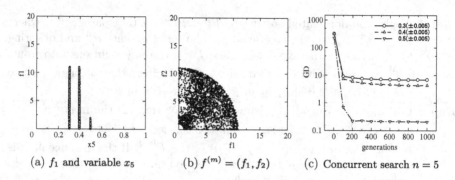

Fig. 3. Proposed method searching concurrently on the original space $f^{(m)} = (f_1, f_2)$, $m = 2$, $n = 5$, and expanded space $f^{(M)}$, $M = 4$. Final generation, DTLZ3 problem.

We run the algorithms 30 times using different random seeds and present average results, unless stated otherwise. The number of generations is set to 1000 generations, parent and offspring population size are $|P_t^A| = |Q_t^A| = 2250$ for the search on the expanded space and $|P_t^B| = |Q_t^B| = 250$ for the search on the original space. As variation operators, the algorithms use SBX crossover and polynomial mutation, setting their distribution exponents to $\eta_c = 15$ and $\eta_m = 20$, respectively. Crossover rate is $pc = 1.0$, crossover rate per variable is $pcv = 0.5$, and mutation rate per variable is $pm = 1/n$.

Figure 3 shows results by the proposed method searching concurrently on the original space $f^{(m)} = (f_1, f_2)$, $m = 2$, and on the expanded space $f^{(M)}$, $M = 4$, for $n = 5$ variables. From Fig. 3(a) it can be seen that the proposed method effectively finds solutions around the two preferred values $x_5 = 0.3$ and $x_5 = 0.4$. In addition it also finds solutions around $x_5 = 0.5$, the value at which solutions become Pareto optimal in this problem. Also, from Fig. 3(b) note that the solutions found are within the threshold distance d established as a condition for solutions desirability.

Figure 3(c) shows the generational distance GD over the generations. GD is calculated separately grouping solutions around the preferred values $x_5 = 0.3$, $x_5 = 0.4$ and optimal value $x_5 = 0.5$. Solutions are considered within a group if the value of x_5 is in the range $[x_5 - 0.005, x_5 + 0.005]$. Note that GD reduces considerably for the three groups of solutions. This clearly shows that the concurrent search on the original space pulls the population closer to the Pareto optimal front and achieves good convergence in addition of finding solutions around the preferred values in variable space. These solutions are valuable for the designer to analyze alternatives that include practically desirable features in addition to optimality.

Figure 4 shows the number of solutions that fall within the desirable area at various generations of the evolutionary process, i.e. solutions located within a distance $d = 10$ of the instantaneous set of Pareto optimal solutions in $f^{(m)}$. Results are shown for DTLZ3 problem with $m = \{2, 3\}$ original objectives

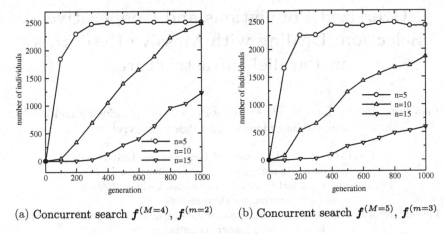

(a) Concurrent search $f^{(M=4)}$, $f^{(m=2)}$ (b) Concurrent search $f^{(M=5)}$, $f^{(m=3)}$

Fig. 4. Number of solutions within the desirable area over the generations, i.e. solutions located within a distance $d = 10$ of the Pareto optimal solutions in $f^{(m)}$ found by the proposed method. DTLZ3 problem with $m = 2$ and $m = 3$ original objectives and two additional fitness functions.

varying the number of variables $n = \{5, 10, 15\}$. Note that the proposed method can effectively find a large number of solutions for any number of variables.

4   Conclusions

In this work we have proposed a method to search practically desirable solutions expanding the objective space with additional fitness functions associated to preferred values of decision variables. The proposed method evolves concurrently two populations, one in the original objective space and the other one in the expanded space using an enhanced ranking and survival selection that favors optimality as well as practical desirability of solutions. Our experiments show that the proposed method can effectively find a large number of practically desirable solutions around preferred values of variables for DTLZ3 problems with 2 and 3 objectives in its original space and 5, 10, and 15 variables.

In the future we would like to test the approach on other kinds of problems, including real world applications.

References

1. Coello, C., Van Veldhuizen, D., Lamont, G.: Evolutionary Algorithms for Solving Multi-Objective Problems. Kluwer Academic, Boston (2002)
2. Deb. K., Agrawal, S., Pratap, A., Meyarivan, T.: A fast elitist non-dominated sorting genetic algorithm for multi-objective optimization: NSGA-II. KanGAL report 200001 (2000)
3. Deb, K., Thiele, L., Laumanns, M., Zitzler, E.: Scalable multi-objective optimization test problems. In: Proceedings of the Congress on Evolutionary Computation 2002, pp. 825–830. IEEE Service Center (2002)

Oversized Populations and Cooperative Selection: Dealing with Massive Resources in Parallel Infrastructures

Juan Luis Jiménez Laredo[1]([✉]), Bernabe Dorronsoro[3], Carlos Fernandes[2,4],
Juan Julian Merelo[4], and Pascal Bouvry[1]

[1] FSTC-CSC/SnT, University of Luxembourg, Luxembourg
{juan.jimenez,pascal.bouvry}@uni.lu
[2] Laseeb, Technical University of Lisbon, Lisbon, Portugal
cfernandes@laseeb.org
[3] Laboratoire d'Informatique Fondamentale de Lille, University of Lille, Lille, France
bernabe.dorronsoro_diaz@inria.fr
[4] Geneura Lab, University of Granada, Granada, Spain
jmerelo@geneura.ugr.es

Abstract. This paper proposes a new selection scheme for Evolutionary Algorithms (EAs) based on altruistic cooperation between individuals. Cooperation takes place every time an individual undergoes selection: the individual decreases its own fitness in order to improve the mating chances of worse individuals. On the one hand, the selection scheme guarantees that the genetic material of fitter individuals passes to subsequent generations as to decrease their fitnesses individuals have to be firstly selected. On the other hand, the scheme restricts the number of times an individual can be selected not to take over the entire population. We conduct an empirical study for a parallel EA version where *cooperative selection* scheme is shown to outperform binary tournament: both selection schemes yield the same qualities of solutions but *cooperative selection* always improves the times to solutions.

Keywords: Selection schemes · Evolutionary algorithms · Parallelization · Execution times

1 Introduction

We seek after a more efficient exploitation of massively large infrastructures in parallel EAs by balancing *population size* and *selection pressure* parameters. To that aim, we assume platforms in which the number of resources can be always considered *sufficient* (see e.g. [5]), i.e. large enough to allow a parallelized population to eventually converge to problem optima. The challenging issue here is to make an efficient use of such resources since a too large population size can be considered oversized: a parametrization error leading to unnecessary wastes of computational time and resources [8]. We show that, in the case of oversized populations, selection pressure can be increased to high values in such a way that

G. Nicosia and P. Pardalos (Eds.): LION 7, LNCS 7997, pp. 444–449, 2013.
DOI: 10.1007/978-3-642-44973-4_47, © Springer-Verlag Berlin Heidelberg 2013

computing time is minimized and the solution quality is not damaged. Hence, the population sizing problem can be redefined into a twofold question that we call the *selection dominance* criterion; a selection scheme A can be said to dominate other selection scheme B if:

a) any arbitrary population size P sufficient to B is always sufficient to A. In our case, we set up the sufficiency criterion to the algorithm performing with a success rate (SR) greater or equal to 0.98 ($SR \geq 0.98$).
b) the execution time due to the pair (A, P) is strictly smaller than the execution time due to (B, P).

Under the selection dominance perspective, it is a good practice to tune the selection pressure to maximum values which still respect the sufficiency criterion. Nevertheless, by doing that we may fall in the following well-known dilemma: On the one hand, a high selection pressure will eventually make the algorithm converge faster but with the risk of losing diversity and getting stuck in local-optima. On the other hand, a low selection pressure will improve the success rate expectations at the cost of a worse parallel execution time.

In the ideal case, a selection operator should be able to self-regulate the selection pressure according to the problem features and the given population size. However, in this paper, we limit the scope of the research to demonstrate a new selection scheme that is better than binary tournament ($s2$) in the sense that $s2$ solutions are always dominated: binary tournament is shown to be outperformed in execution times while both selection schemes have equivalent sizing requirements, i.e. same population sizes are *sufficient* in both cases.

The new selection scheme –introduced in Sect. 2– is inspired by reciprocal altruism, a simple form of natural cooperation in which the fittest individuals decrease their own fitnesses in order to allow breedings of less fitter ones. An experimental analysis is conducted in Sect. 3. Finally, some conclusions and future lines of research are exposed in Sect. 4.

2   Cooperative Selection

The design of new selection schemes is an active topic of research in EC. In addition to canonical approaches such as ranking, roulette wheel or tournament selection [3], other selection schemes have been designed to trade off exploration and exploitation [1] or to be able to self-adapt the selection pressure on-line [2], just to mention a few. Cooperation has been also considered in the design of co-evolutionary EAs [9] in which sub-populations represent partial solutions to a problem and have to collaborate in order to build up complete solutions. However, to the extent of our knowledge, there have been no attempts for designing selection schemes inspired by cooperation.

Cooperative selection, in this early approach, is not more than a simple extension of the classical tournament selection operator. As in the latter, a set of randomly chosen individuals $\vec{s} = \{random_1(P), \dots, random_s(P)\}$ compete for reproduction in a tournament of size s. The best ranked individual is then selected for breeding. The innovation of the new operator consists of each individual having two different measures for fitness. The first is the standard fitness function f which is calculated in the canonical way while the second is the cooperative fitness f_{coop} which is utilized for competing. Since we analyze the operator in a generational scheme context, at the beginning of every generation f_{coop} is initialized with the current fitness value (f) of the individual. Therefore, every first tournament within every generation is performed the same way as with the classical tournament selection. The novelty of the approach relies on the subsequent steps: after winning a competition of a tournament $\overrightarrow{s}$, the f_{coop} of the fittest individual is modified to be the average of the second and the third, which means that, in the following competitions, the winning individual will yield its position to the second. Since each f_{coop} is restarted with the f value every generation, it is likely that fitter individuals reproduce at least once per generation but without taking over the entire population. The details of the cooperative selection scheme are described in Procedure 1.

Procedure 1. Pseudo-code of Cooperative Selection

---

procedure COOPERATIVESELECTION(s)

    #1. Ranking step:
    # Competing individuals in $\overrightarrow{s}$ are ranked according to their cooperative fitnesses f_{coop}^{rank}
    $rank(\overrightarrow{s}) \leftarrow \{f_{coop}^1, f_{coop}^2, f_{coop}^3, \dots, f_{coop}^s\}$

    #2. Competition step:
    # The individual with the highest cooperative fitness f_{coop}^1 is selected
    $winner \leftarrow rank_1(\overrightarrow{s})$

    #3. Altruistic step:
    # After being selected, the *winner* of the competition decreases its own fitness
    $f_{coop}^1 \leftarrow \dfrac{f_{coop}^2 + f_{coop}^3}{2}$

    return *winner*

end procedure

As in *tournament selection*, the only parameter to adjust in *cooperative selection* is the tournament size. We performed preliminary studies in order to tune such a parameter. In those experiments, we found out that a cooperative tournament size of 16 (*Coop s16*) is equivalent to binary tournament $s2$ in terms of selection pressure. Therefore, all experiments will be conducted for such a parameter value.

3   Analysis of Results

In order to analyze the performance of the cooperative selection scheme, we conduct simple experiments in a master-slave GA [4] and tackle an instance of length $L = 200$ of the onemax problem [10]. The GA, in addition to be parallel,

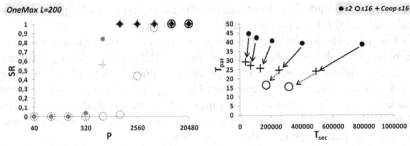

(a) *Scalability of the success rate (SR)*
    as a function of the population size
    (P). Marks in bold show a SR ≥
    0.98.

(b) *Tradeoff between T_{sec} and T_{par} for*
    $SR \geq 0.98$ and different population
    and tournament sizes.

Fig. 1. Scalability of a master-slave GA tackling an instance of the onemax problem
with size L = 200. Parameters s2 and s16 represent different tournament sizes of 2 and
16 respectively and *Coop s16* stands for cooperative selection with a tournament size of
16. Arrows in sub-figure (*b*) indicate a *"dominated by"* relationship between selection
schemes using equal population sizes. Some circles for *s16* are missing in (*b*) as this
setting does not yield *sufficiency* ($SR \geq 0.98$) for population sizes smaller than 10240.

follows a 1-elitism generational scheme. Besides, only a simple point crossover
was considered as breeding operator[1] and three different types of selection parametrization: two of them using tournament selection with tournament sizes of 2
(*s2*) and 16 (*s16*), and one using cooperative selection with a tournament size of
16 (*Coop s16*). Then every setting is analyzed for population sizes scaling so that
the SR can be estimated and the sufficiency criterion met. Our method to estimate optimal population sizes starts with a population size of $P = 40$, doubling
P in every step until $P = 20480$. Each parametrization is run independently 50
times so that a fair estimation of the success rate (SR) can be made.

  For the sake of simplicity, we assume that each individual is sent for evaluation to a single processor so as to apply the following execution time metrics
[6]:

- T_{sec}: is the sequential optimization time and refers to the number of function
  evaluations until the first global optimum is evaluated.
- T_{par}: is the parallel optimization time and accounts for the number of generations it takes to find the problem optimum.

  Figure 1(a) shows the scalability of the SR with respect to the population
size for the three different selection operators. The SR scales in all cases with a
sigmoid shape in which smaller population sizes perform poorly and larger ones
reach the sufficiency $SR \geq 0.98$. The only remarkable difference between para-
meterizations rely on the transition phase of the sigmoid. Such transitions occur

---

[1] We follow here the selectorecombinative approach of Lobo and Lima [7] for studying
the scalability of the population size.

much earlier in $s2$ and *Coop s16* than in $s16$. This allows smaller populations to be sufficient for the formers while not for the latter.

In Fig. 1(b), the analysis of the trade offs T_{sec}/T_{par} shows that, whenever a given population size is sufficient in the three settings, the winning strategy is $s16$, i.e. the maximum selection pressure. Nevertheless, such results do not imply that $s16$ *dominates* $s2$ since $s16$ requires of larger population sizes to achieve better performance. *Coop s16*, however, outperforms parallel and sequential times of $s2$ while having the same population requirements. Therefore, it can be said that *Coop s16* dominates $s2$ under any setting.

4   Conclusions and Future Works

In this paper, we have proposed the *cooperative selection* scheme, an extension of tournament selection that, by implementing an altruistic behavior in winners of the competitions, is able to outperform binary tournament. Without assuming any knowledge on the problem domain, *cooperative selection* is shown to outperform the utilization of parallel resources in a simple test case: given identical population sizes (i.e. same computing platform), *cooperative selection* saves computational efforts with respect to binary tournament and requires of less parallel execution time to yield the same quality in solutions.

As a future work, we plan two major lines of research. The first is the straightforward application of *cooperative selection* to massively parallel EAs as in the case of GPU-based EAs or as in volunteer-computing-based EAs. The second line of research is related to high-dimensional real-parameter optimization. Here we think that *cooperative selection* could perform well since, on the one hand, typical frameworks for benchmarking usually impose restrictions on time and, on the other hand, problems with high-dimensionality require of large amounts of resources in order to minimize optimization errors.

Acknowledgments. This work was supported by the Luxembourg FNR Green@Cloud project (INTER/CNRS/11/03) and by the Spanish Ministry of Science Project (TIN2011-28627-C04). B. Dorronsoro acknowledges the support by the Fonds National de la Recherche, Luxembourg (AFR contract no 4017742).

References

1. Alba, E., Dorronsoro, B.: The exploration/exploitation tradeoff in dynamic cellular genetic algorithms. IEEE Trans. Evol. Comput. **9**(2), 126–142 (2005)
2. Eiben, A.E., Schut, M.C., De Wilde, A.R.: Boosting genetic algorithms with self-adaptive selection. In: Proceedings of the IEEE Congress on Evolutionary Computation, pp. 1584–1589 (2006)
3. Eiben, A.E., Smith, J.E.: Introduction to Evolutionary Computing. Springer, Heidelberg (2003)
4. Lombraña González, D., Jiménez Laredo, J., Fernández de Vega, F., Merelo Guervós, J.: Characterizing fault-tolerance of genetic algorithms in desktop grid systems. In: Cowling, P., Merz, P. (eds.) EvoCOP 2010. LNCS, vol. 6022, pp. 131–142. Springer, Heidelberg (2010)

5. Laredo, J.L.J., Eiben, A.E., van Steen, M., Merelo Guervós, J.J.: Evag: a scalable peer-to-peer evolutionary algorithm. Genet. Program. Evolvable Mach. **11**(2), 227–246 (2010)
6. Lässig, J., Sudholt, D.: General scheme for analyzing running times of parallel evolutionary algorithms. In: Schaefer, R., Cotta, C., Kołodziej, J., Rudolph, G. (eds.) PPSN XI. LNCS, vol. 6238, pp. 234–243. Springer, Heidelberg (2010)
7. Lobo, F., Lima, C.: Adaptive population sizing schemes in genetic algorithms. In: Lobo, F., Lima, C., Zbigniew, M. (eds.) Parameter Setting in Evolutionary Algorithms, vol. 54, pp. 185–204. Springer, Heidelberg (2007)
8. Lobo, F.G., Goldberg, D.E.: The parameter-less genetic algorithm in practice. Inf. Sci. Inf. Comput. Sci. **167**(1–4), 217–232 (2004)
9. Potter, M.A., De Jong, K.A.: A cooperative coevolutionary approach to function optimization. In: Davidor, Y., Männer, R., Schwefel, H.-P. (eds.) PPSN LNCS, vol. 866, pp. 249–267. Springer, Heidelberg (1994)
10. David Schaffer, J., Eshelman, L.J.: On crossover as an evolutionarily viable strategy. In: Belew, R.K., Booker, L.B. (eds.) ICGA, pp. 61–68. Morgan Kaufmann, San Francisco (1991)

Effects of Population Size on Selection and Scalability in Evolutionary Many-Objective Optimization

Hernán Aguirre[1]([✉]), Arnaud Liefooghe[2,4],
Sébastien Verel[3,4], and Kiyoshi Tanaka[1]

[1] Shinshu University, Matsumoto, Japan
{ktanaka,ahernan}@shinshu-u.ac.jp
[2] Université Lille 1, LIFL, UMR CNRS 8022, Lille, France
arnaud.liefooghe@lifl.fr
[3] Université Nice Sophia-Antipolis, Nice, France
verel@i3s.unice.fr
[4] Inria Lille-Nord Europe, Lille, France

Abstract. In this work we study population size as a fraction of the true Pareto optimal set and analyze its effects on selection and performance scalability of a conventional multi-objective evolutionary algorithm applied to many-objective optimization of small MNK-landscapes.

1 Introduction

Conventional multi-objective evolutionary algorithms (MOEAs) [1] are known to scale up poorly to high dimensional objective spaces [2], particularly dominance-based algorithms. This lack of scalability has been attributed mainly to inappropriate operators for selection and variation. The population size greatly influences the dynamics of the algorithm. However, its effects on large dimensional objectives spaces are not well understood. In this work we set population size as a fraction of the true Pareto optimal set and analyze its effects on selection and performance scalability of a conventional MOEA applied to many-objective optimization. In our study we enumerate small MNK-landscapes with 3–6 objectives, 20 bits, and observe the number of Pareto optimal solutions that the algorithm is able to find for various population sizes.

2 Methodology

In our study we use four MNK-landscapes [3] randomly generated with $m = 3$, 4, 5 and 6 objectives, $n = 20$ bits, and $k = 1$ epistatic bit. For each landscape we enumerate all its solutions and classify them in non-dominated fronts. The exact number of true Pareto optimal solutions POS^T found by enumeration are $|POS^T| = 152$, 1554, 6265, and 16845 for $m = 3$, 4, 5, and 6 objectives, respectively. Similarly, the exact number of non-dominated fronts of the landscapes are 258, 76, 29, and 22, respectively.

G. Nicosia and P. Pardalos (Eds.): LION 7, LNCS 7997, pp. 450–454, 2013.
DOI: 10.1007/978-3-642-44973-4_48, © Springer-Verlag Berlin Heidelberg 2013

We run a conventional MOEA for a fixed number of generations. The algorithm uses a population P from which it creates an offspring population Q by recombination and mutation. The population P for the next generation is obtained from the joined population $P \cup Q$ by survival selection. In this work we use NSGA-II as the evolutionary multi-objective optimizer, set with two point crossover with rate $pc = 1.0$, and bit flip mutation with rate $pm = 1/n$.

Once evolution is over, we compare the set of POS^T with the sets of unique non-dominated solutions obtained at each generation after survival selection to determine which are true Pareto optimal solutions, count their number at each generation, and their accumulated number found during evolution.

3   Experimental Results and Discussion

Let us denote by F_1 the set of non-dominated solutions in population P, and F_1^T the set of solutions in by F_1 that are true Pareto optimal solutions. Figure 1 shows the number of solutions in F_1 and F_1^T over the generations for $m = 3$ and 4 objectives, running the algorithm for 100 generations with three different population sizes $|P| = 50$, 100 and 200.

First we analyze results for $m = 3$ objectives. When we set population size to $|P| = 50$ or 100, a value smaller than the number of true Pareto optimal solutions $|POS^T| = 152$, it can be seen in Fig. 1(a.1) and (a.2) that after few generations all solutions in the population are non-dominated, $|F_1| = |P|$. However, not all solutions in F_1 are true Pareto optimal solutions, i.e. $|F_1^T| < F_1 = |P|$. Also, it is important to note that F_1^T fluctuates up and down after an initial increase. On the other hand, when we set the population size to a value larger than the number of true Pareto optimal solutions, $|P| = 200 > |POS^T| = 152$, it can be seen in Fig. 1(a.3) that the instantaneous non-dominated set is a subset of the population, $F_1 \subset P$. Also, note that from generation 35 onwards, all non-dominated solutions in the population are also true Pareto optimal, $F_1 = F_1^T$. In this case, the algorithm finds and keeps in P almost all true Pareto optimal solutions, 147 out of 152, during the latest stage of the search.

It is known that the number of true Pareto optimal solutions $|POS^T|$ increases considerably with the number of objectives. However, this is often ignored and the algorithm is set with a very small population size compared to $|POS^T|$. To study these cases, Fig. 1 (b.1)–(b.3) show results for $m = 4$ objectives setting population size to the same values used for $m = 3$ objectives, which are very small compared to $|POS^T|$, i.e $|P| \leq 200 < |POS^T| = 1554$. Note that these settings of population size magnify the difficulties observed for $m = 3$ with $|P| = 50$ or $|P| = 100$. That is, fewer solutions are true Pareto optimal, although the set of non-dominated solutions of the population quickly contains mutually non-dominated solutions only. Also, larger fluctuations are observed in the number of true Pareto optimal solutions F_1^T.

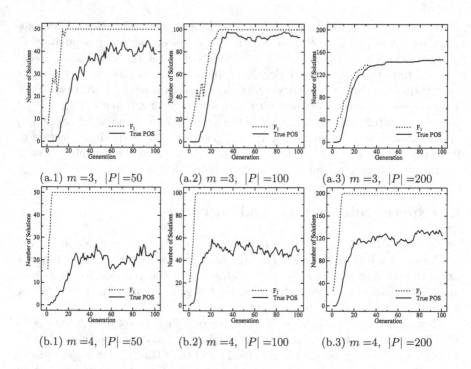

Fig. 1. Number of non-dominated F_1 and actual number of true Pareto optimal solutions F_1^T in the population over the generations. $|POS^T| = 152$, and 1554 for $m = 3$, and 4 objectives, respectively. Population sizes $|P| = 50$, 100, and 200.

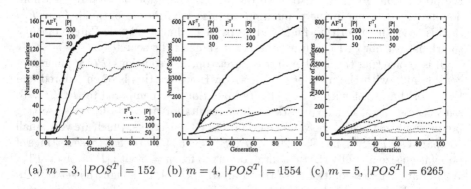

Fig. 2. Accumulated and instantaneous number of true Pareto optimal solutions, AF_1^T and F_1^T, $m = 3$, 4, and 5 objectives. Population sizes $|P| = 50$, 100, and 200.

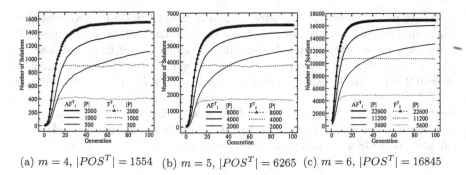

(a) $m = 4$, $|POS^T| = 1554$ (b) $m = 5$, $|POS^T| = 6265$ (c) $m = 6$, $|POS^T| = 16845$

Fig. 3. Accumulated and instantaneous number of true Pareto optimal solutions, AF_1^T and F_1^T, $m = 4$, 5, and 6 objectives. Population sizes 1/3, 2/3 and 4/3 of POS^T.

In general, if $|P|$ is set to a value smaller than $|POS^T|$, the algorithm cannot keep all true Pareto optimal solutions in the population. However, we would expect an ideal algorithm to keep as many true Pareto optimal solutions as the size of its population, $|F_1^T| = |F_1| = |P| < |POS^T|$. This is not what we observe in our results. To explain this behavior, Fig. 2 shows the instantaneous number of true Pareto optimal solutions in the population $|F_1^T|$ and its accumulated number $|AF_1^T|$ over the generations for population sizes $|P| = 50$, 100, and 200. Note that a large number of true Pareto optimal solutions are found by the algorithm. However, not all these solutions remain in the population (except in the case $m = 3$ $|P| = 200$). Some of these solutions are lost from one generation to the next one during the survival selection step of the algorithm. At this step, the algorithm joins the population P with the offspring population Q and ranks individuals with respect to dominance-depth. The best rank is given to true Pareto optimal solutions and also to some others that are not true optimal but appear non-dominated in the combined population. Let us call the set of best ranked non-dominated solutions obtained from $P \cup Q$ as $F_1^{P \cup Q}$. If this set $F_1^{P \cup Q}$ is larger than the population P, a sample of them $P = F^1 \subset F_1^{P \cup Q}$ is chosen based on crowding distance during the survival step. At this point, some true Pareto optimal solutions are dropped in favor of lest crowded non-optimal solutions. Summarizing, $P = F^1 \subset F_1^{P \cup Q}$ and therefore $F_1^T \subset F_1$ is more likely to occur for population sizes smaller than the number of true Pareto optimal solutions $|POS^T|$.

Figure 2(a) and Fig. 3(a)–(c) show results for $m = 3$, 4, 5 and 6 objectives using population sizes that correspond approximately to 1/3, 2/3 and 4/3 of the set POS^T, respectively. From these figures note that increasing population size from 1/3 to 4/3 of POS^T translates into a striking performance scalability of the algorithm, measured on terms of the number of true Pareto optimal solutions found and kept in the population. For population size 4/3 of POS^T the number of $AF_1^T = F_1^T \subset F_1$ and the algorithm can actually find and keep in the population 147 out of 152, 1545 out of 1554, 6248 out of 6265, and 16842 out of 16845 true Pareto optimal solutions for 3, 4, 5 and 6 objectives, respectively.

These results show that the effectiveness of the algorithm in many-objective landscapes depends strongly on the size of the population. However, it should be noted that larger populations demand more computational time and memory. Also, a relatively larger number of solutions need to be evaluated. For example, after 100 generations, using a population size 4/3 of POS^T, the conventional MOEA used in this study evaluates approximately a number of solutions equivalent to 2%, 19%, 76% and 215% of the size of the search space for $m = 3$, 4, 5, and 6 objectives, respectively. In the future, we would like to analyze the efficiency of MOEAs in many-objective landscapes.

4   Conclusions

In this work we analyzed the effects of population size on selection and scalability of a conventional dominance-based MOEA for many-objective optimization. We showed that the performance of a conventional MOEA can scale up fairly well to high dimensional objective spaces if a sufficiently large population size is used compared to the size of the true Pareto optimal set.

References

1. Deb, K.: Multi-Objective Optimization using Evolutionary Algorithms. Wiley, Chichester (2001)
2. Ishibuchi, H., Tsukamoto, N., Nojima, Y.: Evolutionary many-objective optimization: a short review. In: Proceedings of IEEE Congress on Evolutionary Computation (CEC 2008), pp. 2424–2431. IEEE Press (2008)
3. Aguirre, H., Tanaka, K.: Insights on properties of multi-objective MNK-landscapes. In: Proceedings of 2004 IEEE Congress on Evolutionary Computation, pp. 196–203. IEEE Service Center (2004)

A Novel Feature Selection Method
for Classification Using a Fuzzy Criterion

Maria Brigida Ferraro[1], Antonio Irpino[2],
Rosanna Verde[2], and Mario Rosario Guarracino[1,3(✉)]

[1] High Performance Computing and Networking Institute,
National Research Council, Naples, Italy
[2] Department of European and Mediterranean Studies,
Second University of Naples, Caserta, Italy
[3] Department of Informatics, Kaunas University of Technology, Kaunas, Lithuania
mario.guarracino@cnr.it

Abstract. Although many classification methods take advantage of
fuzzy sets theory, the same cannot be said for feature reduction methods.
In this paper we explore ideas related to the use of fuzzy sets and we propose a novel fuzzy feature selection method tailored for the Regularized
Generalized Eigenvalue Classifier (ReGEC). The method provides small
and robust subsets of features that can be used for supervised classification. We show, using real world datasets that the performance of ReGEC
classifier on the selected features well compares with that obtained using
them all.

1 Introduction

In many practical situations, the size and dimensionality of datasets is large and
many irrelevant and redundant features are included. In a classification context,
learning from huge datasets could not work well even if theoretically more features should lead more discriminant power. In order to face with this problem
two kinds of algorithms can be used: *feature transformation* (or extraction) and
feature selection. Feature transformation consists in constructing new features
(in a lower dimentional space) from the original ones. These methods include
clustering, basic linear transforms of the input variables (Principal Component
Analysis/Singular Value Decomposition, Linear Discriminant Analysis), spectral
transforms, wavelet transforms or convolution of kernels. The basic idea of a feature transformation is simply projecting a high-dimensional feature vector onto
a low-dimensional space. Unfortunately, the projection leads a loss of the measurement units of features and the obtained features are not easy to interpret.
Feature selection (FS) may overcome this disadvantages.

FS aims at selecting a subset of features relevant in terms of discrimination capability. It avoids the drawback of the output interpretability, because
the selected features represent a subset of the given ones. FS is used as a pre-
processing phase in many contexts. It plays an important role in applications
that involve a large number of features and only few samples. FS enables data

G. Nicosia and P. Pardalos (Eds.): LION 7, LNCS 7997, pp. 455–467, 2013.
DOI: 10.1007/978-3-642-44973-4_49, © Springer-Verlag Berlin Heidelberg 2013

mining algorithms to run when it is otherwise impossible given the dimensionality of the dataset. Furthermore, it permits to focus only on relevant features and to avoid redundant information. FS strategy consists of the following steps. From the original set of features, a candidate subset is generated and then evaluated by means of an evaluation criterion. The goodness of each subset is analyzed and, if it fulfills the stopping rule, it is selected and validated in order to check whether the subset is valid. Otherwise, if the subset does not fulfill the stopping rule, another candidate is generated and the whole process is repeated.

The FS methods are classified as *filters*, *wrappers* and *embedded*, depending on criterion used to evaluate the feature subsets. Filters are based on intrinsic characteristics of features to reveal their discriminating power and do not depend on predictor. These methods select features by ranking. Different relevance measures can be used. These measures include correlation criteria [1], the mutual information metric [2–4], class similarity measures with respect to the selected subset (FFSEM [5] and filter methods presented in [6,7]) and the separability of neighboring patterns (ReliefF [8]). A filter procedure may involve a forward or a backward selection. Forward selection consists in starting with no features and then, at each iteration, one or more features are added if they bring additional contribution. The algorithm stops when no features among the candidates lead a significant improvement. Backward selection (or elimination) starts with all features. At each iteration, one or more features are removed if they reduce the value of the total evaluation.

Filters present a low complexity but the discriminant power may be not high, since the evaluation criterion can be not associated with the classifier in use. Embedded methods do not separate the learning from the feature selection phase, thus embedding the selection within the learning algorithm. At the time of designing the predictor, these methods pick up the relevant features. Embedded methods include decision trees, weighted naive Bayes (Duda et al. [9]), FS using the weight vector of Support Vector Machines (SVM) (Guyon et al. [10], Weston et al. [11]).

In wrapper methods, FS depends on classifiers. Namely, each candidate subset is evaluated by analyzing the accuracy of a classifier. These methods, unlike the filters, are characterized by high computational costs but high classification rates are usually obtained. Filter algorithms are computationally more efficient, although their performance can be worse than wrapper algorithms.

In a classification framework, data may present characteristics of different classes and can be affected by noise. To cope with this problem, classes may be considered as fuzzy sets and data belong to each class with a degree of membership. Fuzzy logic improves classification by means of overlapping class definitions and improves the interpretability of the results. In the last years, some efforts have been devoted to the development of methodologies for selecting feature subsets in an imprecise and uncertain context. To this extend, the idea of fuzzy set is used to characterize the imprecision. Ramze Rezaee et al. [12] present a method consisting of an automatic identification of a reduced fuzzy set of a labeled multi-dimensional data set. The procedure includes the projection of the original data set onto a fuzzy space, and the determination of the optimal subset of fuzzy features by using conventional search techniques. A k-nearest neighbor (NN) algorithm is used. Pedrycz and Vukovich [13] generalize feature selection method by introducing a mechanism of fuzzy feature selection. They propose to consider granular features, rather than numeric. A process of fuzzy feature selection is carried out and numerically quantified in the space of membership values generated by fuzzy clusters. In this case a simple Fuzzy C-Means (FCM) algorithm is used. More recently, a new heuristic algorithm has been introduced by Li and Wu [5]. This algorithm is characterized by a new evaluation criterion, based on a min-max learning rule, and a search strategy for feature selection from fuzzy feature space. The authors consider the accuracy of k-NN classifier as the evaluation criterion. Hedjazi et al. [14] introduce a new feature selection algorithm, MEmbership Margin Based Attribute Selection (MEMBAS). This approach processes in the same way numerical, qualitative and interval data based on an appropriate and simultaneous mapping, using fuzzy logic concepts. They propose to use the Learning Algorithm for Multivariable Data Analysis (LAMBDA), a fuzzy classification algorithm that aims at getting the global membership degree of a sample to an existing class, taking into account the contributions of each feature. Chen et al. [15] introduce an embedded method. It is an integrated mechanism to extract fuzzy rules and select useful features, simultaneously. They use the Takagi-Sugeno model for classification. Finally, Vieira et al. [16] consider fuzzy criteria in feature selection by using a fuzzy decision making framework. The underlying optimization problem is solved using an ant colony optimization algorithm previously proposed by the same authors. The classification accuracy is computed by means of a fuzzy classifiers.

A different approach is considered in the work proposed by Moustakidis and Theocharis [17]. They propose a forward filter FS based on a Fuzzy Complementary Critrion (FuzCoC). They introduce the notion of fuzzy partition vector (FPV) associated with each feature. A local fuzzy evaluation measure with respect to patterns is used and it takes advantage of fuzzy membership degrees of training patterns (projected on that feature) to their own classes. These grades are obtained using a fuzzy output kernel-based SVM. FPV aims at detecting the data discrimination capability provided by each feature. It treats each feature on a pattern-wise base, thus allowing to assess redundancy between features. They obtain subsets of discriminating (highly relevant) and non-redundant features. FuzCoC acts like a minimal-redundancy-maximal-relevance (mRMR) criterion. Once features have been selected, the prediction on class labels is obtained using a 1-NN.

In the present work, we take inspiration from the above methodology and from [18] to devise a novel wrapper FS method. It can be seen as a FuzCoC constructed by a ReGEC (Guarracino et al. [19]) classification approach. By means of a binary linear ReGEC, a one-versus-all (OVA) strategy is implemented, that allows to solve multiclass problems. For each feature, distances between each pattern and classification hyperplanes are computed, and they are used to construct the membership degree of each pattern to its own class. The sum of these grades represent the score associated with the feature, that is the capability to discriminate the classes. In this way, all features are ranked, and the selection process determines the features leading to an increment of the total accuracy on training set. Hence, only features with highest discrimination power are selected.

The advantage of this strategy is that it takes into account the peculiarity of the classification method, providing a set of features consistent with it. We show that this process fits out a robust subset of features, thus, a change in training points produces a small variation in the selected features. Furthermore, using standard datasets, we show that the classification accuracy obtained with a small percentage of available features is comparable with that obtained using all features.

This paper is organized as follows. In the next section, a description of the forward filter FS SVM-FuzCoC ([17]) is given. Section 3 contains our proposal, FFS-ReGEC, and the novel algorithm is described. In order to check the adequacy of the proposed procedure, in Sect. 4, we present a discussion on the dataset SONAR. Some comparative results on real world datasets are given in Sect. 5. Finally, Sect. 6 contains some concluding remarks and open problems.

2   SVM-FuzCoC

Let $D = \{\mathbf{x}_i, i = 1, \cdots, N\}$ be the training set, where $\mathbf{x}_i = \{x_{ij, j=1, \cdots, n}\}$ (n is the total number of features). The training patterns in D are initially sorted by class labels:

$$D = \{D_1, \cdots, D_k, \cdots, D_M\}$$

where $D_k = \{\mathbf{x}_{i_1}, \cdots, \mathbf{x}_{i_{N_k}}\}$ denotes the set of class k patterns and N_k is the number of patterns included in D_k, with $\sum_{k=1}^{M} N_k = N$ (M is the number of classes). Following the OVA methodology, the authors initially train a set of M binary K-SVM classifiers on each single feature, to obtain fuzzy membership of each pattern to its class. Let x_{ij} denote the feature j component of pattern $\mathbf{x}_i$, $i = 1, \cdots, N$. According to FO-K-SVM, fuzzy membership value $\mu_k(x_{ij}) \in [0, 1]$ of x_{ij} to class k is computed by

$$\mu_k(x_{ij}) = \begin{cases} 0.5 & \text{if } f_k(x_{ij}) = m_{ijk} = 1 \\ \dfrac{1}{1 + e^{\left(\ln\left(\frac{1-\gamma}{\gamma}\right)\right) \cdot \left(\frac{f_k(x_{ij}) - m_{ijk}}{|1 - m_{ijk}|}\right)}} & \text{if } m_{ijk} \neq 1 \end{cases} \quad (1)$$

where $f_k(x_{ij})$ is the decision value of the kth K-SVM binary classifier trained by x_{ij}, $m_{ijk} = max_{l \neq k} f_l(x_{ij})$ is the maximum decision value obtained by the rest $(k-1)$ K-SVM binary classifiers, and γ is the membership degree threshold fixed by the user.

The fuzzy partition vector (FPV) of feature j is defined as

$$G(j) = \{\mu_G(x_{1j}), \cdots, \mu_G(x_{Nj})\} \tag{2}$$

where $\mu_G(x_{ij}) = \mu_{c_i}(x_{ij}) \in [0,1]$, $i = 1, \cdots, N$. Generally, $\mu_G(x_{ij})$ is determined using the general formula (1) by replacing k with c_i, i.e., the class label which pattern x_{ij} belongs to. Each FPV can be considered as a fuzzy set defined on D:

$$G(j) = \{x_{ij}, \mu_G(x_{ij}) | x_{ij} \in D\}, \quad |D| = N, \quad i = 1, \cdots, N$$

where $\mu_G(x_{ij})$ denotes the membership value of x_{ij} to fuzzy set G.

Consider a set of initial features, $S = \{z_1, \cdots, z_n\}$, where, $z_j = [x_{1j}, \cdots, x_{Nj}]^T$. For each feature they construct in advance the associated FPV by means of the FO-K-SVM technique. Let $FS(p) = \{z_{l_1}, \cdots, z_{l_p}\}$ denote the set of p features selected up to and including iteration p. The cumulative set $CS(p)$ is an FPV representing the aggregating effect (union) of FPVs of the features contained in $FS(p)$:

$$CS(p) = G(z_{l_1}) \cup \cdots \cup G(z_{l_p}) \tag{3}$$

$CS(p)$ fits out approximatively the quality of data coverage obtained by the features selected at the pth iteration.

Let z_{l_p} be a candidate feature to be selected at iteration p. $AC(p, z_{l_p})$ denotes the additional contribution of z_{l_p} with respect to the cumulative set $CS(p-1)$ obtained at the preceding iteration, and it is determined by

$$AC(p, z_{l_p}) = G(z_{l_p})| - |CS(p-1)| \tag{4}$$

Feature selection, according to SVM-FuzCoC, follows the algorithm in Fig. 1.

3   Fuzzy Feature Selection ReGEC

The proposed FFS-ReGEC is a wrapper FS, incorporating a FuzCoC. The training patterns in D are initially sorted by class labels. Following the OVA methodology, we initially train a set of M binary linear ReGEC classifiers on each single feature, to obtain fuzzy membership of each pattern to its class. Let x_{ij} denote the feature j component of pattern $\mathbf{x}_i$, $i = 1, \cdots, N$. According to FO-ReGEC (Fuzzy Output ReGEC), a fuzzy membership value $\mu_{c_i}(x_{ij}) \in [0,1]$ of x_{ij} to its own class c_i is computed by

$$\mu_{c_i}(x_{ij}) = f + (1-f) \cdot e^{-\dfrac{\|x_{ij} - c_i\|^2}{dm^2}} \tag{5}$$

where $\|x_{ij} - c_i\|^2$ is the squared distance of x_{ij} from its original class c_i, $dm^2 = min_{l \neq i} \|x_{ij} - c_l\|^2$ is the minimum squared distance of x_{ij} from the other classes and f is the minimum membership (fixed). The fuzzy score s_j of feature j is defined as

$$s_j = \sum_{i=1}^{N} \mu_{c_i}(x_{ij}) \tag{6}$$

Feature selection according to FFS-ReGEC consists of the following steps. From the feature set we select the feature j with the highest score s_j, obtained by (6). Then we consider the set of non-selected features. At each iteration p, we consider a candidate with the highest score among non-selected ones. Let D_p be the dataset obtained considering the features selected at iteration $(p-1)$ and the candidate. We consider a linear Multi-ReGEC algorithm (Guarracino et al. [20]) and we compute the accuracy rate on training set. If the last added feature increases accuracy on training set, we add it to the set of selected features. We iterate the procedure until a candidate leads an increment of the total accuracy. In order to explain better this procedure, the algorithm of the FS is presented in Fig. 2.

Data: The feature set $S = \{z_1, \cdots, z_j, \cdots, z_n\}$
Result: The set $FS(m) = \{z_{l_1}, \cdots, z_{l_m}\}$ of m finally selected features
  Initialization
      Compute the feature FPVs $G(z_j)$, $j = 1, \cdots, n$;
      Set $CS(0) = \emptyset$, $FS(0) = \emptyset$;
  Select the first feature: find feature z_{l_1} such that

$$z_{l_1} = argmax_{l=1,\cdots,n}\{|G(z_l)|\}$$

  Set $CS(1) = G(z_{l_1})$, $FS(1) = FS(0) + \{z_{l_1}\}$;

while $(p < n)$ **do**
    Compute the additional contribution of the remaining features

$$AC(p, z_j) = G(z_j)| - |CS(p-1), \quad j = 1, \cdots, n, \quad j \neq l_1, \cdots, l_{p-1}$$

    Find $z_{l_p} \in S$ such that

$$l_p = argmax_{j=1,\cdots,n}\{|AC(p, z_j)|\}, \quad j = 1, \cdots, n, \quad j \neq l_1, \cdots, l_{p-1}$$

    Calculate the percentage improvement of z_{l_p} with respect to $CS(p-1)$:

$$h_{l_p} = \frac{|AC(p, z_{l_p})|}{|CS(p-1)|} \times 100\%$$

    if $h_{l_p} > e_z$ (e_z is the threshold determined by the designer) **then**
      | $CS(p) = CS(p-1) \cup G(z_{l_p})$;
      | $FS(p) = FS(p-1) + \{z_{l_p}\}$;
    else
      | Terminate FuzCoC procedure at iteration m;
    end
end

Fig. 1. SVM-FuzCoC FS

Data: The feature set $S = \{z_1, \cdots, z_j, \cdots, z_n\}$
Result: The set $FS(m) = \{z_{l_1}, \cdots, z_{l_m}\}$ of m finally selected features
Initialization
     Compute the feature scores s_j, $j = 1, \cdots, n$,
     Set $FS(0) = \emptyset$
Select the first feature: feature z_{l_1} such that

$$z_{l_1} = argmax_{l=1,\cdots,n}\{s_l\}$$

Set $FS(1) = FS(0) + \{z_{l_1}\}$
Compute the accuracy rate $acc(1)$;

while $(p < n)$ *AND* $(acc < 1)$ *OR* $(m < \min(N, perc * n + 1)$ **do**
   | Select the feature z_{l_p} with the highest score among non-selected at the $(p-1)$ th
   | iteration;
   | Let D_p be the dataset obtained considering the features in $FS(p-1)$ and the
   | candidate z_{l_p};
   | Run a MultiReGEC classification algorithm on D_p;
   | Compute the accuracy rate $acc(p)$;
   | **if** $(acc(p) > acc(p-1))$ **then**
   | | $FS(p) = FS(p-1) + \{z_{l_p}\}$;
   | **else**
   | | $FS(p) = FS(p-1)$;
   | **end**
end

Fig. 2. FFS-ReGEC

4 A Case Study

In this section we check the adequacy of FFS-ReGEC by using the dataset
Sonar [21] from UCI. The dataset is characterized by 208 samples, 60 features
and 2 classes. The dataset is composed of 111 samples obtained by sending
sonar signals to a metal cylinder at various angles and under various conditions,
and 97 patterns obtained from rocks under similar conditions. The transmitted
sonar signal is a frequency-modulated chirp, with rising frequency. The data set
contains signals obtained at different angles, spanning $90°$ for the cylinder and
$180°$ for the rock.

The 60 variables values range in $[0, 1]$. Each number represents the energy
within a particular frequency band, integrated over a certain period of time.

We generate 1000 random splits partitioning the original dataset into training
and testing sets, 70% and 30%, respectively. Figure 3 shows the distribution of
the 1000 hold outs according to the number of selected features in the training
step. Furthermore, we report the mean test accuracy and the standard deviation
for each set of partitions having the same number of selected features. The
average number of selected features on 1000 random partitions is 7.42, with an
average test accuracy equal to 76.46%. We also note that in 84.70% of times the

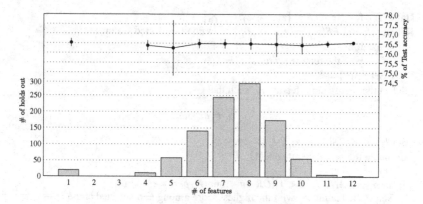

Fig. 3. The distribution of 1000 hold outs according to the number of selected features, the average test accuracy and the standard deviation for each set of splits with the same number of selected features.

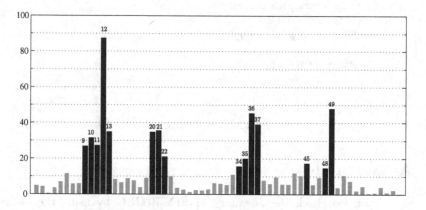

Fig. 4. Percentages of times features are selected in 1000 hold outs

number of selected features is between six and nine. It is also interesting that all 20 partitions in which a single feature is selected, it is always feature 12.

In Fig. 4 we report the number of times (in percentage) features are selected in 1000 hold outs. In the figure, the darkest bars are related to the 15 most selected features.

In Fig. 5 the absolute value of the correlations *corr* among the top 15 most chosen features is shown. We can see that some features are highly correlated (the darkest zones), hence we cluster them by means of the hierarchical clustering derived by the dendrogram depicted in Fig. 6, in which the vertical axis represents the value $1 - |corr|$, the complement to one of the absolute value of the correlation, and by clustering together those features with $|corr| > .6$. We obtain 4 clusters, forming the groups reported in Table 1.

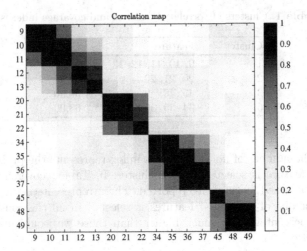

Fig. 5. Correlation plot

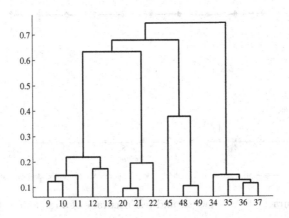

Fig. 6. Dendrogram of the classification of the 15 feature

In each cluster, features are correlated when all data are considered. In the hold out process the discrimination capability might be lower because of presence/absence of some patterns. In this case, it would be desirable that another feature from the cluster of correlated patterns is selected, which is also consistent with the idea that an FS procedure should avoid redundancy in selected features. To verify whether the proposed algorithm has this characteristic, we perform the following test.

Let $A(\mathbf{z}_j)$ be the set of hold outs where the feature $\mathbf{z}_j$ has been selected with cardinality $|A(\mathbf{z}_j)|$, and let C_k be a cluster of features. For each cluster, we compute the following *coverage* index:

$$CI(k) = \frac{|\bigcup_{\mathbf{z}_j \in C_k} A(\mathbf{z}_j)|}{H}$$

Table 1. Clusters of correlated features and covarage indexes.

Cluster	Features	CI (%)
1	9, 10, 11, 12, 13	99.60
2	20, 21, 22	61.40
3	45, 48, 49	60.90
5	34, 35, 36, 37	68.90

where H is the number of hold outs. This index represents the probability that at least one feature is selected from a cluster. In Table 1 coverage indexes of clusters of correlated features are reported. These represent the percentages of times at least one of correlated features is selected in all the considered hold outs. These percentages are much higher than those we would obtain with a random selection.

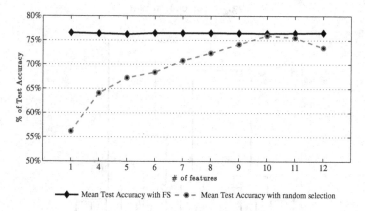

Fig. 7. Test accuracy of FFS-ReGEC and test accuracy of a random feature selection.

Then, we perform a paired t-test in order to validate the difference between the average accuracy of the proposed FS and the average accuracy of a random selection. The null hypothesis is that the random feature selector and FFS-ReGEC have the same mean accuracy. The sample consists of 1000 pairs of accuracy calculated on 1000 hold outs (70-30). For each hold out, we perform the FS, then we compute the test accuracy. On the same hold out, we randomly select a set of features with the same cardinality of the set of the selected features, then we compute the test accuracy with these random features. We obtained a rejection of the null hypothesis with p-value lower than 0.001, hence we conclude that the mean accuracies are different. The accuracy obtained with the proposed FS procedure is significantly greater than test accuracy obtained by a random selection. Finally, in Fig. 7, the values of test accuracy for different numbers of selected features are reported for both FFS-ReGEC and random selector. For each number of selected features, we consider the number of holds out selecting this number of features and we consider the average test accuracy obtained with FFS-ReGEC and that of the random selector.

We conclude that the proposed strategy always select a number of features sufficient to discriminate patterns with almost the same accuracy, whatever are the patterns used to train the classifier.

5   Comparative Analysis

In order to validate the proposed FFS-ReGEC, we consider 9 real-world datasets taken from UCI. All tests have been performed implementing the algorithms in Matlab, and comparing the results with those available in literature ([17]). The details are reported in Table 2, where the name of the dataset, the number of patterns, features, classes are shown.

To be consistent with the validation strategy used in [17], we consider 100 random splits partitioning the original dataset into training and testing sets, 70 % and 30 %, respectively. We check test accuracy (TA) obtained by ReGEC procedure without FS, by 1-NN, by the proposed FS and by SVM-FuzCoC. In Table 3 we also report the mean number of selected features (sf).

We note that the dimensionality reduction obtained using the novel FS method is comparable with that of SVM-FuzCoC, and in most cases it is higher. The selection process achieves a reduction in the number of features of approximately 99 % for genomic datasets, where the number of features is in the order of thousands. The classification accuracy obtained by ReGEC on the reduced sets is comparable with that obtained by the method, when all features are used. Accuracy results of FFS-ReGEC well compare with those of SVM-FuzCoc. In addition, when SVM-FuzCoc reaches classification rates higher than those obtained with FFS-ReGEC, these rates are the highest ones compared with all FS procedures considered in [17], and when classification rates of FFS-ReGEC are higher than those resulting from SVM-FuzCoc, these are the best ones compared with all the other algorithms. This means that on specific problems FFS-ReGEC can be a suitable alternative to existing methods.

Table 2. Dataset details

Dataset	Number of patterns	Number of features	Number of classes
Glass	214	9	6
Page Blocks	5472	10	5
Pen digits	10992	16	10
WDBC	569	30	2
Ionosphere	351	34	2
Sonar	208	60	2
SRBCT	83	2308	4
Leukemia	72	5147	2
DLBCL	77	7070	2

Table 3. Test accuracy (TA) rates for Multi-ReGEC, 1-NN, FFS-ReGEC, SVM-FuzCoC and number of selected features (sf) for the FS procedures

Dataset	Features	Multi-ReGEC	1-NN	FFS-ReGEC		SVM-FuzCoC	
		TA (%)	TA (%)	TA (%)	sf	TA (%)	sf
Glass	9	70.30	70.50	69.05	5.95	**73.36**	6.00
Page Blocks	10	95.53	95.52	93.41	4.40	**95.04**	7.00
Pen digits	16	99.34	99.31	**98.82**	13.09	97.22	12.00
WDBC	30	86.89	86.67	92.31	2.96	**96.48**	7.74
Ionosphere	34	85.98	85.57	**90.36**	7.07	89.46	4.00
Sonar	60	81.39	81.92	**76.46**	7.42	73.17	19.00
SRBCT	2308	90.64	91.29	94.24	24.00	**98.88**	33.00
Leukemia	5147	93.29	91.43	91.80	10.29	**95.71**	12.86
DLBCL	7070	88.92	89.32	82.20	52.32	**93.22**	15.50

6   Concluding Remarks

In this paper we propose a novel fuzzy feature selection technique. It uses the ReGEC algorithm to select the most promising set of variables for classification. In future, we will devise techniques to weight the contribution of the variables in the computation of the classification model, in order to enhance the discrimination capability of the most promising ones.

Acknowledgment. This work has been partially funded by Italian Flagship project *Interomics* and Kauno Technologijos Universitetas (KTU).

References

1. Guyon, I., Elisseeff, A.: An introduction to variable and feature selection. J. Mach. Learn. Res. **3**, 1157–1182 (2003)
2. Battiti, R.: Using mutual information for selecting features in supervised neural net learning. IEEE Trans. Neural Netw. **5**, 537–550 (1994)
3. Peng, H., Long, F., Ding, C.: Feature selection based on mutual information criteria of max-dependency, max-relevance, and min-redundancy. IEEE Trans. Pattern Anal. Mach. Intell. **27**, 1226–1238 (2005)
4. Ooi, C.H., Chetty, M., Teng, S.W.: Differential prioritization in feature selection and classifier aggregation for multi class microarray datasets. Data Min. Knowl. Discov. **114**, 329–366 (2007)
5. Li, Y., Wu, Z.F.: Fuzzy feature selection based on min-max learning rule and extension matrix. Pattern Recogn. **41**, 217–226 (2008)
6. Mao, K.Z.: Orthogonal forward selection and backward elimination algorithms for feature subset selection. IEEE Trans. Syst. Man Cybern. B **34**, 629–634 (2004)
7. Fu, X., Wang, L.: Data dimensionality reduction with application to simplifying RBF network structure and improving classification performance. IEEE Trans. Syst. Man Cybern. B. **33**, 399–409 (2003)

8. Kononenko, I.: Estimating attributes: analysis and extensions of relief. In: Bergadano, F., De Raedt, L. (eds.) ECML 1994. LNCS, vol. 784, pp. 171–182. Springer, Heidelberg (1994)
9. Duda, R.O., Hart, P.E., Stork, D.G.: Pattern Classification. Wiley, New York (2001)
10. Guyon, I., Weston, J., Barnhill, S., Vapnik, V.: Gene selection for cancer classification using support vector machines. Mach. Learn. **46**, 389–422 (2002)
11. Weston, J., Elisseff, A., Schoelkopf, B., Tipping, M.: Use of the zero-norm with linear models and kernel methods. J. Mach. Learn. Res. **3**, 1439–1461 (2003)
12. Ramze Rezaee, M., Goedhart, B., Lelieveldt, B.P.F., Reiber, J.H.C.: Fuzzy feature selection. Pattern Recogn. **32**, 2011–2019 (1999)
13. Pedrycz, W., Vukovich, G.: Feature analysis through information granulation and fuzzy sets. Pattern Recogn. **35**, 825–834 (2002)
14. Hedjazi, L., Kempowsky-Hamon, T., Despénes, L., Le Lann, M.V., Elgue, S., Aguilar-Martin, J.: Sensor placement and fault detection using an efficient fuzzy feature selection approach. In: 49th IEEE Conference on Decision and Control, 15–17 December 2010, Hilton Atlanta Hotel, Atlanta, GA, USA
15. Chen, Y.C., Pal, N.R., Chung, I.F.: An integrated mechanism for feature selection and fuzzy rule extraction for classification. IEEE Trans. Fuzzy Syst. **20**, 683–698 (2012)
16. Vieira, S.M., Sousa, J.M.C., Kaymak, U.: Fuzzy criteria for feature selection. Fuzzy Sets and Syst. **189**, 1–18 (2012)
17. Moustakidis, S.P., Theocharis, J.B.: SVM-FuzCoC: a novel SVM-based feature selection method using a fuzzy complementary criterion. Pattern Recogn. **43**, 3712–3729 (2010)
18. Guarracino, M.R., Cuciniello, S., Pardalos, P.: Classification and characterization of gene expression data with generalized eigenvalues. J. Optim. Theory Appl. **141**, 533–545 (2009)
19. Guarracino, M.R., Cifarelli, C., Seref, O., Pardalos, P.: A classification algorithm based on generalized eigenvalue problems. Optim. Methods Softw. **22**, 73–81 (2007)
20. Guarracino, M.R., Irpino, A., Verde, R.: Multiclass generalized eigenvalue proximal support vector machines. In: 4th IEEE Conference on Complex, Intelligent and Software Intensive Systems, pp. 25–32 (2010)
21. Gorman, P., Sejnowski, T.: Analysis of hidden units in a layered network trained to classify sonar targets. Neural Netw. **11**, 75–89 (1988)

Author Index